COFFEE BIOTECHNOLOGY AND QUALITY

COFFEE BIOTECHNOLOGY AND QUALITY

Proceedings of the 3rd International Seminar on Biotechnology in the Coffee Agro-Industry, Londrina, Brazil

edited by

T. Sera

C.R. Soccol

A. Pandey

S. Roussos

SPRINGER-SCIENCE+BUSINESS MEDIA, B.V.

A C.I.P. Catalogue record for this book is available from the Library of Congress.

DOI 10.1007/978-94-017-1068-8

Printed on acid-free paper

Presentation

This book is essentially based on the papers presented in the 3rd International Seminar on Coffee Biotechnology (III SIBAC), held at Londrina, Parana, Brazil, during May 24-28 1999. This seminar was organised by IAPAR (Instituto Agronomico do Parana)Brazil, IRD (Institut de Recherche pour le Développement) France, and UFPR (Universidade Federal do Parana) Brazil.

At the beginning of the book, the Organizing committee, the Executive committee and the International Scientific committee of the III SIBAC are presented followed by the preface which gives a general scope of Coffee Biotechnology and Quality along with comments on the general layout of the five sections in which the Seminar was structured: General introduction on coffee biotechnology, Coffee breeding, tissue culture and genetics, Pest control, Post harvest technology, and Coffee residues biotechnology.

We like to thank the European Union (DG XII), the Parana State Secretary of Agriculture, Londrina District, EMBRAPA, IRD-DIC, IAPAR and UFPR for their sponsoring. Thanks also go to the members of the International Society of Biotechnology for Coffee Agro-industry for planning this event, to the Local executive Committee at Londrina, and to Fançoise Simoni and Corinne Antonelli for their excellent technical assistance and efficiency in producing the book from the manuscripts.

T. Sera
C.R. Soccol
A. Pandey
S. Roussos

Contents

Pest control

Post-harvest technology

Coffee residues biotechnology

Organizing Institutions and Sponsors

IAPAR

Instituto Agronomico do Parana
Coffee Research Program, (IAPAR), C.P. 481, CEP86001-710, Londrina, PR, Brazil;
Tel. 055- 043 – 37 62 326 and 37 62 000 Fax. 055 – 043 – 37 62 101 and 37 62 102
Contact: T. Sera ; E-mail : tsera@pr.gov.br

IRD

Institut de Recherche pour le Développement
Laboratoire de Microbiologie de l'IRD ; Institut Fédératif de Recherche en Biotechnologie Agro-Industrielle de Marseille IFR-BAIM; Universités de Provence et de la Méditerranée, Case 925, ESIL, 163 Avenue de Luminy, 13288 Marseille Cedex 9, France
Tel. (33) 04 91 82 85 82 ; Fax. . (33) 04 91 82 85 70
Contact : S. Roussos ; E-mail : roussos@esil.univ-mrs.fr

UFPR

Universidade Federal do Parana
Laboratorio de Processos Biotecnologicos, Departamento Engenharia Quimica, (UFPR), 81531-970 Curitiba-PR – Brazil ;
Tel. 055- 041 366 2323 ext. 285 Fax. Tel. 055- 041 266 02 22
Contact : C.R. Soccol ; E-mail : soccol@engquim.ufpr.br

Sponsored by :

Union Européenne DG XII, CBP&D/Café, MDIC/Embrapa, Secretaria Municipal de Agricultura e Abastecimiento Cidade de Londrina, Fundaçao ARAUCARIA, Governo do estado de Parana- SEAB-PR, CIRAD, IRD-DIC.

Organizing Committee of the III SIBAC

Carlos R. Soccol – UFPR – Brasil
Sevastianos Roussos - IRD – France
Tumoru Sera – IAPAR - Brazil

Executive Committee

Coordinator : Tumoru Sera

Scientific programme: Sevastianos Roussos, Tumoru Sera, Carlos R. Soccol, Carlos R. Riede, Luiz F. Pereira, Luiz F. Vieira

Institutional contacts : Florindo Dalberto, Marcos Krieger, Rogério M.L. Cardoso

Financial income: Armando Androciolo, Francisco B. Lima, Luiz F. Ferrari, Samir Cury

Marketing : Marcos V.F. Martins, Michele Peres

Edition : Oswaldo Petrin

Posters : Maia C.L.L. Dias, Maria L. Crochemore

Forein contacts : Luiz F. Kalinowsky

Informatic and computing : Welfrid Stenzel

Reproduction & Grafic : Tadeu K. Sakiyama

Finance control : José Pereira da Silva, Julio César Chaves

« Dia de Campo » : Armano Androcioli, José A. Azevedo, Marcos Pavan, Paulo H. Caramoni, Rodolfo L. de Carvalho

International Scientific Committee :

Alemar B. Rena (CBP&D CAF2/Brasil), Alex Bustillo (CENICAFE-Colombia), Antonio P. Nacif (CBP&D CAF2/Brazil), Ashok Pandey (CSIR-India), Carlos R. Soccol (UFPR-Brazil), Daniel Martinez-Carrera (COLPOS/Mexico), Christopher Augur (IRD-Mexico), Dominique Berry (CIRAD-France), Eduardo Aranda (Inst. Ecologia/ Mexico), Gerardo Saucedo-Castañeda (UAMI-Mexico), German Moreno-Ruiz (CENICAFE-Colombia), Gustavo Viniegra-Gonzalez (UAMI-Mexico), Isabelle Perraud-Gaime (IRD-Mexico), Jaime Zuluaga-Vasco (Kaffesma-Colombia), Joseph Le Bars (INRA-France), Luis F. Pereira (IAPAR/Brasil), Leo Pyle (Univ. Reading/UK), Marc Berthouly (CIRAD-France), Mariano Gutierrez-Rojas (UAMI-Mexico), Moacyr Paschoal (UFLA/Brasil), Maurice Lourd (IRD-Brasil), Ney Sakiyama (UFV/Brasil), Oliveiro Guerreiro (IAC/Brasil), Philippe Lashermes (IRD-France), Robin Duponnois (IRD-Senegal), Ivone Ferrao (INCO-DC-UE/ Belgium), Sevastianos Roussos (IRD-France), Tumoru Sera (IAPAR/Brasil

Secretariat: Françoise Simoni, Emerson Nakamura, Marcos Z. Altéia, Rubens Sacchetto

Preface

Recent advents in Biotechnology have revolutionised the scientific developments of human society. One of the thrust areas related with human welfare has been the agricultural biotechnology and coffee research was not unaffected by this. During the past one and a half decades, a large amount of research directed towards the pre- and post-harvest technological developments on coffee biotechnology have led to change the face of coffee industry. With this back-ground, the third International Symposium on the Biotechnology of Coffee Agro-industry was organised at Londrina during May 25-28, 1999 (III SIBAC).

Coffee is considered as one of the most important beverages in the world. It is an important agricultural product on the world market and is widely grown tropical tree crop in about 50 countries. Economy of several countries such as Colombia, Brazil and Central America depends heavily on coffee production for foreign currency earnings. Presently an estimated area of about 11.2 x 106 ha all over the world is under coffee plantation producing about four millions tons of green beans annually. The supply of coffee consumed world-wide primarily comes from two cultivated species: *Coffea arabica* and *C. canephora*, which account for about 75 and 25% of the world market, respectively. These two varieties are commercially known as "arabica" and "robusta", respectively.

The present book, which is the outcome of the III SIBAC, comprises 48 chapters from the known experts in their respective areas. The entire manuscript has been classified in to five sections, General, Coffee breeding, tissue culture and genetics, Pest control, Post-harvest technology, and Coffee residues biotechnology. There are two chapters under General section. The first chapter by G. Viniegra-Gonzalez, which was the opening lecture of the III SIBAC, deals with the relationship of biotechnology with the coffee production. Asking a very fundamental question- what first, production or marketing, the article discusses various features and implications of application of biotechnology in coffee agro-industry. The second chapter of this section traces the historical developments of coffee industry in India and discusses various developments taken during the past several decades. It is interesting to note that the largest number of races (as many as 30) of arabica variety has been recorded from India.

The second section of the book on Coffee breeding, tissue culture and genetics consists of 17 chapters. The first chapter by Fazuoli et al. deals with the aspects of classical breeding and modern biotechnology of coffee. It emphasises that tissue culture, genetic transformation and molecular marker techniques are the most important modern biotechnology techniques that can be of help in coffee improvement. The second chapter of the section highlights the major achievements of the IAPAR in coffee breeding. It describes the role of biotechnology in developing coffee cultivars using IAPAR model of high-density coffee planting. The genetic improvement of the coffee, especially for the production, has not been easy. Generally it takes several years of serial productions to know the long-term productive capacity of any cultivar. Through the combination of a group of methods, techniques and genetic improvement procedures it is possible to achieve efficiency in improvement programmes. The chapter by Berthouly and Etienne describes the somatic embryogenesis of coffee. The chapter by Pasqual *et al.* also deals with the embryo culture of coffee. There are several articles, which deal with the genetic aspects of coffee. The chapter by Sreenath reports the role of biotechnology for genetic improvement of Indian coffee. India is among the top ten coffee producing and exporting countries along with Brazil, Columbia, Indonesia, Vietnam, Mexico, Ivory Coast, Ethiopia, Uganda and Guatemala. Through the means of biotechnological improvement programmes, good successes have been achieved in developing new high-yielding cultivars. Yet another article from India describes a process for producing synthetic seeds in coffee. Some other articles report various other aspects such as molecular breeding by Lashermes *et al.*, microsatellites in coffee by Rovelli *et al.*, heterosis in coffee by Fontes *et al.*, protoplast fusion by Cordeiro *et al.*, protein in coffee endosperm by Acuna *et al.* Coffee cultivation in the world has benefited greatly from the successful breeding programmes, resulting three to four times increase in the yields. However, yield in several species seems to have reached a plateau, which is hard to overcome. Therefore, one of the great challenges for the breeders is to increase the yield of the present cultivars, which could be met by adopting DNA marker and molecular breeding techniques. This is the theme and content of four chapters by Sakiyama, Ruas *et al.*, Santa Ram and Sreenath, and Dufour *et al.*

The third section of the book on Pest control comprises nine chapters. There are two chapters by Bustillo and Pellegrin *et al.*, which present case studies on pest control in Colombia and South Pacific, respectively. The coffee berry borer is considered as one of the most important pests to coffee all over the world. This insect destroys the coffee berry and causes fall and weight loss of the harvested berry. It also affects the quality of the beverage. Bustillo advocates the need to develop and implement an integrated pest control management programme based on the biological control of the pest instead of use of a chemical insecticide. There are several chapters, which deal with the development and characterisation of isolates as biological agents to control coffee pathogens. These include one on characterisation of two isolates for use against coffee berry borer by Velez *et al.*, NHB as agents for the control of root knot nematodes by Duponnois *et al.*, and others by Jimenez-Diaz *et al.* and Villacorta and Torrecillas. A chapter by Araujo *et al.* describes a technique for studying the mycorrhizal fungi culture parameters on agar media while one another chapter by Roussos *et al.* describes solid state fermentation as a novel approach for producing the spores of fungal biopesticides for insect control.

Section four of the book on Post-harvest technology has six chapters. Generally it has been found that the actual coffee potential is lower than a rational and competitive coffee crop. To improve productivity with lower cost/bag, one of the alternatives is the formation of commercial F1 hybrid seedling. This is the theme of the article presented by Fadelli and Sera. Article by Soares *et al.* describes the species related differences in Brazilian green coffee contaminated by ochratoxin A. Bars and Bars took the case of mycotoxigenesis in grains and describe mycotoxic prevention in coffee.

The remaining two chapters in the section are on the quality aspects of coffee. The article by Petracco deals with the quality aspects of espresso coffee as influenced by the coffee botanical variety and the one by Prete *et al.* reports correlation between the electric conductivity and the quality of coffee beverage and supports it as a measure for quality of coffee beans exudates.

The fifth and last section on Coffee residues biotechnology focuses the recent attempts to beneficially use the by-products generated at various stages of coffee processing through biotechnological means. One of the main focuses has been on the biological detoxification of anti-nutritional and anti-physiological factors from the residual pulp and husk. The first chapter of the section by Roussos *et al.* goes into elaborate discussion on various possibilities to utilise various residues for the production of feed, mushrooms, aroma compounds, gibberellic acid and enzymes etc. There are several other chapters by Perraud-Gaime *et al.*, Fan *et al.*, Brand *et al.*, Machado *et al.*, Soares *et al.*, Gutierrez-Sanchez *et al*, Auguilar *et al.*, Aranda and Barois on the production of value-added products from coffee pulp and coffee husk. A chapter by Woiciechowski *et al.* describes a process to hydrolyse the coffee husk and then to use the fermentable sugar-rich hydrolysate as the substrate for fermentation processes. Martinez and Clifford have presented an excellent review on coffee pulp polyphenols. The last chapter of the section deals with a case study on the production of mushrooms on coffee pulp in Mexico.

Efforts have been made to present each chapter well documented and providing state-of-art information. To make the chapters complete in themselves, some information may be similar in to that in other chapter(s) but considering the subject matter as a whole, we hope that it would not be considered as repetition. We sincerely believe that all the authors have done their best in their contributions. We take this opportunity to thank all the authors for their co-operation and to our reviewers who took time out from their busy schedules to give their opinion on the manuscripts.

May 2000 The Editors

List of Contributors

Acuña Ricardo
Cenicafe, Chinchiná,A.A. 2427 Manizales, Caldas – Colombia
Tel. + 968 50 65 50 ; Fax + 968 504 723
E-mail. fcracu@cafedecolombia.com

Aguilar C.N.
Departamento de Biotecnología, Universidad Autónoma Metropolitana, Unidad Iztapalapa (UAMI), A.P. 55-535, C. P. 09340, México, D.F. – Mexico

Aranda Eduardo
Departmento de Biología de Suelos, Instituto de Ecología, 91000 Xalapa – México
Tel. + 52 (28) 42 18 50 ; Fax. 52 (28) 18 78 09
E-mail. Isabelle@conacyt.edu.mx

Augur Christopher
IRD Mexico, Calle Ciceron 609, Colonia los Morales, 11530 México DF – Mexico ;
E-mail : augur@xanum.uam.mx

Berthouly Marc
CIRAD-AMIS,B.P. 5035, 34032 Montpellier Cedex 1, France
Tel. 00 33(0) 4 67 61 71 07 Fax. 00 33(0) 4 67 61 56 05
E-mail : berthouly@cirad.fr

Brand Débora
Laboratorio de Processos Biotecnologicos, Departamento Engenharia Quimica, (UFPR), 81531-970 Curitiba-PR – Brazil
Fax. + 55 41 266 02 22
E-mail : bodeby@uol.com.br

Bustillo Alex
Disciplina de Entomología, Cenicafé, A. A. 2427, Manizales, Caldas – Colombia
Tel. + 968 50 65 50 ; Fax + 968 504 723
E-mail. fcbus@cafedecolombia.com

Cordeiro A.T.
DBV, Universidade Federal de Viçosa, Av. P.H. Rolfs, s/n-36.571-000, Viçosa-MG - Brazil

Cortez J.C
Divisão de Desenvolvimento Rural, Ministério da Agricultura e do Abastecimento, 13087-000 Campinas (SP) – Brazil ; Email: cortezcafe@hotmail.com

De Araujo Alvaro
Laboratoire de Microbiologie de l'IRD, IFR-BAIM; ESIL Case 925, Universités de Provence, 163, Avenue de Luminy, F-13288 Marseille Cedex 9 ; France
E-mail : alvaro.dearaujo@libertysurf.fr

Dufour Magali
CIRAD-CP, B. P. 5035, 34032 Montpellier Cedex 1, France
Tel. 00 33 (0) 4 67 61 55 88 Fax. 00 33 (0) 4 67 61 57 93

Duponnois Robin
Institut de Recherche pour le Développement (IRD), Laboratoire de Biopédologie, BP 1386, Dakar – Sénégal
Tel. (221) 849 33 33 – 849 33 24 Fax. (221) 832 16 75
E-mail : Robin.Duponnois@ird.sn

Fadelli S.
Universidade Estadual de Londrina (UEL), PIBIC/CNPq/IAPAR; Londrina – Brazil ;

Fazuoli L.C.
IAC – Centro de Café e Plantas Tropicais - CP 28 - CEP 13001-970 – Campinas, SP – Brazil; E-mail : fazuoli@.cec.iac.br

Fontes M. José Roberto
Departamento de Fitotecnia, Universidade Federal de Viçosa (UFV), 36571-000 Viçosa-MG, Brazil
Fax. : + 31 899 26 14
E-mail : jrfontes@alunos.ufv.br

Graziosi Giorgio
Università degli Studi di Trieste, Dipartimento di Biologia, Piazzale Valmaura 9, 34138 Trieste, Italy
Tel. + 39 040 81 18 76 Fax. + 39 040 81 08 60
E-mail. Graziosi@univ.trieste.it

Guerreiro-Filho Oliveiro
Instituto Agronômico de Campinas, Centro de Café, CP 28; 13001-970, Campinas, SP, Brazil. E-mail: oliveiro@cec.iac.br

Gutiérrez-Sánchez Gerardo
Departamento de Biotecnología, Universidad Autónoma Metropolitana, Unidad Iztapalapa (UAMI), A. P. 55-535, C. P. 09340, México, D. F., Mexico

Labat Marc
Laboratoire de Microbiologie IRD, IFR-BAIM; Universités de Provence et de la Méditerranée, Case 925, ESIL, 163 Avenue de Luminy, 13288 Marseille Cedex 9, France; E-mail: labat@esil.univ-mrs.fr

Lagemaat (Van de) Jurgen,
Biotechnology and Biochemical Engineering, Depart Food Science and Technology, University of Reading, PO Box 226, Whiteknights, Reading, RG6 6AP- UK;

Lashermes Philippe
Genetrop-GAP, IRD, BP 5045, F-34032 Montpellier Cedex 1- France
Tel. + 33 (0) 4 67 41 61 85 Fax. + 33 (0) 4 67 41 6181 and 4 6754 78 00
E-mail. Philippe.Lashermes@mpl.ird.fr

Le Bars Joseph
Laboratoire de Pharmacologie-Toxicologie I.N.R.A., BP 3, 31931 Toulouse Cedex, France Tel. 00 33 (0) 5 61 86 27 74

Leifa Fan
Laboratorio de Processos Biotecnologicos, Departamento Engenharia Quimica, (UFPR), 81531-970 Curitiba-PR – Brazil
Fax. + 55 41 266 02 22
E-mail : soccol@engquim.ufpr.br

Machado M.M. Christina
Laboratorio de Processos Biotecnologicos, Departamento Engenharia Quimica, (UFPR), 81531-970 Curitiba-PR - Brazil
Fax. + 55 41 266 02 22
E-mail : soccol@engquim.ufpr.br

Martinez-Carrera Daniel
College of Posgraduados in Agricultural Sciences, Campus Puebla, Mushroom Biotechnology,Laboratory, A.P 701, 72001 Puebla – Mexico
Tel.00 52 22 852 798 Fax. 00 52 22 852 161
E-mail. dcarrera@colpos.colpos.mx

Martínez Díaz C.P.
Disciplina de Entomologia, Cenicafé, AA 2427, Manizales, Colombia
Chinchiná, Caldas – Colombia

Muniswamy B.
Biotechnology Centre, Unit of Central Coffee Research Institute, Coffee Board, Manasagangothri, Mysore - 570 006, Karnataka, India

Pasqual Moacir
Laboratório de Cultura de Tecidos, Departamento de Agricultura da Universidade Federal de Lavras, 37200-000, Lavras – MG, Brazil Fax. + 55 35 829 11 32
E-mail. mpasqual@ufla.br

Pellegrin Fréderic
IRD (ex-ORSTOM), Laboratory of Plant Pathology, BP 5045, 34032 Montpellier, France Tel. 00 33 (0) 4 67 41 62 27 E-mail :pellegrin@mpl.ird.fr

Perraud-Gaime Isabelle
IRD Mexico, Calle Ciceron 609, Colonia los Morales, 11530 México DF – Mexico ;
E-mail : isa@xanum.uam.mx *and gaime@ data.net.mx*

Petracco Marino,
Illycaffe S.p.A., via Flavia, 110 – Trieste - Italy
Tel. 00 39 040 389 03 97 ; Fax. 00 39 040 389 04 90
E-mail : marino.petracco@illy.it

Prete C.E. Cassio
Departamento de Agronomia, Universidade Estadual de Londrina (UEL), caixa postal 6001, 86051-990, Londrina - PR, Brazil

Pyle David Leo
Biotechnology & Biochemical Engineering, Deparement of Food Science Technology, The University of Reading, PO Box 226, Whiteknights, Reading, RG6 6AP, UK
Fax. 44 (0) 118 93 100 80
E-mail : D.L.Pyle@afnovell.reading.ac.uk

Ramirez-Martinez R. José
Laboratorio de Fitoquimica, Universidad Nacional Experimental Táchira (UNET), Apartado 436, San Cristóbal 5001, Táchira, Venezuela
Tel. + 58 76 5304 22 Fax. + 58 76 53 24 54
E-mail :jramir @epsilon.funtha.gov.ve

Roussos Sevastianos
Laboratoire de Microbiologie IRD, IFR-BAIM; Universités de Provence, Case 925, ESIL, 163 Avenue de Luminy, 13288 Marseille Cedex 9, France
Tel. + 33 (0) 4 91 82 85 82 Fax. + 33 (0) 4 91 82 85 70
E-mail : roussos@esil.univ-mrs.fr

Rovelli Paola
Università degli Studi di Trieste, Dipartimento di Biologia, Piazzale Valmaura 9, 34138 Trieste, Italy
Tel. + 39 040 81 18 76 Fax. + 39 040 81 08 60
E-mail. Rovelli@univ.trieste.it

Ruas M. Paulo
Departamento de Biologia General, Universidade Estadual de Londrina, 86051-990 Londrina - PR, Brazil

Sakiyama S. Ney
Departamento de Fitotecnia, Universidade Federal de Viçosa, 36571-000 Viçosa - MG, Brazil *E-mail : sakiyama@mail.ufv.br*

Santa Ram
Biotechnology Centre, Unit of Central Coffee Research Institute, Coffee Board, Manasagangothri -Mysore -570 006, India
E-mail : tisscult@giasbg01.vsnl.net.in

Saucedo-Castañeda Gerardo
Departamento de Biotecnologia; Universidad Autónoma Metropolitana, Iztapalapa, Apartado Postal 55-535, CP 09340 D.F., Mexico
Tel. + 525 724 47 11 Fax. 525 724 47 12
E-mail : saucedo@xanum.uam.mx

Sera Tumoru
Coffee Research Program, Agronomic Institute of Parana State (IAPAR), C.P. 481, CEP86001-710, Londrina, PR, Brazil
Tel. 00 55 (04)3 37 62 326 Fax: 00 55 (04)3 37 62 101
E-mail: tsera@pr.gov.br

Soares Marlene
Laboratorio de Processos Biotecnologicos, Departamento Engenharia Quimica, Universidade Federal do Parana (UFPR), 81531-970 Curitiba-PR – Brazil
Fax. + 55 41 266 02 22
E-mail : soccol@engquim.ufpr.br

Soccol Carlos R.
Laboratorio de Processos Biotecnologicos, Departamento Engenharia Quimica, Universidade Federal do Parana (UFPR), 81531-970 Curitiba-PR – Brazil
Tel.. + 55 41 361 31 91 Fax. + 55 41 266 02 22
E-mail : soccol@engquim.ufpr.br

Sondahl Maru
Fitolink Corp., 6 Edinburgh Lane, Mt. Laurel, NJ 08054 USA

Sreenath L.
Biotechnology Centre, Unit of Central Coffee Research Institute, Coffee Board, Mysore - 570 006,Karnataka, India
E-mail : tisscult@giasbg01.vsnl.net.in

Srinivasan C.S.
Central Coffee Research Institute, Coffee Research Station, 577 117 Chikmagalur Dist, Karnataka, India
Tel. + 91 08 265 430 29 Fax. 91 08 265 43 143

Valencia Jiménez Arnubio
Departamento de Química, Facultad de Ciencias Exactas y Naturales. Universidad de Caldas, Apartado Aereo 275, Manizales-Colombia
Fax. (68) 886 25 20
E-mail. fcaval@cafedecolombia.com

Valente Soares M. Lucia
Universidade Estadual de Campinas, Faculdade de Engenharia de Alimentos. CP 6121, 13081-970, Campinas, São Paulo, Brazil
Fax. + 55 19 788 78 90
E-mail. valente@fea.unicamp.br

Vélez Arango Patricia Eugenia
Disciplina de Entomologia, Cenicafé, AA 2427, Manizales, Colombia
Tel. + 968 50 65 50 ; Fax + 968 504 723
E-mail. fcpvel@cafedecolombia.com

Villacorta Amador
Instituto Agronômico do Paraná-IAPAR, Rod. Celso Garcia Cid. Km 375, CEP 86001-970, Londrina-PR, Brazil
E-mail :

Viniegra-Gonzalez Gustavo
Departamento de Biotecnologia; UAM-I, A.P. 55-535, CP 09340 D.F., Mexico;
Tel. + 525 724 47 11 Fax. 525 724 47 12
E-mail : vini@xanum.uam.mx

Woiciechowski L. Adenise
Laboratorio de Processos Biotecnologicos, Departamento Engenharia Quimica, (UFPR), 81531-970 Curitiba-PR, Brazil
Fax. + 55 41 266 02 22
E-mail : soccol@engquim.ufpr.br

Chapter 1

BIOTECHNOLOGY AND THE FUTURE OF COFFEE PRODUCTION

G. Viniegra-González

Departamento de Biotecnologia; Universidad Autónoma Metropolitana, Iztapalapa, Apartado Postal 55-535, CP 09340 D.F., Mexico.

Running title: Coffee and biotechnology

1. Introduction

Coffee is a crop from tropical countries, which, according to FAO statistics, has annual production around 4 million tons of green beans with sales between 6 to 12 billion dollars (Fig. 1).

Coffee price has followed an oscillatory decreasing trend (Fig. 2). Such oscillatory behaviour of coffee prices has a main harmonic component with an average period, τ = 8.8 years, and an average amplitude close to one US$/lb. In other words, coffee market, as many markets for cash crops, has periodic booms and crashes that make difficult to plan the development of coffee producing economies. Therefore, any further technological and commercial development for coffee production should try to buffer those periodic market crises, by increasing the diversification of coffee outputs (valuable products and by products) and by a significant reduction of production and transformation costs (increasing yields and decreasing production costs).

The world trend to protect the environment is a major concern affecting coffee production and transformation, because many easy and ready techniques are to be discarded due to environmental considerations. For example, pest control using chemical compounds leaving toxic residues or, disposing of waste waters by the simple expedient to dump the untreated discharges in rivers or lakes. But, perhaps the environmental challenge may be a blessing instead of a problem. Specially, if new technologies introduced to meet environmental standards are, at the same time, oriented to increase the quality, diversity and commercial opportunities for coffee derivatives. For example, producing new varieties of organic coffee beans with better price, or giving rise to new high value added products from coffee pulp.

T. Sera et al. (eds.), Coffee Biotechnology and Quality, 1–16.

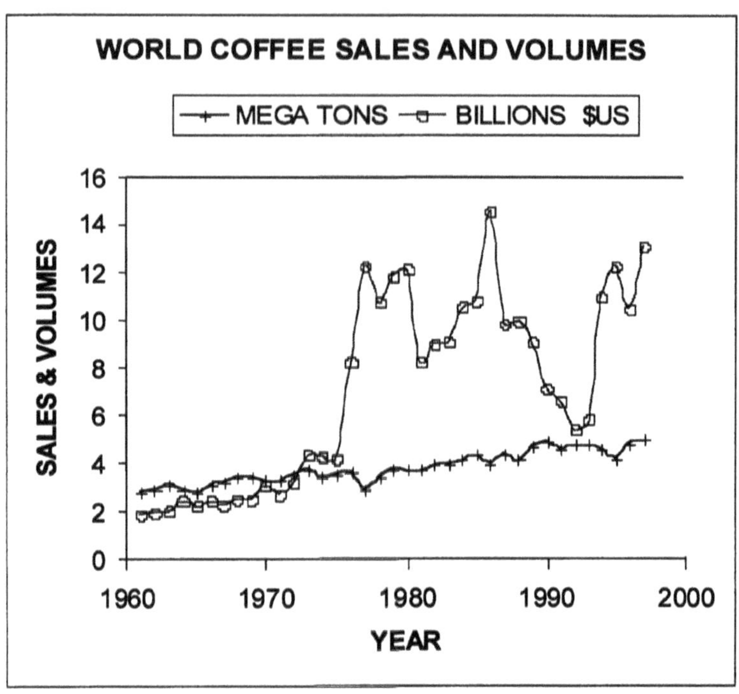

Figure 1. Global trends of coffee market from FAO Statistics (FAOSTAT, Internet). Annual sales are shown in billion dollars (circles). Annual volumes of production are shown as Mton or million of metric tons (crosses).

Another important consideration is the growing concern with "health foods" which is related to the increasing life expectancy around the world. Now, more people die at old age with the, so called, degenerative diseases, such as, cancer, heart and vascular problems. For this reason, certain foods and beverages are being labelled as "bad or junk foods" and others as "good and health foods". Coffee is now in a transition period for their future marketing. In the dark side, caffeine is considered as a "risk factor". Mainly, because of their stimulating effects on the heart, kidneys and brain and there is a strong ongoing debate on the goods and evils of caffeine in our diet. But in the bright side, coffee, tea and other traditional beverages, are being discovered as a natural and abundant source of antioxidant phenols. Such chemicals are considered in the top list of "health foods" because some people claim they have a protective role against degenerative diseases. Unfortunately, research in the field of antioxidant chemicals from coffee cherries is still a neglected field and more is said in the popular journals about green tea, red wine than about coffee derivatives.

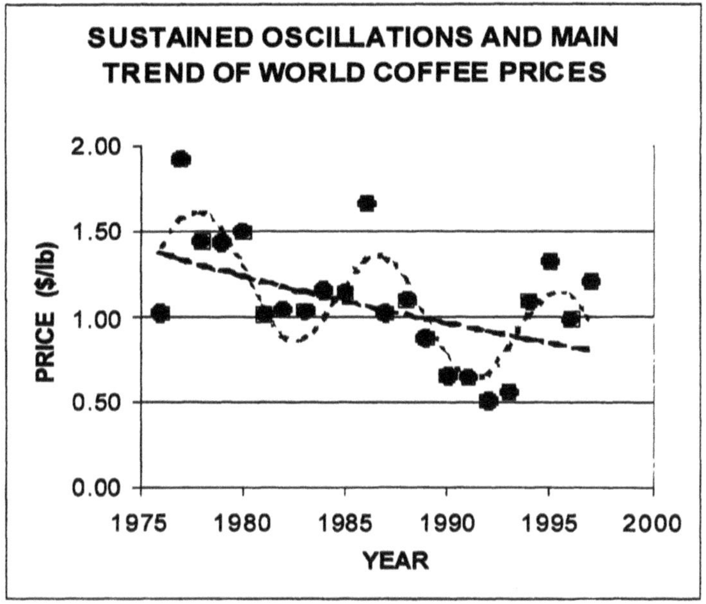

Figure 2. Annual coffee price fluctuations (estimated from Fig. 1). Actual average prices are shown in circles. The linear and the oscillatory trends are shown as continuous plots. They correspond to least square error fits of average prices.

2. The nature of market oscillations in cash crop economics

2.1 A DYNAMIC MODEL FOR COFFEE MARKET

A simple model illustrates some of the remarkable properties of coffee market (Fig. 3).

This is defined by three variables, X = the raw material in the producing region (green beans), Y = the intermediate material transported from the producing to the consuming regions and Z = the stocks of green beans to be processed and consumed.

$$\begin{array}{cccc} V_0 & V_1 & V_2 & V_F \end{array}$$

$$\rightarrow X \rightarrow Y \rightarrow Z \rightarrow$$

The primary production rate, VO, should be sensitive to the stock level, Z, because the price is a decreasing function of the stocks or inventories (Fig. 4).

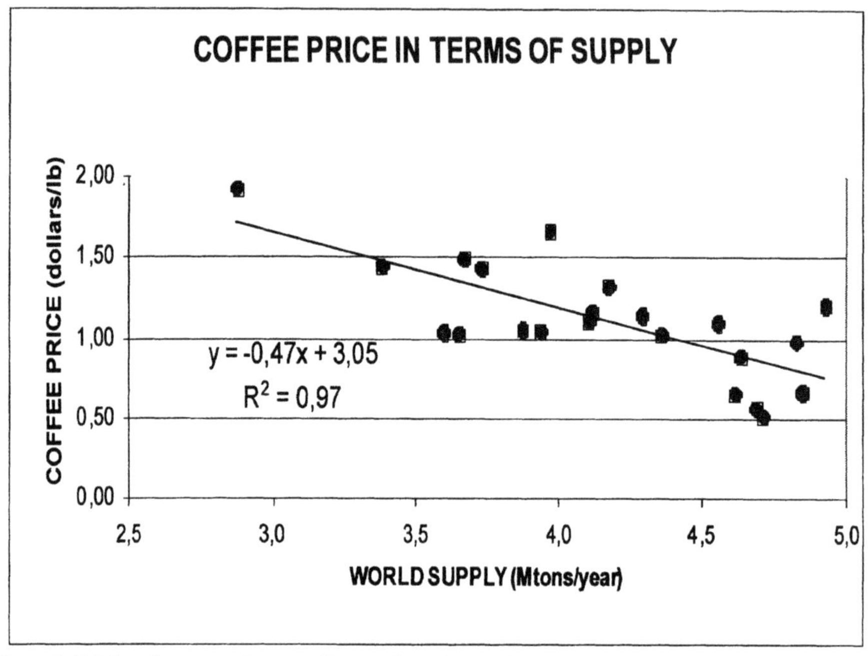

Figure 3. Relations between price vs coffee supply (estimated from Figs. 1 and 2). The equation shown in the graph corresponds to the adjusted linear trend

$$Vo = Vi/[1 + (Z/Ki)^p] \qquad [1]$$

Where, Ki, is saturation constant for the negative feedback between the stocks and the primary production. For small values of Ki, primary production is inhibited by low values of Z. The coefficient, $p \geq 0$, is an indication of the delay that exists between stock accumulation and the inhibition of coffee production. For values, $p > 1$, the producers will have a tendency to neglect small stock rises, Z, and will continue to produce coffee despite some reductions in coffee price. For values, $p < 1$, coffee producers will be very sensitive to small accumulations of Z shutting down primary production.

At the end of the market chain there is the consumption rate, V_F, which is a convex function saturated by the market inventories, Z, (Fig. 5) because there is a maximal consumption rate, V_C, with a saturation constant K_C, as follows.

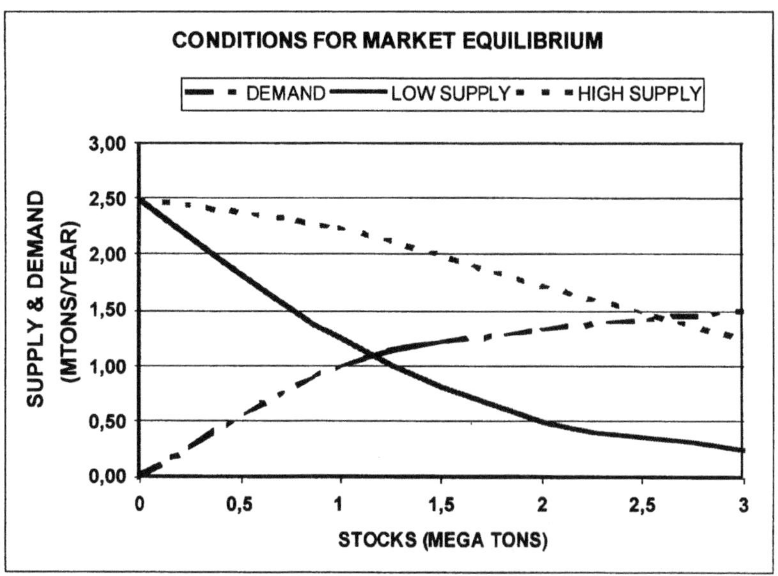

Figure 4. Graphical solution for the steady-state (equilibrium) conditions between coffee supply and demand. Parameters were, $V_0 = 2.5$ Mton/year; $V_F = 2.0$ Mton/year; $K_F = 1$, and, $K_I = 1$ or 3 Mton, for low and high supply curves, respectively

$$V_F = V_C Z/(K_M + Z) \qquad [2]$$

There are two other variables, X, the amount of green coffee beans in the producing farmlands, and Y, the coffee inventory in the hands of the middlemen, to have a complete dynamic system which follow the usual mass balance considerations. Here for simplicity it has been assumed that coffee losses along the market chain are negligible as indicated in Eqs. 3 to 5.

$$dX/dt = V_O - k_1X \qquad [3]$$

$$dY/dt = k_1X - k_2X \qquad [4]$$

$$dZ/dt = k_2X - V_F \qquad [5]$$

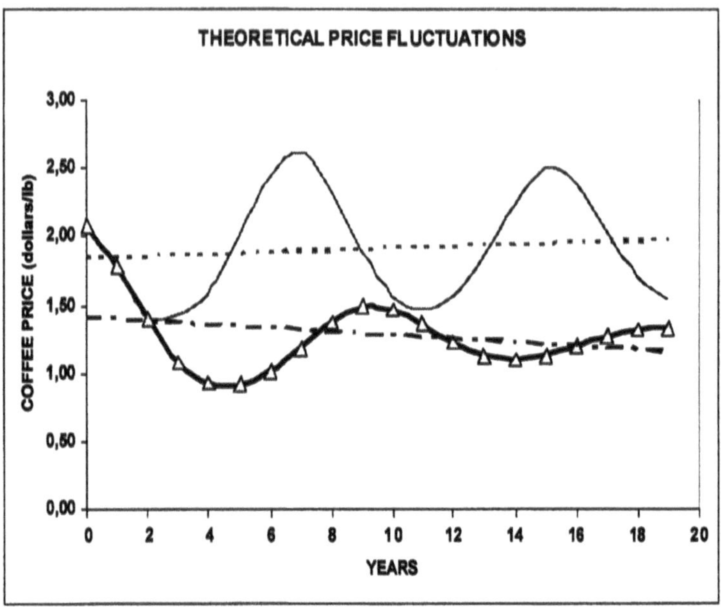

Figure 5. Calculated price fluctuations for two different steady-state (equilibrium) conditions. The upper curve (squares) with steady oscillations corresponds to the equilibrium condition with low coffee supply level ($K_1 = 1$ Mton) and for that reason shows higher price levels. The bottom curve (triangles) with damped oscillations shows lower price levels because of the higher level of coffee supply ($K_1 = 3$ Mton). Simulations were calculated with an Excel spreadsheat using a first order Euler approximation. Differential equations used are shown above. Numerical values of the other parameters were: $V_0 =$ 2.5 Mton/year; $V_F =$ 2.0 Mton/year; $K_F = 1$ Mton; $k_1 = k_2 = 1$/year; $p = 2$.

2.2 BOTTLENECK FOR AGRICULTURAL CASH PRODUCTS

The market would be at equilibrium if each one of the rates of change, dX/dt, dY/dt, dZ/dt vanish together. This amounts to the so-called steady-state conditions which is possible only when V_O (input) = V_F (output). Figure 4 shows the graphical solution for such a condition. In one theoretical case, the saturation level is high (i. e., $K_i = 3$ Mton) and in other theoretical case, the saturation level is low ($K_1 = 1$ Mton). Hence, the market bottleneck is the final rate of consumption, V_F, because there is a maximum amount of coffee that consumers can buy, and also, because the rate of growth of consumers is rather low as compared to the rate of growth of production. Statistical evidence from FAO statistics shows that, in fact, the main price has a decreasing trend (Fig. 2), suggesting that production is overcoming consumption, because there is a definite negative correlation between coffee stocks and prices (Fig. 3).

2.3 OSCILLATING BEHAVIOUR OF SUPPLY AND DEMAND

Price instabilities are a good reason to study market dynamics, specially, when fluctuations are quasi periodic and have wide amplitude. Figure 2, shows unequivocal evidence of sustained coffee price oscillations as commented above. The source of those oscillations usually is a delay between the market final response (accumulation of inventories, Z) and the primary input adjustments (coffee bean production, X). For example, agricultural production may have oscillatory behaviour if there is a delay between planting, culturing and harvesting. Thus, in coffee production the plant becomes mature after four of five years after being planted. This fact of life makes difficult to decide whether to stop and keep going with the production, even though the coffee prices have already fallen one or two years before.
The model represented by the dynamic equations 3 to 5 can be solved numerically as shown in Figs 5 and 6 for the two kinds of market equilibria presented in Fig. 4. The equilibrium point with high level of supply (K_1 = 3 Mton) shows damped oscillations, whereas, the equilibrium point with low level of supply (K_1 = 1 Mton) shows sustained oscillations. Such theoretical results seem to indicate that market behaviour is steadily oscillatory when there is a shortage of inventories. On the other side, market stability with a high level of coffee inventories may be too dear, because stock accumulation is associated with low prices and financial losses for coffee producers. Therefore, the main question is how to stabilise the coffee market without incurring in price crisis. The obvious answer is to develop product and by product diversification which goes beyond the usual production of coffee as a beverage but also as a rich source of high price chemicals, specially, from cheap and polluting materials such as coffee pulp and coffee bean residues.

2.4 SOCIAL IMPORTANCE AND DIFFICULTIES OF PRODUCT DIVERSIFICATION

Product diversification is well developed for crops such as cereals and legumes. For example, the large surplus of maize from U.S.A. is used to produce a wide variety of products such as, animal feeds, fructose syrups, alcohol, fermentation products, edible oil, starch, gluten, plastics, etc. In Brazil, sugar cane uses have been diversified to include the alternative production of gasohol as a way to cope with increasing fuel demands and saturation of sugar market. However, there are many problems in finding successful alternative markets for traditional products such as coffee because of the following two main factors:

a) The low level of R & D investment in tropical countries. This hinders the discovery of new uses for traditional products.
b) The complex web of social, cultural, political and financial arrangements that are associated to traditional crops and which cannot be changed overnight in order to develop new industrial concerns for alternative uses of conventional or waste products.

An interesting example of product diversification is the case of grape and wine industry in France. The traditional main product for French grape yards is wine, but wine is also a product that has found market saturation for more than a century in Europe. To cope with this problem, the French government has issued a number of codes and regulations that help to manage wine surpluses and also help to clean the environment. This comprises a network of distilleries producing liquor from the grape pomace or marc, as it is called the solid grape residue from wine making. The usual arrangement is between a number of wine co-operatives that get together to finance and manage a distillery. The main role of such an enterprise is to produce brandy or "eau de vie" from the residual sugar in the pomace. Surplus amounts of grapes can also be transformed in liquors when the price of wine falls down, working as alternative outlets for crop surplus. This scheme has various advantages:

a) There is no distraction for primary and traditional wine making in the co-operatives.
b) Special factories are dedicated to new products with new ideas derived from R&D.
c) Surplus raw materials (grape) can be transformed into new products, such as wine pigments, grape seed oil and grape sugar syrups, without entering the same market for the traditional or basic product, i.e., wine.
d) Waste materials (grape pomace) can be transformed in new value added products, helping to finance the cost of waste disposal.

It should be stressed that French distilleries were not developed as an answer to conventional market forces in the usual sense of private supply and demand. They were a social adaptation to a long series of political and social problems derived from wine market instability at the end of 19th century and now are becoming a response to the problem of solid grape waste disposal. In other wine making countries different ways of solving the problem of grape over production and waste disposal have been developed, suggesting that product diversification is a problem to be tackled in close relation to the social conditions of each given region.

3. Alternatives for product diversification from coffee derived materials

3.1 COFFEE PULP AS A SOURCE OF NEW MATERIALS (ANTIOXIDANTS AND ANTITUMORAL DERIVATIVES)

The solids consumed as coffee beverage are only 6% of the solids present in the cherry. Most of coffee materials usually are rotten in the field as manure with little if any economic benefit for the coffee producers or transformers. Chemical composition of the coffee pulp shows that it contains sugars, pectin and phenols, etc. In fact, caffeine is 1.5% of the coffee solids and is the only coffee chemical that is extracted and marketed around the world, mainly as a low price additive for cola drinks. Thanks to the pioneer work done at the University of Zulia in Venezuela, we now have a more clear picture of the types of chemicals in coffee cherries and pulp (Clifford, 1979; Ramírez-Martínez, 1988; Clifford and Ramírez-Martínez, 1991; De Colmenares *et al.*, 1994, 1998). It has been found that the main contribution to phenolic compounds in the coffee beans and pulp are, chlorogenic acid, 42.2%, epicatechin, 21.6%, isochlorogenic acid I, 5.7%, isochlorogenic acid II, 19.3% (Ramírez-Martínez, 1988). Also, condensed tannins are abundant in the coffee pulp from fresh cherries but they polymerise rather quickly after drying due to spontaneous oxidation (De Colmenares *et al.*, 1994). Chlorogenic acids can be characterised by gel filtration (Trugo *et al.*, 1991), or by HPLC analysis (Bicchi et al., 1995) and can be obtained from coffee pulp and coffee waste (after roasting the beans) using solvent extraction (Dibert and Cros, 1989; Dibert *et al.*, 1989; Bicchi *et al.*, 1995). This kind of chemicals are now considered as food antioxidants with some evidence that may protect animal cells against somatic mutations associated to cancer (Huang and Ferraro, 1992; Abraham, 1996; Guillot et al., 1996). New chemical derivatives of coffee phenolic compounds can be obtained by selective enzyme esterification (Guyot *et al.*, 1997) and transglycosylation (Millqvist-Fureby *et al.*, 1998). An interesting use of proanthocyanidin from coffee cherries is the inhibition of coffee parasite *Hemileia vastatrix* (De Colmenares *et al.*, 1998). Also, an important part of coffee flavour and fragrance is related to chologenic acids (Bicchi *et al.*, 1995) and is being considered in the manufacture of coffee substitutes (Haffke and Engelhardt, 1986). An interesting possibility is the use of coffee cells in biotransformation processes in order to produce valuable fine chemicals involved in their metabolism (Koge *et al.*, 1992 a, b)

3.2 COFFEE PULP AS A RAW MATERIAL FOR FERMENTATION PROCESSES

Coffee pulp can be used as a substrate for the production of enzymes by microbial fermentation (Antier *et al.*, 1993; Srivastava, 1993; Roussos *et al.*, 1995; Baracat-Pereira *et al.*, 1997). Such enzymes can be used in turn as catalysts for food industries, including some improvements on coffee processing i.e., pectin breakdown using cleaner techniques for coffee depulping.

Traditional coffee fermentation can be improved by the adequate use of micro-organisms (Castelein and Verachtert, 1984) and also as a raw material for producing biogas, compost, single cell protein (Tauk, 1986; Rajasekhar *et al.*, 1983) and citric acid (Shankaranand and Lonsane, 1994).

3.3 COFFEE PULP AS A RAW MATERIAL FOR FOOD, FEED AND FERTILISERS

Mushroom production is an interesting way to dispose of coffee pulp. For example, the edible mushroom *Pleurotus ostreatus* can be grown on plastic bags filled with pasteurised coffee pulp (Martinez, 1984; Calzada et al., 1987). Also *Pleurotus sajor-caju* has been grown on coffee pulp (Wong and Wang, 1991) showing that tannin content is greatly diminished, leaving a spent residue to be used for animal feeding or for composting.

cherries yield 600 kg of pulp and 300 kg of green beans, the potential sales value for both products is of the same order of magnitude. Mushroom market is important around the world but is constrained by habits and uses in different countries. Unfortunately, many of the tropical countries do not have a strong demand for mushrooms in their diet. However, the combined world market of edible mushrooms (mainly *Agaricus bisporus, Pleurotus sp., Lentinus edodes)* has a sales figure similar to the combined market of industrial fermentation products, such as organic acids and enzymes.

Coffee pulp can be used for animal feeding and has been the subject of significant amount of research (Aregheore, 1998; Villagran Blanco, 1981; Okai *et al.*, 1985; Abate, 1988; Abate and Pfeffer, 1986; Porres *et al.*, 1993; Habte, 1989; Mason and Giner, 1979; Fagbenro and Arowosoge, 1991; Rolz *et al.*, 1988; Peñaloza *et al.*, 1985; Larde, 1989, 1990 a, b; Givens and Barber, 1986; Bressani and Gonzalez, 1978; Sikka et al.., 1985; Peñaloza-Izurieta, 1981; Vargas et al., 1982; Abate, 1984). But has three major problems:

a) the final value of the product is low since it has to compete with abundant low price feedstuff,
b) caffeine, tannin and other phenolic compounds have important anti-nutritional value and are difficult to remove and,
c) it is a perishable material that is not worth drying due to the low added value of the final product.

However, if coffee pulp and other coffee residues are the by-products from other more cost-effective processes, such as, mushroom production or chlorogenic, caffeine and tannin extraction, the spent residue free from anti-nutritional compounds, would certainly be a valuable material to support animal production in the tropics.

Composting coffee pulp is an easy process to develop in the farmlands using earthworms. Mechanical composting will be forbidding because of the very low value of the compost. However, the use of organic farming in order to produce coffee seedlings and the need to make an "ecological" or "green" impression on sophisticated buyers, may be a sufficient reason to compost coffee pulp. Specially if the environmental regulations preclude dumping the waste in rivers or require stabilisation before land use as a manure. Fortunately enough, there is now significant amount of research results and practical applications for composting this material (Onsando and Waudo, 1982; Orozco *et al.*, 1996; Uribe-Henao, 1983; Tauk, 1984, 1985 a, b; Wu, 1995). What remains to be done is the practical development of this technology whenever and wherever it becomes feasible.

Anaerobic digestion of coffee residues, specially the digestion of the spent waters from the wet process, has been studied through the past ten years (Villagrán Blanco, 1981; Calzada et al., 1984 a, b, 1986; Leon *et al.*, 1980; Boopathy, 1987; Lanting *et al.*, 1988; Bello Mendoza *et al.*, 1997). This is the best option to clean wastewater from the wet coffee process. But it may be too expensive if there is no previous reduction of the amount of spent water by judicious changes in depulping, material handling and water managing. Experiences in Chinchiná, Colombia, indicated that adequate water management could reduce water consumption to less than 5 L per kg of cherries, when the usual amount is more than four times larger. With such words of caution, anaerobic digestion will be the ideal solution to water pollution in coffee mills using the wet process. In this context, biogas will be a cheap by product to be used locally as a clean fuel for many thermal processes needed, namely, drying coffee beans.

3.4 IMPROVEMENTS OF COFFEE BREEDING AND PEST CONTROL

Plant biotechnology is exploding around the world. Many new techniques for cultivar development are being patented and published. It would be far too ambitious to have an adequate review of this hot research subject. But some reviews are available (Villalobos, 1989; Shand, 1990; Chatterjee et al., 1991; Colton, 1992; Sondahl and Lauritis, 1992; Quesada-Chanto and Jimenez-Ulate, 1996; Carneiro, 1997; Moisyadi *et al.*, 1998). Here some thoughts are given on a difficult but familiar problem. What is first? Production or marketing? Here we assume that product development ought to be based on the market trends. Sophisticated consumers want "organic" products. That means, coffee without obnoxious chemicals, such as pesticide residues. They also want "health" beverages, with safe and powerful antioxidants and less caffeine. Finally, they also want good quality coffee that means distinctive aroma and flavour. Of course, they also want coffee to be sold for the least amount of money.

Coffee producers are not the same. Some produce large amounts of low price coffee. Perhaps to be industrialised and transformed in decaffeinated instant coffee and caffeine. Other sell their best quality product to sophisticated buyers that request special flavours and aroma and request the product to be "organic". Many sell the coffee beans in local markets with low budgets. Obviously, there is no way to have a single type of coffee and many varieties should be developed. But one thing is sure, there is no substitute for good science in coffee breeding. The old fashion techniques of coffee breeding are being transformed by the use of molecular biology techniques. Soon, the genetic maps of many plants will be available and some of those maps may be restricted to company databases. But even if those databases were public, there is a need of basic knowledge to retrieve them by Internet and to read them in a way that is meaningful for the plant breeder. In this case the shortage is in highly trained people in the field of plant molecular biology and also there is a shortage of venture capital because cultivar development is becoming competitive and expensive.

Taking as reference the case of recombinant tomato called "fresh'n savr" produced by the US company called Calgene (now owned by Monsanto). It took nearly 10 years and 20 million dollars to develop it. But after it was licensed, it became a commercial flop, mainly because of image problems with the products of genetic engineering and also because of some unexpected problems of pest control, since the new genetically engineered tomato resulted sensitive to mouldy infections.

Perhaps there is a need for a blend between old and new plant breeding techniques. For example, new phenotypes can be matched with genotypes if adequate molecular markers are known. This in turn, depends from DNA probes and DNA sequences that are known to be related to specific genes. Thus a combination of molecular and agricultural

techniques may be the easiest way to overcome many social objections to the term "genetic engineering". This way, "natural" mutations can be screened and micro propagated in vitro. Helping to overcome the natural time constraints of developing new varieties in the old fashion way. But in order to do so, more information is needed for the coffee phenotypes at molecular level and this requires a sound knowledge of the biochemistry and physiology of the plant, together with the proper commercial information on the convenience of the chosen phenotype.
Unfortunately, these new developments of coffee biotechnology happen when public budgets are being slashed down in tropical countries. Coffee is a tropical plant to be studied and developed in the tropics. However most of the advanced publications on plant physiology and genetics are done in European research institutes devoted to coffee biotechnology which now are retreating from their traditional experimental grounds in Africa. Also, large companies from Switzerland, and US, which have important commercial interests in Latin America, do not have important R&D institutions in this region. Thus, the investment of R&D for coffee biotechnology is not increasing as much as needed by the present predicament of coffee market discussed above.

4. The role of Biotechnology in the new context of coffee production and marketing

As has been suggested at the outset of this article, there is a need to face a serious problem of market instability and a decreasing trend of coffee prices. Therefore, product diversification may be quite an adequate answer to this problem. In this context, biotechnology offers interesting possibilities.
In the first place, there is a need for improving the public image of coffee based in high value added products that are good for the health of most people. Recent results in the analysis of chlorogenic acids present in green beans and coffee pulp seem to be a promising way both to develop new varieties of coffee plants which are richer in antioxidants and poorer in caffeine but with good flavour and aroma.
Another interesting possibility is to transform coffee pulp in high added value products such as the same type of antioxidants, pigments and tannins, to be formulated for a large variety of new food and "nutraceutic" products (fine chemicals that are used as food additives). Surely, many people will be willing to try new food additives if they have the promise to reduce cancer and heart diseases. Hence, there is the need to do basic research on the health uses of coffee derivatives, which warrant an expansion of coffee industry.
Coffee producers have to comply with increasing environmental regulations. Dumping waste on the land or in rivers may be an easy way to dispose of such materials but environmental regulations will force them to cleaner ways. Perhaps the best way is to find new by products such as mushrooms, enzymes, citric acid, compost and biogas, in order to reduce the net cost of the new environmental regulation. In this sense, pest resistant varieties of coffee will be necessary to reduce or eliminate the use of polluting pesticides. In all such cases, biotechnology offers interesting possibilities to find

practical answers to those problems. The brief literature review presented in this paper supports this notion.
There is no substitute for knowledge. Knowing how to do things is an important way to achieve power in this world. Unfortunately, knowledge is also expensive and someone has to pay to obtain the practical gains form this knowledge. Thus the predicament of many tropical countries is how to finance research and development in order to protect and further the development of traditional endeavours such as coffee making and marketing. The author hopes that international co-operation will be a significant pathway to remove many of the hindrances to support and develop the biotechnology applied to this field.

5. References

Abate A. (1984), Investigation of coffee pulp as a feed ingredient for young goats in Kenya. Inaugural-Dissertation Universitat Bonn. Landwirtschaftliche Fakultat

Abate A., Pfeffer E. (1986), Changes in nutrient intake and performance by goats fed coffee pulp-based diets followed by a commercial concentrate, Animal Feed Sci. Technol., 14 (1/2), 1-10

Abate A. (1988), Coffee pulp: some indices of nutritional importance. Bull. Animal Health and Production in Africa (Bulletin de la santé et de la production animale en Afrique), 36 (1), 39-45

Abraham S.K. (1996), Anti-genotoxic effects in mice after the interaction between coffee and dietary constituents, Food Chem. Toxicol., 34 (1): 15-20

Antier P., Minjares A., Roussos S., Raimbault M., Viniegra-Gonzalez G. (1993), Pectinase-hyperproducing mutants of *Aspergillus niger* C28B25 for solid-state fermentation of coffee pulp. Enz. Microbiol., Technol., 15: 254-260

Aregheore E.M. (1998), A review of implications of antiquality and toxic components in unconventional feedstuffs advocated for use in intensive animal production in Nigeria, Veter. Human Toxicol., 40 (1): 35-39

Baracat-Pereira M.C., Minussi R.C., Coelho J.L.C., Silva D.O. (1997), Tea extract as an inexpensive inducer of pectin lyase in *Penicillium griseoroseum* cultured on sucrose, J. Ind. Microbiol. Biotechnol.,18 (5): 308-311.

Bello Mendoza R., Sanchez V,J.E. (1997), Anaerobic filter treatment of wastewater from mushroom cultivation on coffee pulp World J. Microbiol. Biotechnol., 13 (1): 51-55

Bicchi C P., Binello A.E., Pellegrino G.M., Vanni A.C. (1995), Characterisation of green and roasted coffees through the chlorogenic acid fraction by HPLC UV and principal component analysis. J. Agric. Food Chem., 43 (6): 1549-1555

Boopathy R. (1987), Inoculum source for anaerobic fermentation of coffee pulp, Appl. Microbiol. Biotechnol, 26(6): 588-594

Bressani R., Gonzalez J.M. (1978), Evaluation of coffee pulp as a possible substitute for maize in rations for broiler chicks, Arch. Latinoamer. Nutri., 28 (2): 208-221.

Calzada J.F., Arriola, M.C de., Castañeda, H. O., Godoy, J.E. and Rolz, C. (1984a), Methane from coffee pulp juice: experiments using polyurethane foam reactors, Biotechnol. Lett., 6 (6): 385-388

Calzada, J.F., Porres E. Yurrita A., Arriola M.C., Micheo F. (1984b), Biogas production from coffee pulp juice: one and two phase systems, Agric. Wastes, 9 (3): 217-230.

Calzada J.F., Rolz C., Del Carmen de Arriola M., Castañeda H., Godoy J E. (1986), Biogas production from coffee pulp juice using packed reactors: scale-up experiments, MIRCEN J. Appl. Microbiol. Biotechnol., 2(4): 489-492

Calzada J.F., León R. de., Arriola M.C., Rolz C. (1987), Growth of mushrooms on wheat straw and coffee pulp: strain selection. Biol. Wastes, 20(3): 217-226

Carneiro M.F. (1997), Coffee biotechnology and its application in genetic transformation, Euphytica 96 (1): 167-172

Castelein J., Verachtert H. (1984), Coffee fermentation, Biotechnology, 5:587-615

Chatterjee G., Singh G., Thangam P. (1991), Clonal propagation of bamboo, coffee and mimosa. In-Horticulture. New Technologies and Applications (J. Prakash and R. L. M. Pierik, eds.), Bangalore, India, pp. 261-264

Clifford M.N. (1979), Chlorogenic acids their complex nature and routine determination in coffee beans, Food Chem., 4 (1): 63-71

Clifford M.N., Ramirez-Martinez J.R. (1991), Phenols and caffeine in wet-processed coffee beans and coffee pulp, Food Chem., 40(1): 35-42

Colton R.L. (1992), Soluble coffee's new biotechnology. Developments in food science. Food Sci. Human Nutr., 29 341-346

De Colmenares N.G., Ramirez-Martinez J.R., Aldana J.O., Clifford M.N. (1994), Analysis of proanthocyanidins in coffee pulp, J. Sci. Food Agric., 65 (2): 157-162

De Colmenares N.G., Ramirez-Martinez J.R., Aldana J.O., Ramos-Niño, M E., Clifford, M., Pekerar, S. and Mendez, B. (1998), Isolation, characterisation and determination of biological activity of coffee proanthocyanidins. J. Sci. Food Agric., 77(3): 368-372

Dibert K. and Cros, E. (1989), Solvent extraction of oil and chlorogenic acid from green coffee. I. equilibrium data. J. Food Eng., 10 (1): 1-11

Dibert, K., Cros E., Andrieu J. (1989), Solvent extraction of oil and chlorogenic acid from green coffee. II: Kinetic data. J. Food Eng., 10(3): 199-214

Fagbenro O.A., Arowosoge I.A. (1991), Growth response and nutrient digestibility by *Clarias isheriensis* (Sydenham, 1980) fed varying levels of dietary coffee pulp as replacement for maize in low-cost diets. Biores. Technol., 37(3): 253-258

Givens D.I., Barber W.P. (1986), In vivo evaluation of spent coffee grounds as a ruminant feed. Agric. Wastes, 18 (1): 69-72

Guillot F.L., Malnoe A., Stadler R.H. (1996), Antioxidant properties of novel tetraoxygenated phenylindanisomers formed during thermal decomposition of caffeic acid. J. Agric. Food Chem., 44 (9): 2503-2510

Guyot B., Bosquette B., Pina M., Graille J. (1997), Esterification of phenolic acids from green coffee with an immobilised lipase from *Candida antarctica* in solvent free medium. Biotechnol. Lett., 19 (6): 529-532

Habte T.-Y. (1989), Investigations on the nutritional value of coffee pulp. PhD Thesis, Justus-Liebig-Universitat Giessen

Haffke H., Engelhardt U.H. (1986), Chlorogenic acids in coffee substitutes. Chlorogensauren in Kaffee-Ersatzstoffen. Zeit. Lebensmittel-Unter. Forschung, 183(1): 45-46

Huang M.T., Ferraro T. (1992), Phenolic compounds in food and cancer prevention. ACS Symposium Series, (507): 8-34

Koge K., Orihara Y., Furuya T. (1992a), Effect of pore size and shape on the immobilisation of coffee (*Coffea arabica* L.) cells in porous matrices. Appl. Microbiol. Biotechnol., 36(4): 452-455

Koge K., Orihara Y., Furuya T. (1992b), Immobilisation and culture of coffee (*Coffea arabica* L.) cells on novel culture systems using a basket-shaped unit, Biotechnol. Tech., 6(4): 313-318

Lanting J., Jordan J.A., Schone M.T., Kull A., Carey W.W., Kitney B.L. (1988), Thermophilic anaerobic digestion of coffee wastewater. Proc. Ind. Waste Conf., Purdue University. (43rd): 513-524

Larde G. (1989), Investigation on some factors affecting larval growth in a coffee-pulp bed. Biol. Wastes 30(4): 11-19

Larde G. (1990a), Growth of *Ornidia obesa* (Diptera:Syrphidae) larvae on composting coffee pulp. Biol. Wastes, 34(1): 73-76

Larde G. (1990b), Recycling of coffee pulp by *Hermetian lucens* (Diptera: Stratiomyidae) larvae. Biol. Wastes, 33 (4): 307-310

Leon R., Calzada F., Herrera R., Rolz C. (1980), Fungal biomass production from coffee pulp juice for producing compost, biogas, feeds, ethanol, pectin and other chemical compounds. J. Ferment. Technol., 58(6): 579-582

Martinez D. (1984), *Pleurotus ostreatus* cultivation on agricultural wastes. I. Isolation and characterisation of native strains in different solid media in the laboratory. Biotica, 9 (3): 243-248

Mason J.F., Giner L. (1979), "Converting coffee pulp into cattle feed. Food Eng. Internatl., 4(9): 35

Millqvist-Fureby A., Mac Manus D.A., Davies S., Vulfson E.N. (1998), Enzymatic transformations in supersaturated substrate solutions. II. Synthesis of disaccharides via transglycosylation. Biotechnol. Bioeng., 60 (2): 197-203

Moisyadi S., Neupane K.R., Stiles J.I. (1998), Cloning and characterisation of a cDNA encoding xanthosine N7 methyltransferase from coffee (*Coffea arabica*). Acta Hortic., (461): 367-377

Okai D.B., Bonsi M.L.K., Easter R.A. (1985), Dried coffee pulp (DCP) as an ingredient in the diets of growing pigs. Trop. Agric., 62(1): 62-64

Onsando J.M., Waudo S.W. (1992), Effect of coffee pulp on *Trichoderma* sp: in Kenyan tea soils. Trop. Pest Management,.38(4): 376-381

Orozco FH., Cegarra J., Trujillo LM., Roig A. (1996), Vermicomposting of coffee pulp using the earthworm *Eisenia fetida*: effects on C and N contents and the availability of nutrients. Biol. Fertil. Soils, 22(1/2): 162-166

Peñaloza Izurieta W. (1981), Solid fermentation of coffee pulp. MS Thesis, Universidad de San Carlos de Guatemala, Centro de Estudios Superiores en Nutricion y Ciencias de Alimentos

Peñaloza W., Molina M.R., Brenes R.G., Bressani R. (1985), Solid-state fermentation: an alternative to improve the nutritive value of coffee pulp. Appl. Environ. Microbiol., 49(2): 388-393

Porres C., Alvarez D., Calzada J. (1993), Caffeine reduction in coffee pulp through silage. Biotechnol. Adv., 11(3): 519-523

Quesada-Chanto A., Jimenez-Ulate F. (1996), In vitro evaluation of a *Bacillus* sp for the biological control of the coffee phytopathogen *Mycena citricolor*. World J. Microbiol. Biotechnol., 12 (1): 97-98

Rajasekhar T., Udayshankar K., Abraham K.O., Seshadri R., Shankaranarayana M.L. (1983), Studies on the utilisation of coffee by products (Husk, pulp, skin, leaves, twigs, wood) possible uses as manure, fuel, biogas, coffee pulp molasses, protein and caffeine extraction. Indian Coffee, 47 (4): 9-11

Ramirez-Martinez J.R. (1988), Phenolic compounds in coffee pulp: quantitative determination by HPLC. J. Sci. Food Agric., 43(2): 135-144

Rolz C., Leon R., Arriola M.C. (1988), Biological pre-treatment of coffee pulp. Biol. Wastes, 26(2): 97-114

Roussos S., Aquiáhuatl M de los A., Trejo-Hernandez M del R., Perraud I.G., Favela E., Ramakrishna M., Raimbault M.,Viniegra-Gonzalez G. (1995), Biotechnological management of coffee pulp - isolation, screening, characterisation, selection of caffeine-degrading fungi and natural microflora present in coffee pulp and husk. Appl. Microbiol. Biotechnol., 42(5): 756-762

Shand H. (1990), Coffee and biotechnology. Genewatch, 6 (2/3): 12-13

Shankaranand VS., Lonsane BK (1994), Coffee husk: an inexpensive substrate for production of citric acid by *Aspergillus niger* in a solid-state fermentation system. World J. Microbiol. Biotechnol., 10 (2): 165-168

Sikka S.S., Bakshi M.P.S., Ichhponani J.S. (1985), Evaluation *in vitro* of spent coffee grounds as a livestock feed. Agric. Wastes, 13(4): 315-317

Sondahl M.R., Lauritis J.A. (1992), Coffee biotechnology in agriculture. In: Biotechnology of Perennial Fruit Crops (eds. F. A. Hammerschlag and R. E. Litz), pp. 401-420

Srivastava K.C. (1993), Properties of thermostable hemicellulolytic enzymes from *Thermomonospora* strain 29 grown in solid state fermentation on coffee processing solid waste. Biotechnol. Adv., 11 (3): 441-465

Tauk S.M. (1984), Identification of fungi isolated from coffee pulp. Naturalia, 9:57-60

Tauk S.M. (1985a), Effect of the addition of nitrogen and phosphorus on the growth of *Candida utilis* in coffee pulp juice. Rev. Microbiol., 16 (3): 188-194

Tauk S.M. (1985b), Use of fungal inocula and pumice for composting coffee pulp. Agric. Ecosystems Environ., 14 (3/4): 291-298

Tauk S.M. (1986), Effect of concentration of total sugar and aeration on the growth of *Candida utilis* in culture of coffee pulp. Rev. Microbiol., 17 (3): 254-263

Trugo L.C., Maria C.A.B., Werneck C.C. (1991), Simultaneous determination of total chlorogenic acid and caffeine in coffee by high performance gel filtration chromatography. Food Chem., 42 (1): 81-87

Uribe-Henao A. (1983), Influence of composted coffee pulp on coffee yield. Cenicafe, 34(2): 44-58

Vargas E., Cabezas M.T., Murillo, B., Braham, E. and Bressani, R. (1982), Effect of high levels of dehydrated coffee pulp on the growth and adaptation of young steers. Arch. Latinoamer. Nutri.,32 (4): 973-989

Villagrán Blanco L.R. (1981), Evaluación química de tratamientos microbiológicos anaerobicos para reducir los factores antifisiologicos de la pulpa de café. Master's Thesis, Facultad de Ciencias y Farmacia. INCAP T, Universidad de San Carlos de Guatemala

Villalobos V.M. (1989), Advances in tissue culture methods applied to coffee and cocoa. Plant Biotechnologies for developing countries. Proc. International Symp., Technical Centre for Agricultural and Rural Co-operation and the FAO, Luxembourg

Wong Y.S., Wang X. (1991), Degradation of tannins in spent coffee grounds by *Pleurotus sajor-caju*. World J. Microbiol. Biotechnol., 7(5): 573-574

Wu N. (1995), Composting coffee pulp in El Salvador. BioCycle, 36 (11): 82-83

Chapter 2

COFFEE CULTIVATION IN INDIA

C.S. Srinivasan, N.S. Prakash, D. Padma Jyothi, V.B. Sureshkumar, V. Subbalakshmi

Central Coffee Research Institute, Coffee Research Station, Chikmagalur-577 117, India

Running title: Coffee cultivation in India

1. Introduction

Although coffee is not an essential commodity for human consumption, its stimulating properties are too well known to ignore its importance in human health, if taken at moderate levels. In India, coffee has a place of pride among plantation crops grown and is traditionally cultivated on the south - western hill slopes since 150 years. As history indicates, the commercial cultivation of coffee in India was started during early 19th century by the European entrepreneurs (Anonymous, 1995). These early coffee plantations have flourished in South India for nearly a century till the leaf rust epidemics were seen and soon the leaf rust became the disease of concern for arabica coffee in view of its high economic importance. Incidentally, the largest number of its races (as many as 30) has been recorded from India. Thus, the fate of coffee industry, especially arabica has been largely dependent on release of varieties resistant to rust with acceptable yield levels, bean standards and cup quality. The Central Coffee Research Institute (CCRI) established near Balehonnur in Chikmagalur District of Karnataka State has rendered yeoman services in breeding and release of improved arabica and robusta varieties to the growers from time to time since its establishment in 1925 as "Mysore Coffee Experimental Station". The success achieved in developing rust resistant varieties in India during 1940's (Narasimhaswamy, 1961) are pioneering ones in the history of coffee breeding. It is pertinent to mention that the classical work of Mayne (1932, 1936, 1939) on the existence of physiological races in rust pathogen supplemented the breeding efforts not only in India but also elsewhere in the world.

T. Sera et al. (eds.), Coffee Biotechnology and Quality, 17–26.

1.1. GEOGRAPHICAL INFORMATION

In India, the major coffee tracts are distributed in hill ranges of Southern part of the country between 9^0 and 14^0 N latitudes, and 500-1500 m altitudes; the arabica in high elevations and robusta in low lands. The annual rainfall varies between 1000–2500 mm with the temperature ranges between 8- 34^0C.

2. Arabica improvement

In India, improvement of arabica coffee was started even before the commencement of organized research. Some enterprising planters made frantic efforts to develop superior cultivars through selection/hybridization, in order to tackle the rust disease. In this context, hybrids such as 'Hamiltons', 'Jacksons', 'Netrakonda' and selections like 'Coorgs', 'Chicks' and 'Kents' were some important ones, which found place in the pages of coffee breeding history. Nevertheless some of these selections are still found here and there in commercial plantations. 'Kents' is one such selection made by a private planter which served as the major source for planting during the 1920s and is still internationally acclaimed for its liquor quality. In spite of the best efforts of the planting community, plant materials with stable resistance could not be developed due to the ability of the rust pathogen to mutate into virulent races. This necessitated the organized research activities for improvements of arabica coffee in order to evolve superior cultivars with leaf rust resistance, high yielding potential, superior quality and wide adaptability.

2.1. ORGANIZED RESEARCH TOWARDS GENETIC IMPROVEMENT

Since the inception of organized research in coffee in India during 1925, development of improved strains using conventional breeding techniques has been found successful. The success of breeding programme in any crop depends on the exploitation of the available genepool. As a first step towards varietal improvement, initial germplasm with about 250 diverse collections of arabica coffee from different coffee growing estates were established at CCRI during 1925-40. In subsequent years, several exotic wild collections of coffee and cultivated forms of arabica and robusta were introduced from other coffee growing countries to establish a massive gene bank. Presently, the Institute's gene bank comprises of over 350 arabica collections, 15 robusta types and 18 different species of *Coffea* and is considered as one of the recognized germplasm centers for coffee in the world. The individual collections of the gene bank have been assessed and genes conferring rust resistance, bush formation, production potential, quality and drought hardiness have been identified and exploited from time to time in developing 12 superior strains of arabica for commercial cultivation.

2.2. EARLY INDIAN SELECTIONS (1925-1960)

Over the years the simultaneous cultivation of both arabica and other diploid species such as robusta and liberica has resulted in the natural hybridization between these types. The early Indian selections such as S.288 and S. 795 were evolved from some of these natural hybrids identified during the indigenous germplasm Collection.

S. 288 was the first selection released during 1936-37. It was superior to other varieties such as 'Chiks' and 'Kents' under cultivation then, by showing resistance to leaf rust races II and I. Strain S. 288 was developed by pure line selection from S.26 mother plant a putative natural hybrid between liberica and arabica. This selection has the potential to give moderate yields of about 1000 kg/ha but produces defective beans to the tune of 20-30%. To improve the quality aspects of S.288, it was crossed with 'Kents' and S.795 the most popular strain of Indian arabica was developed. Since its release in 1945-46, S.795 attracted the attention of planters and is still the most popular selection occupying large area (60%) under cultivation. The bushes of this variety are tall, very vigorous, wide spreading with profuse growth and high yield potential of 2000 kg/ha. This selection is widely adaptable to different agro-climatic zones. The beans are oblong, bold representing 70% 'A' grade with good cup quality (Narasimhswamy, 1960).

The S_{H3} factor responsible for resistance to several physiologic races of leaf rust pathogen was identified in these strains with possible introduction from *C. liberica*. Thus, these two commercial arabica cultivars of Indian origin differ from the rest of pure arabica cultivated in the Globe elsewhere (Wagner and Bettencourt, 1965). The S.288 and S.795 remained resistant over a period of time till uncommon rust races VIII, XII and XIV became common on these cultivars due to host preference. This necessitated further work to develop resistance to race VIII the most common and virulent race affecting S.795 populations. As some of the exotic introductions from Ethiopia, such as Cioccie, Agaro and S_{12} Kaffa were found resistant to race VIII, Sln.4 was developed following pure line selection in some of these collections. This selection includes three Ethiopian arabica collections viz. Cioccie, Agaro and Tafarikela. Among these, the first two exhibit semi-erect branching, while Tafarikela shows drooping nature. All these are tall types and moderate yielding cultivars. Tafarikela is characterized by an early ripening. Cioccie and Agaro are resistant to common rust races I, II and also to race VIII.

Tafarikela exhibits general tolerance to leaf rust pathogen showing very less rust build up under field conditions, even though susceptible to race I & II under laboratory screening.

2.3. RECENT DEVELOPMENTS (1960-1985)

2.3.1. Tetraploid inter-specific hybrids:

In subsequent years, focus was shifted on development of spontaneous as well as synthetic hybrids. 'Devamachy', a putative robusta x arabica hybrid of indigenous origin, was exploited to develop Selection 5. This selection comprises of two important families derived by crossing Devamachy x Rume Sudan and Devamachy x Indian selection S.333. Devamachy x Rume Sudan line is known for its field tolerance to rust and in recent screening at CIFC reported to be resistant to Coffee Berry Disease also.

Table 1. Bean grades and quality descriptors of S.2931.

Location	Bean grade percentage					Quality descriptors
	A	B	C	Pb	T	
CCRI	70	8	3	18	1	Raw : Greyish, medium size beans Roast : Even roast, good swelling Brown centre cut, a few pales Liqour : Fair body, good acidity
RCRS Thandigudi	68	20	-	10	2	Raw : bluish grey, medium bold beans Roast : Even roast, good swelling Liquor : Fair to good body, fair acidity, sightly unclear

However, a small sized bean is the negative aspect in this line. The out come of S. 333 x Devamachy cross has been found promising and is characterized by semi-drooping habit with good resistance to leaf rust (85% of the population) and bolder bean size (Table 1) (Srinivasan and Ramachandran, 1997). This variety has shown the yield potential of 1300 - 2000 kg/ha under different agro-climatic zones (Table 2).

In one of the trials this line has recorded an average yield of 2048 kg/ha (5 years) maintained without any fungicidal sprays (Raju *et al.*, 1996).

With the experience in exploiting natural hybrids, systematic approach to develop inter-specific hybrids through programmed hybridization between tetraploid vs. diploid and diploid vs. diploid was also taken up at CCRI in order to develop Sln. 6 and Sln. 11. The Sln. 6 was evolved by crossing robusta and arabica followed by back crossing to arabica. While developing this hybrid the genome of 2x & 4x occurring in robusta and arabica, respectively were not altered. The resultant triploids were backcrossed to arabica and the resultant progeny was subjected to pure line selection. S_3 of BC_2 has been released for commercial cultivation as Sln.6.

Table 2. Performance of S.2931 progenies at different locations.

Sl. No.	Location	Progeny	Elevation (m)	Rainfall (mm)	Mean yield (Kg /ha)
	Pulney Hills				
1	Pillavali Estate (1972 planted)	F3	900	1551	1340*
2	RCRS, Thandigudi (1985 planted)	F4	1350	1875	2048
	Chikmagalur				
3	CDF, Arasinaguppe (1980 planted)	F4	970	1290	1332*
	Coorg				
4	Margolly Estate, (1972 planted)	F3	887	1397	1554* 1361

*With Bordeaux spray

While advancing the generations selection was made to minimize the percentage of empty locules which are otherwise high due to hybridity (Sreenivasan, 1983). The plants of Sln.6 are vigorous, wide spreading and tend more towards arabica phenotype with rust resistance and acceptable quality. The yield potential is around 1500 kg/ ha. Thus, the robusta x arabica hybrid of Indian origin differs from arabustas and other similar hybrids produced elsewhere where the tetraploid robusta was crossed with tetraploid arabica to derive viable types, which mostly tend towards robusta phenotypes. Subsequently, crosses between Sln.6 and HDT have also been developed and are under evaluation. These hybrid progenies (S.4369 and S.4375) are very vigorous with uniform branching pattern and good fertility status. Both the populations are manifesting high field tolerance to rust (95-97%). In the development of Sln. 11, liberica and eugenioides the two diploid parents were crossed. Interestingly on this diploid F_1 hybrid, an arabica like amphiploid sucker with 44 chromosome number was spotted which bred true by seed and this was further advanced to develop Sln.11.

This selection is known to exhibit good tolerance to leaf rust under field conditions but with smaller bean size. Incidentally, its resemblance to arabica is indicative of the probable contribution of these two species for the origin of arabica in nature (Vishweshwara, 1975). The Sln 11 was further crossed with other arabica such as Ciociee, Tafarikela, HdeT with multiple objectives of improving bean size, early ripening etc. and the progenies are under evaluation.

Table 3. Performance of F_1 hybrids (HdeT x other arabicas).

F_1 hybrid	Parentage	Mean yield (Kg/ha) (6th to 10th Year Avg.)	Rust resistance %
S.2790	HdeT x Tafarikela	1760	90.00
S.2792	Tafarikela x HdeT	1258	74.08
S.2794	HdeT x Geisha	1950	98.00
S.2795	HdeT x S.1934	994	87.50
S.2800	HdeT x Bourbon	2162	86.67
S.2803	S_{12} Kaffa x HdeT	1733	90.00

In addition to the above, Hibrido de Timor (HdeT), a natural Robusta x Arabica hybrid was introduced from the Coffee Rusts Research Center (CIFC), Portugal, in 1961 and manifests high degree of resistance to rust. After pure line selection, this was released as Sln. 8. Now, HdeT is a major source for rust resistance and widely used in breeding for rust resistance all over the world. Several F_1 hybrids between HdeT and other arabica were made available by CIFC, Portugal as well as developed in India. All these combinations manifest a high degree of resistance and productivity potential (Table 3).

The superior combination, HdeT x Tafarikela was released as Sln.9. The bushes are vigorous, semi-erect to drooping with drought tolerance and early ripening behaviour and come up well in marginal areas and supply positions.

2.3.2. Dwarf and semi-dwarf hybrids:

To meet the planters' demand for dwarf and semi-dwarf varieties suitable for high-density planting, mutants such as 'San Ramon' and 'Caturra' have come in handy. San Ramon, a dwarf mutant susceptible to several races of leaf rust, has been used in a series of crosses with the resistant lines S.795, Agaro and HdeT to produce hybrids, which are found suitable for cultivation in low rainfall areas. The cross of Caturra with S 795 and Cioccie made available by CIFC is found promising and named as Sln.10.

The Catimor hybrid developed as an international collaborative programme between CIFC and other coffee growing countries including India was found promising and released by giving popular name " Cauvery" in 1985. This material became popular in a very short time owing to its precocious bearing nature, high yield potential (2 to 2.5 tonnes / ha) and good response to the intensive cultivation. Though, this material was resistant to leaf rust in initial years, break down in resistance was observed due to the appearance of five new rust races (Rodrigues *et al.*, 1993) which could infect majority of these populations (Prakash *et al.*, 1998). Further, advanced generation of Villa Sarchi x HdeT, Catuai x Catimor (both supplied by CIFC) and Catimor x Sarchimor developed in India are found promising with high uniformity and vigor, rust resistance, bold beans coupled with good cup quality (Srinivasan, 1996). These lines are under multilocation evaluation and may form future materials for commercial release.

3. Robusta improvement

The second important and commercially cultivated species is *C. canephora* ('Robusta'). Hence, due importance has also been given for improvement of robusta in India. Robusta coffee possesses several useful characters like high tolerance to leaf rust pathogen, white stem borer, nematode invasion and potentiality to give consistent yields under irrigation. On the other hand, inability to endure long drought, late cropping as well as late stabilization of yields and inferior quality compared to arabica, are some of the negative aspects of robusta coffee. Keeping these aspects in view improvement of robusta coffee was undertaken by CCRI.

As Robusta is cross-pollinated, breeding methods such as mass selection, clonal propagation followed by pedigree selection and inter-specific hybridization have been used to evolve three commercial types. The seedling progenies of 12 high yielding mother plants selected from private estates were evaluated. Two materials S.270 and S.274 were found superior and released for commercial cultivation as Sln. 1 (R). These types grow into moderately large trees with vigorous growth. The fruits and seeds are bold, borne in large clusters. Liquor is rich in body and the cup is neutral. S.274 is preferred over S.270 on account of its bolder beans and it is performing well under irrigated conditions.

Further, 17 superior plants which have yielded twice or more than the family mean yield were identified from the initial single tree progenies of S.267 and S.278 and named as Balehonnur Robusta (BR selection). Based on the individual performance of these clones, BR 9, BR 10 and BR 11 were found to be superior in production and as such seed mixture of these clones was issued for commercial cultivation as Sln.2 (R). Clones were also issued for the establishment of bi/polyclonal gardens. The clones showed high degree of stability for 'A' grade beans and other characteristics.

Table 4. Performance of C x R hybrid in different agro-climates and bean grades.

Zone	Yield (Kg/ha)	Grade percentages (CxR sibs)		
1. Mudigere	1031	A	-	56.68
2. Koppa	1550	B	-	12.96
3.Coorg	2080	C	-	7.80
		Pb	-	19.46
4. Wynad	1335	BBB	-	3.10

The third robusta line developed in the series is C x R hybrid. This is similar to the "Congusta" hybrid developed in Java earlier (Cramer, 1957). *C. congensis* another diploid species known for its compact bush stature and quality was crossed with S.274 robusta in 1942. Back-crossing to robusta and sib mating was followed to achieve compact stature with bold fruits and seeds (Anonymous 1998). Beans are medium to bold, golden brown and gives soft liquor with flavour nearer to arabica (Table 4).

4. Conclusions

The continuous efforts of Indian institution involved in the research and development in coffee for various improvements have been of great usefulness in improving coffee cultivation in India. It is noteworthy to mention that the arabica varieties developed at CCRI have gradually replaced the old cultivars and presently cover 98% of the total area under arabica coffee in India. However, due to the long economic life, the intake of improved robusta strains is rather slow. In spite of this, the C x R hybrid is gaining popularity owing to its superior quality characteristics and is being preferred over other robusta.

5. Newer challenges

The ever-increasing racial diversity of the leaf rust pathogen (*Hemileia vastarix B. &* Br), sudden appearence of coffee berry borer (Hypothemus hampei Ferrari) in 1991 and increased awareness for quality of the product due to market liberalization pose new challenges in coffee cultivation. Accordingly, a paradigm shift has been taking place in coffee research. Achieving the durability of rust resistence through gene pyramiding approach, development of high yielding strains with a better response to intensive cultivation through multiple crossing programmes, evolving pest resistant and low caffeine lines through biotechnological means and molecular biology approaches are the best considered emerging trends of the day.

6. References

Anonymous (1995), Coffee Guide, Coffee Board Research Department, Govt. of India.

Anonymous (1998), Coffee Board Research Department, CxR Technical Bulletin

Cramer P.G.S. (1957), In- F. L.Wellman, Review of Literature of Coffee Research in Indonesia. SIC Editorial American Institute of Agricultural Sciences, Turrialba, Costa Rica.

Mayne W.W. (1932), Physiologic specialization of *Hemilea vastatrix* B.& Br., Nature, **129**,510

Mayne W.W. (1936), Annual Report of the Coffee Scientific Officer. 1935-36. Mys. Coffee. Exp. Stn. Bull. **14**,21

Mayne W.W. (1939), Annual report of the Coffee Scientific Officer, 1938-39, Mys. Coffee Exp. Stn. Bull. 19.

Narasimhaswamy R L. (1960), Arabica selection S.795. Its origin and performance. Indian Coffee. **24**, 197-204.

Narasimhaswamy R.L. (1961), Coffee leaf rust disease (*Hemileia)* in India. Coffee Turrialba, **3**, 33-39

Prakash N.S., Srinivasan C.S., Naidu R. (1998) Races of coffee rust (*Hemileia vastatrix): Is* India a potential source? PLACROSYM XIII. In press.

Raju T., Srinivasan C.S., Jagadeesan M. (1996) Performance of CCRI released arabica selections in the Pulneys in Tamilnadu. Indian Coffee, **LXV** (3): 17-20.

Rodrigues Jr.C.J., Varzea V.M.P., Godinho I.L., Palma S., Raho R.C. (1993) New physiologic races of *Hemileia vastatrix*. ASIC, 15th Colloque, Montpellier, 1993. 318-321.

Sreenivasan M.S. (1983), Cyto-embryological studies of coffee hybrids. Ph. D. Thesis, University of Mysore.

Srinivasan C.S. (1996), Current status and future thrust areas of research on varietal improvement and horticultural aspects of coffee. J. Coffee Res., 26(1): 1-16.

Srinivasan C.S., Ramachandran M. (1997), Selection 5B. S.2931 (S.333x Devamachy hybrid). An old arabica hybrid rediscovered with promising features. Indian Coffee, 61(11): 4-6.

Vishveswara S. (1975), Coffee: Commercial qualities of some newer Selections. Plant. Chron. 70, 122.

Wagner M., Bettencourt A.J. (1965) Inheritance of reaction to Hemileia vastratrix Berk. Br. In- Coffea arabica L. In progress report 1960-1965, Coffee Rust Research Centre, Oeiras, Portugal.

Chapter 3

BREEDING AND BIOTECHNOLOGY OF COFFEE

L.C. Fazuoli, M. Perez Maluf, O. Guerreiro Filho, H. Medina Filho, M.B. Silvarolla

IAC – Centro de Café e Plantas Tropicais - CP 28 - CEP 13001-970 – Campinas, SP – Brazil fazuoli@.cec.iac.br

Running title: Breeding and biotechnology

1. Classical breeding

1.1. COFFEE SPECIES AND VARIABILITY

The taxonomic position of some *Coffea* species is controversial. Many of the known species have been discovered along harvests made in the tropical forests of Africa since 1940. The literature suggests that the true *Coffea* species are those from central and equatorial regions of Africa, including Madagascar and the neighbouring islands close to Indian Ocean. The other species from Asian regions, previously described as being part of the genus *Coffea*, are no longer considered true *Coffea* species. Chevalier (1947) in his classification, considered 65 species from which 24 belonged to other genera. Cramer (1957), however, suggested the existence of at least 100 species. Purseglove (1968), cited by Wrigley (1988) referred to 50 species of the *Coffea* genus, from which 33 were from Tropical Africa, 14 from Madagascar and 3 from Mauritius and Reunion Islands. As these species are studied, divergences are noticed as for the number of true species from Madagascar. New *Coffea* species were described by Bridson (1982), especially in Eastern parts of Africa, although some of them have not been thoroughly characterized. Leroy (1980) recognized three genera of coffee plants: *Coffea, Psilanthus* and *Nostolachma*. The latter is restricted to Asia and Indonesia. He also distinguished three subgenera of *Coffea* and two of *Psilanthus.*

More recent classifications (Bridson, 1987; Bridson & Verdcourt, 1988; Bridson, 1994) group coffee plants into two genera: *Coffea* and *Psilanthus*. The genus *Coffea* comprises subgenera *Coffea* and *Baracoffea* and the genus *Psilanthus* are divided in two subgenera, *Psilanthus* and *Afrocoffea*. While the *Baracoffea* subgenus is

T. Sera et al. (eds.), Coffee Biotechnology and Quality, 27–45.

represented by seven species, the subgenus *Coffea* holds a total of 80 species, 25 from Continental Africa and 55 from Madagascar.
The difficulty in classifying *Coffea* species lies on the fact that complete collections or even live collections bearing representatives of the described species are not available. The existence of such collections would be extremely beneficial not only from the taxonomic point of view but also for breeding purposes. However, most of the African countries have enforced serious limitations on transportation of native material out of the country, such as the new species described by Bridson (1994). In addition, the species were not deposited in the germplasm collections of their original countries, so it is necessary to collect new material from this eastern region. The establishment of international and national Germplasm Stocks of known species and its derivatives is very important. Since among the wild forms there are highly advantageous characteristics such as resistance to diseases and insects, tolerance to drought and other environmental conditions, and also different characteristics of the plants (root system, leaves, flowers, fruits, etc.). *Coffea* germplasm collections have been established in the Ivory Coast, Madagascar, Costa Rica, Colombia, Brazil and other countries. However, in all of these countries very few accessions of other *Coffea* species are present besides *C. arabica* and *C. canephora.* At Instituto Agronômico de Campinas, São Paulo State, Brazil, it has been possible to gather 18 species of the genus *Coffea* and three of the genus *Psilanthus*. Some species have many accessions such as *C. arabica* and *C. canephora*, but very few specimens of other species are present. In many of the collections, it is very common to see plants that do not represent the original species since they were introduced by seeding, i.e. they are in fact hybrids of different species. Among the known species, the two most important in the international market are *C. arabica* (arabica coffee) and *C. canephora* (robusta coffee). About 70% of the coffee traded in the world is of the arabica type and 30% is robusta. Other species of coffee, depending on their agronomic characteristics, are important in breeding programs.
C. arabica is native of a restricted region, marginal to the other species, in South-western Ethiopia, South-eastern Sudan and Northern Kenya (Chevalier, 1947; Charrier, 1978; Bridson, 1982; Leroy, 1982). *C. arabica* is a noble species that produce good quality coffee. There are several varieties of *C. arabica* and more than 40 single gene mutants have been described (Carvalho *et al.*, 1991). The characteristics of these mutants are highly variable and include, for example, caffeine content, leaf shape or colour, growth and shape of the plant, type of flower and of blooming, shape and colour of fruits and seeds, resistance to diseases, insects and nematodes.
The other species of commercial interest *(C. canephora)* is geographically more widely distributed and can be found in tropical and subtropical areas of western and central regions of the African continent. This wide area corresponds to the Guinea Republic, Liberia, Sudan and Uganda. There is a high concentration of types in the Democratic Republic of Congo (Chevalier, 1947; Charrier & Berthaud, 1985). There is a high variability in this species in relation to the size and shape of the plants, leaves, fruits and seeds. The root system of *C. canephora* representatives are highly developed and

resistance to most diseases and insects is observed very frequently. The levels of caffeine and soluble solids *C. canephora* seeds are higher than in arabica coffee.

Other *Coffea* species are not economically cultivated but the importance of these wild species lies on their great variability for many characteristics especially resistance to diseases, insects and nematodes. Most wild *Coffea* species have reduced flower production, which accounts for the low fruit yield. Some of the species are caffeine-free or have a low level of caffeine and soluble solids in the seeds. Other species have extremely early or late maturation, and at least one of *C. racemosa* is resistant to drought and have deciduous leaves.

1.2. BIOLOGY OF REPRODUCTION AND RELATIONSHIP BETWEEN SPECIES

The species *C. arabica* is tetraploid with 2n=44 chromosomes. It is self-compatible and reproduces mostly by self-fertilization, which occurs in about 90% of the flowers. The cross-fertilization rate has been evaluated along several consecutive years in Campinas by using different recessive mutants. The most appropriate for this determination is the cera mutant, which exhibits yellowish endosperm when self-fertilized. Because of the xenia effect and recessiveness of the allele a cera coffee plant originate greenish seeds when fertilized by pollen of surrounding coffee plants homozygous for the dominant allele for green endosperm. The percentage of natural cross-fertilization has been determined to be around 10% in *C. arabica* (Carvalho & Mônaco, 1962). Successive generations of selfing do not reduce the vigour or productivity of the plants. Fertilization in *C. arabica* happens around 24 hours after pollination and the first cell division of the endosperm occurs 21-27 days after fertilization. The first zygote division occurs 60-70 days after pollination (Mendes, 1958).

C. canephora and the other known species of the genus *Coffea* are diploid, have 2n=22 chromosomes, are self-incompatible and reproduce exclusively by cross-fertilization (Berthaud, 1980; Conagin & Mendes, 1961, Devreux *et al.*, 1959). Only one exception is known, the species *C.* sp. *Moloundou*, which is diploid and self-compatible. The incompatibility in *C. canephora* was determined as being of the gametophytic type, which is also seen in *C. dewevrei* and *C. congensis*. There are evidences that incompatibility in *C. canephora* is conditioned by a S allelic series, in a single locus (Conagin & Mendes, 1961; Devreux *et al.*, 1959). The exact number of genes involved in incompatibility is not known.

As for the diploid species belonging to the *Psilanthus* genus it is known that at least *Psilanthus bengalensi* and *P. travancorensis* are self-compatible. Their floral structure suggests that they are also self-species (Medina Filho et al., 982).

In an attempt to understand the relationships of coffee species, inter-specific hybridizations were performed between 14 Coffea species at IAC by Carvalho *et al.*(1984a). These studies and others, described by Longo (1972), Medina (1972),

Maglio (1983), were discussed by Medina Filho *et al.* (1984) and indicated that the success of hybridizations between species of the same subsection is not necessarily greater than between coffee plants from different subsections. In the same way there is not a general trend relating the success of hybridizations with the geographical distribution of species or even with the morphological similarities among plants. Also, the hybridization success, viability and hybrid fertility may not be predicted on the basis of biochemical similarities or number and behaviour of chromosomes.

1.3. GENETIC ANALYSES OF COFFEA ARABICA AVAILABLE VARIABILITY AND USE OF SOME CHARACTERISTICS FOR BREEDING

Among the perennial coffee plants, *C. arabica* is certainly the best studied species concerning inheritance and dominance relationships of genetic factors. Research on inheritance of the main characteristics of the *C. arabica* varieties has proven to be very important for obtaining new cultivars through genetic recombination. The *C. arabica* Arabica cultivar (*Nacional* or *Típica*) has been used in theses studies as a standard.
Along an observation period of 67 years, 42 mutations were found that allowed genetic analyses. Only the genetic factors *anomala* and *anormalis* showed linkage (Carvalho, 1960). Anomalies were found that were apparently related to cytological traits, and others have cytoplasmic origin. The factors studied affect some characteristics of the plant such as height, type of branching and of leaves, flowers, fruits and seeds (Carvalho *et al.* 1991), in addition to resistance to diseases and insects. Many of these factors are dominant, others are recessive or show incomplete dominance in relation to the alleles of cultivar Arabica (Carvalho, 1958; Krug & Carvalho, 1951). Among the factors that affect plant height Caturra (Ct), São Bernardo (Sb), Vila Lobos (Vl) and San Ramón (Sr) are all dominant and independent (Carvalho *et al.* 1984b). These factors are of great economic interest since they imply in shortening of the internodes of the orthotropic and plagiotropic branches resulting in more compact and smaller coffee trees. The implication is that these characteristics make harvesting and phytosanitary treatments easier, ultimately reducing production costs. An example of use of the Caturra factor was the development of cultivars Catuaí Vermelho and Catuaí Amarelo. These cultivars are short, vigorous, highly productive and are widely cultivated in Brazil and in other parts of the world (Carvalho & Mônaco, 1972; Fazuoli, 1986; Carvalho & Fazuoli, 1993). It has also been observed that some factors have a marked pleiotropic effect, such as the laurina factor (*lr*). This factor affects not only plant size and height, leaf size, shape and size of fruits and seeds but also reduce the amount of caffeine in the seeds (Krug *et al.*, 1959; Carvalho *et al.*, 1986).
The *cera* (*ce*) factor, because of the xenia phenomenon mentioned previously, has its value for demonstrating for the first time that the coffee seed is formed by true endosperm and it has since being utilized for estimating natural cross-fertilization rates (Krug & Carvalho, 1939, 1951). Once *ce* fruits exhibit yellow seeds this mutant also served to demonstrate that the green colour of a normal seed is not due to chlorogenic acids since *ce* seeds have the same constitution as the green ones in relation to those

compounds (Carelli *et al.*, 1974). Along the years, several compilations have been published on the genetic analyses performed in *C. arabica* (Carvalho & Krug, 1949; Carvalho, 1958; Krug & Carvalho, 1951; Carvalho *et al.* 1991).
Sources of resistance to diseases have been detected in *C. arabica.* They include resistance to several pathogens such as rust, "mancha aureolada" (*Pseudomonas syringae* pv. *garcae*), fruit anthracnose or CBD (Coffee Berry Disease), caused by *Colletotrichum coffeanum,* now designated *C. kahawae* (Bettencourt & Carvalho, 1968; Moraes *et al.*, 1974; van der Vossen & Walyaro, 1980). As for resistance to nematodes sources have been found in accessions from Ethiopia (Fazuoli, 1981).
Thus there is reasonable variability for some characteristics in *C. arabica* that has been used in breeding programs. However, for characteristics related to resistance to diseases, insects and nematodes the variability is limited, reason why arabica breeding depends heavily on the sources of genetic resistances present on the diploid *Coffea* species.

1.4. USE OF CHARACTERISTICS OF DIPLOID *COFFEA* SPECIES IN BREEDING OF *C. ARABICA*

The model of informal classification of species proposed by Harlan and De Wet (1971), when applied to the genus *Coffea,* indicated that there is a great potential for breeding of *C. arabica* (primary pool) concentrated in the secondary gene pool (Medina Filho *et al.* 1984). The species *C. canephora, C. congensis*, *C. racemosa* and *C. dewevrei* (the later now designated *C. liberica var. dewevrei*) belong to this secondary pool. They all have been used as important sources of resistance to insects, diseases, nematodes and to adverse environmental conditions. A good example of such use was the development of the Arabica cultivars Icatu Vermelho, Icatu Amarelo and Icatu Precoce at IAC. These cultivars exhibit resistance to *Hemileia vastatrix* of *C. canephora,* introgressed by back-crossings and selections into the arabica cultivars Mundo Novo and Bourbon Vermelho (Mônaco *et al.*, 1974; Fazuoli, 1991; Carvalho & Fazuoli, 1993).
C. canephora has also been exploited indirectly by the use of the Timor Hybrid that is probably derived from a spontaneous backcross of a natural hybrid, between *C. arabica* and *C. canephora* (Bettencourt, 1973). The Timor Hybrid shows *C. arabica* phenotype, has 44 chromosomes in somatic cells, is self-fertile and has stable meiotic behaviour (Rijo, 1974). The cultivars Obatã (IAC 1669-20), Tupi (IAC 1669-33) and IAPAR 59 (also derived from IAC 1669) which are productive and short in size show high levels of resistance to rust derived from the Timor Hybrid, hence indirectly from *C. canephora* (Fazuoli *et al.*, 1996, 1999; IAPAR, 1993). It is noteworthy that certain progeny of *C. canephora* are also sources of resistance to *Colletotrichum coffeanum*, pathogen of CBD. This is an extremely serious disease, very common in African countries and causes fruits to fall upon infection (Carvalho *et al.*, 1976; van der Graaf, 1981; van der Vossen & Walyaro, 1981). Some cultivars of *C. canephora*, like Apoatã (IAC2258) also exhibits resistance to the nematodes *Meloidogyne exigua, M.*

incognita, *M. paranaensis* and *Pratylenchus coffeae*. Apoatã can be used as a nematode resistant rootstock for Arabicas or as a Robusta cultivar (Fazuoli, 1981; Fazuoli *et al.*, 1984, Carvalho & Fazuoli, 1993). This represents a viable solution for the persistence of coffee as a cash crop in several regions of Brazil where nematodes are present (Medina Filho *et al.* 1999).

The species *C. racemosa* that belongs to the secondary gene pool of *Coffea arabica* crosses quite easily with *C. arabica*, producing natural or artificial triploids. *C. racemosa* has been used as a source of resistance to leaf miner (*Perileucoptera coffeella*) and to drought (*Medina Filho et al.*, 1977a). Since the time for maturation of *C. racemosa* in Campinas is very reduced (2 to 3 months) this characteristic has been utilized in the development of early *C. arabica* cultivars.

Another species from the secondary gene pool is *C. dewevrei* or *C. liberica var. dewevrei*, which, besides being a good yielder, exhibits resistance to rust, nematodes and leaf miner. This species is quite late, taking one year from blooming to fruit ripening. The fruit has a thick endocarp that could be a physical barrier for the berry borer (*Hypotenemus hampei*).

C. congensis has been used in its tetraploid form in Campinas. This species, besides having an exuberant root system and good yield, is resistant to the nematodes *M. exigua*, *M. incognita* (Fazuoli *et al.*, 1983) and to *H. vastatrix*. Although referred as *C. congensis* the coffee plants utilized in this work and deposited at IAC' Germoplasm Collection are probably hybrids of *C. congensis* x *C. canephora* for the reasons described below. The observation of recent introductions of *C. congensis* at IAC and detailed information from Charrier (personal comm.) indicate marked differences between the original plants and the recent introductions of this species in Campinas. This confirms the suspicion that the material previously studied would actually be a hybrid of *C. congensis* and another species of the same genus. According to Chevalier (1947), the species *C. congensis* remains submerged in water during certain time along the year, at its native environment. Adaptation to these environments may represent an advantage since in some tropical areas excess humidity in the soil is very damaging to *C. arabica* coffee trees.

Another species that has been used is *C. stenophylla*, which belongs to the third gene pool of *C. arabica*. This species is nearly immune to the leaf miner (Medina Filho *et al.*, 1977b). However its use is limited due to difficulties in hybridizations with *C. arabica* (Carvalho & Mônaco, 1967). In this case, more sophisticated methods such as *in vitro* culture of protoplast fusion products are indicated as an aid for *C. stenophylla* use for genetic improvement of Arabicas (Sondahl *et al.*, 1980).

The species *C. salvatrix*, which belongs to the secondary gene pool, is also used in genetic breeding for resistance to *P. coffeella* at IAC. Tetraploid *C. salvatrix* coffee plants were crossed to *C. arabica* in order to transfer resistance. However, one must be cautious when using *C. salvatrix* in breeding programs since it is extremely attractive to ants (Carvalho & Mônaco 1967; Mazzafera, 1991).

An important consideration is that in every program, where wide hybridizations are used for the improvement of *C. arabica*, quality tests must be effected during the selection cycles, since the *C. arabica* good quality beverage must be ultimately recovered. A number of species evaluated for quality in Campinas showed that with one exception, all give very bad cupping. The exception is *C. eugenioides*, which has a very fine aroma, tasting fruity and clean. Interesting enough is that *C. eugenioidis* is one of the most likely ancestor (together with *C. canephora, C. congensis* or other unknown species) of *C. arabica*. It is quite possible that the superior beverage of *C. arabica* traces back to *C. eugenioides*. If this speculation is true, *C. eugenioidis* represents a "missing link" explaining the origin of the good flavour of *C. arabica* as originated by natural crosses of diploid species supposedly of bad cupping quality . The implication of this for breeding Arabicas and Robustas for flavour improvement is obvious and has been stressed by Medina Filho *et al.* (1984).
Caffeine is a chemical component of great interest due to the stimulating effect that it produces. It is possible to reduce or increase caffeine content of varieties to meet requirements of distinct markets. Caffeine reduction could be achieved by using caffeine-free wild species from Madagascar in hybridizations or by transfer of the recessive allele laurina (*lr*) of *C. arabica* to commercial cultivars, since this factor reduces caffeine levels by half (Carvalho *et al.*, 1965; Carvalho *et al.*, 1983; Charrier & Berthaud, 1975). The development of caffeine-free *C. canephora* has been studied by Ky *et al.* (1999) in crosses with the species *C. pseudozanguebarie* devoided of caffeine. Some species of the genus *Psilanthus*, such as *P. bengalensis* and *P. travancorensis* (previously *Coffea*, then *Paracoffea*) that do not have caffeine in their seeds can also be used in similar programs. The isolation of a hybrid of *C. arabica* with *P. bengalensis* was of particular interest (Charrier, 1978). However, the utilization of species of other genera in coffee breeding is more difficult due to greater genetic barriers. It would be also possible to increase the levels of caffeine by breeding using *C. canephora*, since it has significant variability for this characteristic (Charrier & Berthaud, 1975).
Since *C. arabica* is a tetraploid and the other species of the *Coffea* genus are diploid, difficulties are found for using these diploid species in the breeding of *C. arabica*. Although it is possible to recover the tetraploidy from triploid hybrids through backcrossing with *C. arabica* (Medina Filho *et al.*, 1977a, 1983), a more attractive alternative would be the duplication of the number of chromosomes in diploid species before crossing with *C. arabica*. This strategy has proven appropriate for *C. canephora* and resulted in the *C. arabica* Icatu cultivar (Fazuoli, 1991; Carvalho & Fazuoli, 1993). Treatment of seeds or seedlings with colchicine is used for production of tetraploids from diploid individuals. Triploid individuals should be used as male genitors in crossings or back-crossings. In hybridizations using tetraploid coffee plants as male genitors the transfer of aneuploidy to the progeny is thus decreased.
The several examples of the successful utilization of *Coffea* germplasm in breeding of *C. arabica*, demonstrated that the available gene pool present in the collections still

has an enormous potential to contribute for the development of new cultivars with specific characteristics, circumventing phytosanitary problems or satisfying new demands of consumers.

2. Modern Biotechnology

Among the modern biotechnology techniques that can be of help in coffee improvement emphasis should be given to tissue culture, genetic transformation and molecular marker techniques.

2.1. TISSUE CULTURE

The term tissue culture corresponds to a series of techniques that allow the *in vitro* culture of plant tissues from parts of an adult plant. Tissue cultures have been successfully established for numerous wild and cultivated species from all kinds of tissues and organs, such as leaves, embryos, anthers and pollen. The use of such techniques has allowed the development of several applications such as micropropagation of genotypes that are of interest but difficult to cultivate, easier maintenance of germplasm stocks, production of haploid and diploid homozygotes, genotype transformation, among others. Presently, as more advanced molecular biology techniques are developed the establishment of more efficient tissue culture methodologies is needed for all commercial cultivars. Once established, these techniques will extensively help in genetic breeding programs.

The first attempts of *in vitro* culture were in micropropagation of explants, vegetative nodes or apical meristems. This type of culture is more efficient for small-scale production of plants; for example, the multiplication of special genotypes or the establishment and maintenance of *in vitro* collections.

The culture method must be carefully chosen according to the final objective and experimental conditions previously tested for each species. For instances, for uses of greater demand such as propagation of cultivars the most indicated methodology is somatic embryogenesis. Generally, the culture can be started from tissue or organ fragments that will develop into an amorphous mass of continuous and undifferentiated growth in culture media. This mass, named callus, can be maintained indefinitely in this state through transfers. Differentiation into tissues and organs, and regeneration into adult plants are reached by adding appropriate growth regulators (plant hormones) to the culture medium. A variation of this methodology is protoplast culture. In this case the culture is started from a suspension of plant cells that had their cell walls removed. This type of cells can be used in transformation experiments. Differentiation and regeneration of plants are accomplished afterwards according to somatic embryogenesis procedures. Another type of culture can be started from intact anthers or from microspores. Anther culture is of great importance in improvement programs since it allows the development of haploid plants. In addition, microspores have their

chromosomes duplicated after colchicine treatment, giving rise to homozygous diploid plants.
A big effort has been made to establish the *in vitro* cultivation of several species of the genus *Coffea*. The first successful attempt to regenerate *C. arabica* plants was accomplished in 1975, through direct somatic embryogenesis (Herman & Haas, 1975). Several studies have been undertaken since then in an attempt to establish the ideal culture conditions (reviewed by van Boxtel, 1994). Presently, regeneration of adult plants from embryogenic calluses is successfully obtained (van Boxtel, 1994). In *C. arabica* the frequencies of embryo isolation and plant regeneration from embryogenic calluses are low and depend on the genotype (van Boxtel & Berthouly, 1996). This regeneration, however, was not possible when the culture was started with protoplasts isolated from leaves of *C. arabica* (van Boxtel, 1994). Recently, in another type of experiment, androgenesis has been established for the Catimor variety. In this particular case the culture was started from floral buddings, and anthers and microspores were isolated and cultivated in different culture media. Regeneration of plants from anther cells has also been successfully achieved (Carneiro & Antão da Silva, 1999). These experiments open an important opportunity for coffee improvement since the production of haploid plants represents saving of time and increases in efficiency for generation of hybrids.
The use of tissue culture techniques in coffee can be an essential tool for solving problems encountered in breeding programs. One of them is the low genetic variability for resistance to diseases, insects and nematodes observed in *C. arabica*. This problem can be overcome by the introduction of genes from another species through genetic transformation. Therefore it is essential that an efficient methodology for *in vitro* cultivation of *Coffea* species is established, determining the influence of factors such as type and age of starter tissue, culture medium composition, time period from embryogenesis to regeneration into adult plants, etc. Another aspect that can be solved with this technology is propagation and conservation of germplasms of coffee. At the moment, propagation is done mainly through seeds. The use of somatic embryogenesis for obtaining embryos that could be multiplied in a large scale in bioreactors, and germinated afterwards, the so-called micropropagation, has receiving some attention. In field experiments, development of plants originated from somatic embryos was similar to that of plants obtained from seeds (Sondahl *et al.*, 1999). The lack of biofactories specialized in coffee, the high production cost and nonexistence of data concerning the effect of the process in the quality of the beverage are limiting factors for utilization of micropropagation for coffee.

2.2. GENETIC TRANSFORMATION

Genetic transformation can be understood as the modification of genotypes through introduction of specific genes leading to a programmed alteration of the transformed individual. In transformation of plant tissues the genes introduced can be those related to characteristics of agronomic interest, such as resistance to diseases and insects, control of hormone synthesis and other substances of commercial interest. Nowadays, the existence of transformed or genetically modified individuals is being constantly broadcast, and understanding the potential of these new methodologies is becoming extremely important.

The method of transformation utilized in plants varies according to the morphological and physiological characteristics of the species. The cheapest and most efficient method is done with the bacterium *Agrobacterium tumefaciens*. This bacterium is a natural pathogen of plants and promotes the disordered growth of infected plant cells, leading to formation of galls. This alteration results from the action of genes that code for growth regulators present in a plasmid of the bacterium. During infection these genes are transferred to the plant genome where they are activated. Prior to transformation experiments the *A. tumefaciens* plasmid is modified in the laboratory and it receives the gene of interest to be introduced in the plant as well as marker genes for selection of the transformed material. Transformation using this methodology is normally performed in cultured plant cells, which will subsequently be selected for the presence of the bacterial plasmid and regenerated into transformed plants. In coffee, plants have been efficiently transformed through this methodology. The gene *CryIA(c),* which confers resistance to *P. coffeella* and was isolated from *Bacillus thuringiensis*, has been introduced into *C. arabica* and *C. canephora.* The regenerated plants are being tested for resistance to leaf miner (Leroy *et al.*, 1999).

Another methodology that has been successfully utilized for coffee transformation is biobalistic or bombarding of plant cells. In this case, the marker and the gene of interest are introduced into plant tissues by a strong pressure, exerted by compressed helium gas. This bombarding is done in a special apparatus where DNA associated with micro particles is positioned under the gas jet and directed over the tissue to be transformed. In addition to yield of high transformation efficiency, another advantage of this method is that the transformed tissue does not suffer mechanical damage. Up to this moment, tests performed with various *C. arabica* tissues such as intact leaves, cells in suspension and embryogenic calluses have been promising (van Boxtel, 1994). It has been possible to detect the transient expression of the GUS marker gene in these experiments.

The impulse for development of transformation methodologies came from the increasingly complex knowledge of the molecular structure of the gene. Molecular biology techniques allowed the isolation and cloning of genes, identification of regulatory regions and mechanisms of gene action. In coffee, there are some examples that represent possibilities for utilization of these technologies in improvement programs. In one of these examples, resistance to the leaf-miner can be obtained through the use of genes isolated from *B. thuringiensis.* These genes, related to synthesis of toxins that have insecticide action have been transferred to cells in culture (Leroy *et al.*, 1999). In another type of experiment, the coffee gene coding for xantosyne-N7-methyltransferase, which is an enzyme involved in caffeine biosynthesis, was introduced into coffee cells. The rational behind this experiment was to invert the sense of the introduced sequence so that the produced RNA would be complementary to the normal gene RNA and would pair with it. In this manner, no enzyme would be produced and consequently caffeine synthesis would be blocked (Moisyadi *et al.*, 1999). Current agriculture requires more and more varieties with greater agronomic plasticity. The new molecular biology techniques, associated with efficient tissue culture and transformation systems allow the isolation of very specific genotypes in a controlled manner, i.e. without introducing undesirable characteristics. Biotechnology can contribute considerably in genetic improvement programs of coffee since research done up to this moment has demonstrated that there is a potential for transfer of genes of interest, in a faster way, without the need of back-crossings. Nonetheless, the success of this approaches depends on a strong knowledge of not only the genotype being evaluated but also of the mechanisms of action and type of inheritance of the genes of interest. In addition, a selection and progeny analyses program is of fundamental importance so that the obtained varieties actually have the desired characteristics.

2.3. MOLECULAR MARKERS

The development of electrophoresis techniques, which allow the separation and visualization of different molecules, opened the possibility of knowing and evaluating the genotypes of individuals. Analyses of electrophoresis patterns allow the characterization of gene loci, in relation to their inheritance patterns, and number and frequency of alleles in a population. This way, it is possible to identify molecular markers associated with specific genotypes.

A molecular marker can be defined as any phenotype originated from the expression of a gene or a specific fragment of DNA (Ferreira & Grattapaglia, 1996). The first molecular markers used in genetic analyses were isozymes (Tanksley & Ortom, 1983). Presently, an array of molecular markers has been identified through techniques that allow a wide characterization of the genetic polymorphism present in different plant varieties. These markers are important tools in assisted breeding programs and in identification of varieties. Several breeding programs of cultivated plants have utilized genetic-molecular maps generated by these markers, together with traditional methods, for selection of cultivars of economic interest (Rafalski & Tingey, 1993; Lashermes *et al.*, 1997a).

The first DNA marker developed was RFLP (*Restriction Fragment Length Polymorphism*). The polymorphism observed in RFLP analyses is based on the length of fragments generated by digestion with restriction enzymes (Botstein *et al.*, 1980). The development of the principle of chain polymerization reactions (PCR-Polymerization Chain Reaction) made it possible to identify markers obtained from amplifications of random sequences in the genome. These amplified sequences vary in size and number depending on the genotype that is being analysed. In this manner, populations that show high genetic variability have distinct patterns of markers. On the other hand, populations showing low genetic variability exhibit uniform marker patterns. The most utilized method for obtaining markers in plants are: RAPD (*Random Amplified Polymorphic DNA*), initially employed by Williams *et al.*(1990) and Welsh & McClelland (1990); AFLP (*Amplified Fragment Length Polymorphic*), developed and applied by Vos *et al.* (1995); and microsatellites (Tautz, 1989; Weber & May, 1989). In general, all these methods aim to evaluate the genetic variability by the same principle, i.e. the existence of polymorphisms in regions of the DNA that are conserved or not, in a specific population. The differences between the methods are the specificity and capacity for detecting this polymorphism, and particularly the experimental cost. Thus, RAPD markers are the simplest and cheapest of all of them. However, in some situations, such as the case with species showing low genetic variability, RAPD is not suited to identify polymorphisms that allow the distinction of different genotypes. Microsatellites, on the contrary, have a good amplitude but are costly.

Several experiments have been conducted in an attempt to establish a molecular genetic map for coffee species. However, the genetic variability found in a population of *C. arabica* is not very high, thus the identification of molecular markers necessary for breeding programs is hindered. Up to this moment many *C. arabica* cultivars have been established through classical breeding as the result of crossings with other species of the genus *Coffea* (Sondhal & Lauritis, 1992). These inter-specific crossings associated with an expected rate of 10% of allogamy lead to the conclusion that a certain level of molecular polymorphism exists between *C. arabica* varieties.

Establishment of genetic maps for coffee based on isoenzymes and RFLPs, either from total or chloroplast DNA, was not successful due to the low polymorphism level (Paillard *et al.*, 1996; Lashermes *et al.*, 1996a). The sequence between the regions coding for the 18S and 26S ribosomal units, known as ITS (*internal transcribed spacer*), is another marker that has been used to characterize different coffee species. In the study of Lashermes *et al.* (1997b), several species and varieties of *Coffea* were evaluated for polymorphism in the ITS region. The results indicate that ITS may serve either as an inter or intra-specific marker, since polymorphisms were observed at both levels. Up to now, among all techniques RAPD was the one that best demonstrated the level of genetic diversity between several cultivars belonging to *C. arabica* and *C. canephora* species, since specific markers for the varieties studied were determined (Lashermes *et al.*, 1996b). However, intra-specific variation was not determined for *C. arabica* (Orozco - Castillo *et al.*, 1994) Based on this information it is possible to infer that methodologies such as AFLP and microsatellites that reveal a greater number of loci should be used for identification of polymorphisms in *C. arabica*.

In addition to their use in identification of species and varieties, markers can be important tools in breeding programs. In this case, the identification of molecular markers that co-segregate with desirable agronomic traits, such as resistance to diseases and insects, productivity, beverage quality can be a helpful instrument to help in assisted selection. This type of analysis is important in species that exhibit a long life cycle, such as coffee since the identification of desired progeny can be accomplished in the initial stages of the plant development. So, it is not necessary that all of the progeny are planted and evaluated in experimental fields, which saves time, physical space and resources.

3. Possibilities for the use of Modern Biotechnology in coffee plants

3.1. EXAMPLES OF USES OF TISSUE CULTURE

3.1.1. Leaf culture

Culture of coffee leaves could be utilized for somaclonal variation or micropropagation studies. However, use of direct somatic embryogenesis in temporary or permanent immersion bioreactors, and biofactories would be interesting for propagation of disease-free *C. canephora* clones, special plants (e.g. hybrids with multiple resistance - rust, nematodes and/or leaf miner), *C. canephora* clones resistant to several species and/or races of nematodes.

3.1.2. Anther culture

Anther culture would be interesting for the production of haploid plants. In the case of diploid species, homozygous diploid individuals could then be obtained allowing genetic analyses in these autoincompatible species and production of F_1 hybrids from 100% nematode-resistant *C. canephora.*

3.1.3. Embryo culture

Embryos originated from inter-specific or inter-generic crossings that present difficulties in their natural development could grow and become plants when placed in *in vitro* cultures. This is the case, for instances, in crossings of *C. stenophylla* or *P. bengalensis* with *C. canephora* or *C. arabica* (Carvalho & Mônaco, 1967; Charrier, 1978).

3.1.4. Preservation of cells and propagules

The preservation of cells or propagules is related to the *in vitro* conservation of gene pools of species, inter-specific hybrids, haploids, etc., aiming to maintain a germplasm stock in the laboratory.

3.1.5. Root culture

There are possibilities of developing techniques that allow the selection of coffee plants from axenic root cultures together with nematodes.

3.2. PLANT TRANSFORMATION

Transformation using genes from bacteria, fungi and viruses could be used in coffee especially for transfer of characteristics (genes) that are not found in wild species. An example is resistance to coffee berry borer (*H. hampei*) since, at present, sources of resistance to this insect have not been identified in coffee germplasm. Researchers at CENICAFÉ in Colombia were able to clone genes from *Streptomyces* sp that produce chitinase (Personal communication). This enzyme probably acts in the beetle chitin and hence could be used in the control of coffee berry borer.

Since there are sources of resistance to other insects in diploid species these genes have been transferred to *C. arabica* in classical breeding (crossings, auto-fecundations, and back-crossings). For example, the development of cultivars that are resistant and tolerant to *P. coffeella* (leaf miner) is currently in the final stages of evaluation (Guerreiro Filho *et al.*, 1999). In addition to this Bt genes are being incorporated into coffee cultivars by some researchers (Leroy *et al.*, 1999).

Another use of transformation concerns the isolation of coffee plants tolerant to herbicides. In this case the interest is mostly from the companies that want to sell their chemical products. In all cases of transformation uses for incorporation of genes, attention must be given to the possible genetic interactions, which can result in positive or negative modifications in other plant characteristics.

It is important to emphasize the use of transformation without introduction of foreign genes in the coffee genome such as the genetic manipulation of ethylene synthesis (Neupane *et al.*, 1999), blocking of caffeine biosynthesis through insertion of inverted sequences (Moisyadi *et al.*, 1999), which, in these cases, would be related to the quality of the product.

Finally, the use of transformation in coffee plants will have a great advance whenever it is possible to know the genome of *C. arabica* and diploid species, and to clone specific genes from diploid species that could be transferred to *C. arabica*. The utilization of these cultivars would not represent greater risks and would be advantageous since the damages that introduced bacterial, fungal and viral genes could cause to the environment and human health have not been thoroughly evaluated.

3.3. UTILIZATION OF MOLECULAR MARKERS

The following examples of utilization of molecular markers in coffee can be cited:

a) Characterization of coffee plants (species, hybrids, cultivars, clones, etc.) aiming the knowledge and protection of the material through a safer identification. For characterization of coffee plants descriptors related to morphological characteristics of the plant such as leaves, flowers, fruits and seeds have been used. Other characteristics such as physiological, agronomic, technological and resistance or tolerance to adverse factors (diseases, insects, nematodes and environment) have also been utilized. At present, the use of modern techniques for DNA extraction and the development of molecular markers will allow a better characterization of mutant coffee plants of *C. arabica* and specially the diploid *Coffea* species and other genera.

b) Assisted selection using molecular markers. In this case, one seeks to associate resistance to rust, nematodes, maturation, organoleptic properties, etc., with characteristics of better definition and easier evaluation, such as molecular markers pattern.
c) Genomic analyses and evaluation of genetic diversity. For this purpose one can characterize the genome of *C. arabica* and main diploid species or the genome of some disease agent, such as insects or important nematode.
d) Somatic hybridization between coffee species would also be interesting and possible with the use of molecular marker techniques.

4. Conclusions

Since the coffee cultivars available in the market are highly productive, it can be affirmed that the recent tendencies for improvement are related to resistance to diseases, insects and nematodes, resistance and tolerance to adverse environmental conditions, better root systems and plant architecture, with reduction in the size and production of a better quality drink. In most cases, classical breeding was and still is important. However, the possibility of using modern biotechnology such as tissue culture, molecular markers and transformation is evident. The use of transformation in coffee with genes cloned from bacteria, fungi and viruses should be analyzed under two aspects: the scientific and the community interests. The scientific aspect refers to a unavoidable advance in the next decades. From the community point of view more considerations are required, since at the moment the environment is not suitable for the utilization of this type of methodology. On the other hand, the use of tissue culture and especially of molecular markers will certainly be in the spotlight of studies on coffee breeding and could mean great advances in a very near future.

5. Acknowledgements

This research was supported by a Grant from PNP&D Café/Embrapa. LCZ, OGF and HPMF thank to CNPq and MPM to FUNAPE for a fellowship.

6. References

Berthaud, J. (1980), L'incompatibilité chez *Coffea canephora* méthode de test et détérminisme génétique. Café, Cacao, Thé 24: 267-274

Bettencourt, A.J. & Carvalho, A. (1968), Melhoramento do cafeeiro visando resistência à ferrugem. Bragantia, 27:35-68

Bettencourt, A. J. (1973), Considerações gerais sobre o Híbrido do Timor. Campinas, Instituto Agronômico, Circ. n.º 23. 20 p.p.

Botstein, D.; White, R.L.; Skolnick, M. & Davis, R.W. (1980), Construction of a genetic linkage map in man using restriction fragment length polymorphisms. Am. J. Hum. Genet. 32: 314-321

Bridson, D. M. (1982), Studies in *Coffea* in *Psilanthus* (Rubiaceae subfam. *Cinchonoideae*) for part 2 of Flora of Tropical East África: *Rubiaceae*. Kew Bulletin 36 (4): 817-859

Bridson, D.M. (1987), Nomenclatura notes on *Psilanthus*, including *Coffea* sect. *Paracoffea* (Rubiaceae tribe Coffeae). Kew Bulletin LR (2) 453-460

Bridson D.M. & Verdcourt, B. (1988), "Flora of Tropical East África – Rubiceae (Part 2). Polhill R. M. (eds), 727 pp

Bridson, D.M. (1994), Additional notes on *Coffea* (Rubiaceae) from Tropical East África. Kew Bulletin 49 (2): 331-342

Carelli, M.L.; Lopes, C.R. & Mônaco, L.C. (1974), Chlorogenic acid content in species of *Coffea* and selections of *C. arabica*. Turrialba, 24: 398-401

Carneiro, M.F. & Antão da Silva, C.M. (1999), Androgenesis in some Catimor progenies. 18th International Conference on Coffee Science (Helsinki). Paris: ASIC, PA 741

Carvalho, A. & Krug, C.A. (1949), Genética de *Coffea* XII. Hereditariedade da cor amarela da semente. Bragantia, Campinas, 9 :193-202

Carvalho, A. (1958), Advances in coffee production technology. Recent advances in our knowledge of coffee trees. 2 Genetics. Coffee and Tea Industries, New York, 81:30-36

Carvalho, A. (1960), Genética de *Coffea*. XXV. Ligação genética dos fatores anormais e anômala. Bragantia, Campinas, 19:CXCICXCV. (Nota 38)

Carvalho, A. & Mônaco, L.C. (1962), Natural cross pollination in *Coffea arabica*. Proc XVI Intern. Hort. Congr. Brussels 4 : 447-449

Carvalho, A.; Tango, J. S. & Mônaco, L. C. (1965), Genetic control of caffeine content in coffee. Nature, 205: 314

Carvalho, A. & Mônaco, L.C. (1967), Genetic relationships of selected *Coffea* species. Ciência e Cultura, 191:161-165

Carvalho, A. & Mônaco, L.C. (1972), Transferência do fator Caturra para o cultivar Mundo Novo de *Coffea arabica*. Bragantia, Campinas, 31 (31): 379-399

Carvalho, A.; Mônaco, L.C. & van der Vossen, H.A.M. (1976), Café Icatu como fonte de resistência a *Colletotrichum coffeanum*. Bragantia, 35: 343-347

Carvalho, A.; Sondahl, M.R. & Sloman, C (1983), Teor de cafeína em seleções de café. Congresso Brasileiro de Pesquisas Cafeeiras, Poços de Caldas, MG., 10:111-113

Carvalho, A., Medina Filho, H.P. & Fazuoli, L.C. (1984a), Evolução e melhoramento do cafeeiro. In: Aguiar-Perecin, M.L.R., Martins, P.S., Bandel, G. Eds. Tópicos de Citogenética e Evolução de Plantas. I Colóquio sobre Citogenética e Evolução de Plantas, Piracicaba, São Paulo. Revista Brasileira de Genética. p. 215-234

Carvalho, A., Medina Filho, H.P., Fazuoli, L.C. & Costa, W.M. (1984b), Números de locos e ação de fatores para porte pequeno em *Coffea arabica*., Bragantia, Campinas, 43(2): 421-442

Carvalho, A., Fazuoli, L.C. & Mazzafera, P. (1986), Melhoramento do café laurina. In. Reunião Anual da Sociedade Brasileira para o Progresso da Ciências, 38, Curitiba. Anais. p. 908

Carvalho, A., Medina Filho, H.P., Fazuoli, L.C., Guerreiro Filho, O. & Lima, M.M.A. (1991), Aspectos genéticos do cafeeiro (Genetic Aspects of the coffee tree). Revista Brasil. Genet., 135-183

Carvalho, A. & Fazuoli, L.C. (1993), Café. *In*: Furlani, A.M.C. & Viégas, G.A. eds. O melhoramento de plantas no Instituto Agronômico, Campinas, Instituto Agronômico. Cap. 2, p. 29-76

Charrier, A. (1978), La structure génétique des caféiers spontanés de la region Malgashe (Mascarocoffea). Leurs relations avec les caféiers d'origine africaine (Eucoffea). Memories ORSTOM, Paris, 87, 223 p.p

Charrier, A. & Berthaud, J. (1975), Variation de la teneur en caféine dans le genre *Coffea*. Café, Cacao, Thé 19 (4) : 251-264

Charrier, A. & Berthaud, J. (1985), Botanical classification of coffee. In: Clifford, M.N. & Willson, K.C. Eds Coffee. Botany, Biochemistry and production of beans and beverage. Westport, connecticut. The Avi Publishing, Inc. p. 13-47

Chevalier, A. (1947), Les caféiers du globe. III. Systhématique des caféiers. Maladies et insectes nuisibles. Encyclopédie biologique 28 : 1-256 Fascicule III, Paul Lechevalier, Paris

Conagin, C.H.T.M. & Mendes, A.J.T. (1961), Pesquisas citológicas e genéticas de três espécies de *Coffea* – autoincompatibilidade em *C. canephora*. Bragantia, 20 : 787-804

Cramer, P.J.S. (1957), A review of literature of coffee research in Indonésia. In: Wellman, F. ed. Muc. Publ. N.° 15. Turrialba, Inter. Am. First of Agric. Sci. P. 1-262

Devreux, M., Vallaeys, G., Pochet, P. & Gilles, A. (1959), Researches sur l'autosterilite du caféier Robusta (*Coffea canephora Pierre*), Publ. INEAC 78, 44 p.p

Fazuoli, L.C. (1981), Resistance of coffee to the root-knot nematode species *Meloidogyne exigua* and *M. incognita*. Colloque International sur la protection des cultures tropicales, Lyon, France. p. 57

Fazuoli, L.C.; Costa, W.M., Gonçalves, W. & Fernandes, J.A.R. (1983), Identificação de resistência em *Coffea canephora* e *C. congensis* ao nematóide *Meloidogyne incognita*, em condições de campo. Ciência e Cultura, São Paulo, 35 (7):20 (Suplemento).

Fazuoli, L.C.; Costa, W.M., Gonçalves, M., Lima, M.M.A. & Fernandes, J.A.R. (1984), Café Icatu como fonte de resistência e/ou tolerância ao nematóide *Meloidogyne incognita*, In: Congresso Brasileiro de Pesquisas Cafeeiras, 11, Londrina. Resumos. Rio de Janeiro, IBC/GERCA, P. 247-248

Fazuoli, L.C. (1986), Genética e Melhoramento do Cafeeiro. *In*: Rena, A. B. ; Malavolta, E. ; Rocha, M. ; Yamada, Y. Editores. Cultura do Cafeeiro. Piracicaba. Potafós, p. 87-113.

Fazuoli, L.C. (1991), Metodologias, critérios e resultados da seleção em progênies do café Icatu com resistência a *Hemileia vastatrix*. Campinas, 322 p (Tese de Dourtorado-Universidade Estadual de Campinas – UNICAMP).

Fazuoli, L.C., Medina, H.P., Guerreiro Filho, O., Lima, M.M.A., Silvarolla, M.B., Gallo, P.B. & Costa, W.M. (1996), Obatã (IAC 1669-20) e Tupi (IAC 1669-33), cultivares de café de porte baixo e resistentes a ferrugem. Congresso Brasileiro de Pesquisas Cafeeiras, 22°, Águas de Lindóia, SP, p. 149-150

Fazuoli, L.C., Medina, Filho, H.P., Guerreiro Filho, O., Gonçalves, W., Silvarolla, M.B. & Lima, M.M. A. (1999), Coffee cultivars in Brazil. Association Scientifique Internationale du Café – 18 ème Colloque Helsinki, Finlândia (no prelo).

Ferreira, M.E. & Grattapaglia, D. (1996), Introdução ao uso de marcadores moleculares em análise genética. 2ª.edição. Embrapa. p.p. 220

Guerreiro Filho, O., Silvarolla, M.B. & Eskes, A. B. (1999), Expression and mode of inheritance of coffee resistance to leaf miner. Euphytica, 105(1):7-15

Harlan, J.R. & De Wet, J.M.J. (1971), Toward a rational classification of cultivated plants. Taxon. 20 (4): 509-517

Herman, E. & Haas, G. (1975), Clonal propagation of *Coffea arabica* L. from callus culture. HortScience 10, 588-589.

IAPAR. Café IAPAR 59 (1993), Londrina, PR., IAPAR. Folder

Krug, C.A., Carvalho, A. (1939), Genetical proof of the existence of *Coffea* endosperm. Nature, London, 144 : 515

Krug, C.A., Carvalho, A. (1951), The genetics of *Coffea*. Advances in Genetics, New York, 4: 127-158.

Krug, C.A.; Carvalho, A. & Antunes Filho, H. (1959), Genética de *Coffea*. XXI. Hereditariedade dos caracteres de *Coffea arabica* L. var. laurina (Smeathman). DC. Bragantia, Campinas, 13: 247-255

Ky, C-L.; Louarn, J., Akaffou, S., Guyot, B., Chrestin, H., Charrier, A., Hamon, S. & Noirot, M. (1999), The use of genetic resources for coffee cup taste improvement. In International Seminar on Biotechnology in the Coffee Agroindustry, 3. Londrina, p. 28. (Resumos).

Lashermes, P.; Cros, J.; Combes, M.C.; Trouslot, P.; Anthony, F., Hamon, S. & Charrier, A. (1996a), Inheritance and restriction fragment lenght polymorphism of chloroplast DNA in the genus *Coffea* L. Theor. Appl. Genet. 93: 626 -632.

Lashermes, P., Trouslot, P., Anthony, F., Combes, M.C. & Charrier, A. (1996b), Genetic diversity for RAPD markers between cultivated and wild accessions of *Coffea arabica*. Euphytica 87: 59 - 64.

Lashermes, P.; Agwanda, C.O.; Anthony, F., Combes, M.C., Trouslot, P. & Charrier, A. (1997a), Molecular marker-assisted selection: a powerful approach for coffee improvement. 17th International Conference on Coffee Science (Nairobi). Paris: ASIC, 474-480.

Lashermes, P., Combes, M.C., Trouslot, P. & Charrier, A. (1997b), Phylogenetic relationships of coffee-trees species (*Coffea* L.) as inferred from ITS sequences of nuclear ribosomal DNA. Theor. Appl. Genet. 94: 947 - 955.

Leroy, J.F. (1980), Evolution et taxogénèse chez les caféiers. Hypothèse sur leur origine. Comptes Rendus de l'Académie des Sciences, Paris, 291 : 593-596.

Leroy, J.F. (1982), L'origine Kenyane du genre *Coffea* L. et la radiation des espèces à Madagascar. Colloque Scientifique International sur le Café, Salvador, Bahia, Brasil. 10:413-442.

Leroy, T.; Henry, A.M., Philippe, R., Royer, M., Deshayes, A., Frutos, R., Duris, D., Dufour, M., Tesserreau, S., Jourdan, I., Bossard, G.; Lacombe, C. & Fenouillet, C. (1999), Genetically modified coffee trees for resistance to coffee leaf miner. Analysis of gene expression, insect resistance and agronomic value. 18th International Conference on Coffee Science (Helsinki). Paris:ASIC, B 209.

Longo, C.R. L. (1972), Estudo de pigmentos flavonóides e sua contribuição á filogênia do gênero *Coffea*. Piracicaba, 91 p. p. (Tese Doutorado-Esc. Sup. Agric. Luiz de Queiroz – Universidade de São Paulo, Brasil).

Maglio, C.A.F.P. (1983), Morfologia dos cromossomos nucleolares em fase de paquiteno no gênero *Coffea* L. Campinas, 92 p.p. (Tese de Mestrado – Universidade Estadual de Campinas – UNICAMP).

Mazzafera, P. (1991), Análises químicas em folhas de cafeeiros atacados por *Atta* spp. Revista de Agricultura, Piracicaba, 66(1):33-45.

Medina, D.M. (1972), Caracterização morfológica de híbridos interespecíficos de *Coffea*. Piracicaba, 111 p.p. (Tese de Doutorado – Esc. Sup. Agric. Luiz de Queiroz – Universidade de São Paulo.

Medina Filho, H.P., Carvalho, A. & Medina, D.M. (1977a), Germoplasma de *Coffea racemosa* e seu potencial no melhoramento do cafeeiro. Bragantia 36, nota n.° 11: 43-46.

Medina Filho, H.P., Carvalho, A. & Mônaco, L.C. (1977b), Melhoramento do cafeeiro XXXVII. Observações sobre a resistência de cafeeiro ao bicho mineiro. Bragantia 36: 131-137.

Medina Filho, H.P., Carvalho, A., Fazuoli, L.C. (1982), *Coffea bengalensis* e *C. travancorensis*: espécies diplóides, autocompatíveis e autógamas de *Coffea*. Ciência e Cultura. Reunião Anual da SBPC, Campinas, São Paulo, 34: 713 (Resumo).

Medina Filho, H.P., Carvalho, A., Gonçalves, W. & Levy, F.A. (1983), Melhoramento do cafeeiro visando resistência ou tolerância ao bicho mineiro (*Perileucoptera coffeella*). Congresso Brasileiro de Pesquisas Cafeeiras, Poços de Caldas, MG. 10:84-86

Medina Filho, H.P., Carvalho, A., Sondahl, M.R., Fazuoli, L.C. & Costa, W.M. (1984), Coffee breeding related evolutionary aspects. In: Plant Breeding Reviews (J. Janick, ed.). Avi Publish. Co. Connecticut, USA V. 2 p. 157-193

Medina Filho, H.P., Fazuoli, L.C., Guerreiro Filho, O., Gonçalves, W., Silvarolla, M.B., Lima, M.M.A. & Thomaziello, R.A. (1999), Increasing robusta production in Brazil: the potential of 200 thousand ha in São Paulo state. Association Scientifique Internationale du café. 18 ème Colloque. Helsinki, Finlândia. (no prelo)

Mendes, A.J.T. (1958), Avances in Coffee Production Technology: Recent advances in our knowledge of coffee trees. 3 – Cytology, Coffee and Tea Industries 81: 37-41

Moisyadi, S.; Neupane, K.R.; Stiles, J.I. (1999), Cloning and characterization of xanthosine-N7-methyltransferase, the first enzyme of the caffeine biosynthetic pathway. 18th International Conference on Coffee Science (Helsinki). Paris:ASIC, B 214

Mônaco, L.C.; Carvalho, A.; Fazuoli, L.C. (1974), Melhoramento do cafeeiro. Germoplasma do café Icatu e seu potencial no melhoramento. *In*: Congresso Brasileiro de Pesquisas Cafeeiras, 2, Poços de Caldas, Resumos. Rio de Janeiro, IBC/GERCA, p. 103

Moraes, S.A., Sugimori, M.H., Thomaziello, F.M. & Carvalho, P.T.C. (1974), Resistência de cafeeiros a *Pseudomonas garcae*. In: Congresso Brasileiro de Pesquisas Cafeeiras, 2., Poços de Caldas – MG, 1974. Resumos Rio de Janeiro, IBC, p. p. 183

Neupane, K, R., Moisyadi & Stiles, J. I. (1999), Cloning and characterization of fruits-expressed ACC synthase and ACC oxidase genes from coffee. 18 th International Conference on Coffee Science. (Helsinki) Paris : ASIC, B 213

Orozco - Castillo, C., Chalmers,K.J., Waugh, R. & Powel, W. (1994), Detection of genetic diversity and selective gene introgression in coffee using RAPD markers. Theor. Appl. Genet. 93: 934- 940.

Paillard, M., Lashermes, P. & Pétiard, V. (1996), Construction of a molecular linkage map in coffee. Theor. Appl. Genet. 87: 43 - 47

Rafalsky, J.A. & Tingey, S.V. (1993), Genetic diagnostics in plant breeding - RAPDs, microsattelites and machines. Trends Genet. 9: 275-280

Rijo, L. (1974), Observações cariológicas no cafeeiro Híbrido do Timor. Portugaliae Acta Biologica, Lisboa, 13: 157-168

Sondahl, M.R., Chapman, M. & Sharp, W.R. (1980), Protoplast liberation, cell wall reconstitution and callus proliferation in *Coffea arabica* callus tissues. Turrialba, 30 (2): 161-165

Sondahl, M.R. & Lauritis, J.A. (1992), Coffee. In: Biotechnology of perennial fruit crops. Biotechnolgy in agriculture 8: 401- 420

Sondahl, M.R., Sondahl, C.N. & Gonçalves, W. (1999), Micropropagação do cafeeiro. I Encontro sobre Produção de Café com Qualidade. (Viçosa) Ed. Laércio Zambolin. 228 - 240

Tanksley, S.D. & Orton, T.J. (1983), Isozymes in Plant Genetics and Breeding. Elsevier, Amsterdam, Part A, pp.516, Part B, pp.472

Tautz, D. (1989), Hypervariability of simple sequences as a general source of polymorphic DNA markers. Nucleic Acid Res. 17: 6463- 6471

Teixeira, A.A., Carvalho, A., Fazuoli, L.C. & Mônaco, L.C. (1974), Avaliação da bebida de alguma espécies e híbridos interespecíficos de *Coffea*. Ciência e Cultura, 26(7): 565

van Boxtel, J. (1994), Studies on genetic transformation of coffee by using electroporation and the biolistic method. Tese apresentada na Universidade de Wageningen. pp125

van Boxtel, J. & Berthouly, M. (1996), High frequency somatic embryogenesis from coffee leaves. Plant Cell Tissue & Organ Culture 44: 7-17

van der Graaf, N.A. (1981), Selection of Arabica coffee types resistant to coffee berry disease in Ethiopia. Meded. Landbouwhogeschool. Wageningen, 81: 11

van der Vossen, H.A.M. & Walyaro, D.J. (1980), Breeding for resistance to coffee berry disease in *Coffea arabica* L. II. Inheritance of the resistance. Euphytica, 29:777-791.

van der Vossen, H.A.M. & Walyaro, D. J. (1981), The coffee breeding programme in Kenia: A review of progress made since 1971 and plan of action for the coming years. Kenya Coffee. 46(541)

Vos, P., Hogers, R., Bleeker, M., Reijans, M., van de Lee, Hornes, M., Fritjers, A., Pot, J., Peleman, J., Kuiper, M. & Zabeau, M. (1995), AFLP: a new technique for DNA fingerprinting. Nucleic Acid Res. 23: 4407- 4414

Weber, J.K. & May, P.E. (1989), Abundant class of human DNA polymorphisms which can be typed using the polymerase chain reaction. Am. J. Hum. Genet. 44: 388-397

Welsh, J. & McClelland, M. (1990), Fingerprinting genomes using PCR with arbitrary primers. Nucleic Acid Res. 18: 7213 -7218

Williams, J.G.K.; Kubelif, A.R.; Livak, K.J.; Rafalsky, J.A. & Tingey, S.V. (1990), DNA polymorphisms amplified by arbitrary primers are useful as genetic markers. Nucleic Acid Res. 18: 6531-6535.

Wrigley, G. (1988), Coffee. (Tropical Agriculture series). Logman Scientific & Technical. USA. 612 p

Chapter 4

DEVELOPMENT OF COFFEE CULTIVARS IN REDUCED TIME BY USING BIOTECHNOLOGY IN THE "IAPAR MODEL FOR HIGH DENSITY PLANTING"

T. Sera[1]

[1]*Coffee Research Program, Breeding and Genetics Area, Agronomic Institute of Parana State (IAPAR), CEP86001-970, Londrina-PR, Brazil*

Running title: Coffee breeding

1. Introduction

The genetic improvement of the coffee aiming at high productivity is not an easy task. This is due to the existence of the juvenile period, to the need of evaluating several years of serial production to quantify the long term productive capacity, and to the existence of the pronounced annual oscillation of production where the coffee plantations are grown under full sun and without shade (Stevens, 1949). Efforts have been continuously made to improve the coffee productivity through research and development in coffee breeding and crop techniques. The program started in 1933 by Agronomic Institute of Campinas (IAC) obtained increases of up to 240% of productivity (Mundo Novo variety) in comparison to the first coffee brazilian cultivar denominated Arabica (Fonseca *et al.* 1978). The improvement methods normally applied in Brazil, consisted in the selection of individual plants, followed by the progenies evaluation, hybridization and genealogical selection. In addition, the backcross method has been applied for the transference of specific characters, such as short internodes and resistance to pests and diseases, present in other cultivars or in related species of *Coffea arabica* L. The selection of individual plants can be accomplished either in a commercial plantation, or in experimental fields especially planted with the purpose of genetic improvement. In the

T. Sera et al. (eds.), Coffee Biotechnology and Quality, 47–70.

latter, the production is registered by several successive years for the identification of superior genotype (Medina *et al.*, 1984).

The deep knowledge of the cultivation environment, processing, commercialization and consumption is of fundamental importance in the successful development of a cultivar. The use of the "Technological model for the production of coffee in Paraná " ("IAPAR Model of high density planting") as an environment for the development of coffee cultivars has been taken as an example in the presentation of the theme. It consists of the following points:

- production model built up with more than 80 technologies;
- plantation in dense spacing, between 8 and 12 thousand plants/ha;
- potential productivity up to 4800kg/ha/year in the best coffee farms;
- profit of 50ha in traditional system obtained in 10ha, by increasing the efficiency (productivity, quality, lower cost/bag);
- survival to very low prices and profit with bad ones;
- creation of one employment for each ha of coffee cultivated at the cost US$5000/ha;
- qualified and productive workers as a solution;
- economic reactivation in more than 200 municipalities of Paraná state-Brazil;
- adaptation to small, medium and large farms;
- initial return of investment within 2.5 years from planting;
- economic stability with integrated and moderated diversification in the property;
- planting scaled yearly during 4 to 10 years, distributed quartely and half-yearly within each year, using own resources;
- gradual and continuous adoption of new technologies and cultivars;
- suitable for descapitalized farmers;
- independence to financing sources;
- ecological technology with soil, pests and diseases management;
- co-existence with nematodes through the integrated management of control techniques;
- smaller cost of fertilizing per bag based on rational soil and nutrition management;
- co-existence with frosts, recovering production potential after two years;
- regional, local and topographic recommendation of cultivar, adapted to climate and specific soils of the farm;
- harvest in parts by using 4 groups of cultivars from early to super-late maturity, to distribute the harvesting period;
- quality coffee with at low cost, by using simple technologies;
- coffee is a need for the urban consumers and consumption is in ascension.

The main objective of IAPAR has been to support technologically the coffee growers of Paraná state under the "Paraná Coffee Plan" increasing the profitability and the economic stability of producers, through the economic efficiency at the property. Efficiency will be reached based on:

a) Increase of productivity;
b) Reduction of the production cost per bag;
c) Improvement of product quality;
d) Integrated and moderate diversification in the property.

2. Needs of genetic improvement for the coffee

To reach the economic efficiency in the production of quality coffee at lower cost, there is need to develop several research projects:

a) The potential of productivity of 80 clean coffee bags/ha/year for excellent producers is not accessible for most. It is necessary to overcome several critical points through cultivars and its appropriate management;
b) The accomplishment of the potential of quality coffee production faces several difficulties, including uniform maturation, distribution for harvesting period by using cultivars from early to super-late maturation groups and of larger size grains;
c) The cost of production of US$80-100/bag of the traditional coffee system and more than US$50/bag for high density coffee are still high, and the crop harvest is the costly component. There is a need to make the diversification of cultivars by maturation groups, and to develop cultivars with small canopy and large fruit size, besides resistance to parasites;
d) The vulnerability to environmental oscillations is high, needing more resistant cultivars to frost, drought and to poor soils;
e) The control of the rust disease Hemileia vastatrix is difficult in dense spacing, making it more expensive in cost and depreciating the quality and increasing the susceptibility to frosts. It is necessary to develop new cultivars with more genes for resistance and tolerance, more adapted to this cultivation system;
f) The root knot nematodes, especially Meloidogyne incognita with its 5 well-known races is limiting in the infested property, needing to intensify the development of resistant cultivars;

g) The bacterial disease Pseudomonas syringae pv. garcae is becoming a limiting disease in altitude areas and in south and east faces, needing to develop resistant cultivars;
h) The leaf miner (*Perileucoptera coffeella)* is becoming more harmful and of difficult control, needing to intensify the development of resistant cultivars;
i) The high cost of control of the weeds of the coffee crop in formation justifies the development of resistant cultivars to post-emergent herbicides;
j) There is the need to prevent the introduction of antracnosis of the fruits and to begin the evaluation of resistance of the available germ-plasm;
k) The fruit borer control with incorporation of resistant genes by genetic transformation is necessary;
l) There is the need to reduce the time spent in the development of the cultivars.

3. Reduction in the time spent for coffee cultivar development

a) The methods of genetic improvement of plants usually used in annual species can be used for coffee, a perennial one, since respected its characteristics. However, the improvement of perennial species as the coffee plant presents some very important particularities, needing to adopt special procedures for not having a low efficient program:
b) Usually there is a long period until flowering for the accomplishment of the crossings and for the production of seeds.
c) It is important to evaluate all the agronomic characteristics as well, and not only productivity and quality, due to relatively high cost of substitution of the cultivars.
d) High cost of the field evaluation due to the need of large areas and time for experimentation.
e) Need of evaluation of the precocity and productive longevity.
f) Need of evaluation of the annual oscillation of the production.
g) Flowering and annual production on the same plant

4. Factor 'annual genetic gain'(Gsa) in the coffee cultivar selection

Attempts are constantly being made to identify new procedures that can increase the efficiency in the development of cultivars, and Gsa has proved efficient in this regard. Gsa is estimated by the increment in the average performance of a progeny that is accomplished in each selection cycle. A selection cycle includes the establishment of a segregating population, genotype evaluation, selection of superior genotypes and use of the genotypes selected as parental, to form a new population for a new selection cycle. The time consumed in each selection cycle can vary considerably (Fig 3). In the comparison of alternative strategies, Gsa needs to be expressed in annual base. One needs to trace strategies that uses human resources, space, materials, and available time rationally to propitiate the largest possible gain a year (Sera and Alves, 1999).

GSa values limit the adoption of some improvement methods normally used in annual crop in a classic way. The consequence is a smaller efficiency of a perennial species improvement program if adopt the same procedures as used for the annual plants (Fig.1 and Fig. 2), specially for seed propagated species. This low efficiency is measured by the low genetic gain of the selection in a year. To proceed, it will be defined and compared by the expression of genetic gain per year (GSa) until the obtaining cultivars in annual and perennial plants propagated sexual or asexually.

$$GSa = (DS \times h^2) \div (F_n \text{ x years per cycle})$$

Where,

GSa = annual genetic gain;

DS = selection differential;

h^2 = heritability = genetic variance (σ^2_G) / phenotypic variance (σ^2_F) = σ^2_G / σ^2_F;

F_n = number of necessary selection cycles to obtain cultivars.

Annual plants. h^2 depends on the selective outline, and they generally carries a selection cycle per year repeating several generations.

$$GSa = DS \text{ x } \frac{\sigma^2_G}{\sigma^2_F} \div F_n$$

Where,

σ^2_G = genetic variance;

σ^2_F = phenotypic variance

Coffee cultivar type clone. If the coffee plant is propagated vegetatively, the $\sigma^2_G = \sigma^2 a + \sigma^2 d + \sigma^2 i$ and then the genetic gain expected a year it would be:

$$GSa = DS \text{ x } \frac{\sigma^2 a + \sigma^2 d + \sigma^2 i}{\sigma^2 a + \sigma^2 d + \sigma^2 i + \sigma^2 e} \div \text{Years per generation}$$

Where,
$\sigma^2 a$ = addictive genetic variance;
$\sigma^2 d$ = dominant genetic variance;
$\sigma^2 i$ = epistatic genetic variance;
$\sigma^2 e$ = environmental variance.

All σ^2_G could be taken as advantage and a cultivar, generally heterozigotic, with heterosis is propagated vegetatively for the later generations, resulting in higher GSa compared to cultivars propagated by seed due to larger h^2 and Fn = 1.

Coffee cultivars type lineage propagated by seeds. In conventional coffee cultivars propagated by seeds, seeking the selection of homozigous genotypes in most of the loci, GSa it would be:

$$GSa = DS \times \frac{\sigma^2 a}{\sigma^2 a + \sigma^2 d + \sigma^2 i + \sigma^2 e} \div (\text{Years per generation} \times F_n)$$

Only $\sigma^2 a$ it could be taken as advantage in this case and, necessarily, we would have to carry population selection trough a few generations until homozygous. Resulting GSa is much smaller than the propagated vegetatively, or obtained in plants of annual cycle.
Thus, it is evident that depending on the method of propagation and of the type of cultivar to obtain, a longer or shorter time could be required for obtaining the superior genotypes that could be used commercially by the producers. The adoption of some special strategies in the genetic improvement and anticipated releasing of new coffee genotypes for farmers can reduce the time spent in the development and adoption of new cultivars. Besides, we can get a great reduction in the experimental area, resulting in high economy and substantial increase in the efficiency of the breeding program.

5. Reduction of the time spent in the development of coffee cultivars

In the development of a cultivar, the reduction of the time spent in a generation or selection cycle, or in the cultivar development is of fundamental importance to increase the efficiency of an improvement program, where the first significant flowering occur only after 2 or even more years after sow (Sera, 1987). Very frequently this is a limiting factor in the development of a genetic breeding project, within a reasonable period.

Besides, one of the desirable characteristics of a coffee cultivar is that it begins to produce earlier and, at the same time, continue producing for many years (Medina *et al.*, 1984). For that, it is necessary to evaluate the production during the productive useful life of the plant or its progeny, to make sure that such selection presents this characteristic.

After the juvenile period, a single selection field demands many years of observation during which the experimental area can not be taken used for other purposes. This way, a lot of space, time and resources are consumed, while the number of plants that can be measured is limited. Thus, the possibility to carry a program of rational genetic coffee improvement (Fig. 1 and Fig. 2) involving multiple objectives using, for example, a scheme of back-crossing outline or the genealogical method, depends largely on the reduction of the number of generations and of making the anticipated selection for production with less number of years per generation. Other less complex characteristics such as resistance to pests, diseases and environmental adversities can be evaluated and selected during the juvenile period (Sera, 1984).

There are several methods and alternative techniques which could be applied in the field, during process of obtaining cultivars (Sera and Alves, 1999):

a) reduction of the juvenile period;
b) selection in the juvenile period;
c) anticipated selection for production in the initial years;
d) selection in early generations;
e) methods that don't evaluate the production (bulk, SSD and back-crossing);
f) use of biotechnological techniques and genetic engineering;
g) obtaining of double-haploid cultivars;
h) use of the heterosis of hybrid cultivars F1, F2 and synthetic;
i) vegetative propagation for use of clone type cultivars;
j) anticipated release of cultivars.

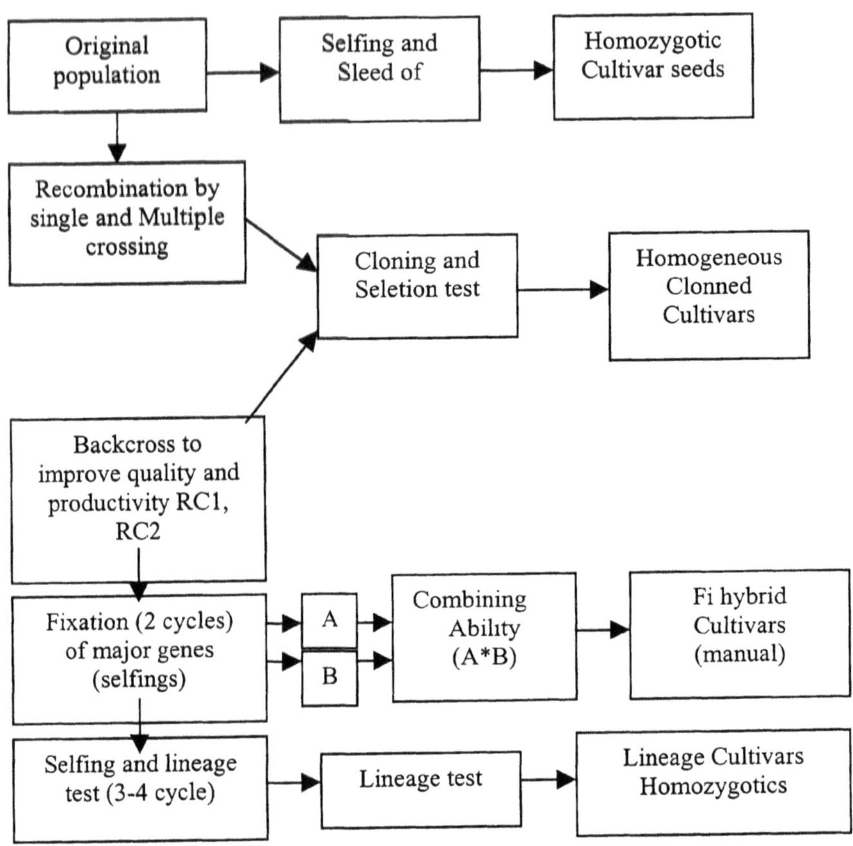

Figure 1. Breeding of an autogamous perennial plant, arabica coffee (*Coffea arabica* L.). Source: Van der Vossen, 1985.

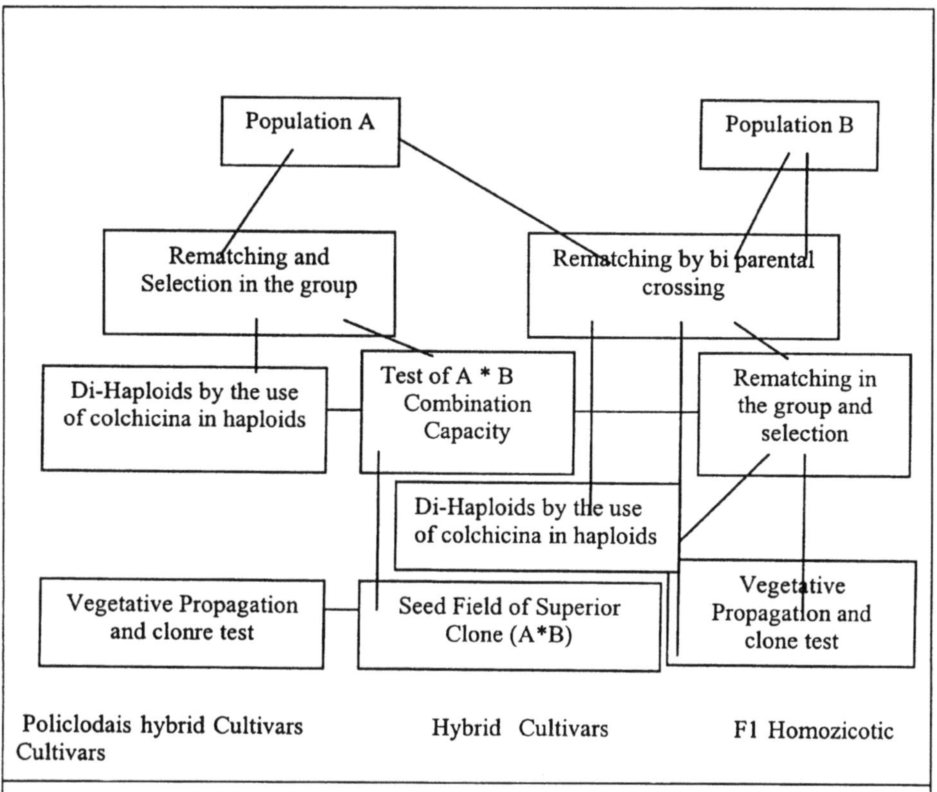

Figure 2. Breeding of an allogamous self-incompatible perennial plant, robusta coffee (*Coffea canephora* Pierre). Source: Van der Vossen, 1985.

Each one of the methods, techniques and available procedures for the development of cultivars presents advantages and disadvantages. The probability to obtain superior cultivar can be increased by the increase of the number of genotypes number that can be tested every year, but this will increase costs that need to be minimized by alternative strategies, using special techniques and procedures, with adaptation for perennial plants. Appropriate strategies that facilitate the fast progress of generations, cloning, double-haploid and anticipated cultivar releasing can promote great impact in the efficiency of a coffee breeding program.

5.1. REDUCTION OF THE JUVENILE PERIOD

In general, a project of lineage cultivar development involves several selection generations. Consequently, the length of a reproduction generation is very often the most limiting factor to complete a project in a reasonable period.

- **Selection for early flowering genotype**: It happens in the coffee plant in the same way as it happens in other perennial plants such as apple, pear, grapevine, fig and citrus. (Janick and Moore, 1975; Sera, 1984).
- **Immediate sowing after the harvest seeds**: It is common to wait ideal natural conditions for sowing. Since germination is carried out under controlled conditions and in green-house, a time reduction of 6 months can be obtained for the coffee plant. If the coffee seedlings are planted during the month of October, they can be expected to save one agricultural year of the coffee. However, if is planted during April-May, one could expect to have lost one year per cycle (Sera, 1984).
- **Good cultivation conditions**: By maintaining the conditions of nursery such as irrigation, temperature and light intensity, application of manure, control of pests, etc., as close as possible to fields, we can reduce the juvenile period (Janick and Moore, 1975; Söndhal *et al.*, 1972; Sera, 1984). This also applies for cultivation in vitro (Söndhal et al, 1981) or in the field (Monaco *et al.*, 1978). In coffee, it is possible to produce flowers with 6 months in the field by using 6 month old seedlings. So it is possible to have first seed production with 20 months beginning from sowing.
- **Dwarfing graft**: The grafting of vegetative shoot or buds of woody perennial plants in weak grafts, old or dwarf has frequently been indicated, as in the case of apple tree, to reduce the juvenile period to less of half of the time (Janick and Moore, 1975). This could be tried in coffee plants.
- **Environmental floral induction**: Continuous light exposition is appropriated to reduce the flowering time of long day plants. For plants of short days, less exposition to light is useful but the effect of short days depends on the optimal photo-period. For the tropical plants, 8 hours could be more appropriate (Wellensieki, 1962). Nitrogen or water deficiency can induce early flowering.
- **Environmental reduction of flowering-maturation period**: The deficiency of water can induce the reduction of time from flowering to the maturation. The deficiency of nitrogen during the filling of the grains leads to small seeds and to early maturation. The increase of the phosphorus tends to reduce the maturation period. High temperature can induce up to three month precocity in maturation in

some late maturation cultivars of coffee plant, especially if combined with nitrogen or water deficiency (Sera, 1979), allowing the sow in May and plantation in the field in October.

5.2. EVALUATION AND JUVENILE SELECTION FOR PRODUCTIVITY

The juvenile selection for productivity can be practiced for the qualitative and quantitative characteristics of high herdability which are highly correlated with production. The selection can be made in the seedlings level, both in the nursery or in the field, before the first harvest.

5.2.1. Selection in seedlings before transplant to nursery recipient:

Characteristics such as root volume and vegetative vigor (Ramos, 1980) can be selected during the period of emission of cotyledonary leaves up to two definitive leaves, before the transplant to the plastic bag recipient, saving area occupied in the nursery.

5.2.2. Selection in seedlings in the nursery:

Yield correlated traits such as resistance to insects, diseases, nematodes, drought, aluminum and frost can be selected before planting in the field, saving space and time for the evaluation and computing data. This would liberate more space, resources and time for more valuable material.

5.2.3. Juvenile selection in the field:

Characteristics such as ramification, diameter of the crown, resistance to wind, resistance to extreme temperatures and number of flowers per node can be selected in field conditions, before the first harvest. Three or more plants can be planted per hole and one can be selected (Valencia, 1973; Srinivasan, 1980; Mawardi and Hartobutoyo, 1981; Srinivasan and Vishveshwara, 1981). Based on initial field evaluations, some treatments can be eliminated some treatments by evident agronomic inferiority, thus increasing the efficiency of the selection and saving time and resources.

5.3. ANTICIPATED SELECTION FOR PRODUCTIVITY AND PRODUCTION LONGEVITY IN INITIAL YEARS

The productivity is a quantitative characteristic, difficult to work in cultivar development programs that incorporate new characteristics such as resistance to parasites, difficult environmental conditions and change in the cultivation system, or quality improvement demanded by consumers.

The understanding of the production components and their correlation can aid in the anticipated selection of the promising genotypes for productivity. It is common negative correlation between number and size of the grains, between production of leaves and seeds, between size of plants and possible number of plants per area, etc. (Gifford and Evans, 1981).

In perennial plants, once production starts, it continues for several years, usually for 10-20 years, and therefore initial production cannot represent the total production (Carvalho, 1952; Rocha *et al.*, 1980). Thus, genotypes that present production precocity can provide initially larger production, which could be smaller when a longer period of time is considered that is, precocious senescence. Genotypes could provide late production but with production longevity. The ideal genotype would be the one that begins to produce early with high productivity and stay productive for a larger period of time (Medina *et al.*, 1984).

Thus, it is necessary to make a premature evaluation system for production (Waryaro and Van der Vossen, 1979; Srinivasan, 1981) that could guarantee a fast progress in each selection generation, resulting in quick obtaining of new cultivars with larger genetic progresses (Fig. 3).

5.3.1. Selection index

The premature identification of more productive genotypes with larger precision than is reached during the production of the initial years is highly desirable in the genetic improvement of the coffee plant. According to Medina *et al.* (1984), good cultivars should begin to produce early and maintain high productions for several years. The best progenies can be chosen based on the first six serial harvests to represent 15 to 20 years of production (Carvalho, 1952; Fazuoli, 1977). However, a perennial crop in such a period of time that begins to produce significantly only after three years of vegetative period is still considered time consuming, besides consuming space and resources. The coefficient of determination (R^2) indicates the percentage in which the real production is estimated by the multiple regression equation. Sera (1987), using multiple regression estimated these coefficients for selection index composed by measurable characteristics, in the first three years of production. He obtained R^2=0.615 for the first 3 productions, R^2=0.489 for the first 2 productions and R^2=0.070 for the first production. The last one is considered very low. Following, the estimates of the best production indexes are presented, taking into account other characteristics correlated with production, for the available characteristics in the third, second and first year of crop harvests, as well as without the production.

First three years: R^2=0.791 using the size and production of grains ÷ production of fruits, height of the canopy, diameter of the canopy, production of the 1st. biennium, annual increment of the production in the first three years of production and annual oscillation of production in the first three years.

First two years: R^2 = 0.717 using the size of the grains, production of grains ÷ production of fruits, height of the canopy, diameter of the canopy and production of the first biennium.

First year: R^2 = 0.571 using the size of the grains, production of grains ÷ production of fruits, height of the canopy, diameter of the canopy and 1st. production.

Without production. R^2 = 0.454 using height and diameter of the canopy.

The R^2 obtained by using production and correlated traits is always larger than just using the production. Using only height and diameter of the canopy, the correlation coefficient was almost equal to one using the first two harvests, which agrees with the results obtained by Walyaro (1983). Sera (1987) obtained the classifications by comparing the progenies with base total productions, and the estimated ones anticipated by indexes were verified to include the seven better progenies of 72 (i = 10%). It was necessary to select the 13 best for the first three years, the 20 best for the first two years and the 50 best for the first year of harvest. Therefore, it would be viable to anticipate the selection with reasonable safety, having production data and correlated characters obtained in the first two years of cropping. The use of evaluations until first harvest ($R^2 < 0.5$) only is for discard of inferior selections or pre-selection, as indicated by Araújo (1978) and Srinivasan (1982), but the reduction in 50% is very significant and important to reduce human and material resources that can be used for other activities.

Thus, anticipated selection for the reduction of the time spent in the development of cultivars is feasible (Table 1) by using indexes of production that compose production components and production to guarantee that the superior progenies are included in those anticipatedely selected. It is advisable to use low selection intensities, as much between as within the progenies.

Table 1. Comparison of time spent in the obtaining lineage type cultivars by using alternative breeding procedures: **(A)** Classic genealogical selection, **(B)** early generations selection, **(C)** early generations selection + anticipated selection, **(D)** generation advance without yield evaluation by using SSD + early generations + anticipated selection and **(E)** generation advance without yield evaluation by using back-cross of a dominant gene + anticipated selection in arabic coffee (*Coffea arabica* L.), an autogamous perennial plant propagated by seed.

Table1. Comparison of time spent in the obtaining lineage type cultivars by using alternative breeding procedures:

Year	Alternative breeding procedures				
	(A) Classic genealogical selection	(B) Early generations sel.	(C) Early generation + Anticipated selection	(D) S.S.D. + Early gener. + Anticipated sel.	(E) BC + Early gener. + Anticipated Sel.
1	Fl Plant	Fl Plant	Fl Plant	Fl Plant	Fl RC1 Plant
2	Harvest & Planting	Harvest & Planting	Harvest & Planting	Harvest & Planting	Flowering and RC_2
3	Flowering	F_2 and Flowering	F_2 and Flowering	F_2 and Flowering	Planting F1, RC_2
4	1st harvest	1st harvest	1st harvest	Harvest and planting	Flowering and RC_3
5	2nd harvest & Planting	2ndharvest&Planting	2nd harvest &Planting	F_3 and Flowering	Planting and F_1, RC_3
6	F_2 and Flowering	F_3 and Flowering	F_3 and Flowering	Harvest and planting	Flowering and ⊗
7	1st harvest	1st harvest	1st harvest	F_4 and Flowering	F_2 and homozigotic
8	2nd harvest and planting	2nd harvest&Planting	2nd harvest&Planting	1st Regional test	**CULTIVARS**
9	F_4 and Flowering	F_4 and Flowering	F_4 and Flowering	2nd Regional test	Basic Seed
10	1st harvest	1st Regional test	1st Regional test	**CULTIVARS**	Flowering
11	2nd harvest and planting	2nd Regional test	2nd Regional test	Basic Seed	1ª. Production
12	F_5 and Flowering	3rd Regional test	**CULTIVARS**	Flowering	RELEASE
13	1st harvest	4nd Regional test	Basic Seed	1ª. Production	Commercial Seeds
14	2nd harvest and planting	5rd Regional test	Flowering	RELEASE	Flowering
15	F_6 and Flowering	6nd Regional test	1ª. Production	Commercial Seeds	*COFFEE CROP*
16	1st Regional test	**CULTIVARS**	RELEASE	Flowering	
17	2nd Regional test	Basic Seed	Commercial Seeds	*COFFEE CROP*	
18	3rd Regional test	Flowering	Flowering		
19	4nd Regional test	1ª. production	*COFFEE CROP*		
20	5rd Regional test	RELEASE			
21	6nd Regional test	Commercial Seeds			
22	**CULTIVARS**	Flowering			
23	Basic Seed	*COFFEE CROP*			
24	Flowering				
25	1ª. Harvest				
26	RELEASE				
27	Commercial Seeds				
28	Flowering				
29	1ª harvest				
30	*COFFEE CROP*				

5.3.2. Production longevity and annual oscillation evaluation

Two of the main constraints to get fast advance of generation selection in coffee breeding is longevity and oscillation of productivity. This is possible by using correlation estimates (Sera, 1980; Sera, 1984; Sera, 1987; Sera and Alves, 1999)

For longevity, it is possible to use selected mother plants of initial generations while the generations are advanced.

For production, annual oscillation has a correlation between productivity and vegetative vigor in the same year. More stable genotypes have better vigor after a year of high yield.

5. 4. SELECTION IN EARLY GENERATIONS

This approach can be used for autogamous and allogamous species, to select individuals inbred lines or populations at the early stage of endogamy. The objective should be to eliminate lines or populations that don't present merit for posterior selection. Repeated tests of segregant populations F_2 or F_3 would supply the medium to evaluate the performance of the different crossings. This method is useful to identify superior F_2 plants through progeny tests with replications and then to evaluate and select homozygotic individuals inside of derived lines of F_2

This procedure can identify homozygotic progenies in generations as early as F_3, reducing the time spent in the development of cultivars (Fig. 1), although the cost of the materials and evaluation conducted increases a lot. In perennial plants the time is restriction factor; it is favorable, because individuals, lines or inferior populations are early discharged in the autogamy process. Characteristics of high herdability such as early maturation can be selected in F_2 generations. Characteristics highly correlated with the production such as vegetative vigor, ramification, volume of the canopy and size of the grains can be selected with success. Characteristics of low herdability such as production, can also be selected in F_2, since the genetic variability of the germ-plasm, and the population size are high and the environmental control is rigorous (Sera, 1980; Mawardi and Hartobutoyo, 1981; Mawardi *et al.*, 1983).

5. 5. GENERATIONS ADVANCE WITHOUT PRODUCTION SELECTION IN INITIAL GENERATIONS

Improvement methods such as " Single-Seed-Descent " (SSD), population and back-cross, can take fast progress in the segregant generations towards the homozygotic, or to have high similarity to the recurrent parental. They allow the progress of breeding, works in certain stage without accomplishing the evaluation of the productivity, which are expensive and time consuming operations. The advantage of this breeding methods is to advance the initial generations without selection for the production (as exemplified using SSD in Table 1).

The genealogical method has been popular for endogamous population of annual plants for situations when only one generation is advanced per year. Nowadays, it has been substituted with other methods such as SSD, in which it is possible two to three generations in one year using greenhouse, stove or nursery. A comparison of the genealogical methods, with selection in early generations, with and without anticipated selection for productivity, SSD and back-cross, applied to the coffee plant, is presented in Fig. 3.

5. 6. USE OF BIOTECHNOLOGICAL TECHNIQUES

Biotechnological techniques mentioned below have opened new perspectives for the genetic improvement of perennial species such as coffee (Söndhal *et al.*, 1981; Söndhal and Loh, 1988).

5.6.1. Culture of embryos and meristems:

Those techniques already has practical applications in research programs, seeking to increase the efficiency of the breeding programs (Söndhal and Loh, 1988; Etienne *et al.*, 1999).

5.6.2. Microsporos and ovule culture:

The microsporos or ovule culture which can result in the production of large amount of haploid plants. Those plants can be duplicated to obtain double-haploid lines completely homozygous in the F2 generation (Carneiro, 2000), presenting a great potential in the near future.

5.6.3. Somatic embryogenesis starting from leaves:

The direct or indirect somatic embryogenesis (Etienne, *et al*, 1997; Etienne-Barry *et al.*, 1999) has been applied aiming the reduction of the time spent in the development of new cultivars. This time reduction is important when uses clone type cultivars of double hybrids, or simple hybrids, containing several agronomic characteristics in a genotype with heterosis (Bertrand *et al.*, 1997; Bertrand *et al.*, 1999). Nowadays, the production cost of seedlings from somatic embryogenesis is near conventional seedlings (Söndhal *et al.*, 2000).

5.6.4. Genetic transformation:
The genetic transformation of *Coffea* by vectors like *Agrobacterium* and by biobalistic, for incorporating foreign DNA in the coffee genoma is becoming reality (Leroy *et al.*, 1997). Characteristics of wide use in cultivars as the architecture of the plant genes (e.g., compact, erect and conic), male sterility, precocious and late maturation, resistance to nematodes, etc., are expected to be incorporated in near future in commercial cultivars.

5.6.5. Assisted selection by DNA markers:
The use of molecular markers for assisted selection can reduce a time spent in the development of coffee cultivars in several ways. The selection in the juvenile phase, accomplished in laboratory, for qualitative characteristic such as architecture of the plant, color of the fruit, percentage of fruits with empty grain and resistance to diseases and pests will bring great benefits in the efficiency of an improvement program (Lashermes *et al.*, 1997).

5.6.6. Other techniques.
Other techniques such as asexual hybridization through protoplast fusions, selection *in vitro*, "DNA fingerprinting" will help a lot to increase the selective efficiency. Those techniques have the tendency to be incorporated as routine tool for breeders in the near future. The breeding that began as an empiric way had evolved to the scientific breeding, and is in evolution to the biotechnological breeding/genetic engineering.

5.7. USE OF DOUBLE-HAPLOID CULTIVARS

For the autogamous species as the arabic coffee, where the commercial product is a grain, the most economic for the farmers is the use of lineage type cultivars propagated by seeds. The time spent for the development of lineage by the usual method is long, generally around 25 years (Medina *et al.*, 1984) and this period can be reduced for something as eight years.
Multi-line type cultivars are the other type of cultivars in that lineage which are used to give larger phenotypic stability for the mechanical mixture of the several lineage in well-known proportion. It is also used to reduce the time spent in the development of cultivars with resistance to several physiologic races of pathogens. It usually needs a long time to

get to a homozigotic lineage. To reduce this time, there is the alternative of getting the lineage starting from duplicated haploid plants (Nitzsche and Wenzel,1977).
In the usual procedures of improvement, the lineage type cultivars are obtained through successive generations of selfings. Double-haploid are homozygotic plants obtained by the duplication of the chromosome of cells or haploid tissues. The production of diploid homozigotic lines by duplication of the chromosome complement of haploid individuals is an alternative method of development in a very short period. There are several conventional or biotechnological methods to obtain double-haploids, that can be used in perennial plants. In coffee (Carneiro, 2000) has presented important advances aiming haploids plants, specially for selections of Catimor and Sarchimor germ-plasm.
There is the possibility of using haploid plants obtained by conventional, natural or induced methods (bulbosum method) in the process of homozygotic lineage obtaining in coffee tree.

5. 8. USE OF HETEROSIS OF THE HYBRIDS

Hybrid cultivars are used in several cultures seeking to take advantage of the heterosis observed in the hybridizations. In perennial plants, the cost of the hybrid seed can be muffled by several years of production (Sera and Alves, 1999). If a hybrid, for example, produces 20% more per year, in five years, an additional one will have production equivalent to an extra harvest. In perennial plants, the genetic emasculation and the self-incompatibility can be used. The emasculation and the manual pollination can be viable in several crops, especially in plants that produce many seeds per fruit as the passion flower, cocoa tree and *Anonaceae* plants. The use of hybrids can also represent a great economy of time, allowing the breeder to test several combinations of lineage, to select the most promising and to release new cultivars (Bertrand et al, 1997; Bertrand et al, 1999). In coffee plants, where each flower gives less than two seeds, besides hand emasculation, it can be tried gameticides, thermal emasculation and genetic male sterility. Different types of hybrids can also be used.

5.8.1. F1 hybrid type cultivars:

Usually obtained by the hybridization of homozygotic lineage, being viable for perennial plants that produce many seeds per fruit. The readiness of male-sterility genes and self-incompatibility, as well as of chemical and thermal techniques of the male-sterility induction can also facilitate the use of this alternative, especially in plants that produce just a few seeds per fruit.

5.8.2. F2 hybrid type cultivars:
If the sale of hybrid seeds Fl are not economically viable, in some cases, one can cultivate F2 hybrid plants commercially, especially for perennial plants as coffee that produces a few seeds per fruit. In this case, it can take advantage of half of the heterosis of the F1. Some characteristics as the harvest time and the size of the plant should be uniform, even so in some cultivation variability can be admitted for this and other agronomic characteristics.

5.8.3. Synthetic type cultivars:
These types of cultivars are used to preserve the vigor of hybrid without obtaining the hybrid seeds. A population base of hybrid combinations with heterosis of allogamous plants is used to form the seed production fields of cloned superior genotypes, allowing to reach larger productivity than lineage type or population. It can be used in robusta coffee (*Coffea canephora*).

5.9. USE OF CLONE TYPE CULTIVARS

They are cultivars propagated vegetatively, that is to say, all the descendants are genetically the same as the mother-plant. The vegetative propagation for conventional means as the grafting, or un-conventional as *in vitro* micro-propagation allow to obtain quickly superior cultivars (Fig.4). The vegetative propagation allows the use of the whole available genetic variance and the selection usually relapses on highly heterozygotic plants.
The plants of great economic value (fruit and ornamental) are in general propagated vegetatively and they constitute clone type cultivars with high heterosygosity. In plants for which vegetative propagation is not yet economically feasible, the researches can be addressed seeking to improve the techniques to reduce the costs of this type of propagation.
Micro-propagation has been used in several crops (sugar-cane, banana, pineapple, strawberry, eucalyptus, etc) and it presents as advantage, besides the fast propagation, the possibility of obtaining materials free from pathogens, especially virus. On the other hand, during the process of *in vitro* micro-propagation, the possibility exists of finding inherited genetic variations, economically important in the progenies (somaclonal variation).
The use of clone type cultivars is already a reality in Brazil (Paulino *et al.*, 1985; Silveira and Fonseca, 1995), through the use of conventional cutting obtained from Conilon coffee scion (*Coffea canephora var kouillou)*. Although it has quadruplicated

cost, the largest productivity and uniformity allied to other characteristics of cloned cultivars, it has maintained the interest of many coffee growers to implant commercial fields.

The use of cloned cultivars, starting from leaves of hybrid superior genotypes by somatic embryogenesis using bioreactors (Etienne *et al*, 1997), is indicating enough genetic stability and economic viability (Söndhal and Söndhal 2000) to immediate application of this technology in the breeding programs and coffee farms.

5.10. ANTICIPATED RELEASING OF CULTIVARS

In perennial crops several years are required to advance the selection generation. It is interesting to adopt an appropriate outline for anticipation of the cultivar release to increase the efficiency of the coffee breeding program (Fig. 4). For that, it is necessary to spend more in terms of experimental area, evaluation and multiplication of a larger number of selections, as well as the installation of a larger area and of a larger number of potential cultivars in a field of basic seed to save time. In arabic coffee breeding it is possible to save 10 years by the simultaneous conduction of regional evaluation experiment and of basic seed field, since identified the origin by molecular genetic markers ("DNA fingerprinting") of potential cultivars (Sera and Alves, 1999).

5.10.1. Anticipated regional evaluation of selections:

All the promising materials should be tested regional and simultaneously with last generation selection in order to save one stage of the total process from the crossing to the commercial plantation. It consists of evaluating a very larger number of promising progenies, to include among them the first five selections that can become cultivars, saving something as four years in time reduction to release cultivars (Sera, 1987).

5.10.2. Anticipated installation of basic seed field:

To save at least four more years, we should begin concomitantly the installation of regional performance experiments, basic seed field, including all the promising genotypes. The portions of the selections not confirmed for the recommendation should be eliminated and completed with recommended materials.

5.10.3. Anticipated release of cultivars:

The anticipated release of cultivars in early generations, with certain variability for productivity, size of the fruit or grain and for other agronomic characteristics of minor importance, could allow to release unfinished versions with characteristics of high agronomic value four years earlier (e.g., for pests and key diseases, quality, etc,). This would have high economic-social return for the farmers. The complementary selection allows posterior release of improved versions.

Table 2. Time consumed to develop the experimental cultivars and their adoption in large scale by farmers, considering different development alternatives of cultivar and cultivar releasing.

Alternatives	Experimental cultivars	Regional Evaluation	Production Fields (Seed)	Production in farms
Pure Line Conventional	14	22	26(Release)	30
Line+Early generation	8	16(Release)	20	24
Line+Early+Anticipated	8	12(Release)	16	20
Line+Early+SSD	7	10(Release)	14	18
Line+BC+Anticipated	7	8(Release)	12	16
F_2 Hybrid	6	10(Release)	14	18
F_1 Hybrid	2	8 (Release)	8	12
Clone cultivar	2	8 (Release)	-	10

6. Conclusions and futuristic approach

Through the combination of a group of methods, techniques and genetic improvement procedures, it is possible to obtain considerable gains in the efficiency of a breeding program for coffee, a perennial species that time is the main constraint to obtain a reasonable genetic gain per year until the obtaining and planting of an improved cultivar. The adaptation and use of solutions adopted for other perennial species can give a great contribution in terms of saving time, turning the work less tiresome and more enterprising. A strategic plan for a perennial coffee breeding program is vital and decisive to a successful breeder. It is necessary, to have a deep knowledge of all the aspects of the crops, to avoid that life time work results in very little success. Joint research among research breeding programs from different institutions and among different experts, as well as with farmers, is of fundamental importance to increase the efficiency and to reduce the time to complete a cultivar development and releasing. Besides, a deep knowledge of all the aspects of the crop that he deals with, from the origin and evolution of the species to the commercial competitiveness, would also be necessary. A wrong strategy of a coffee breeder is very frustrating, because one doesn't have much time to begin again, or to improve the program, since frequently one spends more than 20 years from first crossing until developing a successful cultivar. The diagnosis of the crop with its problems and potentialities, and the practical knowledge of the crop for intervention has to be as well-planned as possible facing larger priorities inside of the highest priorities. The time factor for the development of a coffee cultivar can be minimized, from about 20 years to 10 years, by changing some simple procedures in the cultivar development strategy and improvement methods, adding new tools such as biotechnological techniques, informatics and bio-statistics. The genetic gain in a year can increase a lot with the availability of human resources, materials, time, and space.

6. References

Araújo Neto K. de. (1978), Avaliação preliminar de progenies de *Coffea arabica* pela estimativa simultânea de vigor vegetativo e carga pendente. Paper presented at 6th Congr. Bras. de Pesq. Cafeeiras, Ribeirão Preto, São Paulo, Brazil. pp.110-113

Bertrand B. (1999), El mejoramiento genético en América Central. In-Desafios de la Caficultura en Centroamerica, B. Bertrand and B. Rapidel (ed.), Editorial Agroamerica, San Jose, pp. 407-456

Carneiro M.F. (2000), Avanços na obtenção de di-haplóides em cultivares de *Coffea arabica*. Paper presented at 3rd SIBAC, May 14-28, Londrina, Paraná, Brazil

Bertrand B., Aguilar, G., Santacreo R., Antonhy F., Ethienne H., Eskes A., Charrier A. (1997), Comportamento d'híbrides F1 de Coffea arabica pour la vigueur, la production et la fertilité en Amérique Centrale. Paper presented at 17th ASIC, July 20-25, Nairobi, Kenya

Carvalho A. (1952), Melhoramento do cafeeiro: vi. estudo e interpretação para fins de seleção de produções individuais na variedade bourbon. *Bragantia*, 12(6): 179-200

Etienne H., Bertrand B., Anthony F., Côte F., Berthouly M. (1977), Lémbriogenèse somatique: un outil pour l'amélioration génètique du caféier. Paper presented at 17th ASIC, July, 20-25, Nairobi, Kenya

Etienne-Barry D., Bertrand B., Vasquez N., Etienne H. (1999), Direct sowing of *Coffea arabica* somatic embryos mass-produced in a bioreactor and regeneration of plants. *Plant Cell Reports, 19*:111-117

Fazuoli L.C. (1977), Avaliação de progenies de café 'Mundo Novo'(*Coffea arabica L.*). MS. Thesis, Universidade de São Paulo, Piracicaba, São Paulo, Brazil

Fonseca M E.S., De Araújo P.F.C., De Pedroso I.A. (1978), *Retorno social aos investimentos em pesquisa na cultura do café*, Badesp, São Paulo, São Paulo, Brazil

Gifford M.R., Evans L.T. (1981), Photosynthesis, carbon paritioning, and yield. In-*Ann.Rev.Plant. Physiol*, 32: 485-509

Janick J., Moore J.N. (1975), Advances in Fruit Breeding, Purdue Univ. Press, West Lafayette, Indiana, U.S.A

Lashermes P., Agwanda C.O., Anthony F., Combes M.C., Troust P., Charrier A. (1997), Molecular marker assisted selection: a powerful approach for coffee improvement. Paper presented at 17th ASIC, July 20-25, Nairobi, Kenya

Leroy T., Royer M., Paillard M., Berthouly M., Spiral J., Tessereau S., Legavre T., Altosaar I. (1997), Introduction de génes d'intérêt agronomique dans l'espèce *Coffea canephora* Pierre par transformation avec *Agrobacterium sp.* Paper presented at 17th ASIC, July 20-25, Nairobi, Kenya

Mawardi S., Hartobutoyo S. (1981), Correlation study of yield and branch characteristics of robusta coffee (*Coffea canephora* Pierre var. *robusta* Cheval.) on F_1 hybrid population. *Menara Perkebunan*, 49 (5), 115-20

Mawardi S., Iswanto A., Hartobutoyo S. (1983), Selection of the F_2 population of arabica coffee: II. a preliminary study for individual selection by applying selection indexes. *Menara Perkebunan*, 51 (5), 128-32

Medina H.P., Carvalho A., Söndahl M.R., Fazuoli L.C., Da Costa W.M. (1984), Coffee breeding and related evolutionary aspects. In- *Plant Breeding Reviews*, J. Janick (ed.), Avi, Westport, pp. 157-194

Mônaco L.C.; Medina F°, H.P., Söndahl M.R., Lima M.M.A. de (1978), Efeito de dias longos no crescimento e florescimento de cultivares de café. *Bragantia*, 37 (4): 25-30.

Nitzsche W., Wenzel, G.(1977), Haploids in plant breeding. In-Advances in Plant Breeding (suplements to Journal of Plant Breeding). Verlag, Berlin. pp. 101

Paulino A.J., Matiello J.B., Paulini A.E. (1985), *Produção de mudas de café Conilon por estacas.* Ministério da Indústria e do Comércio – Instituto Brasileiro do Café, Rio de Janeiro, Rio de Janeiro, Brazil

Ramos L.C. (1980), Desenvolvimento da plântulas de quatro cultivares de café. *Bragantia*, 39:215-218

Ramos L.C., Lima M.M.A., Carvalho A.(1982), Crescimento do sistema radicular e da parte aérea em plantas jovens de cafeeiros. *Bragantia*, 41: 93-99

Rocha T.R., Carvalho A., Fazuoli L.C. (1980), Melhoramento do cafeeiro: XXXVIII. Observações sobre progênies do cultivar Mundo Novo de *Coffea arabica* na estação experimental de Mococa. *Bragantia*, 39(15): 148-60

Sera T. (1979), Zoneamento de cultivares de café para o estado do Paraná. Paper presented at 7th Congr. Bras. Pesq. Cafeeiras, Dec. 7-9, Ribeirão Preto, São Paulo, Brazil

Sera T. (1980), Estimação dos componentes da variância e do coeficiente de determinação genotípica da produção de grãos de café (*Coffea arabica* l.). M.S. Thesis, 62 pp. Universidade de São Paulo, Piracicaba, São Paulo, Brazil

Sera T. (1984), Seleção precoce para produção no melhoramento genético de plantas perenes. Monography, Universidade de São paulo, Piracicaba, São paulo, Brazil

Sera T. (1987), Possibilidade de emprego de seleção nas colheitas iniciais de café (*Coffea arabica* L. cv. Acaiá). PhD Thesis, Universidade de São Paulo, Piracicaba, São Paulo, Brazil

Sera T., Alves S.J. (1999), Melhoramento genético de plantas perenes. In- Melhoramento Genético de Plantas, D. Destro & R. Montalván (ed.) Editora UEL, Londrina, pp. 369-422

Silveira J.S.M., Fonseca, A.F.A. (1995), *Produção de Mudas Clonais de Café Conilon em Câmara Úmida sob Cobertura de Folhas de Palmeira.* Empresa Capixaba de Pesquisa Agropecuária – EMCAPA, Vitória, Espírito Santo, Brazil

Söndahl M.R., Mônaco L.C., Carvalho A., Fazuoli L.C. (1972). Efeito de dias longos em progenies de café (*Coffea arabica),* tratados com benziladenina e giberelina. *Ciência e Cultura*, 24 (sup.): 336-339

Söndahl M.R., Mônaco L.C., Sharp W.R. (1981), *In vitro* methods applied to coffee. In-*Plant Tissue Culture - Methods and Applications in Agriculture,* Thorpe (ed.), Academic, New York, USA, pp. 325-348

Söndahl M.R., Loh W.H.T. (1988), Coffee biotechnology. In- *Coffee (v.4): Agronomy,* R. J. Clarke and R. Macrae (ed), Elsevier, London, pp. 236-262

Söndhal M.R., Söndhal C.N., Gonçalves W. (2000), Custo comparativo de diferentes técnicas de clonagem em café. Paper presented at 3rd SIBAC, May 14-28, Londrina, Paraná, Brazil

Srinivasan C.S. (1980), Association of some vegetative characters with initial fruit yield in coffee (*Coffea arabica L.*). *Journal of Coffee Research*, 10 (2): 21-7

Srinivasan C.S. (1982), Pre-selection for yield in coffee. *Indian Journal of Genetics*, 42 (1):15-19

Srinivasan C.S., Subbalakshmi, V. (1981), A biometrical study of yield variation in some coffee progenies. *Journal of Coffee Research,* 11 (2): 26-34

Srinivasan C.S., Vishveshwara S. (1981), Variability and breeding value of some characters related to yield in a world collection of arabica coffee. *Indian Coffee,* 15 (5): 119-28

Stevens W.L. (1949), Análise estatística do ensaio de variedades de café. *Bragantia*, 9 (5-8): 103-123

Valencia, G. (1973), Relation entre el indice de area foliar y la productividad del cafeto. *Cenicafé*, out/dez: 79-89

Van der Vossen H.A.M. Coffee selection and breeding. In- *Coffee*; *Botany, Biochemistry and Production of Beans and Beverage,* M. N. Clifford and K. C. Willson (ed..), AVI, Westport, pp. 48-96

Walyaro D.J., Van der Vossen, H.A.M. (1979), Early determination of yield potential in arabica coffee by applying index selection. *Euphytica,*, 28:465-472

Wellensiek S.J. (1962), Shortening the breeding cycle. *Euphytica,* 11:5-10

Walyaro D.J. (1983), Considerations in breeding for improved yield and quality in arabica coffee (*Coffea arabica L.*). Phd Thesis, Agricultural University, Wageningen, Netherland

Chapter 5

SOMATIC EMBRYOGENESIS OF COFFEE

M. Berthouly[1], H. Etienne[2]

[1]*CIRAD-CP, 2477 av. du Val de Montferrand- B. P. 5035, 34032 Montpellier Cedex 1, France;* [2]*CATIE, A. P. 11-7170, Turrialba, Costa Rica*

Running title: Somatic Embryogenesis

1. Introduction

Coffee is one of the most important agricultural products on the world market, and is the most widely grown tropical tree crop. Various Latin American countries such as Colombia, Brazil and Central America depend heavily on coffee production for foreign currency earnings. More generally, the economies of about 50 countries depend on this crop, which is grown on 11.2 x 10^6 ha and which produced 5.3 x 10^6 tons of green coffee in 1992 and 5,603 metric tons during 1995. The supply of coffee consumed world-wide primarily comes from two cultivated species: *Coffea arabica* and *C. canephora*, which account for 75 and 25% of the world market, respectively. These two varieties are commercially known as "arabica" and "robusta", respectively.
C. arabica is grown in high-altitude tropical regions (500-2,500 meters). It is the only tetraploid (2n = 44 chromosomes) and autogamous species of the *Coffea* genus. *C. arabica* is well suited to high-altitude regions, and produces high quality coffee with low caffeine content. However, this species is susceptible to main diseases such as leaf rust (*Hemileia vastatrix* Berk & Br.) and coffee berry disease (CBD) (*Colletotrichum coffeanum*).
C. canephora is a diploid (2n = 22 chromosomes), allogamous species that produces coffee with a higher caffeine content, which is considered poor in quality as compared to the quality of *C. arabica*. It is grown in low-altitude tropical regions (0-100 m). With respect to genetic improvement, this species offers advantages of its genes showing resistance to rust, CBD, and nematodes (Bettencourt and Rodrigues, 1988). Given its allogamy, the *C. canephora* species comprises genetically heterogeneous individuals. The progenies of cross-fertilization, therefore, have a substantial genetic variability.
In general, *C. arabica* is more susceptible to disease than *C. canephora*. This is due to the genetic base and greater variability of the latter, given its allogamy. Two major

T. Sera et al. (eds.), Coffee Biotechnology and Quality, 71–90.

fungal diseases affect coffee plants, which are leaf rust caused by *H. vastatrix*, and CBD, caused by *C. coffeanum*. Other pathogens such as *Rhizoctonia*, *Fusarium*, etc. can also affect coffee trees (Haarer, 1969). The word "pest" can have a very broad meaning, ranging from insects to nematodes via weeds (Haarer, 1969). We intend to limit the scope of this article for two main pests that currently cause significant damage in plantations: the coffee berry borer (CBB *Hypothenemus hampei*) and nematodes (root parasites). There seems to be no natural resistance to the CBB, and therefore, possible genetic transformation is the only possible solution. In the case of nematodes, although this approach could be considered, there is a degree of natural resistance to certain types, notably in species such as *C. canephora*. Nematodes are of deep concern world-wide in *C. arabica*.

1.1. CONVENTIONAL PROTOCOLS FOR IMPROVEMENT OF COFFEE

The *C. arabica* breeding method is applied to autogamous species. It is based on producing pure lines by pedigree selection after recombining parental characters brought by the parents. Heterosis has been reported in *C. arabica* in India (Srinivasan and Vishveshwara, 1978), Kenya (Van der Vossen and Walyaro, 1981) and Ethiopia (Ameha, 1983), and more recently in Costa Rica (Bertrand *et al.*, 1997). Trials of vegetatively propagated F_1 hybrids proved unsuccessful, as the conventional propagation methods resulted in poor multiplication rates (Van der Vossen, 1985).
There are generally two ways of breeding *C. canephora* (Bouharmont and Awemo, 1979; Charrier, 1985): a) by seeds, which produce heterogeneous population, and b) vegetatively, which can be summed up as the choice of two clones with the desirable characters. The selected trees are then vegetatively propagated. Polyclonal plantings are necessary to ensure cross-fertilization.
Generally, only *C. arabica* is distributed or sold in seed form after a relatively long pedigree selection process (almost 30-35 years). This process produces stable and homozygous lines with desirable characters. Once the variety is obtained, it is sufficient to establish a seed garden to produce seeds genetically identical to the mother plant, since the plant is autogamous.
Until now, varietal propagation of *C. canephora* was done by vegetative means to benefit from the higher productivity of clones in relation to varieties grown from seeds (Capot, 1997) and from the heterozygosity of the plant. In recent decades, given the complementary characteristics of *C. arabica* and *C. canephora* (quality of the first one and disease resistance of the later), breeding has been greatly appreciated in creating inter-specific hybrids (Icatu). These hybrids are multiplied asexually. Vegetative propagation, therefore, plays an important role in coffee breeding. It enables the propagation and exploitation of heterozygous genetic material. It can also reduce the number of breeding cycles. In fact, with coffee, a perennial crop, it takes at least 35 years to develop a variety. Thus, vegetative propagation can be applied to select *C. canephora* clones, inter-specific hybrids such as arabusta, or *C. arabica* intra-specific hybrids such as Ruiru 11.

Cuttings are the most commonly used vegetative propagation procedure in coffee. However, given the dimorphism of the vegetative axes, the number of orthotopic cuttings that a coffee tree can produce is extremely limited. For large-scale multiplication, the availability of limited number of cuttings can be time consuming. Moreover, horticultural cuttings call for propagation plots, which involve various constraints such as upkeep and required land area (Deuss and Descroix, 1984).

1.2. NEED FOR NEW MASS PROPAGATION TECHNIQUES

Over the past several years, the progress made in biotechnology (micropropagation, molecular biology) has opened up new prospects for research programmes. In this context, the development of *in vitro* multiplication techniques is extremely useful for increasing the multiplication rate enabling the rapid distribution of the hybrids and clones produced. In the past few years, several authors (Staritsky, 1970; Sondahl and Sharp, 1977; Dublin, 1980, 1982; Pierson et al., 1983; Yasuda *et al.* 1985; De Garcia and Menendez 1987; Berthouly *et al.,* 1987) have shown that micro-cutting and somatic embryogenesis can be applied to coffee.
Somatic embryogenesis appears to be an inexpensive alternative for multiplying selected plants. In addition to providing a worthwhile model for fundamental studies, its most promising application is in large-scale plant propagation (Ammirato, 1977). Various studies on coffee have revealed the suitability of the species for somatic embryogenesis and for embryo development into plantlets. However, the multiplication rate is still very low for its large-scale use. In fact, De Garcia and Menendez (1987) working on coffee somatic embryogenesis on a solid medium, reported 60 to 65 somatic embryos per explant after 14 weeks culturing as a record figure.
To exploit the substantial potential of somatic embryogenesis, it is necessary to use a liquid medium and hence, cost of plantlets production can be reduced. Indeed, by culturing cell suspensions, mass embryo production can be scaled-up in a bioreactor. The results obtained with other species are highly promising. Petiard *et al.* (1987) reported a production capacity of 80,000 carrot embryos per litre per day in a ten-litre bioreactor. Currently, there is no *in vitro* propagation method available for commercial coffee production.

2. Micropropagation by somatic embryogenesis

2.1. BACKGROUND

Somatic embryogenesis has been widely used for many years for micropropagation of both herbaceous and woody plants Jain et al., 1995). It has been tested on different types of tissue (leaves, stems, embryos, etc.). In several *Coffea* species and genotypes, somatic embryogenesis is well documented (Staritsky, 1970; Sondhal and Sharp, 1977;

Pierson *et al.*, 1983; Dublin, 1984; Yasuda *et al.*, 1985; Berthouly and Michaux-Ferriere, 1996)

2.2. TYPES OF SOMATIC EMBRYOGENESIS

Since the first work by Staritsky (1970), two types of somatic embryogenesis have been described (Sondhal, 1977):

a) *Low frequency*: the somatic embryos are obtained on only one medium without producing calli. Quite often a small number of somatic embryos (a few to 100 per explant).
b) *High frequency using two medium*: at induction medium for primary callogenesis, and a secondary regeneration medium to produce embryogenic friable calli comprised of solely embryogenic cells producing a large number of somatic embryos (several hundreds to thousands per g of callus), but also the use of a liquid medium for the regeneration phase.

2.3. PROSPECTS FOR SOMATIC EMBRYOGENESIS IN COFFEE

There are three main reasons for using somatic embryogenesis in coffee:

a) To shorten the *C. arabica* breeding cycle by true-to-type micropropagation of hybrids (F_1, F_2, back-crosses);
b) Raid multiplication of *C. canephora* genotypes having multiplication problems (medium-altitude regions);
c) As vehicle for genetic transformation.

2.4. INDUCTION OF EMBRYOGENIC CALLI

2.4.1. Materials and methods:

Leaves were collected from the selected mother plants in the field. All samples were disinfected by immersion in 10% calcium hypochlorite solution containing 1% Tween 80 for 20 min, followed by 8% calcium hypochlorite solution for 10 min, then rinsed 4-times with sterile water.

Explants (1-cm^2) were cut without the main vein and placed on a initial basal medium containing half-strength Murashige and Skoog (1962) salts, 88mM sucrose, 100 casein hydrolysate, 400 malt extract, 10 thiamine, 1 nicotinic acid, 1 pyridoxine, 1 glycine, 100 myoinositol (all in mg.l^{-1}), 2.2 2,4-dichlorophenoxy acetic acid (2,4-D), 2.4 Indol Butyric Acid (*IBA*) and 9.8 6-y-y-dimethylallyamino purine (2iP) (all in µM), solidified with gelrite (2-gl^{-1}) in glass tubes, or small Petri dishes. The pH was adjusted to 5.6 before autoclaving. The cultures were incubated at 25 ±1°C in dark.

After four weeks, explants with callus were transferred to a second medium to promote high frequency rate of embryogenic callus formation. The second basal medium contained (mg.l^{-1}) 200 casein hydrolysate, 800 malt extract, 20 thiamine, 20 glycine, 40 cysteine, 60 adenine sulphate and 200 myoinositol. The cultures were incubated at

26±1°C in dark. After 10-12 weeks, yellow friable embryogenic calli were obtained, which were transferred to either solid or liquid media to proliferate, or regenerate somatic embryos.

2.4.2. Optimum conditions for high frequency embryogenic calli:

Different concentrations of plant growth regulator were used in two successive culture media, and during subculture on the first medium (Table1). The effect of temperature was also tested.

Table 1. Highly embryogenic calli (%) of *Coffea canephora* obtained after 3 or 4 weeks on initial basal medium for conditioning and 10 weeks on a second medium according to PGR ratios.

PGR concentrations in the first medium (μM)			Age of Culture (weeks)	High frequency mbryogenic calli (%) PGR concentration in the second medium (μM)		
2,4-D	IBA	2iP		2,4-D	44	4.4
				BA 35.6	17.8	0 17.8
2.2	2.4	9.8	3	83b	100a	100a
2.2	4.9	9.8	4	100a	100a	100a
4.4	2.4	9.8	3	100a	100a	100a
4.4	4.9	9.8	4	100a	8b	63de
			3	0g	40f	100a
			4	75bc	71cd	60e
			3	80b	100a	33f
			4	100a	100a	92a

Number of explants for each experiment=150.Data followed by the same letter are not significantly different at 5% level (X^2 test). (pgr = plant growth regulateur)

As shown in Table 1, the percentages of high frequency embryogenic calli was dependent upon the age of the culture on the first medium and varied from 0 to 100 according to the different auxin-cytokinin ratios in the medium. The X^2 test at 5% level showed that the percentages of high frequency embryogenic calli were higher when first culture had grown for four weeks (87%) as compared to a first culture grown for three weeks (78%). When the friable calli were obtained from explants on the second medium with only 6-Benzyl aminopyrine (BAP), they contained very small globular embryos (~100μM in diameter). The absence of 2,4-D in the medium could explain the rapid development of embryogenic cells of the calli into globular embryos. Addition of 2,4-

D in the maintenance medium is very important for strong regeneration capacity and to inhibit globular embryo formation.

2.4.3. Genotype effect:

Genotype effect on embryogenic explant production has been seen in many plants, and coffee is no exception (Berthouly and Michaux-Ferriere, 1996; Van Boxtel and Berthouly, 1996). For *C. canephora*, the percentages of highly embryogenic calli from different genotypes varied from 21 to 98% on first medium containing 2.2 µM 2,4-D, 4.9 µM-IBA, and 9.8 µM-2iP, and from 55 to 100 % when the first culture medium was supplemented with 2.2 µM 2,4-D, 2.4 µM-IBA, and 9.8 µM 2iP (Table 2).

Table 2. Highly embryogenic calli (%) obtained on the second induction medium (basal medium + 4,4 µM 2,4-D + 17,8 µM BA) according to genotype and to the population of PGR in the first conditioning medium.

GENOTYPES	High frequency embryogenic calli (%)	
	PGR 1st conditioning medium (µM) 2.2 2,4-D 2.4 IBA 9.8 2iP	PGR 1st conditioning medium (µM) 2.2 2,4-D 4.9 IBA 9.8 2iP
3554 A	96.3bc	48.7ij
3554 B	81.7de	69fg
3751 A	91.6cd	20.8sk
3751 B	100a	90d
3752 A	64.8gh	37.7j
3752 B	88.7d	98.3ab
3754 A	55hi	75,7ef
3754 B	64.9gh	65,3gh
3756 A	100a	96.3bc
3756 B	56hi	83,7de

Number of explants for each experiment= 150.Data followed by the same letter are not significantly different at 5% level (X^2 test)

This variability in obtaining embryogenic calli depended on the genotypes used and has also been observed in *C. arabica*, in both varieties and hybrids (Table 3). Although, genotype effect was observed in both *C. arabica* and *C. canephora*, it was possible to obtain embryogenic calli irrespective of the genotype, more importantly with fairly high response levels. This characteristic makes coffee a plant, which responds well to somatic embryogenesis. Either no or poor response of certain genotypes has been reported for most species, and this is one of the main obstacles to the routine use of this techniques in breeding or mass production schemes.

Table 3. Percentage of embryogenic calli and weight of embryogenic calli obtained from different C. arabica F1 hybrid families

C. arabica Hybrids	N of explants introduced	% embryogenic explants (EC)	Weight of embryogenic calli per explant (ECP mg)
Family 1/hybrid 1	642	38c	282a
Family 2/hybrid 1	504	6d	72b
Family 2/hybrid 2	641	24cd	179a
Family 3/hybrid 1	866	42bc	224a
Family 3/hybrid 2	623	17cd	117ab
Family 4/hybrid 1	639	52abc	262a
Family 4/hybrid 2	384	74ab	284a
Family 4/hybrid 3	385	33cd	278a
Family 4/hybrid 4	382	32cd	211a
Family 4/hybrid 5	285	79a	325a

2.4.4. Seasonal effects:

In addition to the genotype effects, there are also seasonal effects in coffee. Embryogenic callus can depend on the collection time of explant from the mother plant in the field. For example, culturing two genotypes of *C. canephora* using the same medium, the percentages of high frequency of embryogenic calli was obtained, varying from 60 to 100% depending on time of the year under the similar climatic and soil conditions (Table 4).

Table 4. Highly embryogenic calli (%) obtaneid from leaf explants of 2 *Coffee canephora* genotypes collected throughout the year and cultured 4 weeks on the first medium (2,2 μM IBA + 9,8 μM 2iP and second medium (4,4 μM 2,4-D + 17,8 μM BA)

Month	Genotype 3751 A	Genotype 3751 B
January	65efg	60g
February	62fg	63efg
April	100a	98a
May	100a	100a
June	100a	88b
August	100a	76cd
October	73e	71def
November	88b	83cd

Seasonal conditions can influence both percentage of embryogenic explants, as well as the quantity of callus produced per explant. Climatic conditions directly affect contamination levels after transfer *in vitro*. It is obvious preferable to collect vegetal material in dry weather. Lastly, it is important to note that the quantity of calli produced per explant depends on the genotype and culture conditions (Tables 2 and 3).

2.5. EMBRYOGENIC CELL SUSPENSIONS

Within 3-4 months (depending on genotype), friable embryogenic calli developed on explants cultured on a solid medium. At that time a piece of calli was sub-cultured to the liquid medium containing half strength MS minerals, 5 thiamine-HCl, 0.5 pyridoxine-HCl, 0.5 nicotinic acid, 10 L-cysteine, 50 myo-inositol, 100 casein hydrolysate, 200 malt extract, (all in $mg.l^{-1}$), 15 $g.l^{-1}$ sucrose, pH 5.6 and a mixture of 2,4-D and kinetin. The resulting cell suspension cultures were cultured in 250-ml Erlenmeyer flasks at 27°C on a rotary shaker at 100 rpm. Stable embryogenic cell suspension cultures were obtained in 2-3 months. For sub-cultures only aggregates smaller than 1-mm in diameter were collected using a narrow mouth pipette (Falcon).
Stable embryogenic cell suspension cultures of several coffees species and genotypes obtained by this procedure were used for more detailed testing of liquid culture variables, including 2,4-D concentration, photosynthetic photon flux PPF, subculture interval and initial culture density.

2.5.1. Effect of 2,4-D:

Only two concentrations of 2,4-D were tested (4.5 and 9 µM). No significant dose effect was observed. Thus, 4.5 µM 2,4-D and 4.6µM kinetin were chosen for further testing of other liquid media.

2.5.2. Effect of different sub-culture intervals:

Sub-culture intervals of 3, 6, 14 days were tested. Less tissue proliferation was observed when sub-cultured at 3-day intervals rather than 6 or 14-day intervals. No difference was observed between the 6 and 14-day subculture intervals. To minimize culture handling without adversely influencing cell suspension growth, a 14-day subculture interval was regularly used for rapid growth of tissue.

2.5.3. Effect of culture density:

Cell suspension cultures were sub-cultured without readjustment to their initial density. Growth of cell suspensions 20-g l^{-1} and 10-g l^{-1} culture density slowed down between 6 and 8 weeks. In 0.5-g l^{-1} culture density and to a lesser extent in 2-g l^{-1} cell culture density, somatic embryos were observed. No somatic embryos formed in embryogenic cell cultures with cell density 5, 10 and 20-g l^{-1}. Thus, a minimum culture density seemed to be necessary to avoid somatic embryo formation. Calculation of the multiplication rate (MR) demonstrated the exponential growth during the first week of

culture for each density tested. During the second week, the MR of all cultures decreased to a level between 1.0 and 1.8 ($g.l^{-1}$ fresh mater). During the first week, it was between 3-15 $g.l^{-1}$ fresh mater.
Thus, by culturing small amounts of embryogenic callus at a density of 5-g l^{-1}, a relatively low MR can be maintained over several weeks. Under these conditions, sub-culturing in order to avoid a growth plateau is less often necessary. When cultures were sub-cultured weekly with readjustment to their initial density of 10, 15, or 20-g l^{-1}, the MR over 4-week culture period was 2.0, 1.8, and 1.5, respectively. The MR of 10-g l^{-1} cultures was significantly higher than that of 20-g l^{-1} culture. Thus, weekly readjustment of culture density to 10-g calli l^{-1} could be to used accelerate growth of cell suspension cultures.

2.5.4. Nutrient availability in the medium:
Absorption of some macro- and micro-salts by embryogenic cell suspensions was monitored by regularly analysing the used amount of nutrients in the culture medium. After 12 days of culture, 60-70% of the initial concentration of nutrients was still present. After three weeks, macro-salts and nitrogen sources had decreased by 50-70%, and in which the amount of available magnesium was the lowest (30% of initial concentration). These results suggested that amount of MS macro- and micro-minerals at half strength in the culture medium did not deplete in coffee embryogenic cell suspension culture for at least three weeks of culture.
To optimize long-term maintenance of embryogenic cultures, instead of sub-culturing in a fresh medium every two-weeks, a part of the embryogenic callus was discarded and concentrated nutrients were added to the conditioned culture. In all cases, growth was nevertheless inhibited after two months of culture. Therefore, medium renewal remained necessary at regular time intervals in order to sustain the growth of embryogenic suspensions, presumably by eliminating growth-inhibiting substances present in the conditioned medium.

2.6. REGENERATION IN LIQUID MEDIUM

Using a liquid medium for embryo production provide substantial guarantees to improve productivity. It enables the automation of liquid culture methods and reduces the cost as compared to solid medium. Somatic embryogenesis using liquid media for cell suspensions in Erlenmeyer flasks or in bioreactors is limited to the quantity embryogenic to somatic embryos produced, and these embryos have to be transferred to a gel medium several times to ensure their development into plantlets. However, these operations are extremely costly, as they are labour-intensive in terms of individual selection and handling of embryos, and gel culture medium production.
Using a liquid medium for embryo production/or plant procedure and combining the multiplication of embryogenic aggregates in Erlenmeyer flasks make such a technology applicable on an industrial scale. A temporary immersion system has been developed at CIRAD on different plants, and is now being applied on a large scale for coffee. The principle behind it and the conditions for its use are described in section 3. However, as

seen previously, once embryogenic calli are obtained and suspension cell cultures are established, regeneration is possible in the liquid medium according to conventional procedures. The efficacy of regeneration largely depends on certain parameters such as cytokinin levels and the initial quantity of fresh matter.

2.6.1. Effect of benzyladenine (BA) and inoculum density:
Globular somatic embryo formation was inversely related to inoculum density and the earliest abundant somatic embryo formation was mostly observed with an inoculum density of 1-g l^{-1} and 4.4 µM or 17.6 µM BA. Zamarripa (1993) reported the presence of certain substances that were inhibiting to somatic embryogenesis in coffee calli cultured at high density. Addition of BA in the culture medium appeared to neutralize the influence of inhibitory substances and allowed for somatic embryo formation.

3. Development of a new culture system - *Automatic Temporary Immersion System* (RITA^R)- to improve somatic embryogenesis

A new temporary immersion system, known as RITA^R, has been developed at CIRAD over the past eight years (Alvard *et al.*, 1993; Teisson and Alvard, 1994; Berthouly *et al.*, 1995; Etienne *et al.*,1997 b) to avoid transfer to a solid medium during the final stages of somatic embryogenesis, i.e. embryo development, germination and conversion into acclimatizable plants. The device has already proved its efficacy, i.e. its superiority to semi-solid media for producing microcuttings of coffee, banana and *Hevea*, for somatic embryogenesis on coffee (Berthouly *et al.*, 1995; Etienne *et al.* 1997a), banana (Alvard *et al.* 1993), *Hevea* (Etienne *et al.* 1997 b) and citrus species (Teisson and Alvard 1994), and for producing potato microtubers.

3.1. OPERATION

In the beginning, the device used for the initial experiments was a readily available autoclavable filter unit, modified by linking the upper and lower compartments by a glass tube. The Biotrop *in vitro* culture laboratory (CIRAD) has now developed a small, economical device that is easy and flexible to use. The unit is operated in the opposite way to filtration unit. The plant material is placed in the upper part and the liquid medium at the bottom. When the lower compartment is pressurized, using a small laboratory air pump, the medium is pushed up into the upper compartment. A simple timer switch is used to control the frequency and duration of pump operation, hence of immersion. When the pressure is released, the liquid flows back down again. The air blown around the container throughout the immersion period renews the internal atmosphere. All the air-flow is sterilized by 0.2µ hydrophobic vents. Each system is independent and can be moved individually.

3.2. ADVANTAGES OF THE SYSTEM

This container has the same advantages as bioreactors but without following limitations:

Physiological limitations: Permanent contact with the culture medium results invitrification and constant movement of the plant material, that prevents the establishment of polarity and hinders somatic embryo development, oxygenation problems.

Technical and financial limitations: Exorbitant bioreactor cost, their lack of flexibility, complicated technology and fragility in the event of power cuts where the plant material dies within a few hours due to lack of oxygen.

Currently, all these drawbacks, making it impossible to use bioreactors for mass production, particularly in developing countries, are resolved with temporary immersion systems. Temporary immersion enables the development of plantlets from embryogenic cell suspension cultures. Moreover, this development from cell to plants takes place in same container, without handling the plant material; it is sufficient to only change the medium. The explants are in contact with the liquid culture medium for just a few minutes per day, and the duration of contact can be adjusted to prevent any risk of vitrification. The container is ventilated naturally and passively by vents, with pulsed air during immersion to renew all the air in the container.

On a technical level, the RITA^R is small, autoclavable, easily accessible (US$ 55), portable (it can be transported easily without risk of contamination), and easy to operate. In the event of power cuts, plant material is not affected and can withstand power cuts lasting over 10 days. The liquid medium in the lower section is not in contact with the cultures but it maintains a high moisture level. Another advantage for mass production is that laboratories can easily have several units of RITA^R with which one can produce several cultivars at once and adapt production volumes to demand throughout the year.

3.2.1. Regeneration by temporary immersion:

Acclimatate germinated embryos are produced in three stages, each lasting two months:

a) Pro-embryo regeneration,

b) b) Somatic embryo development,

c) Germination.

The whole of the regeneration phase takes place in the light, in the same container and without handling the plant material. However, the medium has to be changed at the end of each stage. Immersion frequency and density are the key factors in this stage. As for regeneration in Erlenmeyer flasks, the optimum density is 1-g of embryogenic clumps per litre of culture medium. The plant material and the medium are in contact for just 2 x 1 min.day^{-1} during the regeneration and embryo development stages and 4 x 5 min.day^{-1} during germination. This very limited exposure to the liquid medium rules out any risk of vitrification, since problems start to arise with immersion frequency/duration combinations of over 4 x 15 min.day^{-1}.

It is possible with most genotypes to obtain very high outputs of around 12,000 somatic embryos with RITAR, i.e. 60,000 embryos/g of cell suspension. All these embryos are all morphologically normal (differentiated cotyledonary embryos) and highly uniform. The conversion rate into plants on gel media is around 90%. However, for the time being, it is apparently impossible to achieve satisfactory subsequent development at such densities. The main aim - obtaining directly acclimatizable material - means working at lower densities of around 3,000 embryos per RITAR. Depending on the genotype, the number of acclimatizable germinated embryos obtained per RITAR can vary between 1,000 and 9,000 (Table 5).
The uniformity and quality of these embryos are better than those obtained on semi-solid media.

Table 5. Performance of *C. arabica* F1 hybrids regenerated by temporary immersion. The tests of embryo conversion into plantlets each considered 200 somatic embryos. The figures quoted are the means of at least three replicates, corresponding to different containers.

C. arabica hybrids	Number of somatic embryos per RITA^R container	Conversion into plantlets of embryos germinated in RITA^R units	
		on gel medium *in vitro*	on horticultural substrate *ex vitro*
Family 1 / hybrid 1	5 000 a	95 a	68 a
Family 2 / hybrid 1	3 809 b	85 bc	50 c
Family 2 / hybrid 2	2 678 c	89 b	61 b
Family 3 / hybrid 1	2 663 c	85 bc	45 d
Family 3 / hybrid 2	2 038 d	83 c	52 c

It is, therefore, possible to achieve sufficiently advanced development for the somatic embryos to acclimatize directly, with success rates already more than 50%. The conversion rate of these somatic embryos into plants *in vitro* is around 90%, which gives one an idea of the scope for improvement in regard *to ex vitro* conversion rates. As far as we know, this is the first time that direct somatic embryo conversion into plants on a horticultural substrate has been observed for a woody plant in the greenhouse. Current research on embryo preparation for direct acclimation via a maturation phase, and on adapting acclimation conditions to this material without leaves lead us to hope that acclimation rates of over 70% should soon be possible.

Embryo conversion into plants can also be obtained directly in RITA[R] units, provided the density does not exceed 500 somatic embryos per container. This operation is clearly pointless as compared to direct acclimation of germinated embryos, which can be produced in much larger quantities. However, this result effectively confirms the importance of the density factor. An initial technical evaluation has been made of the merits of using temporary immersion in a liquid medium for germination and somatic embryo conversion into plants alone. The results are shown in Table 6. This study was carried out with the family one per hybrid 1 (cf. Table 5) and for the production of 9,000 acclimatate somatic embryo.

Table 6. Comparison of the efficacy of somatic embryo germination and conversion into plants using temporary immersion and gel media.

Characteristics of embryo germination and conversion into plants	Data obtained with temporary immersion	Data obtained with semi-solid media
Duration (weeks)	16 ± 1	16 ± 3
Culture density (No embryos/container)	5 000	200 (4 wks) / 40 (4 wks) /6 (4 wks)
No containers	3	50 / 250 / 1670
Area taken up by containers (m^2)	0.3	0.42 / 2.1 / 14
Germination (%)	98 ± 0	97 ± 3
Embryo conversion into plants (%)	60 ± 4 (in greenhouse)	84.3 ± 5.0 (*in vitro*)
Acclimation (%)	ditto	95
N of acclimated plants	9 000	9 000
Operations for 9 000 acclimated plants	2 changes of liquid medium + acclimation	3 transfers to solid media + root induction + acclimation
Plant material handling time (h)	6 (*in vitro*) 22 (acclimation)	81 (*in vitro*) 126 (Root induction + (acclimation)
	total: 30	total: 207

An analysis of the technical data largely favours using the temporary immersion system. It was possible to produce 9,000 acclimatizable germinated embryos using just three RITA units and 1.2 l of liquid media, whereas it took 1,670 pots, each containing 30 ml, i.e. 50-l of gel medium in all to produce the same quantity on a semi-solid medium, which costs more, given the high price of Phytagel. A comparison of the areas used also shows that semi-solid media are less efficient than liquid media, but above all, a comparison of the work times involved proves the merits of culture in

a liquid medium (30 h) in relation to the traditional system (207 h). Production times and costs, which also favour using a liquid medium, were not considered.
These results indicated that temporary immersion system should substantially reduce the cost of producing coffee plants by somatic embryogenesis. This advantage would have been even clearer without limiting to studying embryo germination and conversion into plants, although these are the most costly stages of somatic embryogenesis. The system also ensures greater productivity at the embryogenic expression and embryo development stages, with all four phases taking place in the same container. The results obtained proved that temporary immersion enabled the mass propagation of *C. arabica.* It should be possible to use similar culture conditions successfully for other coffee species, but also for other woody and herbaceous species. Until now, although some authors had described the use of cell suspension or bioreactors to initiate the regeneration process, these systems had not removed the need to plant embryogenic clumps or young somatic embryos on a semi-solid medium to achieve their development into acclimatizable plants (Zamarripa *et al.*, 1991; Neuenschwander and Baumann, 1992).

4. Examples of the use of somatic embryogenesis for coffee improvement and mass propagation

4.1 MASS PROPAGATION OF *C. CANEPHORA* IN UGANDA

Uganda has had a tissue culture laboratory since 1994. Its aim has been to multiply six selected clones of *C. canephora* on large-scale by somatic embryogenesis. The laboratory has three culture chambers. Two are equipped with the temporary immersion system (800 per chamber). There are plans to produce *in vitro* coffee plantlets by somatic embryogenesis each year to achieve large-scale coffee production.

4.2 IMPROVEMENT AND MASS PROPAGATION OF *C. ARABICA* AND *C. CANEPHORA* IN CENTRAL AMERICA

Since 1992, there has been an Arabica coffee genetic improvement project in Central America, linking the *Programa regional para la proteccion y modernizacion de la cafeicultura en America Centrale* (PROMECAFE), CATIE and French scientific support organizations (CIRAD, ORSTOM, and the Ministry of Foreign Affairs). The project has three components: a) study of available germplasm, b) varietal creation, and c) biotechnology consisting in micropropagation which is of crucial importance, since the improvement strategies chosen mean using *in vitro* plantlets.
In effect, varietal creation is based on creating and breeding F_1 hybrids, for which the only possible large-scale multiplication method is micropropagation. The aims of micropropagation under this project are twofold:

a) To introduce *in vitro*, multiply and distribute to project member countries improved material produced by varietal creation; and,
b) To develop a mass propagation method by somatic embryogenesis and transfer it to laboratories of project member countries.

4.2.1. Examples of multiplication of C. canephora parents of the nemaya stock variety:
On various occasions, a hybrid between two *C. canephora* has proved highly resistant and in some cases immune to different nematodes, e.g. *Meloidogyne* and *Pratylenchus* (Anzueto, 1989, Bertrand *et al.*, 1995). The hybrid is T3751 (1-2) x T3561 (2-1), known as the 'Nemaya variety' (F_1 hybrid). Because the demand for this stock is huge, the best way of producing it on a large scale is to set up seed gardens where the two parents are grown side by side. Each tree would thus produce Nemaya variety seeds. As *C. canephora* is allogamous, vegetative propagation is required. Horticultural propagation is not possible as large quantities of each parent would have to be produced and there is only one individual in the CATIE Collection. The number of orthotopic axes available for cuttings or buddings is, therefore, too small.
The decision was, therefore, taken to multiply the two parents by tissue culture, particularly somatic embryogenesis, as the region's requirements are substantial (for instance, 7 ha of seed gardens for El Salvador or Guatemala). The ease with which the material could be introduced in_vitro - a simple square of leaf that could easily be disinfected - was also a determining factor in favour of somatic embryogenesis. Thirty thousand plants of the two parents have been produced at the CATIE laboratory in 1996-1997 and are now being distributed and acclimated in the countries affected by nematodes, with objective of setting up seed gardens.

4.2.2. Example of multiplication of C. arabica F_1 hybrids of agronomic worth:
The arabica coffee genetic improvement programme is original in which is based on hybridizing local varieties (Caturra, Catuai, Catimor, Sarchimor) with wild origins (Ethiopia, Sudan). The programme objectives are to increase the adaptability and productivity of the varieties and improve their disease and pest resistance, while conserving good cup quality.
In vitro micropropagation techniques are the best way of rapid multiplication of selected F_1 hybrids. With its high multiplication potential and the possibility of eventually producing low-cost plants, somatic embryogenesis has been considered useful for large-scale multiplication of the best cultivars. Moreover, using this technique reduces the breeding cycle from 30 to 10 years, and makes it possible to offer growers different candidates for the wide variety of situations in Central America. Several cultivars have been introduced and multiplied on a large scale. Clone performance trials have also been set up. Distribution of these hybrids to the countries participating in the project began in 1997. Around 40 cultivars were targeted for selection, multiplication and testing in clone trials under different soil and climatic conditions in the six project member countries. This research could lead to commercial production of the best hybrids by the year 2003.

4.3. PILOT PROJECT IN TANZANIA

Coffee research in Tanzania has achieved the creation of thousands of hybrids plants established in Lyamungu Research Station. Their average level of yield, without any fungicide spraying, in Lyamungu, is more than double the commercial varieties. A rapid multiplication of the best individual trees is urgently required to make these hybrids available to the coffee growers. It cannot be done through normal cuttings as the rate of multiplication is too low. The pilot propagation project that CIRAD is developing in Tanzania is:

- to start multiplying some promising trees (10) to establish real clones in a wide multi-site comparison trial,
- to train scientists and technicians in tissue culture in coffee propagation,
- to install a small research laboratory in Lyamungu for future studies on other selected trees,
- to help for commercial mass propagation activities in future. Contacts required with private sector

5. Field performance and conformity

5.1. DEFINITION OF VARIATION

Since the work of Larkin and Scowcroft (1981), it is known that tissue culture can cause modifications of the genome known as somaclonal variations. This notion covers three types of variation: (1) Heritable, stable, irreversible changes, (2) Unstable, reversible changes and (3) non-heritable, temporary changes known as epigenetic variations, which affect genome expression (Karp, 1991). Most of this variability is either due to chromosomal alterations, i.e. cuts, translocations, deletions, aneuploidy, polyploidy, gene amplification, transpositions, somatic cross-overs and one-off mutations (Evans and Sharp, 1986), or, in the case of epigenetic mechanisms, to under- or over-methylation of the DNA. In the latter case, the variant phenotype is unstable. For instance, in oil palm, 50% of sterile palms revert to normality after seven years flowering (Duval *et al.*, 1997).

Somaclonal variations are put down to excessive use of phytohormones in the culture medium, particularly auxin, and to over-rapid cell proliferation. In general, somaclonal variations involve a disorganized cell growth phase such as callus or adventive meristem multiplication (Karp, 1991). They affect most plant species, in addition to algae and fungi. Lastly, they are cultivar-dependent (Mohmed and Nabors, 1990; Freytag *et al.*, 1989). This has also been demonstrated in coffee (Sondhal and Lauritis, 1992).

5.2. THE CASE OF COFFEE

For coffee, little is known on the field performance and conformity of plants produced by somatic embryogenesis (Sondhal and Bragin, 1991; Sondhal and Lauritis, 1992). A variability of 10% was recorded out of 12,176 plants of *C. arabica*, which were evaluated. Among the variants modifications affecting leaf and branch morphology, plant size, fruit colour and shape, increased yields, variable ripening precocity and susceptibility to/tolerance of certain diseases were observed. Nestle is known to have set up large-scale trials in Ecuador using somaclonal variation as a source of genetic variability and not true-to-type reproduction. It is widely accepted that in species for which breeding systems are a limiting factor and in primitive species that have not undergone intensive selection - for instance coffee - somaclonal variation can release a new source of variability (Karp, 1991). The percentage of variant plants remains very low - around 1% (Personal communication), and suggests relatively substantial genetic variability in coffee. For instance, the Arabusta, an inter-specific hybrid with an unstable genome, apparently has more somaclonal variations. These results contradicted those of Sondhal and Bragin (1991), since they suggested much lower rates of variants.

5.3. WORK UNDER WAY

In order to remedy the lack of information, PROMECAFE decided to set up an extensive multi-site network with at least 20 sites in four Central American countries and a structure covering 50,000 plants. Moreover, specific studies are being carried out on the effect of certain culture conditions, such as suspension age (from 0 to 2 years), or genotype on the conformity of the plants regenerated. At present, 1,100 *in vitro* plantlets, including 650 somaplants from embryogenic calli of four hybrids are being evaluated in field trials, with 450 microcuttings of the same hybrids as controls (as microcuttings are known not to affect conformity). No difference in plant morphology has yet been observed, and most are about to flower.

One thousand two hundred other somaplants from high frequency embryogenic calli of five other hybrids are currently in the nursery and no variations in morphology have yet been observed either. By the end of 1998, 3,200 somaplants obtained from calli or embryogenic suspensions of various ages, were set to be planted in Costa Rica. By the end of 1999, 60,000 plants of *C. arabica* obtained from somatic embryos, were planned to be planted in Central America. These plants were produced by temporary immersion and under clearly defined culture conditions wholly compatible with the commercial production.

These trials should provide definitive information about the conformity of the material produced by somatic embryogenesis under defined culture conditions. It is important to bear in mind that 5% non-true-to-type plants are a commercially accepted figure. However, if serious fruiting problems arise, it would not be disastrous, as various strategies could be adopted to bring the figure below 5%, or even to eliminate all somaclonal variations. For instance, this was done with banana micropropagation (Cote *et al.*, 1993).

6. Conclusions and prospects

The results presented here show that somatic embryogenesis procedures are applicable to coffee on a large scale, irrespective of the species and genotype. In particular, the technique can be used to modify the conventional Arabica breeding strategy. Until recent years, development of new varieties of this species was geared towards pedigree breeding to produce a homozygous variety that could be distributed as seed. The process is relatively long and takes as long as 30-35 years. The development of a vegetative propagation technique such as somatic embryogenesis should allow:

1. Orientation of genetic improvement programmes towards a new strategy: the creation and use of F_1 hybrids that can be obtained after a 10 to 15-year selection cycle; this will enable breeders to shorten the selection cycle and also to make optimum use of genetic variability, above all by favouring heterosis. Such a programme was launched seven years ago in Central America, involving CIRAD and IRA.
2. Consideration of mass production, including use of the temporary immersion technique developed by CIRAD. Using such a technique to regenerate plants removes the need for isolation and manual transfer of somatic embryos required by the conventional *in vitro* procedure on a gel medium.
3. Consideration of new strategies, such as genetic transformation. In effect, it is essentially the rehabilitation of plant regeneration from cells if such a strategy is to be implemented. This is now possible as the technique has been fully developed.

It is clear that on the brink of the third millennium, thoughts can now be given to new coffee growing strategies based on more rational use of existing biodiversity, and on providing a more rapid response to the various constraints inherent to the crop.

7. Acknowledgements

Our special thanks to Mrs. N. Michaux-Ferrière (CIRAD) for her help in histological studies of the somatic embryogenesis procedure and also for writing this section. We would also like to thank Daniel Alvard and Marc Lartaud (CIRAD) for their help and collaboration on this work.

8. References

Alvard, D., Cute, F., Teisson, C. (1993), Comparison of methods of liquid medium culture for banana propagation. Effects of temporary immersion of explants. Plant Cell Tissue Org. Cult., 32: 55-60

Ameha, M. (1983), Heterosis in crosses of indigenous coffee selected for yield and resistance to coffee berry disease. I. First bearing stage. Acta Horticulture, 140: 155-161

Ammirato, P.V. (1977), Hormonal control of somatic embryo development from cultured cells of caraway: Interactions of abscisic acid, zeatin and gibberellic acid. Plant Physiol., 59: 579-586

Anzueto, F. (1989), Recherche de la resistance aux nematodes dans une collection de Coffea spp. D. E. A,Univ. de Rennes-ENSAR, p. 29

Berthouly, M., Guzman, N., Chatelet, P. (1987), Micropropagation *in vitro* de differentes lignees de *Coffea arabica* v. Catimor. In *12Ème Colloque Scientifique International sur le Cafe*, ASIC, Paris, pp. 462-467

Berthouly, M., Dufour, M., Alvard, D., Carasco, C., Alemanno, L., Teisson, C. (1995), Coffee micropropagation in liquid medium using temporary immersion technique. ASIC, 16eme Colloque,

Berthouly, M., Michaux-Ferriere, N. (1996), High frequency somatic embryogenesis in *Coffea canephora*: induction conditions and histological evolution. Plant Cell Organ Culture, 44: 169-176 Kyoto 9-14 Avril, Vol. II, pp. 514-519

Bertrand, B., Anzueto, F., Pena, M.X., Anthony, F., Eskes, AB. (1995), Genetic improvement of coffee for resistance of root-knot nematodes (Meloidogyne sp.) in Central America. In- XI Colloque Scientifique International sur le cafe, ASIC, Kyoto, Japan, pp. 630-636

Bertrand, B., Santacreo, R., Anthony, F., Etienne H., Charrier A., Eskes, A. B. (1997), Behaviour of *Coffea arabica* F_1 hybrids for resistance, production and technological qualities in Central America. ASIC, Nairobi

Bettencourt, A.J., Rodrigues, C. J. J. R. (1988), Principles and practice of coffee breeding for resistance to rust and other diseases. In- Coffee: Agronomy, (eds.) R. J. Clarke, R. Macrae, Elsevier Applied Science Publishers Ltd., England, pp. 199-234

Bouharmont, P., Awemo, J. (1979), La selection vegetative du cafeier robusta au Cameroun, Café Cacao Thé, 23: 227-237

Capot, J. (1977), L'amélioration du caféier robusta en Côte d'Ivoire. Café Cacao Thé, 21: 233-244

Charrier, A. (1985), Progrès et perspectives de l'amélioration génétique des caféiers. In: *11eme Colloque Scientifique International sur le café*, ASIC, Paris: 403-425

Cote F.X., Sandoval J., Marie P., Auboiron E. (1993), Variations in micropropaged Bananas and Plantains : ligature survey. Fruits, 48: 15-23

De Garcia, E., Menendez, A (1987), Embryogénesis somática a partir de explantes foliares del cafeto "catimor". Café Cacao Thé, 31 (1): 15-22

Deuss, J., Descroix, F. (1984), Le bouturage du caféier robusta dans le programme de replantation de la caféière au Togo. Café Cacao Thé, 28 (3): 165-178

Dublin, P. (1980), Multiplication végétative *in vitro* de l'arabusta. Café Cacao Thé, 24 (4): 281-289

Dublin, P. (1982), Culture de tissus et amélioration génétique des caféiers cultivés. In: *10eme Colloque Scientifique sur le café*, ASIC, Paris, 433-459

Dublin, P. (1984), Techniques de reproduction végétative *in vitro* et amélioration génétique chez les caféiers cultivés. Café Cacao Thé, 28 (4): 231-244

Duval, Y., Amblard, P., Rival, A., Konan, E., Gogor, S., Durand-Gasselin, T. (1997), Progress in oil palm tissue culture and clonal performance in Indonesia and the Côte d'Ivoire. International Planters Conference, Kuala Lumpur, pp. 291-307

Etienne, H., Bertrand, B., Anthony, F., Cute, F., Berthouly, M. (1997 a), L'embryogénèse somatique, un outil pour l'amélioration génétique du caféier. ASIC, 17eme Colloque, Nairobi

Etienne, H., Lartaud, M., Michaux-Ferrière, N., Carron, M. P., Berthouly, M., Teisson, C. (1997 b), Improvement of somatic embryogenesis in Hevea brasilensis (M ll. Arg.) using the temporary immersion technique. *In vitro* cellular and Development Biology.

Evans, D.A., Sharp, W.R. (1986), Somaclonal and gametoclonal variation. In- Handbook of Plant Cell Culture; Techniques and Applications. (eds.) D.A. Evans, W.R. Sharp and P. V. Ammirato, Macmillan Publishing Company, New York, pp. 97-132

Freytag, A.H., Rao-Arelli, A.H., Anand, S.C., Wrather, J.A., Owens, L.D. (1989), Somaclonal variation in soybean plants regenerated from tissue culture. Plant Cell Rep. 8: 199-202.

Haarer, A.E. (1969) Producción moderna de café .Instituto del libro. Edición revolucionaria, La Habana.

Jain S.M., Gupta P.K., Newton R.J.(eds.) (1995), Somatic embryogenesis in woody plants. Vol. 1-3, Kluwer Academic Publishers, The Netherlands

Karp, A. (1991), On the current understanding of somaclonal variation. Oxford surveys of Plant, Molecular Cell Biol., 7: 1-58

Larkin, P.J., Scowcroft, W.R. (1981), Somaclonal variation-a source of variability from cell cultures improvement. Theo. Appl. Genet., 60:197-214

Mohamed, A.S., Nabors, M W (1990), Somaclonal variant plants and wheat derived from mature embryo explants of three genotypes. Plant Cell Rep., 8: 558-560

Monaco, L., Sondhal, M., Carvalho, A., Crocomo, O., Sharp, W. (1977), Applications of tissue culture in the improvement of coffee. In: Applied and Fundamental Aspects of Plant Cell, Tissue and Organ Culture, Reinert J. and Bajaj Y. (eds.) Springer-Verlag, Berlin, pp. 109-129

Murashige, T., Skoog, F. (1962), A revised medium for rapid growth and bioassays with tobacco tissue cultures. Physiol. Plant., 15: 473-497

Neuenschwander, B., Baumann, T. (1992), A novel type of somatic embryogenesis in *Coffea arabica*. Plant Cell Rep., 10: 608-612

Petiard, V., Courtois, D., Masseret, B., Delaunay, P., Florin, B. (1987), Existing limitations and potential applications of plant cell cultures. In: Proceedings 4th European Congress on Biotechnology (Amsterdam), O. M. Neylssl, R.R. Van der Meer, K.C.A.M. Lyuben (eds.) Elsevier Science Publisher B. V., pp. 157-169

Pierson, E.S., Van Lammeren, A.A.M., Schel, J.H.N., Staritsky, G. (1983), *In vitro* development of embryoids from punched leaf discs of *Coffea canephora*. Protoplasma, 115: 208-216

Sondhal, M.R., Sharp, Wr. (1977,) High frequency induction of somatic embryos in cultured leaf explants of *Coffea arabica* L. Z. Pflanzenphysio. Bd. S. 81: 395-408.

Sondhal, M.R., Bragin, A. (1991,) Somaclonal variation as a breeding tool for coffee improvement. In: ASIC, 14eme Colloque Scientifique International sur le Café, San Francisco, pp. 701-710

Sondhal, M.R. and Lauritis J.A. (1992), Coffee. In: Biotechnology of perennial fruit crops. Biotechnol. Agric., 8: 401-420

Srinivasan, C.S., Vishveshwara, S. (1978), Heterosis and stability for yield in arabica coffee. Indian Genet. Plant Breed., 38: 416-420

Staritsky, G. (1970), Embryoid formation in callus tissues of coffee. Acta. Bot. Neerl., 19 (4): 509-514

Teisson, C., Alvard, D. (1994), A new concept of plant *in vitro* cultivation liquid medium: temporary immersion proceedings of the VIII. Proc. 1st Inter. Congr. Plant Tissue Cell Culture, Florence, Italy, pp. 105-109.

Van Boxtel, J., Berthouly, M. (1996), High frequency somatic emnryogenesis in coffeeleaves: Factors influencing embryogenesis, and subsequent proliferation and rgeneration in liquid medium. Plant Cell, Tissue Organ Culture, 44:7-17

Van der Vossen, H.A.M., D.J. Walyaro, D.J. (1981), The coffee breeding programme in Kenya: a review of progress made since 1971 and plan of action for the coming years. Kenya Coffee, 46 (541): 113-130

Van der Vossen, H.A.M. (1985), Coffee selection and breeding. In-Coffee, botany, biochemistry and production of beans and beverage, M. N. Clifford, K. C. Willsons (eds.) Crom Helm Ltd, pp. 48-96

Vasil, V. and Vasil K. (1986), Plant regeneration from friable embryogenic callus and cell suspension cultures of *Zea mays* L. J. Plant Physiol., 124: 399-408

Yasuda, T., Fujii, Y. and Yamaguchi, T. (1985), Embryogenic callus induction from *Coffea arabica* leaf explants by benzyladenine. Plant Cell. Physiol., 26 (3): 595-597

Zamarripa, A., Ducos, J.P., Tessereau, H., Bollon, H., Eskes, A.B., Petiard, V. (1991), Développement d'un procédé de multiplication de masse du caféier par embryogénèse somatique en milieu liquide. In 14ème Colloque Scientifique International sur le café, ASIC, San Francisco, pp. 392-402

Zamarripa, A. (1993), Etude et développement de l'embryogénèse somatique en milieu liquide du caféier (*Coffea canephora* P., *Coffea arabica* L. et hybride Arabusta). PhD Thesis, Ecole Nationale Supérieure Agronomique, Rennes, France

Chapter 6

BIOTECHNOLOGY FOR GENETIC IMPROVEMENT OF THE INDIAN COFFEE

H.L. Sreenath

Biotechnology Centre, Unit of Central Coffee Research Institute, Coffee Board, Mysore - 570 006, India

Running title: Genetic improvement of Indian Coffee

1. Introduction

India is among the top ten coffee producing and exporting countries along with Brazil, Columbia, Indonesia, Vietnam, Mexico, Ivory Coast, Ethiopia, Uganda and Guatemala. In India, both arabica and robusta coffee are cultivated almost in equal areas. Coffee production in India has reached 2.3 lakh tonnes in 1998 from around 19 thousand tonnes in 1950, while the average productivity has increased to over 850 kg/ha from 204 kg/ha, during the same period. The estimated production for 1999-2000 is 2.85 lakh tonnes (Anonymous, 1999).

Coffee is grown in India at inaccessible heights and mountain ranges, especially in South India. These mountain ranges are known for rich biological diversity, both in flora and fauna, constituting a complex ecosystem. During the last 400 years, since coffee was introduced to India, Indian coffee grower has adopted coffee cultivation to suit this complex ecosystem. Traditionally, Indian coffee is grown in the shade of forest trees, making it not only eco-friendly but also superior in quality. Adopting an integrated disease and pest management approach, rather than depending solely on chemicals controls pests and diseases. Chemical fertilizers are used only to a limited extent. Indian coffee is also unique in using solar energy for drying. Thus coffee cultivation in India could be a very good example of sustainable agriculture. It has checked erosion of

T. Sera et al. (eds.), Coffee Biotechnology and Quality, 91–100.

biological diversity as well as soil erosion. It has prevented migration of weaker sections of society, like women and tribals to cities in search of jobs, by providing 4.9 lakh jobs in coffee plantations. It has also provided employment to thousands of people in coffee processing and trade. Ninety eight percent of 1.4 lakh Indian coffee growers are categorised as small growers, whose average holding is just 1.4 hectare.

1.1. MAJOR CHALLENGES TO COFFEE CULTIVATION IN INDIA

Managing the existing biodiversity, tackling major biotic and abiotic stresses, quality improvement and protection of environment are the major challenges to be addressed to make coffee cultivation in India more sustainable.

The most important disease affecting Indian coffee is the leaf rust caused by the fungus *Hemileia vastatrix.* The disease, first recorded from India in 1869, has developed into the most challenging disease of arabica coffee in India. In the fifties, 73% of the planted area in India was covered by arabica producing the superior quality coffee and remaining 27% by robusta, out of a total area of 95,523 ha. However, in 97-98 arabica constituted only 47%, while robusta constituted 53% of a total of 3,05,902 ha. (Anonymous, 1999). Leaf rust is one of the major causes for decline in the area under arabica. Central Coffee Research Institute has done pioneering work in leafs rust management and breeding varieties resistant to this pathogen. However, the pathogen is also evolving rapidly and overcoming the resistance.

The major pests attacking Indian coffee are, the coffee berry borer *(Hypothenemus hampei)*, white stem borer *(Xylotrechus quadripes)*, mealy bugs *(Planococcus citri* and *P. lilacinus)*, shot hole borer (*Xylosandrus compactus)* and root nematode (*Pratylenchus coffeae*) (Anonymous, 1998). Water stress and oxidative stress are the most important abiotic stresses affecting the coffee plants and responsible for severe yield losses. With the expansion of coffee into non-traditional areas, high temperature stress also needs attention.

Effective chemicals have been developed to control the various diseases and pests of coffee. But the small coffee grower who is the backbone of Indian coffee cannot afford them. Also these chemicals are causing serious environmental concerns. The cost of cultivation also needs to be kept at a minimum by reducing input requirement. Then only India can sell its coffee at a competitive price in the International market, since India exports two thirds of the coffee produced. The best option is to develop resistant cultivars with improved quality.

2. Coffee improvement by conventional breeding

Employing conventional plant breeding methods of individual plant selection followed by progeny evaluation, hybridization and pedigree selection, coffee plants with high yield and vigorous plant growth have been developed. In addition to these, back cross methods have been used to transfer specific traits such as short internodes, and disease and pest resistance to *Coffea arabica* from other cultivars or related species. During the last 60 years, Central Coffee Research Institute (CCRI), India, has developed 12 improved cultivars of arabica and 3 improved cultivars of robusta, through conventional breeding (Anonymous, 1996). A few more promising selections are in the advanced stages of evaluation (Srinivasan, 1996). In spite of tremendous contribution to coffee genetic improvement, conventional breeding suffers from the limitations like, the genetic barrier of chromosome number (diploid vs. tetraploid), autoincompatible alleles (diploid species), lack of genetic understanding and long breeding cycles. As a result of these factors, transfer of genetic traits from wild outbreed species of the genus to the cultivated species is quite difficult.

3. Biotechnology for coffee improvement

Biotechnology can supplement the efforts of coffee breeders with additional tools, which can overcome the limitations of traditional plant breeding to a great extent. There is tremendous scope for application of crop biotechnology tools for genetic improvement of Indian coffee. Coffee improvement through biotechnology can focus on three different areas of applications: agronomy, processing industry and consumers. Agronomic benefits should focus on reducing direct and indirect farming costs. To reduce coffee-farming costs, the new technologies can address fertilizer efficiency, disease and pest resistance and crop management aspects that will reduce labour utilization for e.g., herbicide tolerance and mechanized harvesting. The overall objective of biotechnology should be to make Indian coffee cultivation more sustainable.

Various biotechnological approaches such as micropropagation, embryo rescue, anther culture, cell line selection, somaclonal variation, protoplast culture, *in-vitro* preservation, marker aided selection and genetic transformation have tremendous potential for genetic improvement of Indian coffee (Sreenath, 1998a). Ever since the first report on tissue culture regeneration of coffee by Staritsky (1970), there has been a steady progress in coffee biotechnology research (Sondahl and Loh, 1988; Sreenath and Naidu, 1997).

4. Coffee biotechnology research in India

Practically, all research work on coffee biotechnology in India has been done at CCRI. Tissue culture research has been conducted at CCRI with the short term and long term objectives of micropropagation and genetic improvement and germplasm preservation, respectively (Muniswamy and Sreenath, 1999). For achieving these objectives, research efforts were focussed on stem, leaf, integument, apical bud and node cultures for micropropagation, anther culture, embryo culture, endosperm culture, protoplast culture, genetic transformation and molecular markers for genetic improvement and *in-vitro* preservation techniques, including cryopreservation, for germplasm preservation. Significant achievements made in these areas are given below.

4.1. MICROPROPAGATION

In coffee, micropropagation is possible by microcuttings as well as somatic embryogenesis. Various experiments conducted to develop micropropagation protocols for Indian cultivars of coffee have resulted in plant regeneration through somatic embryogenesis (Fig.1) from stem, leaf and integument tissues in more than 20 genotypes/clones of arabica, robusta and hybrid selections (Raghuramulu *et al.,* 1987; Babu *et al.*, 1993; Jayashree *et al.,* 1995; Muniswamy and Sreenath, 1995b; Sreenath, 1998c). Genotype differences were found in respect of callus induction, somatic embryogenesis and plant regeneration in arabica coffee (Naidu *et al.*, 1999). Clear differences were seen between arabica and robusta cultivars in somatic embryogenesis and germination of somatic embryos (Babu et al., 1993). Studies were conducted to standardize encapsulation techniques for developing synthetic seeds (Muniswamy and Sreenath, 1995a).

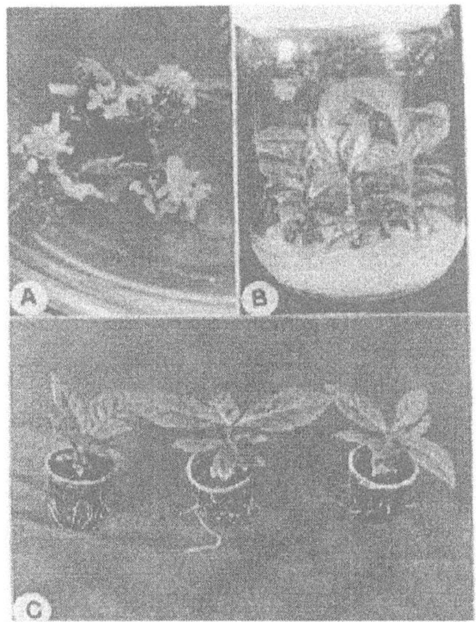

Figure 1. Plant regeneration through somatic embryogenesis. A: Somatic embryos formed on leaf explant; B: Plans regenerated from somatic embryos. C: Somatic embryo derived plantlets hardened in net pots.

Protocols were developed initially for small scale hardening (Muniswamy *et al.*, 1994). Later facilities were developed for medium scale hardening (unpublished results). Trial plots of tissue cultured plants are being established in different agro-climatic zones for comparison with the seedling progeny (Sreenath, 1998c). Three improved selections viz., Cauvery, Sln.9 and CxR are being used for large-scale field evaluation of micropropagated plants. Vegetative growth and cropping were normal in the micropropagated plants. Plant regeneration was achieved from the apical bud and nodal explants (Ganesh and Sreenath, 1997), but further refinement of this technique is required. Effect of TIBA and BAP was tested on integument cultures of *C. canephora*. In this study, the integument tissue was found to be aseptic, with high callusing ability and the cultures were phytohormone autotrophic (Babu *et al.*, 1997). For the first time in any plant species, plant regeneration was achieved from integument tissues of CxR (Sreenath *et al.*, 1995) and Cauvery cultivars (unpublished results).

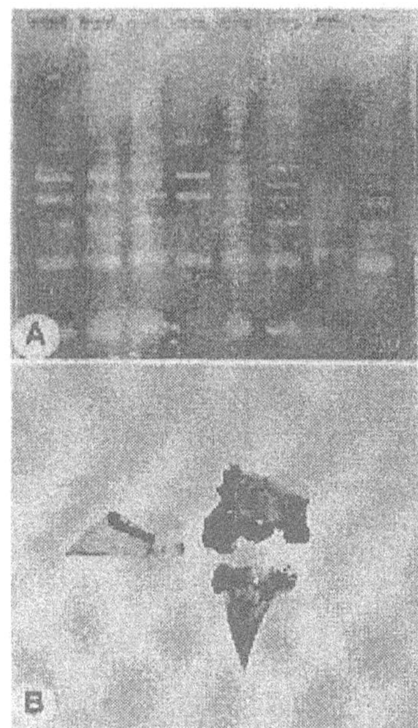

Figure 2. Plant regeneration through somatic embryogenesis. A: RAPD marker profile of coffee; B: Coffee explant showing the gus expression (blue stained callus).

4.2. GENETIC IMPROVEMENT

With the objective of using the tissue culture technology for genetic improvement, research was done on embryo culture, anther culture and endosperm culture. The research has resulted in plant regeneration through embryo culture by direct germination, as well as, somatic embryogenesis (Muniswamy *et al.*, 1993; Sreenath *et al.*, 1989; Muniswamy and Sreenath, 1997). Plants produced from the embryo culture have been established in the field. In addition to this, plants derived from embryo culture of inter-specific crosses between cultivated *C. canephora* (robusta) and three indigenous wild species, viz., *C. bengalensis*, *C. travancorensis* and *C. wightiana* and from five intervarietal crosses have been established in the field and are being evaluated (Sreenath *et al.*, 1992; unpublished results).

Plantlets were regenerated from the anther culture of CxR cultivator (Muniswamy and Sreenath, Unpublished work) and a few of the plants are established in soil. This was the first achievement in a diploid species of coffee. Callus cultures were established from endosperm tissues (Raghuramulu, 1989) and plantlets were regenerated through embryogenesis from the endosperm calli of S. 2803 and BR clones (Muniswamy and Sreenath, Unpublished work).

Protocols have been optimized for isolation of protoplast from embryogenic calli (Mamatha and Sreenath, 1998). DNA markers are becoming powerful tools in the hands of coffee breeders (Sreenath 1998b, 1999). Protocols have been developed for isolation of DNA and producing RAPD (Ram and Sreenath, Unpublished work) (Fig. 2A) and AFLP markers (Sreenath *et al.*, Unpublished work). The RAPD markers are being used for genetic fingerprinting of leaf rust differentials.

Genetic transformation can be very useful for engineering disease and pest resistance in coffee (Sreenath, 1998d) and for improving yield and cup quality (Sreenath, 1998e). Genetic transformation protocols are being developed for genetic engineering of coffee. Transient expression of *gus* gene was achieved in leaf tissues of CxR cultivars after co-cultivation with *Agrobacterium tumefaciens* (Naveen and Sreenath, Unpublished work)

4.3. *IN VITRO* PRESERVATION

In vitro preservation of zygotic embryos achieved under slow growth condition up to two years in *C. arabica* (Naidu and Sreenath, 1999). Successful cryopreservation of zygotic embryos is achieved in three species (Krishna and Sreenath, unpublished work).

5. Conclusions and futuristic approaches

For India, coffee earns around 450 million dollars in foreign exchange, which is very important for the economy of the country. Due to sustained research efforts of CCRI, the production and productivity of Indian coffee have improved steadily during the last five decades. However, further intensive research efforts are needed to make Indian coffee cultivation more sustainable in the next millennium. Biotechnology is identified as a thrust area for meeting this challenge. In India, research efforts on coffee biotechnology are focussed on micropropagation, genetic improvement and germplasm preservation. Research efforts at CCRI have resulted in successful plant regeneration through somatic embryogenesis in more than 20 selections. For the first time in any plant species, plant regeneration was achieved from integument tissues. Protocols have been optimized for direct and indirect plant regeneration from embryo culture. Plant regeneration was achieved for the first time from the anther culture in the diploid CxR cultivars of coffee. Plant regeneration was also achieved from endosperm cultures of *C. arabica* (S. 2803), which is again a first achievement in any coffee species.

There is great potential for application of biotechnology for genetic improvement of Indian coffee cultivars and growing coffee in an eco-friendly way. Results obtained so far are encouraging. Currently, the efforts are focussed on converting the vast potential of biotechnology into reality. Molecular plant breeding is one more step in the long evolution of our crop improvement practices and it is expected to play a positive role in meeting the challenges of coffee cultivation in India. This new branch of crop improvement is made possible by the recent quantum leap in our ability to transform genes to plants from unrelated organisms and our understanding on how plant genes function and confer specific desirable traits. The gene transfer technology will be useful to Indian coffee for protecting plants from pests and diseases, (Sreenath, 1998d) optimizing plant growth in normal, as well as stressful environment, improving yield and producing better cup quality (Sreenath, 1998e) and developing speciality coffees like caffeine free varieties (Sreenath, 1997).

Developing molecular markers linked to leaf rust resistance, characterization of coffee germplasm with DNA markers, engineering leaf rust resistance in coffee with anti-fungal genes from heterologous sources, isolation of genes of leaf rust resistance from resistant cultivars, developing *Bt* based biopesticides to combat major pests, engineering insect resistance using Bt and proteinase inhibitor genes, isolating caffeine degradation genes for manipulation of caffeine content in the beans, are identified as thrust areas of coffee biotechnology research in India. It may be concluded that biotechnology will help to create coffee varieties that will fit into the eco-friendly coffee cultivation system in India and make coffee cultivation more sustainable in the new millennium.

6. References

Anonymous, (1996), Coffee Guide, Central Coffee Research Institute, Chikmagalur

Anonymous, (1998), A compendium on pests and diseases of coffee and their management in India, Central Coffee Research Institute, Chikmagalur

Anonymous, (1999), Database on Coffee, July 1999, Coffee Board, Bangalore

Babu K.H., Shanta H.M., Ganesh D.S., Sreenath H.L. (1993), Plant regeneration through high frequency somatic embryogenesis from leaf tissues in two cultivars of coffee. J. Plant. Crops, 21: 307 –312

Babu, K.H., Jayashree, G., Naidu, M.M., Ganesh, D.S. and Sreenath, H.L. (1997), Effect of TIBA and BAP on integument cultures of CxR cultivar of robusta coffee *(Coffea canephora)*. In Biotechnological Applications of

Plant Tissue and Cell Culture, G.A. Ravishankar and L.V. Venkataraman (eds.), Oxford and IBH Publishing Co. Pvt. Ltd., New Delhi, pp. 296-300

Ganesh, D.S., Sreenath, H.L. (1997), Shoot regeneration from apical bud and node cultures in four clones of robusta coffee *(Coffea canephora)*. In- Biotechnological Applications of Plant Tissue and Cell culture, G.A. Ravishankar and L.V. Venkataraman (eds.), Oxford and IBH Publishing Co. Pvt. Ltd., New Delhi, pp. 301-306

Jayashree, G., Ganesh, D.S., Sreenath, H.L. (1995), Effect of kinetin on germination of somatic embryos of CxR cultivar of coffee. J. Coffee Res., 25(2): 102-105

Mamatha, H.N., Sreenath, H.L. (1998), Optimization of conditions for isolation and culture of coffee protoplast from embryogenic calli. Paper presented at the XIII PLACROSYM, December 14-18, Coimbatore, Tamilnadu, India

Muniswamy, B., Naidu, M.M., Sreenath, H.L. (1993), Somatic embryogenesis and plant regeneration from cultured immature zygotic embryos of *Coffea canephora* Pierre. J. Plant. Crops, 21 (suppl.): 346-350

Muniswamy, B., Naidu, M.M., Sreenath, H.L. (1994), A simple procedure for hardening of *in-vitro* raised plantlets of *Coffea canephora*. J. Coffee Res., 24: 49-53

Muniswamy, B., Sreenath, H.L. (1995a), Standardization of encapsulation techniques for producing synthetic seeds in coffee. J. Coffee Res., 25: 24-29

Muniswamy, B., Sreenath, H.L. (1995b), High frequency somatic embryogenesis from cultured leaf explants of *Coffea canephora* on a single medium. J. Coffee Res., 25: 98-101

Muniswamy, B., Sreenath, H.L. (1997), Genotype specificity of direct somatic embryogenesis from cultured embryos in coffee. In - Biotechnological Applications of Plant Tissue and Cell Culture. G.A. Ravishankar and L.V. Venkataraman (eds.), Oxford and IBH Publishing Co. Pvt. Ltd., New Delhi, pp. 204-209

Muniswamy, B., Sreenath, H.L. (1999), Tissue culture studies in coffee and its applications. Indian Coffee, LXIII (3), 7-8

Naidu, M.M., Sreenath, H.L. (1999), In *vitro* culture of coffee (*Coffea arabica* L) zygotic embryos for germplasm preservation. Plant Cell Tissue Org. Cult., 55: 227-230

Naidu, M.M., Ganesh, D.S., Jayashree, G., Sreenath, H.L. (1999), Effect of genotype on somatic embryogenesis and plant regeneration in arabica coffee. In - Plant Tissue Culture and Biotechnology, Emerging Trends. P.B. Kavi Kishor (ed.), Universities Press India Ltd., Hyderabad, pp. 90-95

Raghuramulu, Y. (1989), Anther and endosperm cultures of coffee. J. Coffee Res., 19: 71- 81

Raghuramulu, Y., Purushotham, K., Sreenivasan, M. S., Ramaiah, P.K. (1987), *In-vitro* regeneration of coffee plantlets in India. J. Coffee Res., 17: 57-64

Sondahl, M. R., Loh, W.H.T. (1988), Coffee biotechnology. In - Coffee Agronomy Vol. 4, R. J. Clarke and R. Macrae (eds.), Elsevier Applied Science, London, pp. 235-262

Sreenath, H.L. (1997), Development of caffeine free coffee varieties. Indian Coffee, LXI (10): 13-14

Sreenath, H.L. (1998a), Biotechnology approaches for coffee improvement. Indian Coffee, LXII (1): 18-19

Sreenath, H.L. (1998b), DNA marker techniques come to the help of coffee breeders. Indian Coffee, LXII (2), 11-13

Sreenath, H.L. (1998c), Micropropagation of superior coffee genotypes. Indian Coffee, LXII (4): 17-19

Sreenath, H.L. (1998d), Potential and strategies for genetic engineering of coffee plants for disease and pest resistance. Indian Coffee, LXII (6): 7-11

Sreenath, H.L. (1998e), Enhanced carbohydrate accumulation by genetic engineering can lead to higher yield and rich brew in coffee, Indian Coffee, LXII (8): 21-22

Sreenath, H.L. (1999), Use of molecular markers in coffee. In - Plant Tissue Culture and Biotechnology, - Emerging Trends. P.B. Kavi Kishor (ed.), Universities Press India Ltd., Hyderabad, pp. 289-293

Sreenath, H.L., Muniswamy, B., Naidu, M. M., Dharmaraj, P.S., Ramaiah, P.K. (1992), Embryo culture of three inter-specific crosses in coffee. J. Plant. Crops, 20 (suppl.): 243-247

Sreenath, H.L., Muniswamy, B., Raghuramulu, Y., Ramaiah, P.K. (1989), Immature embryo culture in coffee. Paper presented at XIII Plant Tissue Culture Conference, December 18-20, Shillong, Meghalaya, India

Sreenath, H.L., Naidu, R. (1997), Quarter century of research in coffee biotechnology – a review. In - Biotechnological Applications of Plant Tissue and Cell Culture, G.A.

Ravishankar and L.V. Venkataraman (eds.), Oxford and IBH Publishing Co. Pvt. Ltd., New Delhi, pp. 14-27

Sreenath, H.L., Shantha, H.M., Babu, K.H., Naidu, M. M. (1995), Somatic embryogenesis from integument (perisperm) cultures of coffee. Plant Cell Rep., 14: 670-673

Srinivasan, C.S. (1996), Current status and future thrust areas of research on varietal improvement and horticultural aspects of coffee. J. Coffee Res., 26: 1-16

Staritsky, G. (1970), Embryoid formation in callus cultures of coffee. Acta Bot. Neerl., 19: 509- 514

Chapter 7

MOLECULAR BREEDING IN COFFEE (*COFFEA ARABICA* L.)

P. Lashermes[1], M. C. Combes[1], P. Topart[2], G. Graziosi[3], B. Bertrand[4], F. Anthony[2]

[1]*IRD, Montpellier BP 5045, 34032 Montpellier Cedex 1 France;* [2]*CATIE, Turrialba, Costa Rica;* [3]*Università deli Studi di Trieste, Italy;* [4]*IICA/PROMECAFE, Costa Rica*

Running title: Molecular breeding in coffee

1. Introduction

The cultivated coffee *Coffea arabica* L. (2n=4x=44) is an allotetraploid species native to Africa, containing two diploid genomes that originated from two different diploid wild ancestors (2n=2x=22), *C. canephora* and *C. eugenioides* or ecotypes related to those species (Lashermes *et al.*, 1999). It is the only polyploid species in the genus and is self-fertile while other *Coffea* species are diploid and generally self-incompatible. *C. arabica* is characterised by a very low genetic diversity (Fig. 1), which is attributable to its origin, reproductive biology, and evolution. In addition, most cultivars are derived from the few trees which survived various efforts to spread arabica growing world-wide (Van der Vossen, 1985; Lashermes *et al.*, 1996a). It is believed that the encountered agro-morphological variation which gave rise to so many named varieties, results from few major-gene spontaneous mutations conditioning plant, fruit and seed characters (Carvalho, 1988). The cultivars, therefore, present a homogeneous agronomic behaviour characterised by a high susceptibility to many pests and diseases, and very low adaptability (Bertrand *et al.*, 1999).

Enlarging the genetic base and improvement of arabica cultivars have become priorities for researchers. Spontaneous accessions collected in the primary centre of diversity as well as wild relative *Coffea* species constitute a valuable gene reservoir for breeding purposes (Anthony *et al.*, 1999).

T. Sera et al. (eds.), Coffee Biotechnology and Quality, 101–112.

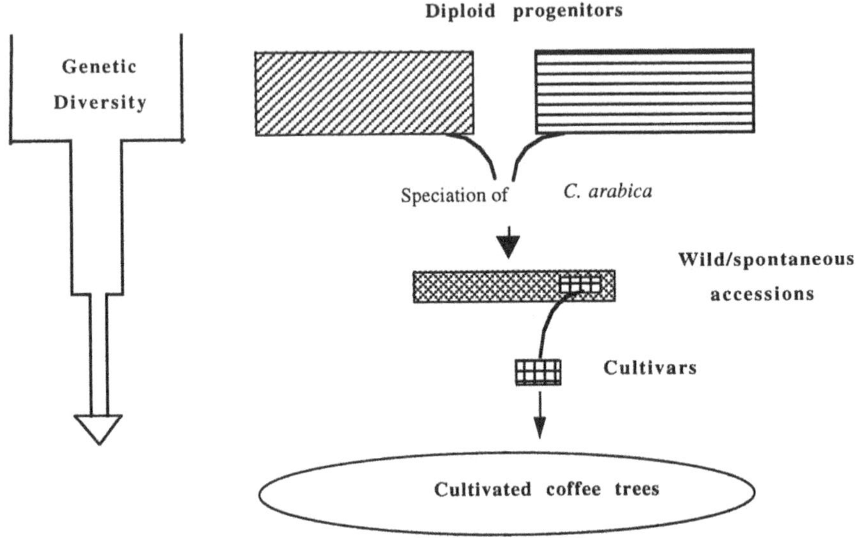

Figure 1. Genetic base of the cultivated Arabica coffee trees.

To date, *C. canephora* provides the main source of disease and pest resistance traits not found in *C. arabica* including coffee leaf rust (*Hemileia vastatrix*), Coffee Berry Disease (CBD) caused by *Colletotrichum kahawae* and resistance to root-knot nematode (*Meloidogyne sp.*). Likewise, other diploid species present considerable interests in this respect. For instance, *C. racemosa* constitutes a promising source of resistance genes to coffee leaf miner (Guerreiro Filho *et al.*, 1999).

The Timor Hybrid is an atypical tree which was identified in a *C. arabica* field, planted in 1927, on the island of Timor (Bettencourt, 1973). Based on information relating to the coffee germplasm introduced into Timor at the beginning of the century, the limited fertility of the original plant, characteristics of disease resistance, and preliminary molecular investigations, it is believed that the Timor Hybrid originated from a spontaneous inter-specific cross between *C. arabica* and *C. canephora* (Bettencourt, 1973; Goncalves and Rodrigues, 1976; Lashermes *et al.*, 1993). Progenies of the Timor Hybrid have been distributed world-wide, and, when observed, showed 2n = 44 chromosomes (Rijo, 1974). In recent decades, they have been used intensively in coffee breeding programmes as the main source of resistance to pests and diseases (Charrier and Eskes, 1997). Exploitation of Timor Hybrid populations has so far relied on conventional procedures in which a hybrid is produced with an outstanding arabica genotype, and the progeny is sledded (or, back-crossed) and selected over at least 3-4 generations. Undesirable genes from the resistance-donor parents are expected to be gradually eliminated. However, conventional coffee breeding methodology faces considerable difficulties in so doing. In particular, strong limitations are due to the long generation time of coffee tree (five years), the high cost of field trial, and the lack of accuracy of current strategy. One can estimate that a minimum of 25 years after hybridisation (five back-cross generations) is required to restore the genetic background of the recipient cultivar and there by ensure good quality of the improved variety. Combining various genes of resistance without reducing coffee quality appears therefore as a very difficult task in an acceptable time-frame through traditional breeding approaches.

In recent years, DNA-based genetic markers have gained widespread applications in many fields of plant genetics and breeding. In particular, the development of marker-assisted selection (MAS) programmes promises to overcome present limitations of conventional coffee breeding. General principle of MAS is that if a gene(s) conferring a trait of interest is linked to an easily identifiable molecular marker, it may be much more efficient to select for the marker than for the trait itself.

2. Molecular analysis of arabica coffee introgression lines

Introgressed arabica genotypes derived from the Timor Hybrid were analysed for the presence of *C. canephora* genetic material using the amplified fragment length polymorphism (AFLP) approach (Vos *et al.*, 1995). In order to gain insights into the mechanism of introgression in *C. arabica*, Lashermes *et al.* (2000) estimated the amount of introgression present in such material. The Timor Hybrid-derived genotypes included in this analysis consisted of two accessions representing two different progenies of the Timor Hybrid (progenies 832-1 and 1343), and 19 introgression arabica lines (BC_1F_4) derived from different hybrids between accessions of various Timor Hybrid progenies (832-1, 832-2, and 1343) and commercial arabica cultivars. The Timor Hybrid-derived genotypes were evaluated using 42 different AFLP primer combinations, and compared to 23 accessions of *C. arabica* and 8 accessions of *C. canephora*.

A total of 1062 polymorphic fragments were scored among the 52 accessions analysed (Fig. 2). The number of polymorphic bands was much higher within the canephora accessions (i.e. 945) than within the accessions of *C. arabica* (i.e. 109). The group constituted by the Timor Hybrid-derived genotypes was distinguished from the accessions of *C. arabica* by 178 markers consisting of 109 additional bands (i.e. introgressed markers) and 69 missing bands. AFLP markers identified in Timor Hybrid-derived genotypes were considered as introgressed markers when not detected

in any of the accessions of *C. arabica* and observed in at least one of the canephora accessions analysed in this study.
AFLP, therefore, seemed to be an extremely efficient technique for DNA marker generation in coffee trees as well as for introgression detection in *C. arabica*. The genetic diversity observed in the Timor Hybrid-derived genotypes appeared approximately double to that in *C. arabica*. Although representing only a small proportion of the genetic diversity available in *C. canephora*, the Timor Hybrid obviously constitutes a considerable source of genetic diversity for arabica breeding. Analysis of genetic relationships among the Timor Hybrid-derived genotypes suggested that introgression was not restricted to chromosome substitution but also involved chromosome recombination. Furthermore, the Timor Hybrid-derived genotypes varied considerably in the number of AFLP markers attributable to introgression (Fig. 3).

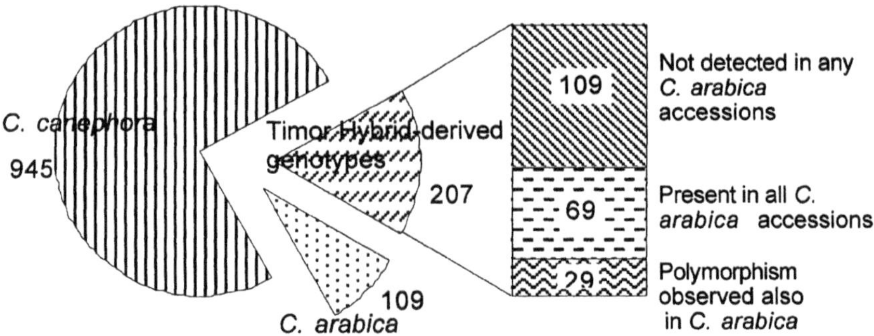

Figure 2. Pie charts depicting the numbers of polymorphic AFLP bands observed among individuals within each group constituted by the accessions of *C. arabica*, *C. canephora* and introgressed Timor Hybrid-derived genotypes, respectively. For the introgressed material, the polymorphic markers either attributable to the Arabica parent or associated with the introgression of *C. canephora* chromosome segments were distinguished (Lashermes *et al.* 1999).

In this way, the introgressed markers identified in the arabica coffee introgressed genotypes were estimated to represent from 8% to 27% of the *C. canephora* genome. Nevertheless, the amount of alien genetic material in the introgression arabica lines remains always substantial and should justify the development of adapted breeding strategies.

3.1.QUARANTINED PATHOGENES

If a virulent pathogen does not occur naturally in the test environment, artificial inoculation is prohibited for safety reasons. For instance, CBD is still restricted to the continent of Africa, and the availability of markers linked to the resistance gene(s) as reported for the *T* gene (Agwanda *et al.*, 1997), could allow pre-emptive breeding in countries (Asia, Latin America) where quarantine barriers are still effective.

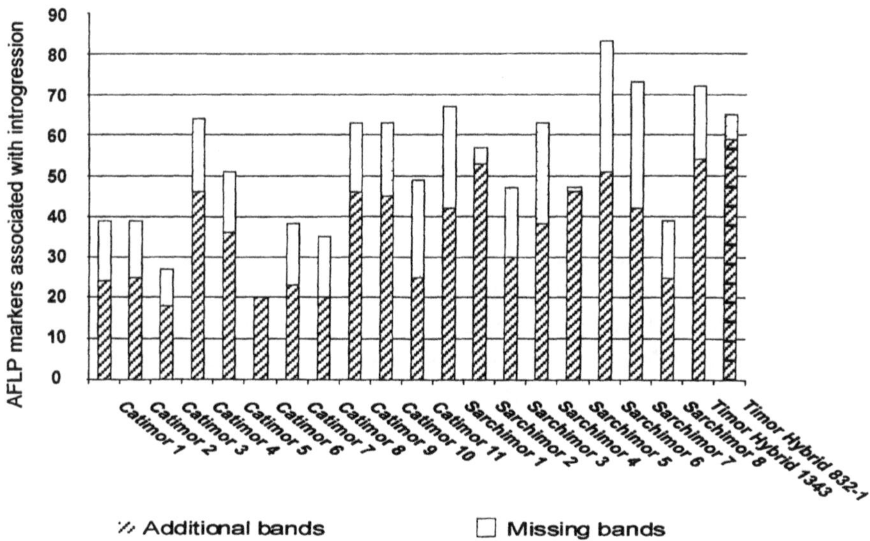

Figure 3. Numbers of AFLP polymorphic bands attributable to introgression detected in Timor Hybrid-derived genotypes (Lashermes *et al.* 1999).

3.2. RELIABILITY/LIMITATION OF DIRECT TESTING

Conventional selection progress could be hampered by the difficulty to ensure reliable test for the resistance trait. Seedling test could also present strong inconvenient. For instance, the present test for evaluation for root-knot nematode is destructive leading to important difficulties in the utilisation of identified plant resistance sources

(Bertrand *et al.*, 1997). In addition, expression of many resistance genes can be strongly influenced by environmental conditions.

3.3. DEVELOPMENTALLY REGULATED CHARACTER

Early selection based on the marker genotype of young seedlings would be particularly beneficial for late expressed traits.

3.4.TRANSFER OF RECESSIVE RESISTANCE GENES

The classical procedure of transferring a recessive resistance gene includes a progeny test after each back-cross generation to determine the presence of the desired allele. With MAS, the transfer can be accomplished without interruptions leading to an important time saving.

3.5. PYRAMIDING OF RESISTANCE GENES/COMBINING VALUABLE TRAITS

Pyramiding of resistance genes has been suggested as a strategy to provide durable resistance (i.e. coffee leaf rust). However, conventional breeding is complicated by the fact that, it is difficult or often impossible to distinguish the various resistance genotypes. Once the different genes conferring resistance to the same pathogen are tagged by tightly linked marker, they could be relatively easily be accumulated into a single genotype via marker-facilitated selection. Comparable advantages versus conventional are procured when trying to combine simultaneously resistance genes to different disease/pests.

4. Molecular-assisted back-cross breeding

Repeated back-crossing simultaneously accomplishes two essential goals: 1) allows segregation to remove donor parent chromosomes unlinked to the target gene, and, 2) allow recombination to remove donor parent segments which are linked to the target gene. Both objectives could be considerably facilitated by the use of molecular markers.

4.1.GENOME SELECTION

Beside the target trait, it is important to consider the complete genome of individuals. Chromosomal segments are segregating within back-cross progenies and the individuals show various contents of the desired parental genome (Fig. 4). A genome selection could, therefore, be performed by the use of markers scattered throughout the genome, resulting in a reduction in number of back-cross generations required to restore the genetic background of the recipient cultivar.

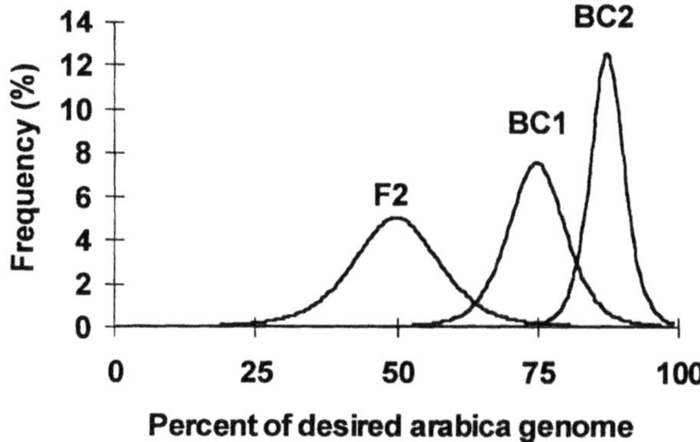

Figure 4. Frequency of individuals in F2, BC1, and BC2 having various contents (%) of the desired parental genome.

Values were estimated (Fig. 5) for a hypothetical arabica genome of 22 chromosome pairs of, on average, 100 cM each (Total genome of 2200 cM), using equations developed by Hillel *et al.* (1990) and Hospital *et al.* (1992). In the absence of selection, parental donor DNA was only removed by a factor of two in each generation. Simulations were given for MAS programme in which the either 10, or 2% best (in terms of percent recurrent parent genome) individuals in each generation were used as the parent for the next generation. Results equivalent to BC5 generation without selection was obtained after only two marker-assisted BC generations allowing a considerable time saving.

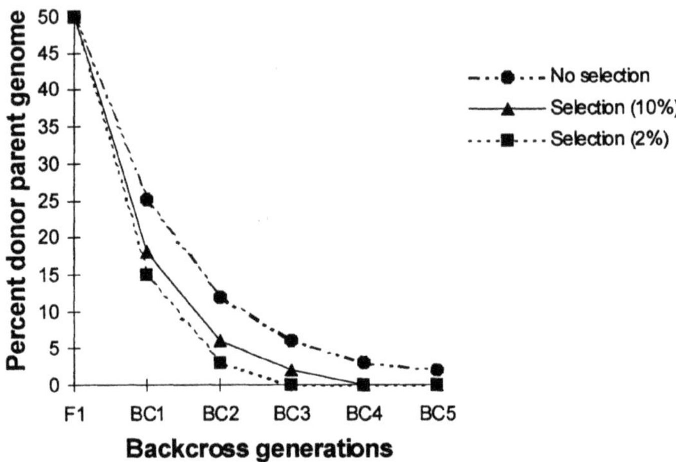

Figure 5. Average content (%) of the donor parental genome in back-cross generations under various intensities of genomic selection.

Table 1 presents the results of various levels of selection imposed on BC1 and BC2. Apparently the moderate levels of selection in both BC1 and BC2 resulted in individual, almost identical to the recipient variety. Achieving a similar result by a unique genome selection in one of these two generations requires extremely intensive selection effort.

Table 1. Percentage of the recipient arabica genome under various intensities selection in BC_1 and BC_2.

Selection in BC_2	Selection in BC_1 (%)			
(%)	100	30	10	2
100	87.5	89.7	91.0	92.0
30	90.9	92.4	93.2	94.1
10	92.5	93.6	94.5	95.1
2	94.1	94.9	95.5	96.1

4.2. REDUCING LINKAGE DRAG

Removing of the linked donor segment could take many generations (Stam and Zeven 1981). Many examples of "linkage drag" are known in which undesirable traits that are closely linked to a target gene, are carried out along during breeding programme (Zeven *et al.*, 1983, Young and Tanksley, 1989). For instance, in Arabica, even after 6 back-cross generations, a region of 32cM flanking a target gene is expected to persist (Fig. 6).

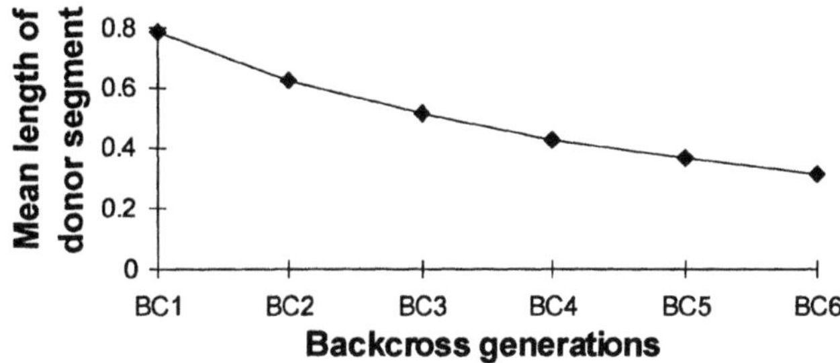

Figure 6. Mean length of donor segment surrounding the target gene after various numbers of back-cross generations The length is expressed as a proportion of the carrier chromosome (chromosome of 100 cM long) (after Stam and Zeven 1981).

In most plant genomes 32cM is enough DNA to contain hundreds of genes. DNA markers can be used to eliminate, or at least significantly reduce, linkage drag by allowing the identification of rare recombinant individuals, which are usually only selected by chance in classical breeding (Paterson *et al.*, 1991). In approximately 150 back-cross plants, there could be 95% chance that at least one plant would have experienced a crossover within 1 cM on one side, or, the other of the gene being selected. With one additional back-cross generation of 300 plants, there would be a 95% chance of a crossover within 1 cM of the other side of the gene, generating a segment surrounding the target gene of less than 2 cM.

5. Genetic mapping

Some utilization of MAS presupposes the existence of a detailed linkage map, which represents the relative order of genetic markers, and their relative distances from one to another along each chromosome of an organism (Paterson *et al.*, 1991).

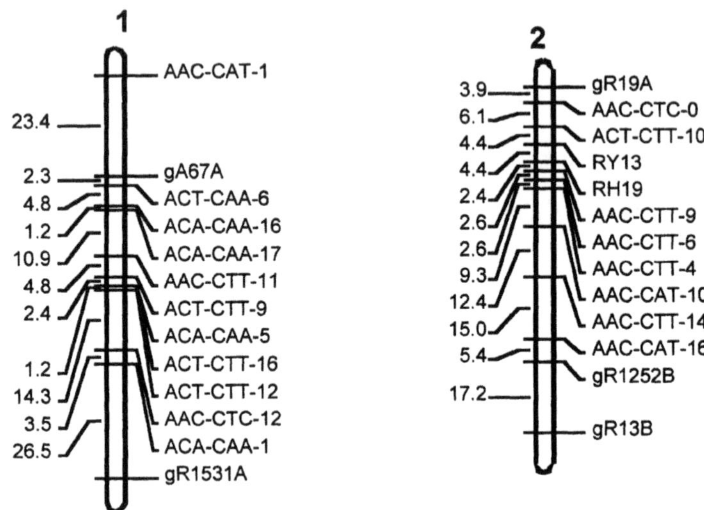

Figure 7. Linkage groups 1 and 2 of the genetic map of *C. canephora*. Map distances in cM are indicated on the left side of linkage groups and marker locus name is on the right side.

A first genetic map of the coffee genome was established using a random population derived from the clone IF200 of *C. canephora* (Paillard *et al.*, 1996). So far, more than 150 markers have been placed on the 11 linkage groups/chromosomes (Fig. 7). Additional linkage maps on different coffee species including *C. arabica* are being constructed. In so doing, the development of microsatellite PCR-based markers will be particularly useful (see chapter by Rovelli *et al.* in this book). These genetic linkage maps are considered useful in providing important information on coffee genome and chromosomal organisation. In particular, one might use it to map important genes as done for the S-locus controlling self-incompatibility (Lashermes *et al.*, 1996b).

6. Conclusions and prospects

The development of molecular markers for coffee trees has opened a new perspective in breeding. The conventional selection of self-, or back-crossed coffee tree progenies for further breeding is extremely laborious and time-consuming. The implementation of MAS could, therefore, be very promising. In particular, the integration of MAS in coffee breeding promises to drastically increase the efficiency of breeding programmes by:

1) allowing for selection at an early stage and on a large number of breeding lines,
2) reducing the number of back-cross cycles required to restore the quality of the traditional cultivars,
3) combining in one-step, selection for various traits or genes of resistance.
Furthermore, new findings from genome research indicate that there is tremendous genetic potential locked up in wild and cultivated germplasm resources that can be released only by shifting the paradigm from searching for phenotypes to searching for superior genes with the aid of molecular linkage maps (Tanksley and McCouch, 1997). In addition, ongoing technological developments, including automation, allele-specific diagnostics and DNA chips, will make MAS approaches based on large-scale screening much more powerful and effective.

7. Acknowledgements

This work was supported in part by the European Community through the International Scientific Co-operation Programme (Inco-Dc Contract Erbic 18CT970181).

8. References

Agwanda, C., Lashermes P., Trouslot P., Combes M.C., Charrier, A. (1997), Identification of RAPD markers for resistance to Coffee Berry Disease, *Colletotrichum kahawae*, in Arabica coffee. *Euphytica*, 97:241-248

Anthony, F., Astorga C., Berthaud, J. (1999), Los recursos genéticos : Las bases de una solución genética a los problemas de la caficultura latinoamericana. In *Desafíos de la caficultura centroamericana*, B. Bertrand & B. Rapidel (eds.), IICA, San José, pp. 369-406

Bertrand, B., Aguilar G., Bompard E., Rafinon A., Anthony, F. (1997), Comportement agronomique et résistance aux principaux déprédateurs des lignées de Sarchimor et Catimor au Costa Rica. *Plantation, Recherche, Développement*, 5:312-321

Bertrand, B., Aguilar G., Santacreo R., Anzueto, F. (1999), El mejoramiento genético en América Central. In *Desafíos de la caficultura centroamericana*, B. Bertrand & B. Rapidel (eds.), IICA, San José, pp. 407-456

Bettencourt, A. (1973), Cosideraçoes gerais sobre o "Hibrido de Timor". Instituto Agronomico de Campinas, Circular No. 31 (285p), Campinas, Brazil

Carvalho, A. (1988), Principles and practices of coffee plant breeding for productivity and quality factors: *Coffea arabica*. In: *Coffee vol. 4: Agronomy*, R. J. Clarke and R. Macrae (eds.). Elsevier Applied Science, London, pp 129-165

Charrier A., Eskes A.B. (1997), Les caféiers. In: *L'amélioration des plantes tropicales*, Charrier A, Jacquot M, Hamon S, Nicolas D (eds.). Collection Repères, CIRAD et ORSTOM, Paris, pp 171-196

Goncalves, M. M., Rodrigues, M. L. (1976), Estudos sobre o café de Timor. II. Nota sobre as posibilidades de producao do hibrido de Timor no seu habitat natural. Lisboa, Portugal. Missao de Estudos Agronomicos do Ultramar, Portugal. Comunicacoes, 86:31-72

Guerreiro Filho, O., Silvarolla M. B., Eskes, A. B. (1999), Expression and mode of inheritance of resistance in coffee to leaf miner *Perileucoptera coffeella*. *Euphytica, 105*: 7-15

Hillel, J., Schaap T., Haberfeld A., Jeffreys A.J., Plotzky Y., Cahaner A., Lavi, U. (1990), DNA fingerprints applied to gene introgression in breeding programs. Genetics, 124: 783-789

Hospital, F., Chevalet C., Mulsant, P. (1992), Using markers in gene introgression breeding programs. *Genetics. 132*: 1199-1210

Lande, R., Thompson, R. (1990), Efficiency of marker-assisted selection in the improvement of quantitative traits. *Genetics, 124*: 743-756

Lashermes, P., Cros J., Marmey P., Charrier, A. (1993), Use of random amplified DNA markers to analyze genetic variability and relationships of *Coffea* species. *Genet. Res. Crop. Evol., 40*: 91-99

Lashermes, P., Trouslot P., Anthony F., Combes M.C., Charrier, A. (1996a), Genetic diversity for RAPD markers between cultivated and wild accessions of *Coffea arabica*. *Euphytica*, 87: 59-64

Lashermes, P., Couturon E., Moreau N., Paillard M., Louarn, J. (1996b), Inheritance and genetic mapping of self-incompatibility in *Coffea canephora* Pierre. *Theor. Appl. Genet.*, 93: 458-462

Lashermes, P., Combes M.C., Robert J., Trouslot P., D'Hont A., Anthony F. and Charrier, A. (1999), Molecular characterisation and origin of the *Coffea arabica* L. genome. *Mol. Gen. Genet., 261*:259-266

Lashermes, P., Andrzejewski, A., Bertrand, B., Combes, M.C., Dussert, S., Graziosi, G., Trouslot, P. and Anthony, F. (2000), Molecular analysis of introgressive breeding in coffee (*Coffea arabica* L.). *Theor. Appl. Genet., 100*: 139-146

Melchinger, A.E. (1990), Use of molecular markers in breeding for oligogenic disease resistance. Plant *Breed.*, 104: 1-19

Paillard, M., Lashermes P., Pétiard, V. (1996), Construction of a molecular linkage map in coffee. Theor. *Appl. Genet., 96*: 41-47

Paterson, A.H., Tanksley S.D. and Sorrells, M. E. (1991), DNA markers in plant improvement. *Adv. Agronomy, 46*: 39-90

Rijo, L. (1974), Observações cariologicas no cafeeira "Hibrido de Timor". *Portugaliae Acta Biologica*, serie A, *vol. XIII*: 157-168

Stam, P., Zeven, A. C. (1981), The theoretical proportion of the donor genome in near-isogenic lines of self-fertilizers bred by back-crossing. *Euphytica, 30*: 227-238

Tanksley, S.D., McCouch, S. R. (1997), Seed banks and molecular maps: unlocking genetic potential from the wild. *Science, 277*: 1063-1066

Van der Vossen, H.A.M. (1985), Coffee selection and breeding. In: *Coffee: Botany, biochemistry and production of beans and beverage*, M.N. Clifford and K. C. Wilson (eds.), Croom Helm, London, Sydney, pp. 48-97

Vos, P., Hogers R., Bleeker M., Reijans M., Van der Lee T., Hornes M., Frijters A., Pot J., Peleman J., Kuiper M. and Zabeau, M. (1995), AFLP: a new technique for DNA fingerprinting. *Nucleic Acids Res., 23*: 4407-4414

Young, N.D., Tanksley, S.D. (1989), RFLP analysis of the size of chromosomal segments retained around the Tm-2 locus of tomato during back-cross breeding. *Theor. Appl. Genet.*, 77: 353-359

Zeven A.C., Knott D.R., Johnson R. (1983), Investigation of linkage drag in near isogenic lines of wheat by testing for seedling reaction to races of stem rust, leaf rust and yellow rust. *Euphytica*, 32: 319-327

Chapter 8

STUDY OF COMBINING ABILITY AND HETEROSIS IN COFFEE

J.R.M. Fontes[1], A.A. Cardoso[1], C.D. Cruz[2], A.A. Pereira[3], L. Zambolim[4], N.S. Sakiyama[1]

[1]Departamento de Fitotecnia, [2]Departamento de Biologia Geral, [4]Departamento de Fitopatologia, Universidade Federal de Viçosa (UFV), 36571-000 Viçosa-MG, Brazil; /UFV; [3]Empresa de Pesquisa Agropecuária de Minas Gerais (EPAMIG/ CTZM Viçosa, Brazil

Running title: Heterosis in coffee

1. Introduction

Combining ability analysis identifies the parents able to transfer their desirable traits to their descendants. It suggests the best hybrid combination and supplies data on the type of gene action, which controls the different agronomic traits. Combining ability in its most general form refers to the behaviour of lines, or cultivars when crossed in direct or reciprocal combinations to obtain the hybrids. The General Combining Ability (GCA) of a line refers to the behaviour of the line in a series of crosses based on the mean value of the resulting F_1 hybrids. The performance of a particular cross may deviate from the mean of the GCA, and this deviation is known as the Specific Combining Ability (SCA) (Allard, 1971).

The terms GCA and SCA were originally defined by Sprague and Tatum, quoted by Melo (1987), using a system of diallel crosses as an experimental method. The GCA is considered as the mean behaviour of a line in a series of crosses. The SCA is the positive, or negative deviation of a determined cross from the mean GCA of its parents. Considering the gene action involved could differentiate the type of GCA. Thus, GCA is associated with genes with mainly additive effects, but also with dominance and epistatic effects (additive x additive). The SCA, however, depends basically on genes with dominance and epistatic and several kinds of interactions.

Fazuoli and Carvalho (1987) studied the SCA of *C. canephora* hybrids. They found that the best combinations were Kouillou x Robusta, BP46 x Kouillou, Uganda x

T. Sera et al. (eds.), Coffee Biotechnology and Quality, 113–121.

SA158 and the progeny from free pollination of Uganda 1646-4. The authors discarded several hybrid combinations, which had no yield, or produced only a few seeds with poor germination. These effects were attributed to self-incompatibility, which is typical of the *C. canephora* species. Cilas *et al.* (1998) showed that lines with low yield might have excellent performance as parents. On the other hand, good varieties might be bad parents. Therefore, the knowledge of both, general and the SCA, is considered essential for the choice of parents to cross in a breeding program.

In heterosis studies, greater emphasis has been given to allogamous species, particularly for maize, where the heterosis is exploited commercially. Recently, however, attention has been given on autonomous species also due to two main reasons: a) the possibility of developing effective systems to commercially exploit the heterosis through F_1 hybrids in these species; and, b) because of the potential use of the yield data of F_1 hybrid and other materials in initial generations, allowing the breeder to give priority to the best crosses (Sarawat *et al.*, 1994).

Heterosis value is frequently expressed in percentages, thus eliminating the unit used to determine the trait. From a practical point of view, the hybrid vigour is interesting when the F_1 hybrid value is greater than the superior parent is (heterosis compared with the superior parent = High-parent heterosis - HPH). Frequently, a cross may show heterosis because it could be more productive than the mean of the parents (Mid-parent heterosis – MPH), but if it is not superior to one of the parents, it may not be of practical interest (Paterniani, 1975).

For a long time in various plant species it has been observed that crosses among genetically divergent parents have greater heterosis than crosses among parents from the same group (Miller and Marani, 1963). Therefore, not all hybrid combinations show heterosis. Greater genetic diversity among the parents increases their probability of contributing with different alleles, resulting in greater heterozygosis in the hybrid. This gives greater vigour, either due to the heterozygosis itself (heterotic genes), or due to the complementation of the genes for vigour (dominant or partially dominant genes for vigour). On the other hand, related parents will have many alleles in common, which reduce the possibilities of heterosis. Thus, lines or cultivars chosen for hybridization should be genetically divergent and have highly desirable traits (Paterniani, 1975; Galvêas, 1988). Bellachew *et al.* (1993) studied six wild *C. arabica* materials crossed in a partial standard diallel. The F_1 hybrids and their parents were assessed for seven vegetative traits to obtain heterosis and GCA estimates. The F_1 hybrids had positive heterosis compared with the best parents (HPH) for all the characteristics studied, with values varying from 3 to 18%. The heterosis for root volume in the P_1 x P_4 cross was 69%. The greatest HPH for all studied traits was observed in certain crosses where the parents had distinct origin and morphological

traits. Srinivasan and Vishveshwara (1978) studied heterosis for grain yield in thirteen crosses among seven *C. arabica* parents during six harvests. Five crosses showed positive heterosis compared with the parental mean (MPH) and seven crosses showed HPH in the mean of the six harvests. The heterosis values were up to 183.3% for the parent mean, and up to 100.41% for the best parent in the F_1 obtained from the cross between *Agaro* x 2045. The *Agaro* and Choeche parents, which were part of a cross second highest in heterosis, were both directly introduced from Ethiopia. The most heterotic F_1 hybrid also had excellent stability and may allow an increase in yield and yield stability in Arabic coffee. Bellachew (1997) studied 2,789 wild and exotic *C. arabica* introductions from Ethiopia and found 279 promising materials. The author reported 69% heterosis in F_1 hybrids for the grain traits and 60% for the yield.
The aim of the present study was to estimate the GCA and SCA among Catuaí lines and descendants of the Híbrido de Timor, and heterosis (HPH and MPH) in their F_1 hybrids.

2. Experimental details

A synthesis of new genetic combinations with resistance to coffee tree rust, using rust resistant coffee trees selected from the germplasm introduced at the Federal University of Viçosa (UFV) was started in 1974. These new combinations were the result of intercrossing of resistant coffee trees and their hybrids with coffee trees of the Catuaí, Mundo Novo and Bourbon commercial varieties to synthesize genotypes with increased resistance and transference of the resistance factors, respectively. The Catuaí Vermelho and Catuaí Amarelo varieties which took part in the crosses were part of the line experiment implanted in Viçosa in 1978, where 12 Catuaí progenies were assessed (six Catuaí Vermelho and six Catuaí Amarelo progenies). These combinations were assessed for rust through artificial inoculation and planted in Campos de Seleção de Híbridos (CSH) for evaluation of their behaviour under cropping conditions. Presently, F_1 plants are being studied in the CSH1 to CSH10, while more advanced generations are evaluated in other fields.
A study of the yield of the existing different hybrids was started, aiming to reorganize the CSH. A total of 74 F_1 hybrids from the crosses among 14 Catuaí lines (Vermelho and Amarelo) with 12 descendants from the Híbrido de Timor were studied. This material was the part of the different fields (CSHs) of the UFV/EPAMIG Breeding Program, at the São José do Triunfo Research Station, Viçosa district, Minas Gerais. The coffee trees were assessed by the mean cherry coffee accumulated yield in grams. Three Catuaí cultivar lines, UFV2145, UFV2114 and UFV2154, were used as controls. Heterosis compared with the mean of the parents (Mid-parent heterosis – MPH) and

heterosis compared with the superior parent (High-parent heterosis - HPH) were calculated as deviations of the F_1 generation from the parental mean (MPH; mp = 100) and from the best parent (HPH; hp = 100), to better assess the practical value of the heterotic behaviour.

$$MPH = \frac{F_1}{(P_1 + P_2) / 2} \cdot 100 - 100$$

MPH = Heterosis (%) compared with the mean of the parents
F_1 = Mean yield of the hybrid
P_1 e P_2 = Mean yield of the parents

$$HPH = \frac{F_1}{hp} \cdot 100 - 100$$

HPH = Heterosis (%) compared with the superior parent
F_1 = Mean yield of the hybrid
hp = Mean yield of the superior parent

The GCA yield values were obtained using the yield mean values of all crosses where the studied parent participated. A descriptive analysis of the data of the first four harvests was made. Some hybrids were also assessed in the sixth, eighth and even tenth years of harvesting.

3. Results and Discussion

The accumulated mean yield of the majority of hybrids was superior to the controls. The UFV427 hybrid showed the same accumulated yield as the best control (UFV2114) in the first four years (13,613g), but was 34.3% superior in the first six years (29,047g). H427 showed a 46.0% greater accumulated yield than the other best control (UFV2145) in the first eight years (39,280g). UFV341-11 was the most productive hybrid derived plant studied in the first four, six, and eight years of accumulated production. In the first four and six years, the yield with this plant was higher by 28.7% and 19.1%, respectively than the best control plant (UFV2114-14). It was also 43.2% superior to the best plant of the other best control (UFV2145-4) in the first eight years.

Several authors have observed the superiority of hybrids compared to the used control varieties, which are normally commercial varieties. Bertrand et al. (1997) studied F1 hybrids from various crosses among traditional varieties (Caturra, Catuaí), some derived from the Híbrido de Timor (Catimor, Sarchimor) and wild coffee trees from Ethiopia and the Sudan, and observed that the hybrids were more vigorous and productive than the best existing varieties.

Table 1 shows the heterosis (MPH and HPH) based on the accumulated yield of the F_1 hybrid plants, which had the best yield during the first ten years of data collection.

4 Years			8 Years			10 Years		
(UFV)	MPH	HPH	(UFV)	MPH	HPH	(UFV)	MPH	HPH
287-3	267	182	427-2	377	205	429-1	339	330
507-3	240	109	429-1	374	281	418-6	323	217
498-11	214	173	506-3	339	230	506-3	281	230
504-7	184	163	430-1	339	229	430-1	264	254
428-7	155	133	415-1	323	251	505-1	240	168
505-9	153	146	511-1	308	246	506-9	234	190
518-6	153	139	429-2	300	221	419-8	232	205
499-1	142	95	506-9	297	198	419-10	230	204
518-8	133	121	429-5	292	215	430-8	218	209
518-5	126	113	427-1	282	145	430-7	204	195
511-4	123	91	419-10	279	230	429-3	195	189
416-6	121	86	506-4	279	185	514-6	191	162
416-4	104	72	419-8	272	223	426-5	187	166
341-11	77	30	423-6	269	132	505-2	187	126
423-3	75	39	498-4	255	123	498-7	186	144
335-5	55	36	418-6	251	221	426-3	161	142
322-6	50	11	428-1	242	224	423-6	161	117
348-5	39	-7	322-6	241	187	506-10	160	125
341-6	38	1	423-10	239	114	419-5	154	134
342-2	16	9	416-6	202	140	423-10	152	110
342-8	12	4	416-4	197	136	498-6	150	113
347-4	4	-16	341-11	176	125	510-5	133	121
337-10	-	-	426-5	170	167	421-3	115	97
513-5	-	-	438-3	139	129	423-2	90	58
338-1	-	-	341-6	124	83	513-3	-	-

The values were grouped in two-year periods. The MPH in 60% of these best plants was greater than 100%. Van Der Vossen (1985) showed considerable evidence of hybrid vigour due to the effect of complementary epistatic genes for yield, especially in F_1 hybrids from crosses among varieties of diverse origin. The author presented accumulated data from the three first harvests of Arabic coffee varieties and F_1 hybrids in two planting densities. The F1 hybrids produced up to 120% more than the parental mean (Laurina x Hfbrido de Timor) for the 3,333 plants/ha density, and up to 109%

more (SL28 X Rume Sudan) for the 6,667 plants/ha density. The best yields, however, were obtained from the Pandang x SL34 cross (15.4% heterosis) and Pandang x Erecta cross (38% heterosis) for the lowest and highest planting densities, respectively. Pandang was the most productive of the varieties in the two densities studied.

The progenies were assessed in up to eight years of production, and the following hybrids were outstanding: a) UFV416-6, UFV416-4, UFV322-6, UFV341-11 and UFV341-6, for their highest MPH at the four, six and eight year assessments; b) UFV418-6 and UFV419-8, for their best MPH at the six, eight and ten year assessments; and, c) UFV287-3, which showed 267% MPH at the four year assessment (period when it was assessed). The UFV504-7, UFV505-9, UFV518-6, UFV518-5, UFV342-2 e UFV342-8 hybrids also stood out with considerable MPH, based on four and six years accumulated yield. Although the UFV427-2 hybrid did not appear among the best during the first four years, it had excellent MPH in the two following two-year period, especially at eight year evaluation with 377% heterosis (the greatest MPH). Best yielding progenies were also found among those with best MPH: for example, UFV341-11, UFV341-6, UFV504-7, UFV342-8, UFV518-6, UFV416-6 e UFV416-4. The means yields of the parents of these progenies are low, indicating that the high yield obtained by the progenies was due to the heterotic effect displayed by them and not because they had parents with high yield capacity.

When the F_1 hybrid yield was compared with the best parent (HPH), which was usually Catuaí, the difference was reduced. This showed that the yield of the Híbrido de Timor parent reduced the mean value. However, the superiority of several F_1 progenies, mainly in the first years of production, was confirmed regardless of the low yield of one of the parents.

The HPH obtained in this study confirmed the superiority of the progenies mentioned earlier, but possible alterations in their classification might occur if this classification has to be based on the best HPH. Thus, the HPH values should also be taken into consideration as in practice, since what is needed is a hybrid, which also outmatches the most productive parent (practical value). Martinez *et al.* (1988) observed that heterosis (in relation to the parental mean) varied from 105 to 169% with a mean of 123%. The mean heterosis value was 102% when compared to the superior parent.

Table 2 shows the GCA data based on mean accumulated yield. The UFV2144-35 cultivar/line stood out among the Catuaí parents in the first four and six years when it was assessed, and the UFV2145-113 cultivar/line stood out at six, eight and ten year assessment. Both parent cultivars were Catuaí Vermelho. For the Híbrido de Timor parents, the UFV378-33 stood out in the first four and six years of production when it was assessed, and the UFV445-46 stood out in the four, six and eight production years, but it was not assessed in the tenth.

Table 2: Ranking of the Catuaí and Híbrido de Timor parents, in decreasing order, based on their GCA, mean accumulated yield and grams of cherry coffee, evaluated during a ten year period, at the São José do Triunfo – UFV-EPAMIG Experimental Center.

CATUAÍ							
(UFV)	**4 Years**	**(UFV)**	**6 Years**	**(UFV)**	**8 Years**	**(UFV)**	**10 Years**
2144-35	13063	2144-35	22664	2145-113	33217	2145-113	43437
2147-295	11417	2145-113	22607	2148-57	31931	2145-79	40720
2143-235	11006	2143-235	20716	2145-79	31336	2143-235	40061
2144-36	10999	2147-295	20011	2143-235	29557	2143-236	36150
2145-79	10906	2148-57	19698	2144-141	29511	2148-57	33336
2145-113	10652	2145-79	19060	2143-236	27459	-	-
2148-57	10421	2144-141	18774	2145-307	26966	-	-
2144-32	10370	2143-236	18753	2147-295	26236	-	-
2144-141	10014	2154-344	18655	2144-32	23826	-	-
2154-344	9662	2144-32	18641	-	-	-	-
2144-71	9325	2145-307	18337	-	-	-	-
2145-307	8775	2144-36	16388	-	-	-	-
2143-236	8661	2144-71	14923	-	-	-	-
2246-139	8633	-	-	-	-	-	-
Mean	10,278.9		19,171.3		28,893.2		38,740.8
					General Mean		**20,655.46**

HÍBRIDO DE TIMOR							
(UFV)	**4 Years**	**(UFV)**	**6 Years**	**(UFV)**	**8 Years**	**(UFV)**	
378-33	13636	445-46	23088	445-46	31488	446-8	37107
445-46	12220	378-33	22846	430-19	31188	439-2	36308
832-1	11701	440-10	20623	446-8	30490	-	-
440-10	11310	450-61	20091	440-10	29047	-	-
446-8	11056	446-8	19888	832-2	28334	-	-
450-61	10748	439-2	19635	450-61	26409	-	-
439-2	10459	430-19	18919	439-2	24336	-	-
832-2	9917	832-2	18436	450-63	23716	-	-
430-19	9742	832-1	16491	449-62	21518	-	-
450-63	8911	450-63	15516	-	-	-	-
449-62	8458	449-62	14717	-	-	-	-
376-2	6779	376-2	13147	-	-	-	-
Mean	10,411.4		18,616.4		27,391.8		36,705.5
					General Mean		**19,093.57**

Regarding SCA, some of the possible hybrid combinations were not assessed in this study because they were not represented in the studied population. Thus, it could possible that the absent hybrid combinations might be superior to the progenies mentioned as superior in this study, because some of the Catuaí parents with good yield did not have progenies which represented them in crosses with the good yield Híbrido de Timor parents, and vice versa. The SCA study based on the mean yield of the F_1 progenies showed that the UFV2144-36 x UFV439-2 cross in the first four and six harvest years and the UFV 2148-57 x UFV 439-2 cross at eight years were the best. Crosses UFV2246-139 x UFV430-19 at the first four years, UFV2144-36 x UFV439-2, UFV2144-35 x UFV445-46 and UFV2144-141 x UFV439-2 at six years, UFV2143-235 x UFV445-46 at eight years, and UFV2145-79 x UFV446-8 at ten years were the best in the evaluation based on SCA obtained from accumulated yield of F_1 progenies. According to Cruz and Regazzi (1994), the best hybrid was that which has the greatest SCA estimate and at least one of the parents has a high GCA estimate. Thus, the hybrid from the UFV2144-35 x UFV445-46 cross, which was outstanding at six years accumulated production, should be considered as the best. It showed an excellent SCA estimate and both parents have high GCA estimate.

4. Conclusions

From the results obtained, it could be concluded that:

1. The majority of hybrids were superior to the Catuaí lines controls;
2. (2) The UFV341-11 was the most productive hybrid, and the UFV427-2 hybrid showed the greatest heterosis (377%);
3. (3) The UFV2144-35 and UFV2145-113 cultivars/lines among the Catuaí, and UFV378-33 and UFV445-46 among the Híbrido de Timor showed the best GCA;
4. (4) The UFV2144-36 x UFV439-2 cross in the first four and six harvest years, and the UFV2148-57 x UFV439-2 cross at eight years were the best in the evaluation based on SCA.

5. References

Allard R.W. (1971), Princípios do Melhoramento Genético das Plantas. Trad. Almiro Blumenschein, Ernesto Paterniani, José T. do Amaral Gurgel e Roland Vencovsky. Agência Norte-Americana para o Desenvolvimento Internacional-USAID e Edgard Blücher Ltda. São Paulo. p. 381

Bellachew B., Ameha M., Mekonnen D. (1993), Heterosis and combining ability in coffee (*Coffea arabica* L.). In: Colloque Scientifique International sur le Cafe, 15º, Montpellier, p.A9

Bellachew B. (1997), Arabica coffee breeding in Ethiopia: a review. In: Colloque Scientifique International sur le Café, 17º, Nairobi, 1997. Association Scientifique Internacionale du Café (ASIC). pp. 415-423

Bertrand B., Aguilar G., Santacreo R., Anthony F., Etienne H., Eskes A.B., Charrier A. (1997), Comportement d'hybrides F_1 de *Coffea arabica* pour la vigueur, la production et la fertilité en Amérique Centrale. In: Colloque Scientifique International sur le Café, 17º, Nairobi, 1997, pp.415-423.

Cilas C., Bouharmont P., Boccara M., Eskes A.B., Baradat Ph. (1998), Prediction of genetic value for coffee production in *Coffea arabica* from a half-diallel with lines and hybrids. *Euphytica*, 104(1): 49-59

Cruz C.D. Regazzi A.J. (1994), Modelos biométricos aplicados ao melhoramento genético. Viçosa, UFV.p 390

Fazuoli L.C. Carvalho A. (1987), Capacidade de combinação de híbridos de *Coffea canephora*. In: Congresso Brasileiro de Pesquisas Cafeeiras, 14, Campinas, 1987, IBC. Pp.93-95

Galvêas P.A.O. (1988), Características Agronômicas de Sete Cultivares de Pimentão (*Capsicum annum* L.) e Heterose dos Seus Híbridos F_1. Tese M. S., Viçosa, UFV. p. 83

Martínez O., Torregroza M., Roncallo E.J. (1988), Heterosis en cruzamientos varietales de maiz de amplia diversidad genética y geográfica. *Agronomía Colombiana*, 5(1-2): 48-52.

Melo P.C.T.de. (1987), Heterose e Capacidade Combinatória em um Cruzamento Dialélico Parcial Entre Seis Cultivares de Tomate (*Lycopersicon esculentum* Mill.). Tese D. S., Piracicaba, ESALQ. p. 108

Miller P.A., Marani A. (1963), Heterosis and combining ability in diallel crosses of upland cotton, *Gossypium hirsutum* L. *Crop Science*, 3(5): 441-444.

Paterniani E. (1975), Estudos recentes sobre heterose - Projeto Nacional de Feijão. Fundação Cargill. Boletim Técnico, p. 36

Sarawat P., Stoddard F.L., Marshal D.R., Ali S.M. (1994), Heterosis for yield and related characters in pea. *Euphytica*, 80: 39-48.

Srinivasan C.S., Vishveshwara S. (1978), Heterosis and stability for yield in Arabica coffee. *Indian Journal of Genetics & Plant Breeding*, 38(3): 416-420.

Van Der Vossen H.A.M. (1985), Coffee Selection and Breeding. In: M. N. Clifford & K. C. Wilson (Eds.) Coffee, Botany, Biochemistry and Production of Beans and Beverage. Croom Helm, London & Sydney. pp. 48-96.

Chapter 9

MICROSATELLITES IN *COFFEA ARABICA* L.

P. Rovelli[1], R. Mettulio[1], F. Anthony[2], F. Anzueto[3], P. Lashermes[4], G. Graziosi[1]

[1]*Università degli Studi di Trieste, Dipartimento di Biologia, Italy;* [2]*CATIE,Turrialba, Costa Rica; 3IICA/PROMECAFE, Costa Rica;* [4]*IRD, BP 5045; 34032 Montpellier cedex 1, France*

Running title: Microsatellites in coffee

1. Introduction

DNA polymorphism has became a widespread tool in biotechnology; in fact, they are frequently used for a number of technical approaches as, for example, in agronomic traits identification, variety characterisation, and marker-assisted breeding programmes. *Coffea arabica* is expected to show polymorphic DNA sequences as any other species and indeed some polymorphisms have been described. However, it has been reported that restriction fragment length polymorphism (RFLP) (Lashermes *et al.*, 1996a) and polymorphism based on polymerase chain reactions (PCR, RAPD) (Orozco-Castillo *et al.*, 1994; Lashermes *et al.*, 1996b) have a relatively low degree of polymorphism (Paillard *et al.*, 1993, 1996). To-date, a high degree of polymorphism has been found only through AFLP (Lashermes *et al.*, 2000). As this species of coffee is autogamous and has a restricted genetic base, its heterozygosity is expected to be relatively low and the probability of finding a polymorphism is correspondingly reduced.

Microsatellites, also known as simple sequence repeats (SSRs), are produced by tandem repetition of sequences from 1 to 6 bp long and constitute highly informative markers. Indeed, they are abundant and are distributed uniformly and randomly in the euchromatin of eukaryotic genomes (Tautz and Renz, 1984; Wang *et al.*, 1994). They are inherited in a codominant Mendelian manner, are somatically stable and highly polymorphic. The polymorphisms are the result of the variation in the number of the

T. Sera et al. (eds.), Coffee Biotechnology and Quality, 123–133.

repeated monomers. Primers can be designed on the single sequences flanking the microsatellite and then used to amplify, by PCR, the locus in various genotypes. Simple gel electrophoresis reveals polymorphic variations in the size of the amplification product.

Microsatellites are considered as the class of genetic elements with generally highly polymorphic sequences, which might provide a reasonable number of genetic markers. They have been found useful as genetic markers in a number of plants such as soy (Akkaya *et al.*, 1992; Morgante and Olivieri, 1993; Morgante *et al.*, 1994), *Arabidopsis* (Bell and Ecker, 1994), barley (Saghai-Maroof *et al.*, 1994), rice (Wu and Tanksley, 1993), Cucurbitaceae (Katzir *et al.*, 1996), corn (Ma *et al.*, 1996), and tomato (Broun and Tanksley, 1996).

We made an attempt to identify the polymorphic microsatellites in *C. arabica*, starting from two genomic libraries, one enriched in $(ATC)_n$ microsatellite and the other in $(TG)_{n,}$ and describe the results in this article.

2. Experimental

2.1. PLANTS AND DNA EXTRACTION

DNA from *C. arabica* var. Caturra was used to construct the genomic libraries. To identify polymorphisms, the following genotypes were analysed: *C. arabica* variety Caturra, *C. arabica* ET-30 (Ethiopia), 12 plants belonging to F_2 generation (*C. arabica* var. ET-30 x *C. arabica* var. Caturra, IRD, Montpellier) and a number of plants of the cultivars Mundo Novo and Bourbon. DNA was extracted from lyophilised leaves following the method described by Murray and Thompson (1980) and Orozco-Castillo *et al.* (1994), as modified by Vascotto et al. (1999).

2.2. CONSTRUCTION AND SCREENING OF THE GENOMIC LIBRARIES

Two genomic libraries were constructed, one enriched in the microsatellite $(ATC)_n$ and one enriched in $(TG)_n$ (Morgante *et al.*, 1998; Rafalski *et al.*, 1996). It involved enrichment prior to cloning and the creation of fragments of DNA with known sequence extremities obtained by binding adapters to them (Karagyozov *et al.*, 1993; Kandpal *et al.*, 1994). The genome of *C. arabica* var. Caturra was partially digested with the enzyme *Tsp509I* (New England Biolabs, USA). Enrichment of the genomic libraries was obtained by selecting fragments containing a microsatellite on magnetic beads covered by strepavidin (Boehringer), conjugated with the biotinilated oligonucleotides $(ATC)_{10}$, or $(TG)_{13}$ The fragments of 200-600 bp, thus selected were cloned at the site of *EcoRI* of the vector Lambda ZAP II (Stratagene, La Jolla, California).

In the case of the genomic library enriched in the microsatellite $(ATC)_n$, the phage plaques were transferred on a nylon membrane Biodyne Plus (Pall) and screened by a ^{33}P- $(ATC)_{10}$ probe in a solution of 5xSSC, 1x blocking reagent (Boheringer Mannheim), 0.1% laurilsarcosina, 0.02x SDS. Two washing (5 min each) with 0.5x SSC and 0.1% SDS at room temperature were carried out, followed by further two washing for 15 min each with the same solution at 45°C. In the case of the TG-enriched genomic library, the phage plaques were screened by a 5'-$(DIG)_3(TG)_{13}$ (Oswel) probe in the same solution as the other probe. Two washing of 5 min each with 2x SSC and 0.1% SDS at room temperature were carried out, followed by further two washing for 15 min each with 1xSSC and 0.1% SDS at 60°C.

2.3. DNA SEQUENCING AND DESIGN OF PRIMERS

The positive clones were sequenced using an automatic sequencer, ABI 373A (Perkin Elmer), following a full scan method. A Thermo Sequence dye terminator cycle sequencing pre-mix kit (Amersham Pharmacia Biotec) was used for the sequencing, following the manufacturer's instructions. Primers were designed on the single sequences flanking the microsatellites using either the Primer3 programme (Whitehead Institute for Biomedical Research of Cambridge, Massachusetts, USA) or *Primers!* for the world wide web (Williamstone Enterprises). Where ever possible, all primers were designed in such a way as to have a T_m of around 58°C, a dimension of 20-22 bp as to obtain an amplification product of 100-300 bp. The primers were designed with the same characteristics in order to amplify all the sequences containing microsatellites under the same amplification conditions. A constant KS tail (5'-TCGAGGTCGACGGTATC-3') was added to one of the two primers for each primer pair. The primers were synthesised by Genset.

2.4. PCR AMPLIFICATION OF THE MICROSATELLITES

Each microsatellite was amplified with a "touchdown" PCR to increase the reaction specificity (Don *et al.,* 1991; Hecker and Roux, 1996; Mellersh and Sampson, 1993). The amplification product was fluorescently labelled via a 3-primer system: primer, 1) was locus specific primer; 2) was the other locus specific primer with the KS constant

tail primer; 3) was the KS primer whose 5' end was conjugated to a fluorescent label (either 6-FAM, 6-carbossifluoresceina, or JOE, 2',7'-dimetossi-4',5'-dicloro-6-carbossifluoresceina). During the first cycles of amplification the pair of locus-specific primers amplified the microsatellite containing sequence and inserted the KS complementary tail, which was primed by the fluorescent KS primer during the successive amplification cycles.
The amplification reaction mix consisted of primer, 100 nM each; fluorescent KS primer, 22 nM; dNTP 200 _M each, Taq-polymerase (Genenco) 0.625 U, 50 ng of genomic DNA, buffer 1x (Tris-HCl 100 mM pH 9.0, KCl 500 mM, TritonX-100 1%), $MgCl_2$ 1.5 mM in a total volume of 25 $^{-1}$. Amplification conditions were: 6 cycles of denaturation 45 sec. at 94°C; elongation for 45 sec at 72°C; annealing for 45 sec, gradually reducing the temperature by 1°C every cycle from 60 to 55°C, followed by a further 34 cycles of denaturation at 94°C for 30 sec, annealing at 55°C for 30 sec elongation at 72°C for 30 sec. The reaction ended with an elongation at 72°C for 8 min. The PCR was carried out in a PTC-200 (MJ Research) thermocycler.

2.5. ANALYSIS TO IDENTIFY POLYMORPHISMS

The amplification products were analysed with an ABI 373A automatic sequencer using the GENESCAN 672 (Perkin Elmer) programme on sequencing gel containing 4.75% polyacrylamide (acrylamide/bisacrylammide 19:1), urea 8.3 M and TBE 1x. The run lasted 10 hours. The volume of the sample analysed in the gel varied from 1-15^{-1} depending on the efficiency of the amplification

3. Results

3.1. ANALYSIS OF THE TWO GENOMIC LIBRARIES

Approximately 3,400 clones of the ATC genomic library were screened, 189 clones were positive (c. 6%) and 111 were sequenced. As regards (TG)n microsatellites, approximately 4,400 clones were screened and 503 found positive, of which 236 were sequenced. A total of over 100 Kbp was sequenced. Approximately 70% of the clones contained a microsatellite sequence. The genome of *C. arabica* contained microsatellite sequences just as any other species.
A total of 69 $(ATC)_n$ microsatellites and 180 $(TG)_n$ microsatellites were found. The ATC microsatellites had an average length of seven repetitions, and a maximum length

of 14 repetitions (Table 1). With regard to the structural characteristics of these microsatellites, $(ATC)_n$ repetitions were rarely compound [in these cases the sequence repeated in tandem adjacent to the $(ATC)_n$ microsatellite was GAC, GCC, GAA and ATT], and generally there was a prevalence of sequences repeated in perfect tandem. The average length of the microsatellites tended to increase as the length of the repeated unit decreased. The average length of the $(TG)_n$ microsatellites was found to be 10 repetitions with a maximum of 27. The most frequent sequence repeated in tandem adjacent to the $(TG)_n$ microsatellite was TC.
Table 1 shows the main characteristics of the microsatellites sequenced. Only 20% of the clones were suitable for primer design: 22 and 24 pairs of primers for $(ATC)_n$ and $(TG)_n$ microsatellite loci respectively. The primers were all initially tested on genomic DNA of *C. arabica* var. Caturra to confirm that they were able to give an amplification product of the desired size.

Table 1. Main characteristics of the ATC and the TG microsatellites.

Library	Number of repeats				Simple		Compound	
	≤ 6	7-10	11-14	>14	Perfect	Imperfect	Perfect	Imperfect
ATC	27	37	5	-	33	25	9	2
TG	21	58	38	63	78	59	24	19

Table 2 reports the sequences of the microsatellites and the size of the amplification products obtained. The two genomic libraries differed greatly in amplification efficiency: only a third of the primers gave an amplification product in the ATC genomic library whereas the primers designed for the TG library gave positive results in approximately 85% cases.

3.2. IDENTIFICATION OF POLYMORPHISMS

Two criteria were applied to distinguish the real alleles from any aspecific amplification product:

a) consideration was given only to clearly defined peaks of high intensity which,
b) showed an effective segregation in F_2. Table 3 shows the sequences of the primers, the number of alleles and of genotypes recorded for each polymorphic locus. The number of alleles and genotypes should, however, be considered provisional. The analyses of other cultivars should lead to further alleles and genotypes being identified. Almost all the polymorphic loci analysed showed one or two alleles per plant (Fig.1); presumably they were homozygotes, or heterozygotes, respectively. Only the locus E12-3CTG showed a profile with 3 or 4 peaks, three always being

present while one was variable. It is possible that in this case the pair of primers amplified a number of independent loci. Overall, $(TG)_n$ microsatellite loci proved to be more polymorphic than $(ATC)_n$; 12 out of 20 $(TG)_n$ loci proved to be polymorphic whereas only one out of 12 $(ATC)_n$ loci analysed was polymorphic.

Table 2. Microsatellite repeat and expected size of the amplification product.

TG Library			ATC Library		
Locus	Repeat	Product (bp)	Locus	Repeat	Product (bp)
14-2CTG	$(CA)_7$	130	10-1CATC	$(ATC)_6$	187
17-2CTG	$(TC)_{14}(CA)_{11}(CA)_{16}$	217	2-1CATC	$(ATC)_8$	205
25-2CTG	$(CT)_{16}(TG)_{10}$	138	6-1CATC	$(GAT)_6$	195
28-2CTG	$(CA)_{15}$	155	A2-2CATC	$(GAT)_7$	152
30-2CTG	$(CA)_{11}$	232	B4-2CATC	$(ATC)_7$	199
32-2CTG	$(CA)_{12}$	128	B7-2CATC	$(ATC)_8$	246
38-2CTG	$(TG)_{11}(GA)_6$	100	B9-2CATC	$(GAT)_6$	290
4-1CTG	$(TG)_8$	117	C2-2CATC	$(ATC)_{14}$	234
7-1CTG	$(TG)_{20}$	191	C4-2CATC	$(ATC)_6$	304
E10-3CTG	$(CA)_7$	136	D10-2CATC	$(GAT)_8$	286
E11-3CTG	$(CA)_8$	175	D4-2CATC	$(ATC)_8$	192
E12-3CTG	$(CA + TA)_{38}$	150	D9-2CATC	$(GAT)_{10}$	307
E5-3CTG	$(CA)_{13}$	211	E1-2CATC	$(GAT)_6$	421
E6-3CTG	$(TG)_{16}$	341	F6-2CATC	$(ATC)_6$	126
E7-3CTG	$(CA)_{10}$	215	F7-2CATC	$(ATC)_4$	182
E8-3CTG	$(CA)_{14}$	198	F8-2CATC	$(GAT)_6$	233
F1-3CTG	$(CA)_{10}$	201	F9-2CATC	$(ATC)_{14}$	172
F9-3CTG	$(CT)_7(CA)_{11}$	137	H10-2CATC	$(ATC)_{14}$	221
G1-3CTG	$(TG)_{13}$	198	H12-2CATC	$(GAT)_6$	236
I2-3CTG	$(TG)_{17}$	173	H6-2CATC	$(ATC)_6$	220
I5-3CTG	$(TG)_{17}$	150	H9-2CATC	$(ATC)_9$	223
I6-3CTG	$(TG)_{15}$	139	I3-2CATC	$(GAT)_{11}$	217
I7-3CTG	$(TA)_5(TG)_{17}$	142			
I9-3CTG	$(TG)_{21}$	212			

Table 3. Primer sequences and alleles

Locus	Foreward primer (5'→3')	Reverse primer (5'→3')	N.of alleles	N. of genotypes
C2-2CATC	CTCTCCCTCAGTCAATTCCA	CTTGGTCTCCCTCCTTTTTC	3	3
4-1CTG	AAAAAGCTGGTCCATGTCAA	GGGGCGTTCAGTTATAAACA	2	3
14-2CTG	TTTTCTTGCTAATCTTTGAGGA	ACTCTAATGGGGTCATGTGG	3	3
17-2CTG	AGGCCTTCATCTCAAAAACC	AGCGTTACTTGAGGCAAAGA	3	2
32-2CTG	AAGGGGAGTGGATAAGAAGG	GGCTGGATTTGTGCTTTAAG	4	4
E6-3CTG	CTGGGTTGGTTCTGATTTTG	GGTTCCCAGAGATTCTCTCC	5	4
E7-3CTG	TGACATAGGGGGCTAAATTG	TTAATGGTGACGCTTTGATG	4	3
E8-3CTG	CACTGGCATTAGAAAGCACC	GGCAAAGTCAATGATGACTC	2	2
E10-3CTG	ATGCCAAGTCGGAAAAGAA	GGCAAGCTCTAGCCTTTGA	2	3
E11-3CTG	AGTGATCTTCGCAGCCATT	TCTTTTTGTGACTGGGCTTC	2	2
E12-3CTG	TGCTTAGGCACTTGATATAGGA	CACGTGCAAGTCACATACTTTA	4	2
G1-3CTG	TGTTGCTGAACTGTGTTGCT	TCCAGAGAAATGTCGGAAGT	2	2
I9-3CTG	TGGCCGTGATAATAAACAGC	ATGTGGCAATCTAAAGCCAA	3	3

4. Discussion

The results reported here were obtained from two DNA genomic libraries of *C. arabica* var. Caturra, enriched in the microsatellite $(ATC)_n$ or $(TG)_n$. They are the first genomic libraries of this type constructed for this species. From the results of the library screening, it is not possible to estimate the frequency of the $(ATC)_n$, or the $(TG)_n$ microsatellites in the *C. arabica* genome. Studies have been conducted on genomic libraries in which the inserts of DNA fragments were not random. The enrichment was certainly substantial but, in the absence of a non-enriched reference genomic library, a precise estimate was not possible.

Although the two libraries allowed the design of almost the same number of primers (22 and 24 pairs of primers), approximately only 50% of the primer pairs tested for amplification of an $(ATC)_n$ microsatellite gave an amplification product. Much better results were obtained with the $(TG)_n$ microsatellites in which 85% of the primers gave an amplification product. Trinucleotide repetitions seemed to be more difficult to analyse, similarly to what was found in rice (Panaud *et al.*, 1996). The cause has still to be ascertained; it might be partially due to the use of a single set of amplification conditions for all the microsatellites and it could be possible that by varying the reaction conditions, an amplification product might be obtained from at least some of these primers. We considered it more important to maintain the same amplification conditions rather than to optimise amplification reactions, which would the development of multiplex, i.e. simultaneous amplification and analysis of several loci.

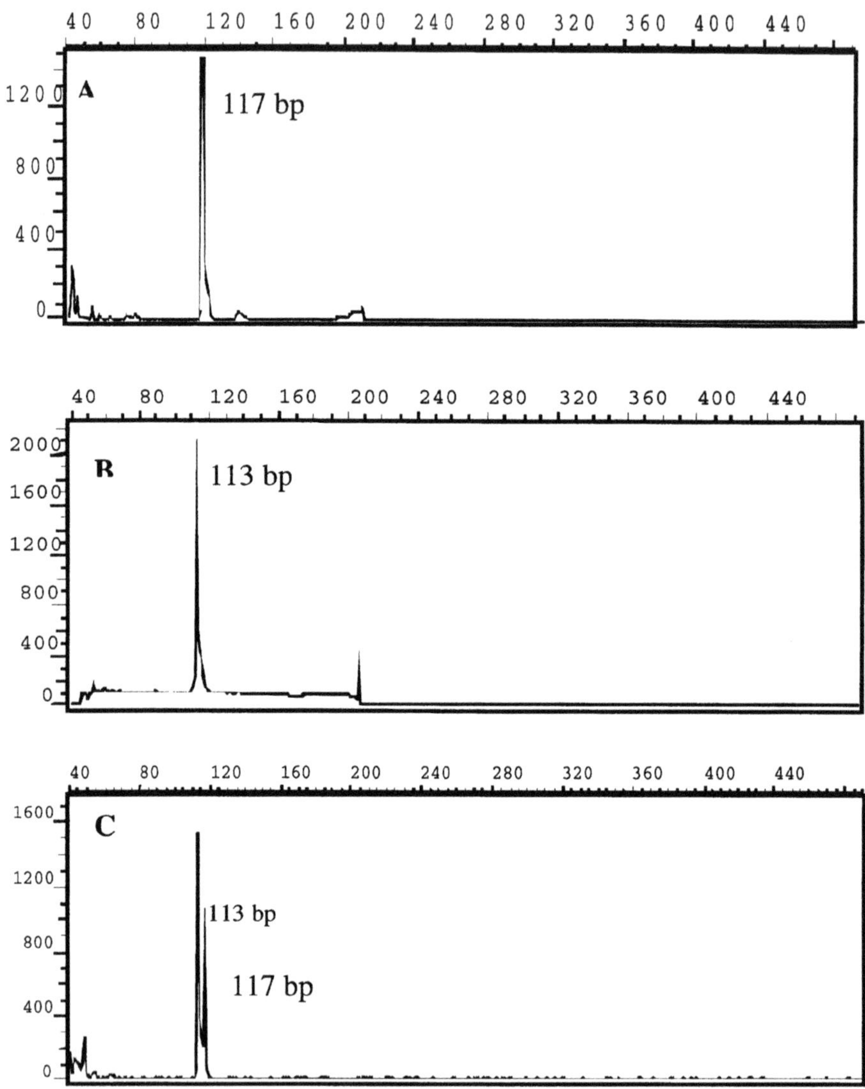

Figure 1. Electropherograms of three segregating plants for the 4-1CTG locus: Pane A: F_2-19; pane B: F_2-8; pane C: F_2-13. Both the parental Caturra and ET-30 showed the same alleles as for F_2-13. The numbers close to the peaks indicates the allele expressed in bp.

The amplification product of each locus gave sometimes several bands of variable intensity, most probably because of the Taq-polymerase slippage during the amplification reaction (Levinson and Gutman, 1987; Schlotterer and Tautz, 1992). However, on the basis of the intensity of the fluorescent signal and of its reproducibility, it was possible to identify the alleles. The segregation of the peaks in the F_2 plants greatly simplified the identification of real alleles.
As *C. arabica* is a tetraploid species, one particularly interesting question concerns the total number of alleles which may be recorded in one single plant, following the amplification of a single microsatellite. In theory, from one (in homozygotes) to two alleles (in heterozygotes) per plant might be expected, if our primers were locus specific for one of the two ancestral genomes. We might have been able to obtain even four bands per plant, if one pair of primers had simultaneously amplified both the pairs of ancestral chromosomes. In reality, the majority of the polymorphic systems analysed showed single bands in at least one plant out of the 12 F_2 plants. A typical example was the 4-1CTG locus in which both the parental lines had two alleles (113bp and 117bp) whereas one F_2 plant showed the 113bp allele only, two F_2 plants gave the 117bp allele only an and all the others gave both alleles. In this case, it is clear that the 4-1CTG locus shows a diploid-type segregation and a similar behaviour has been observed in a further 6 polymorphic loci. In the remaining polymorphic genetic systems it was not possible to distinguish with certainty between diploid and tetraploid behaviour.
It is also interesting to compare the two genomic libraries in relation to the number of bands simultaneously present in the single plants regardless of polymorphisms. The vast majority of ATC loci was monomorph and revealed a single band, whereas a substantial proportion of TG loci revealed two bands, which were present in all the plants analysed even for loci classified as monomorphs. Given the limited number of cultivars analysed we cannot say whether these latter cases were heterozygotes for 1 locus or homozygotes for 2 loci. This doubt should be resolved by extending the analysis to a reasonable number of different varieties.

5. Conclusion and perspectives

The genome of *C. arabica* proved to carry polymorphic microsatellite sequences, as expected. Furthermore some of the primer pairs here described appeared to amplify single loci of the homologous chromosome pair; *i.e.* they were able to discriminate between the two chromosome sets derived from the diploid donor ancestral plants. This result, when confirmed by a consistent behaviour of other loci, could offer an interesting key for studying the origin of the two sets of chromosomes, the actual contribution of

the two diploid progenitors as well as the possible functional evolution of the homologous loci within this organism.
The apparent diploid behaviour of some of the microsatellites introduces an optimistic note in the project of constructing a first genetic map of *C. arabica*. Indubitably, this task would be facilitated if a single locus could unambiguously be attributed to a single linkage group. The hypervariability and co-dominant Mendelian heredity of the microsatellites described here could offer a number of advantages in marker assisted breeding programmes.

6. Acknowledgements

This research project was supported by the European Community grant: INCO-DC Contract no ERBIC 18CT970181.

7. References

Akkaya M.S., Bhagwat A.A., Cregan P.B. (1992), Length polymorphisms of simple sequence repeat DNA in soybean. *Genetics*, 132: 1131-1139

Bell C. J., Ecker, J.R. (1994), Assignment of 30 microsatellite loci to the linkage map of *Arabidopsis*. *Genomics*, 19: 137-144

Broun, P., Tanksley, S.D. (1996), Characterisation and genetic mapping of simple repeat sequences in the tomato genome. *Mol. Gen. Genet.*, 250: 39-49

Don, R.H., Cox, P. T., Wainwright, B.J., Baker, K., Mattick, J.S. (1991), "Touchdown" PCR to circumvent spurious priming during gene amplification. *Nucleic Acids Res.*, 19: 4008

Hecker, K.H., Roux, K.H. (1996), High and low annealing temperatures increase both specificity and yield in touchdown and stepdown PCR. *Biotechniques*, 20: 478-485

Kandpal, R.P., Kandpal, G., Weissman, S.M. (1994), Construction of libraries enriched for sequence repeats and jumping clones, and hybridisation selection for region-specific markers. *Proc Natl Acad. Sci USA*, 91: 88-92

Karagyozov, L., Kalcheva, I.D. Chapman, M. (1993), Construction of random small-insert genomic libraries highly enriched for simple sequence repeats. *Nucleic Acids Res.*, 21: 3911-3912

Katzir, N., Danin-Poleg, Y., Tzuri, G., Karchi, Z., Lavi, U., Cregan, P.B. (1996), Length polymorphism and homologies of microsatellites in several Cucurbitaceae species. *Theor Appl Genet.*, 93: 1282-1290

Lashermes, P., Cros, J., Combes, M.C., Trouslot, P., Anthony, F., Hamon S., Charrier A. (1996a), Inheritance and restriction fragment length polymorphism of chloroplast DNA in the genus *Coffea* L. *Theor. Appl. Genet.*, 93: 626-632

Lashermes, P., Trouslot, P., Anthony, F., Combes, M.C., Charrier, A. (1996b), Genetic diversity for RAPD markers between cultivated and wild accessions of *Coffea arabica*. *Euphytica*, 87: 59-64

Lashermes, P., Andrzejewski, S., Bertrand, B., Combes, M.C., Dussert, S., Graziosi, G., Trouslot, P., Anthony, F. (2000), Molecular analysis of introgressive breeding in coffee (*Coffea arabica* L.). *Theor Appl Genet*, 100: 139-146

Levinson, G., Gutman, G.A. (1987), High frequencies of short frameshifts in poly-CA/TG tandem repeats borne by bacteriophage M13 in *Escherichia coli* K-12. *Nucleic Acids Res.*, 15: 5323-5338

Ma, Z.Q., Roder, M., Sorrells, M. E. (1996), Frequencies and sequence characteristics of di-, tri-, and tetra-nucleotide microsatellites in wheat. *Genome,* 39: 123-130

Mellersh C., Sampson, J. (1993), Simplifying detection of microsatellite length polymorphisms. *Biotechniques,* 15: 582-584

Morgante M., Olivieri A.M. (1993), PCR-amplified microsatellites as markers in plant genetics. *Plant J.*, 3: 175-182

Morgante, M., Rafalski, A., Biddle, P., Tingey, S., Olivieri, A.M. (1994), Genetic mapping and variability of seven soybean simple sequence repeat loci. *Genome*, 37: 763-769

Morgante, M., Pfeiffer, A., Jurman, I., Paglia, G., Olivieri, A.M. (1998), Microsatellite markers in plants. In- *Molecular tools for screening biodiversity. Plants and animals.* Karp, A., Isaac, P. G., Ingram, D.S. (eds.), Chapman and Hall, pp. 206-208 & 288-296

Murray, M.G., Thompson, W.F. (1980), Rapid isolation of high molecular weight plant DNA. *Nucleic Acids Res.*, 8: 4321-4325

Orozco-Castillo, C., Chalmers, K.J., Waugh, R., Powell, W. (1994), Detection of genetic diversity and selective gene introgression in coffee using RAPD markers. *Theor Appl Genet.*, 87: 934-940

Paillard, M., Duchateau, N., Petiard, V. (1993), Diversité génétique de quelques groupes de caféiers: utilisation des outils moléculaires, RFLP et RAPD. Paper presented at ASIC, 15e Colloque, Montpellier, pp.33-40

Paillard, M., Lashermes, P., Petiard, V. (1996), Construction of a molecular linkage map in coffee. *Theor Appl Genet.*, 93, 41-47

Panaud, O., Chen, X., McCouch, S. R. (1996), Development of microsatellite markers and characterisation of simple sequence length polymorphism (SSLP) in rice (Oryza sativa L.). *Mol. Gen. Genet.*, 252: 597-607

Rafalski, J.A., Vogel, J.M., Morgante, M., Powell, W., Andre, C., Tingey, S.V. (1996), Generating and using DNA markers in plants. In- *Nonmammalian genomic analysis: a practical guide,* Birren, B., Lai, E. (eds), Academic Press, London New York, pp.75-134.

Saghi Maroof, M. A., Biyashev, R.M., Yang, G. P., Zhang, Q., Allard, R.W. (1994), Extraordinarily polymorphic microsatellite DNA in barley: species diversity, chromosomal locations and population dynamics. *Proc Natl Acad. Sci USA*, 91: 5466-5470

Schlotterer, C., Tautz, D. (1992), Slippage synthesis of simple sequence DNA. *Nucleic Acids Res.,* 20: 211-215

Tautz, D., Renz, M. (1984), Simple sequences are ubiquitous repetitive components of eucaryotic genomes. *Nucleic Acids Res.*, 12: 4127-4138

Vascotto, F., Degli Ivanissevich, S., Rovelli, P., Anthony, F., Anzueto, F., Lashermes, P., Graziosi, G. (1999), Microsatellites in *Coffea arabica*: Construction and Selection of Two Genomic Libraries.Paper presented at III SIBAC, 24-28 May, Londrina, Brazil

Wang, Z., Weber, J.L., Zhong, G., Tanksley, S.D. (1994), Survey of plant short tandem DNA repeats. *Theor Appl Genet.,* 88: 1-6

Wu, K.S , Tanksley, S.D. (1993), Abundance, polymorphism and genetic mapping of microsatellites in rice. *Mol. Gen. Genet*, 241: 225-235

Chapter 10

STANDARDIZATION OF ENCAPSULATION TECHNIQUE FOR PRODUCING SYNTHETIC SEEDS IN COFFEE

B. Muniswamy, H.L. Sreenath

Biotechnology Centre, Unit of Central Coffee Research Institute, Coffee Board, Manasagangothri, Mysore - 570 006, Karnataka, India

Running title: Synthetic seeds

1. Introduction

Seeds are the primary means of propagation in many crops. The use of genetically uniform tissue culture derived plants could be considered advantageous for crops such as coffee, in which vegetative propagation is difficult to practice. Micropropagation technique is greatly constrained due to high cost involved and inadequate standardization of hardening procedure. During transfer of plants from laboratory to field, generally high mortality rates result that further lower the efficacy of results achieved *in-vitro* (Mathur *et al.*, 1989). For these reasons, efforts have been made to overcome the constraints by encapsulating somatic embryos (Kitto and Janick, 1985; Redenbaugh *et al.*, 1987) in different matrices and to grow them on different media. Studies on somatic embryogenesis and related encapsulation of somatic embryos have been reported by Redenbaugh *et al.* (1993). The first successful examples of synthetic seed technology were in alfalfa and celery. Fujii *et al.* (1989) grew plants at low frequency from synthetic seeds of alfalfa, planted directly in the field. Kirin Brewery Co Ltd., Japan and Plant Genetics Inc., USA conducted field trails in Japan with 20,000 F_1 synthetic seeds in 1988 (Sanada *et al.*, 1993). Recently, Rao *et al.* (1998) reviewed the concept, methods and

T. Sera et al. (eds.), Coffee Biotechnology and Quality, 135–141.

micropropagation as well as germplasm conservation through preservation of encapsulated zygotic and somatic embryos in liquid nitrogen. In coffee, somatic embryogenesis has been achieved from different tissues (Sondahl and Loh, 1988; Sreenath and Naidu, 1997; Muniswamy and Sreenath, 1997). However, reports on encapsulation and subsequent plant regeneration for producing synthetic seeds in coffee are very much limited. Here we report plant regeneration from encapsulated zygotic and somatic embryos of coffee.

2. Materials and Methods

The somatic embryos were regenerated directly from the cultured zygotic embryos of CxR cultivar on MS medium (Murashige and Skoog (1962), supplemented with 1-Naphthaleneacetic acid (NAA) and 6-Benzylaminopurine (BAP). The zygotic embryos of *C. arabica* S.4348 and S.2794 were isolated from mature green fruits. These embryos were cultured on MS medium with ABA (1 mg/l) for 30 days for embryo maturation and then used in encapsulation study.

2.1. PREPARATION OF ENCAPSULATION MATRIX AND BEADS:

Different concentrations of sodium alginate and calcium chloride were tested to standardize bead formation. Sodium alginate solution (2-6%) was prepared in MS medium supplemented with 3% sucrose, with or without growth regulators. pH of the solution was adjusted between 5.6 and 5.8 and sterilized at 121°C for 20 min.
The embryos were mixed with sodium alginate solution and dropped through a Pasteur pipette into calcium chloride solution (2%) held in a conical flask with continuous stirring on a magnetic stirrer. The resulting calcium alginate beads containing entrapped embryos were left in the calcium chloride solution for 30 minutes to complete complexion. The beads were then removed from the solution by using sterile stainless steel sieve. The beads ranging in diameter between 4-6 mm were placed on different media for germination of embryos.

2.2. CULTURE MEDIA:

Encapsulated embryos were cultured *in-vitro* on nutrient medium and *ex-vitro* in potted vermiculite for germination. For *in-vitro* germination, MS medium supplemented with 0.1, or 0.5 mg/l of Kinetin (Kn) or 0.1 mg/l of BAP with 3% sucrose was used. The pH

of the medium was adjusted between 5.6-5.8, solidified with 0.8% bacteriological agar and then
sterilized at 121°C for 20 min. The cultures were incubated at room temperature with 70-80% relative humidity under a photo period of 12 h (~ 2000 lux) using cool white fluorescent tubes. For *ex-vitro* germination of the encapsulated embryos, finely sieved sterile vermiculite filled in the plastic pots was used. They were watered with 1/8 strength MS salts solution on alternate days and sprayed with 0.1% Bavistin once in a week to avoid fungal attack.

2.3. HARDENING:

Plantlets with 2-5 pairs of leaves and a good root system were taken out from culture vessel without damaging the roots and planted in potted vermiculite for hardening. These were watered with 1/8 strength MS salts on alternate days. After growing in net pots for two months, hardened plantlets were planted in polybags, filled with soil mixture.

3. Results and Discussion

3.1. BEAD FORMATION:

Six percent sodium alginate and two percent calcium chloride solutions were optimal to get firm round beads (Fig. 1A) (data not shown). Complexion period of 30 minutes was good. The same conditions were used in further experiments.

3.2. GERMINATION AND HARDENING:

Somatic embryos obtained directly from the zygotic embryos of CxR cultivar at high frequency on 0.1 mg/l of NAA and 1 mg/l of BAP were used for encapsulation. The details of media and comparison of plant development from non-encapsulated and encapsulated embryos are given in the Table 1.
The non-encapsulated embryos cultured on half-strength MS medium with 0.1 mg/l of Kn and 0.1 mg/l of BAP started germinating and developed into plantlets with long hypocotyls, green cotyledonary leaves and roots. However, the embryos encapsulated in half strength MS salts solution, with or without a cytokinin started germinating with

elongation of hypocotyl and enlargement of the cotyledonary leaves (Fig. 1B) after 30-40 days when cultured on half-strength MS medium supplemented with Kn (0.1 or 0.5 mg/l). Later, these developed 11-mm long, thick green hypocotyl and 6-12 mm wide, dark green cotyledonary leaves. Most of these plantlets produced first pair of leaves and roots after three months' culture.

Table 1. Germination of encapsulated and non-encapsulated embryos of coffee (data taken after three months' culture).

Emb.	Medium	Number Embryos cultured	Number Embryos germinated	Hypocotyl leght (mm)	Breadth Cotyledonleaves (mm)	Number plantlets producing 1rst pair of leaves	Number plantlets producing roots
*	1/2 MS+Kn (0.1 mg/l)	30	24 (80)	6	3-4	1 (3.3)	11 (66.7)
**	1/2 MS+Kn (0.5 mg/l)	30	24 (80)	8	6-9	3 (10)	13 (43.3)
**	1/2 MS+Kn (0.5 mg/l)	25	20 (80)	8	7-12	5 (20)	10 (40)
**	Potted vermiculite	25	0	0	0	0	0
***	1/2 MS+Kn (0.1 mg/l)	30	21 (70)	9	6-9	3 (10)	9 (30)
***	_MS+BAP (0.5 mg/l)	20	(85)	9	6-7	5 (25)	2 (10)
***	1/2 MS+Kn (0.5 mg/l)	30	23 (76.6)	0.8	7-12	6 (20)	8 (26.7)
***	Potted vermiculite	25	0	0	0	0	0

Emb.= Embryos ; * Non encapsulated; ** Encapsulated in 1/2 MS basal medium;
*** Encapsulated in 1/2 MS+Kn (0.1 mg/l). Figures in parentheses indicate percentage.

After 2-3 subcultures at 30-45 days interval, on half-strength MS medium supplemented with Kn (0.1 or 0.5 mg/l) and BAP (0.1 mg/l), plantlets developed further and produced shoots with 2-5 pairs of leaves and good roots. The plantlets were hardened with a good survival rate.

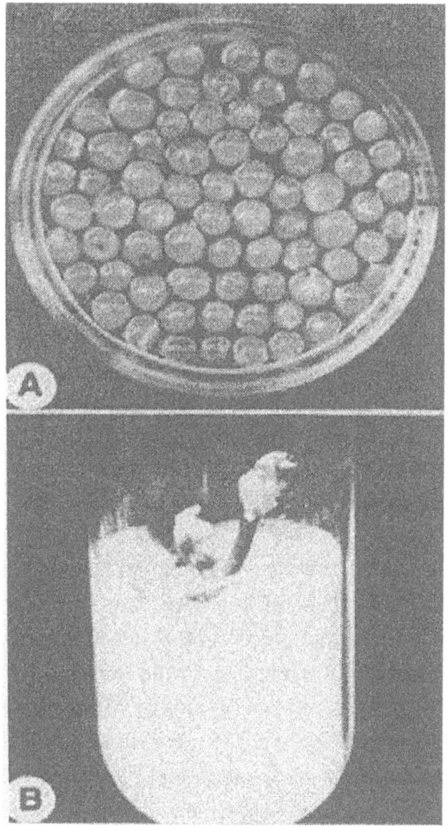

Figure 1. Encapsulation of coffee embryos   A: Encapsulated zygotic embryos;
B: Encapsulated zygotic embryos germinated on MS medium with BAP (0.1 mg/l)

There were clear differences between non-encapsulated and encapsulated embryos in respect of germination and plantlet development. Though, germination and plantlet development were achieved in both encapsulated and non-encapsulated embryos, the encapsulated embryos germinated faster and developed into better plantlets with dark green leaves. Encapsulated embryos responded better when they were cultured initially on half-strength MS medium with 0.5 mg/l Kn for one month and sub-cultured subsequently in presence of reduced concentration of Kn (0.1 mg/l). The time taken to get plantlets suitable for hardening was reduced by two to three months, when encapsulated embryos were cultured as against non-encapsulated embryos.

Encapsulated embryos that were sown directly in potted vermiculite did not germinate. They turned brown and died within 2 weeks. This showed that encapsulation technique was useful for producing synthetic seeds in coffee by using somatic embryos. This would help in scaling up micropropagation and for germplasm conservation through cryopreservation.

In trees species, development of synthetic seeds is valuable because of the need for vegetative propagation and lack of cost effective methods (Redenbaugh *et al.*, 1988). Kamada (1985) also suggested that application of synthetic seeds would depend on the viable, efficient *in-vitro* culture system. Synthetic seed is a new concept in seed biotechnology research, and is mainly considered for propagation and delivery of tissue cultured plants in a more economical and convenient way (Rao *et al.*, 1998).

4. Summary and Conclusions

Somatic embryos were obtained directly from cultured zygotic embryos of CxR cultivar on MS medium supplemented with NAA and BAP. Zygotic embryos of *C. arabica* were isolated from green fruits. They were encapsulated in sodium alginate (6%) in MS salts solution by dropping them into calcium chloride solution (2%). Resulted calcium alginate beads with embryos were cultured *in-vitro* on MS medium and *ex-vitro* in potted vermiculite. Non-encapsulated embryos were cultured on the medium of same composition for comparison. Fast germination and plantlet development was achieved *in-vitro* on half-strength MS medium, supplemented with 0.1 mg/l or 0.5 mg/l of Kn or 0.1 mg/l BAP. When the plantlets produced 3-4 pairs of leaves with a good root system, they were hardened in a polytunnel with a good survival rate. Encapsulated embryos that were sown directly *ex-vitro* in vermiculite could not be regenerated into plantlets. Encapsulation technique could be considered important for producing synthetic seeds in coffee.

5. References

Fujii, J., Slade, D., Redenbaugh, K. (1989), Maturation and green house planting of alfalfa synthetic seeds. *In-vitro* Cell. Develop. Biol., 25: 1179-1182

Kamada, H., (1985), Synthetic seeds. In - Practical Technology on the Mass Production of Clonal Plants, R. Tanaka (ed.), CMC Publishers, Tokyo, p. 48

Kitto, S. , Janick, J. (1985), Production of synthetic seeds by encapsulating asexual embryos of carrot. Am. Soc. Hort. Sci., 110: 277-282

Mathur, J., Ahuja, P.S., Nandalal and Mathur, A.K. (1989), Propagation of *Valeriana wallachi* DC using encapsulated apical and axial shoot buds. Plant Sci., 60: 111-116

Muniswamy, B., Sreenath, H.L. (1997), Genotype specificity of direct somatic embryogenesis from cultured zygotic embryos in coffee. In-Biotechnological Applications of Plant Tissue and Cell Culture, G.A. Ravishanker and L.V. Venkataraman (Eds.), Oxford & IBH Publishing, New Delhi, pp.204 - 209

Murashige, T., Skoog, F. (1962), A revised medium for rapid growth and bioassays with tobacco tissue cultures. Physiol. Plant. 15: 473-497

Rao, P.S., Suprasanna, P., Ganapathi, T.R., Bapat, V.A. (1998), Synthetic seeds: concept, methods and applications. In- Plant Tissue Culture and Molecular Biology, P.S. Srivastava (ed.), Narosa Publishing House, New Delhi, pp. 607-619

Redenbaugh, K., Viss, P., Slade, D., Fujii, J.A. (1987), Scale-up: artificial seeds. In- Plant Tissue and Cell Culture. Alan R, Liss Inc, New York, pp. 473-493

Redenbaugh, K., Fujii, J.A., Slade, D. (1988), Encapsulated plant embryos. In- Biotechnology in Agriculture, A. Mizrahi (ed.), Alan R. Liss, New York, pp 225-248

Redenbaugh, K., Fujii, J.A., Slade, D. (1993), Hydrated coating for synthetic seeds. In- Synseeds: Applications of Synthetic Seeds to Crop Improvement, K. Redenbaugh (ed.), CRC Press Inc., Boca Raton, USA, pp. 35-46

Sanada, M., Sakamoto, Y., Hayashi, M., Mashiko, T., Okamoto, A., Ohnishi, N. (1993), Celery and Lettuce. In- Synseeds: Applications of Synthetic Seeds to Crop Improvement, K. Redenbaugh (ed.), CRC Press, Boca Raton, USA, pp. 305-327

Sondahl, M. R. , Loh, W.H.T. (1988), Coffee biotechnology. In- Coffee Vol.4, Agronomy, R. J. Clarke and R. Macrae (eds.), Elsevier Applied Science, London, pp. 235-262

Sreenath, H. L., Naidu, R. (1997), Quarter century of research in coffee biotechnology- a review. In - Biotechnological Applications of Plant Tissue and Cell Culture, G.A. Ravishankar and L.V. Venkataraman (eds.), Oxford and IBH Publishing Co. Pvt. Ltd., New Delhi, pp. 14-27

Chapter 11

FIELD TESTING OF ARABICA BIOREACTOR-DERIVED PLANTS

M.R. Söndahl[1], C.N. Söndahl[1], W.Goncalves[2]

[1] *Fitolink Corp., 6 Edinburgh Lane, Mt. Laurel, NJ 08054 USA* [2]*Instituto Agronomico, Cx. Postal 28, CEC, Centro de Café, Campinas, SP 13100-000 Brazil*

Running title: Bioreactor-derived plants

1. Introduction

Seeds are the most common form of plant propagation and should be used when the species is autogamous (homozygous), or when reliable production of hybrid seeds can be made in large quantities. Vegetative propagation applies when the plant species does not produce seeds, or hybrid seeds cannot be commercially utilized. Rooting and grafting are the most common methods of vegetative propagation. Micropropagation is utilized when (a) there is difficulty to apply traditional propagation methods; (b) it is important to start from "disease-free" planting materials; or (c) there is a need to produce very large numbers of plants in a short period of time. Micropropagation can be achieved by different in vitro multiplication methods: (1) growth of pre-existing axillary buds; (2) production of shoots via organogenesis; or (3) plantlet production via somatic embryogenesis.

The great majority of commercial micropropagation protocols are based in axillary bud multiplication on solid medium (Kurtz *et al.*, 1991). Micropropagation via organogenesis is very limited in use (Litz and Gray, 1992), probably due to the lack of specific protocols, or because suspicion of high rates of somatic variation. Historically, micropropagation via somatic embryogenesis has been used only in the cases where an axillary bud multiplication is not applicable. The classical examples are the use of somatic embryos for oil palm and date palm propagation (Jones and Hughes 1989).

T. Sera et al. (eds.), Coffee Biotechnology and Quality, 143–150.

somatic embryos for oil palm and date palm propagation (Jones and Hughes 1989). However, somatic embryogenesis is getting a great deal of attention lately, due to its enormous multiplication rates and the relative genetic stability of the resulting plants. Indeed, micropropagation via somatic embryos could bring more efficiency to the industry since it requires reduced physical space and labour, which should translate in reduced unit costs of propagated units.

In vitro propagation is carried by biofactories (or micropropagation labs), and plants are produced in five distinctive phases: inoculation (phase 1), multiplication (phase 2), shoot and root growth (phase 3), hardening (phase 4) and final development (phase 5). Phases 1-3 are under sterile conditions and phases 4-5 take place in *ex vitro* conditions, under greenhouses and/or shaded houses. Plants at phase 5 are ready for field transplant and this is the most common form of selling propagating plants. However, in special cases, hardened plants still in speeding trays or cones (phase 4) are also commercialised. The sale of sterile phase 3 plants is more rare since it requires customers that have hardening facilities. It also has the advantage of facilitating exporting to other countries because of sterility properties and complete lack of soil or potting mixes.

There are more than 100 species in the Genus *Coffea*, but commercial coffee production only relies in two species: *C. arabica* and *C. canephora* (Robusta). These two coffee species share about 70 and 30% of the world coffee market, respectively. Arabica coffee is a self-pollinated species; hence 98% of homozygous individuals can be obtained after six consecutive generations, which requires a minimum of 24 years of selfing and selection. Robusta coffee is an auto-crossing species being propagated by heterozygous seeds. Both Arabica and Robusta would greatly benefit from the availability of mass propagation protocols to quickly put into production elite segregating plants.

Coffee plants could be vegetative propagated by cuttings, grafting and *in vitro* methods. Cuttings has not been found applicable for arabica, but could be used for most robusta genotypes. Grafting of arabica shoots on robusta rootstock is being used at the commercial levels in large number (about 1-2 million/year). *In vitro* methods include (a) axillary bud multiplication by nodal cultures, or (b) plantlets derived from somatic embryos. The propagation method with most potential for large-scale multiplication is via somatic embryo production.

Propagation of coffee via nodal cultures was first proposed by Custer *et al.*, (1980), followed by studies by Söndahl *et al.* (1984). This method yields very limited numbers of cloned individuals (low multiplication rate) after long periods of culture and consequently, very expensive costs per unit. It is suitable for research activities to amplify germplasm collection and to establish clonal gardens.

Coffee somatic embryogenesis was first described with robusta shoot cultures by Staritsky (1970). High-frequency somatic embryogenesis from mature leaf explants of arabica coffee was reported by Söndahl and Sharp (1977). An auxin-free protocol for arabusta leaves was proposed by Dublin (1981). Yasuda *et al.* (1985) reported somatic embryos from cultures of young leaves of arabica. The first report on mass production of robusta somatic embryogenesis, using Erlenmeyer and bioreactor vessels, was presented by Zamarripa *et al.* (1991). A protocol for mass production of arabica somatic embryos was provided by Noriega and Söndahl (1993). Another protocol for mass propagation of coffee somatic embryos was reported by Berthouly *et al.* (1995), using a unique apparatus in a temporary immersion culture.
We made field test evaluation of arabica coffee plants derived from somatic embryos cultivated in liquid bioreactor vessels. Plants derived from solid-agar cultures were also compared with seed-derived plants, which were used as controls. The aim was to evaluate the growth pattern and stability of somatic-embryo derived plants, growing under normal coffee cultivation conditions.

2. Experimental

2.1. PLANT MATERIALS

Fully developed leaves from adult plants of *C. arabica* var. Laurina, line #2165 were used for establishing the *in vitro* cultures. Seeds of the same line were used for raising the control seedlings.

2.2. TISSUE CULTURE

Leaf explants were set onto a primary medium (MS I) for six weeks and then sub-cultured to a secondary medium (MS II) according with the Söndahl's protocols (Söndahl and Sharp 1977; Sondahl *et al.*, 1984). Colonies of friable embryogenic tissues (FET) were isolated after 4-6 months of secondary culture. FET colonies were maintained on solid medium through periodic subcultures.
FET liquid cultures were established from fresh FET colonies using 250 ml Erlenmeyer flasks in rotary shaker at 100 rpm at 25°C, charged with fresh MSII liquid two times a week. Embryogenic suspension cultures were well established after three months and after that cultures were split every three weeks to maintain a packed cell volume (PCV)

of about 5-10 ml/100ml. These suspensions contained only FET cells in cluster aggregates of 0.5-1.0 mm in diameter (Noriega and Söndahl 1993).
Fresh FET suspension cultures were transferred to bioreactor vessels and a cell density of 1-5 PCV/litre was maintained by periodic removal of excess of tissue. Two bioreactor models were used: (a) 7-liter capacity Aplikon vessel equipped with stirring blades (50-80 rpm) and (b) 5-liter capacity Ono vessel equipped with magnetic stirring (70-120 rpm). Complete production protocol and time course on somatic embryo production has already being reported by Noriega and Söndahl (1993).
Somatic embryos derived from solid and liquid bioreactor cultures were plated for germination on MSII medium. Isolated plantlets (phase 3) were transferred to 35x145 mm tubettes for hardening in greenhouse, under 50% light and fog conditions. Coffee plantlets with 3-4 pair of leaves (phase 4) were shipped from New Jersey greenhouse to the Quarantine Service in Brazil (Cenargen). After released, plants were transferred to a local coffee farm nursery. Simultaneously, seedlings of Laurina line 2165 were placed in sand beds and transferred to tubettes for subsequent growth under 50%-30% shading nursery at the same coffee farm.

2.3. EXPERIMENTAL AREA

Coffee plants were identified according their origin as follows: B1 plants from bioreactor Aplikon model, B2 plants from bioreactor Ono model, TC plants from solid agar cultures, and SE plants from seeds. Planting was made at spacing of 2.0 x 1.0 m during the period of March 2-4, 1996. A total of 220 plants were used from liquid cultures, 230 plants from solid cultures and 500 plants from seeds. These plants were established into 35 plots with 10-45 plants/plot in a randomized planting.

3. Results

First field evaluations were made after the first crop in May 1998 and no visible difference could be detected among the treatments. Second evaluation was made at the time of second crop when plants had more than three years of age (May 1999). Several morphological parameters were measured and data are shown in Table 1. Using the values of Control plants as 100%, comparisons were made with plants of other treatments.

Table 1. Morphological evaluation of coffee plants derived from bioreactor cultures (B1, B2) and solid cultures (TC). Seed-derived plants (SE) are control plants. Data from 3-year old plants growing at field conditions.

Treatments	Plant height	Plant diameter	Thickness of main axis	Length of Lateral Branches					
	(m)	(m)	(cm)	LB 1	BL 2	LB 3	LB 4	LB 5	Average
Average B1	1.8	1.2	5.5	63	66	69	65	62	65
% control	93	107	106						100
Average B2	1.9	1.2	5.2	71	71	71	67	69	70
% control	97	108	101						107
Average TC	1.8	1.2	5.1	63	68	63	60	60	63
% control	94	106	98						96
Average SE control plants	2.0	1.1	5.2	64	68	66	64	64	65

For plant height, the *in vitro* plants had values slightly below than the control: B1 (93%), B2 (97%) and TC (94%). The plant diameter values of *in vitro* plants were a little above the control plants: B1 (107%), B2 (108%) and TC (106%). The values of main axis thickness ranged between 98 – 106% in comparison to control plants. Similar data were observed for length of lateral branches (96-107% of the controls).
Screening for alterations on plant phenotype (somaclonal variation) was also made during May/99 and the following data was recorded:

Bioreactor plants (B1 + B2) =	01 variegated type
Solid medium plants (TC) =	02 with larger leaf area
Seed-derived plants (SE) =	02 murta type + 01 angustifolia type

The frequency of these variants types was very low: bioreactor plants (01/220=0.4%), solid medium plants (02/230=0.9%) and seed-derived plants (03/500=0.6%).

4. DISCUSSION

Robusta embryogenic suspension cells (clone R2) were charged at a rate of 1.0 mg fw/litre to a 3-liter bioreactor, model Setric SGI, running at 60 rpm at 26°C (Zamarripa *et al.*, 1991). Under this system, embryogenic tissue proliferated up to a yield of 200,000 embryos/litre on the 49th day of culture, of which about 20% where torpedo embryos (40,000/liter). The conversion rates of robusta somatic embryos to plantlets ranged from 50-70%, which translated into a final yield of 60,000 plantlets per bioreactor/2-months. In greenhouse, the success of hardening was 80-95% after five months. This demonstrated that a 3-liter bioreactor culture of robusta embryogenic cells could provide ca. 48,000 cloned plants every 2-months, which was equivalent to 20 ha of coffee at 2,500 plants/ha.

In the arabica bioreactor cultures system (Noriega and Söndahl, 1993), the cell density was maintained at 1-5 PCV per litre by removing the excess of tissue at each medium exchange. So, the final yield obtained was lower than the robusta (about 9,000 embryos/litre, of which 25% were torpedo-, 45% heart- and 30% globular-shaped embryos). The conversion rate to plantlet was ca. 50% and the hardening survival about 85%. This work illustrated a yield of 19,125 cloned plants per 5-liter bioreactor culture. If five of such units were running at one time, it would be sufficient to produce sufficient cloned plants for 25 ha at a density of 3,800 plants/ha.

Five robusta elite plants were selected for mass propagation via somatic embryogenesis using the bioreactor culture system (Ducos *et al.*, 1999). Cloned plants have been planted in the following countries to test growth behaviour and yield (4,000/site): Philippines, Thailand, Mexico, Nigeria and Brazil. Based on visual inspections of 8,000 plants (over 2-years old) under field conditions, all robusta micropropagated plants had normal vegetative aspect and developed normal flowers and fruits (Ducos *et al.*, 1999).

A population of 20,000 cloned arabica F_1 hybrids have been produced for field-testing in four Central American countries (Etienne *et al.*, 1999). The objective has been to evaluate the performance of the embryo-derived plants under local farming conditions. Out of 4,000 *in vitro* plants under field and nursery conditions evaluated so far, no somaclonal variation has been observed.

Comparative data on bioreactor vs. solid medium x seed-derived plants growing under normal field conditions at second crop stage showed that after four years under field conditions, it was not possible to detect differences among *in vitro* plants and seed-derived plants either for vegetative parameters or for yield (data not shown). The demonstration plot showed that somatic embryo-derived plants could be used for mass propagation with this genotype. The frequency of *in vitro* variation was highly dependent on the genotype (Sondahl and Lauritis, 1992) and so, each elite plant or hybrid selected for mass propagation must be field tested before large plantation areas are established.

Based on average recovery rate of *ex vitro* plantlets equal to 37%, Ducos *et al.*, (1999) provided first comparative production costs as US$ 0.169/somatic embryo-plant versus US$ 0.158 for a robusta cutting. Based on a planting density in the Philippines of 1,600 plants/ha, this difference was only US$ 18.40/ha. *In vitro* production system has many aspects for cost-reduction improvement and the cutting system was already operating close to its maximum efficiency. Considering 50% overhead costs and a profit margin of 30%, the above production costs would be equivalent to a market value of US$ 0.30 for an embryo-derived plant and US$ 0.28 for one cutting.

The use of grafting propagation of arabica varieties on robusta rootstock is being used in areas with high nematode infestation. Very few selected nematode-resistant robusta clones are available for this propagation. In Brazil, such graft planting materials have an average market value of US$ 0.20/unit (W. Gonçalves, personal communication). In consultation with a bio-factory in Brazil (Bionova, personal communication), the price quoted was US$ 0.47 per cloned plant produced via axillary bud method for minimum orders of 5,000 units/genotype. For arabica embryo-derived plants, the quoted price was US$ 0.32/unit of phase 3 plantlets for minimum orders of 50,000/genotype.

5. Summary and futuristic approach

Micropropagation of coffee via mass production of somatic embryos seems to be the most efficient method for scaling-up elite plants into commercial plantations. Multiplication systems using liquid embryogenic cultures, either in Erlenmeyer or bioreactor vessels, are a technology that has competitive costs with other propagation methods. Somatic embryogenesis is envisaged to be the coffee cloning method of the future because it can provide sufficient number of plants for establishing commercial plantations and also it has room for cost reduction due to production scale and process improvement.

Micropropagation could provide the coffee farmers with the ability to quickly adopt superior elite plants being produced by coffee improvement programs. The ability for rapid response and flexibility of changing planting materials may offer immediate gains to farmers and at the same time, it would bring the opportunity for the coffee industry to more closely work with the green coffee production sectors. Elite cloned populations would provide not only agronomic benefits, but also could open the doors for processing and consumer benefits.

This new mass propagation cloning methods for coffee would shorten the release time of improved genotypes from breeding programs and also should give the opportunity of a more frequent turn-over of old to new planting materials as they become available. Since vegetative propagation would be based on heterozygous elite plants (natural or synthetic hybrids) and multiple clone lines would be always present at plantation sites, this technology may help the preservation of heterozygosity and plasticity to environmental changes.

At smaller scale, coffee micropropagation methods would assist improvement programmes for establishing Clonal Testing Fields in different growing environment and to assist germplasm preservation efforts *in vivo* and *in vitro* systems.

6. References

Berthouly M., Dufour M., Alvard, D., Carasco C., Alemanno L., Teisson C. (1995), Coffee micropropagation in a liquid medium using the temporary immersion technique. In: 16th Intl. Scientific Colloquium on Coffee, Kyoto. ASIC, Paris, pp.514-519

Custer J.B.M.., Van Ed G.,. Buijs L.C. (1980), Clonal propagation of *Coffea arabica* L. by nodal culture. IX Intl *in vitro*. Scientific Colloquium on Coffee, London. ASIC, Paris, pp. 589-596

Dublin P. (1981), Embryogénèse somatique directe sur fragments de feuilles de caféier Arabusta. Café Cacao Thé, Paris, 25: 237-241

Ducos J.P., Gianforcaro, M., Florin, B., Petiard, V., Deshayeves, A. (1999), Canephora (Robusta) clones: somatic embryogenesis. In: 18th Intl. Scientific Colloquium on Coffee, Helsinki. ASIC, Paris

Etienne H., Barry-Etienne D., Vasquez N., Berthouly M. (1999), Aportes de la biotecnología al mejoramiento genético del café: el ejemplo de la multiplicación por embriogenesis somatica de híbridos F1 en America Central In: B. Bertrand and B. Rapidel (eds), Desafíos de Caficultura en Centroamerica, IICA, San José, pp.457-493

Jones L.H., Hughes W.H. (1989), Oil Palm (*Elaeis guineensis* Jacq.). In: Biotechnology in Agriculture and Forestry, Vol.5, Trees II. Y.P.S. Bajaj (ed.), Springer-Verlag, Heidelberg, pp. 176-202

Kurtz S.L., Hartman R.D., Chu I.Y.E. (1991), Current methods of commercial micropropagation. In: Scale-up and Automation in Plant Propagation. I.K.Vasil (ed) Academic Press, New York, pp. 7-34

Litz R.E., Gray D.J. (1992), Organogenesis and Somatic Embryogenesis. In: Biotechnology of Perennial Fruit Crops, F.A.Hammerschlag ,R.E.Litz (eds), CAB Intl.,UK, pp. 3-34

Noriega C., Sondahl M.R. (1993), Arabica coffee micropropagation through somatic embryogenesis via bioreactors. In: 15th Intl. Scientific Colloquium on Coffee, Montpellier. ASIC, Paris, pp. 73-81

Söndahl M.R., Sharp W.R. (1977), High frequency induction of somatic embryos in cultured leaf explants of *Coffea arabica* L. Z. Pflanzenphysiol. 81: 395-408

Söndahl M.R., Nakamura T., Medina-Filho H.P., Carvalho A., Fazuoli L.C., Costa W.M. (1984), Coffee. In: P.V. Ammirato *et. Al.*, (eds), Handbook of Plant Cell Culture, Crop Species, Vol.3. MacMillan, New York, pp.564-590

Söndahl M.R., Lauritis J.A. (1992), Coffee. In: F.A. Biotechnology of Perennial Fruit Crops, F.A. Hammerschlag ,R.E.Litz (eds), CAB International, UK, pp. 401-420

Staritsky G. (1970), Embryoid formation in callus cultures of coffee. Acta Bortanica Neerlandica 19: 509-514

Yasuda T., Fujii Y., Yamaguchi T. (1985), Embryogenic callus induction from *Coffea arabica* leaf explants by benzyladenine. Plant ,Cell Physiol. 26: 595-597

Zamarripa A., Ducos J.P., Tessereau H., Bollon H., Eskes A.B., Petiard V. (1991), Développement d'un procédé de multiplication en masse du caféier par embryogénèse somatique en milieu liquide. In: 14th Intl. Scientific Colloquium on Coffee, San Francisco. ASIC, Paris, pp. 392-402

Chapter 12

INTER-SPECIFIC PROTOPLAST FUSION IN *COFFEA*

A.T. Cordeiro[1], L. Zambolim[2], V. Pétiard[3], J. Spiral[3]

[1]DBV and [2]DFP, Universidade Federal de Viçosa, Av. P.H. Rolfs, s/n-36.571-000, Viçosa-MG, Brazil; [3]Centre de Recherche Nestlé, 101, Av. Gustave Eiffel, 37.390-Notre Dame d'Oé, France

Running title: Protoplast fusion in coffee

1. Introduction

Coffee cultivation showing high yields, low production costs and environmental preservation requires employment of genetic materials with high fruit production capacity, uniform fruit maturation and pest and disease resistance. Lack of these traits in the *Coffea arabica* species has directed research aiming to their transference from other donor C*offea* species to arabica materials. However, the genetic and chromosome instability of the inter-specific crosses and the relatively long selection cycle of the coffee breeding programs have hindered the results. The use of protoplast fusion technique on the other hand may overcome these problems. The use of the technique in two different protoplast populations has produced non-fused parents, homo-fusion and hetero-fusion, either symmetric (Gaikwad *et al.*, 1996), or asymmetric (Fahleson *et al.*, 1997). These hetero-fusions which generally represent less than 10% of the whole fused mixture (Waara and Glimelius, 1995), should be screened from the initial population of approximately 5 x 10^5 protoplasts. This can be carried out by fusing metabolically-inactivated parent protoplasts, each one with an inhibitor with different action (Böttcher et al., 1989), allowing thus an screening by metabolic completion which freed only the heterofused protoplasts from developmental impairment.

T. Sera et al. (eds.), Coffee Biotechnology and Quality, 151–156.

2. Experimental

Protoplasts were obtained through enzymatic digestion of cell aggregates of embryonic liquid cultures of canephora and arabica genotypes. The canephora genotype was the Apoatã IAC LC 2258 cultivar, having resistance to *Meloidogyne exigua* (Fazuoli, 1986) and the races 2 (Mazzafera *et al.*, 1989) and 3 (Lima *et al.*, 1989) of *Meloidogyne incognita.* The arabica genotypes assayed were the pathenogenetic diploid DH_3 and the tetraploid Catimor and Catuaí Amarelo.
An enzymatic solution, as described by Spiral and Pétiard (1991), was used to obtain the protoplast (SEOP), except when the protoplast yield (Rp) was examined in response to digestion time and to cellulase and pectinase enzyme concentration. The osmotic pressure of the solutions as well as of all those used with protoplast was fixed at –1.537 MPa (Söndahl *et al.*, 1980), obtained with the variation of mannitol concentration only. The metabolic inhibitors, rhodamine and iodoacetamide, were added before the final 30 minutes of digestion. After purification, the protoplast, whether exposed, or not to electrofusion in a fusion chamber made of a multielectrode system (Sihachakr *et al.*, 1988) were cultivated in the culture medium with agarose (Cordeiro, 1999). The treatments were evaluated through partial, or total counting of the protoplast or the micro-calli regenerated from the plated protoplast.

3. Results and Discussion

The growth kinetics of liquid cultures of the diploid arabica DH_3 and canephora Apoatã genotypes were studied in order to identify the post-subculture period when most cell aggregates was in the division process. Both, the Apoatã and the DH_3 were shown to be in an exponential phase of growth up to an inflection point (t_i), immediately following subculture. From this point on the growth rates on a fresh matter basis, were progressively decreased, until an asymptotic phase of growth was attained (Fig 1A). The initial exponential stage led to the suggestion that immediately following subculture, only a small fraction of cells took part in the cell suspension growth, a trend also observed by Vasil and Vasil (1986). Beyond the inflection point some other factor, probably a component from the medium, or released by cultured cells, started limiting cell growth. Coincidentally, lower protoplast yields (R_p's) in the cell suspension of the both genotypes were also observed for post-subculture times beyond t_i (Fig.1B). Yields about 10^5 (Acuña and Pena, 1991; Söndahl *et al.*, 1980) and 10^7 (Tahara *et al.*, 1994) have been reported in the literature.

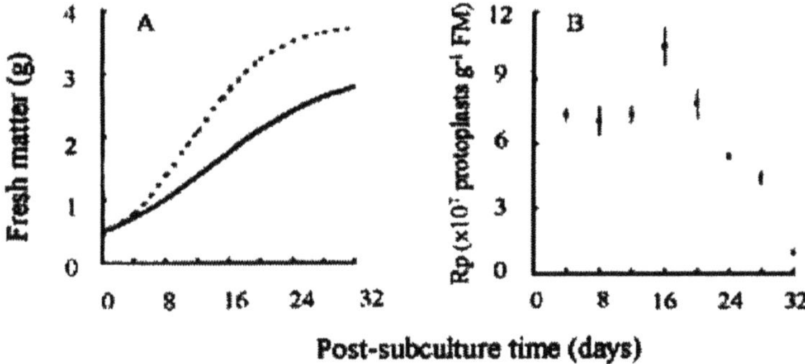

Figure 1-Fresh matter accumulation by cell aggregates (A) of DH_3 (-----) and Apoatã (—) and Rp (B) of Apoatã cell aggregates in relation post-subculture time. Aggregates were cultured in liquid medium at the initial density 10 gL^{-1}. Equations for fresh matter (FM) f (time) for DH_3 and Apoatã are respectively: $FM=3806.58/(1+7.9789e^{-0.1903t})$ $R^2=0.994$ and $FM=3098.83/(1+5.3780e^{-0.1227t})$ $R^2=0.995$. Bars represent the standard error of three replications.

Ninety-six inter-specific electrofusion assays were carried out enrolling the parental genotypes Apoatã with the diploid arabica parental. DH_3 and the tetraploid Catuaí Amarelo and Catimor (Table 1).

Table 1. Numerical attributes displayed by the assays involving different parental combinations in Coffea in relation to their successes, failures and their causes such as contamination, inadequate rhodamine concentration (IRC), and poor quality protoplasts (PQP).

Parental genotypes	Success	Failure		
		Contamination	IRC	PQP
Apoatã + diploid DH_3	6	14	31	1
Apoatã + Catuaí Amarelo	6	2	5	12
Apoatã + Catimor	11	2	6	-
Total	23	18	42	13

Among them, 42 (44%) were unsuccessful because of the inadequate rhodamine concentration employed in the control of the parental protoplast development, especially DH_3 and Apoatã. Twenty-three (24%) were successful, which could be attributed to a differential sensitivity of the arabica towards the inhibiting action of rhodamine. Thus, the higher sensitivity of the tetraploid arabica coffee to that inhibitor rendered the larger success percentage among the three parental combinations assayed. The adequate concentrations of the inhibitors in the control of protoplast development, mainly regarding rhodamine, could be known only at the end of the experimental period, when time was no longer enough to carry out extra electrofusions. Just this knowledge alone allowed an increase in the number of successful trials, improving the efficiency of the control of parental protoplast and protoplast raised from homofusion from 23 (24%) to 65 (~70%).

Figure 2 shows the results from four, out of the 23, successful experiments on the developmental control of parental protoplast and those raised from homo-fusion. Their efficiency and inferiority in the number of micro-calli regenerants from parental protoplast treated with metabolic inhibitors as compared to microcalli regenerants from the same electrofused suggested that the observed difference in the microcalli was due to the inter-specific protoplast fusion, which developed as a consequence of the metabolic completion. The alive regenerants calli are transferred in culture aiming at their differentiation. The embryos produced are sequentially transferred to the M4 and M5 media employed by Zamarripa *et al.* (1991) in order to regenerate plants. These plants are likely to allow for the assessment of efficiencies of the screening method used as well as the potential of protoplast fusion for the production of somatic hybrids in *Coffea*.

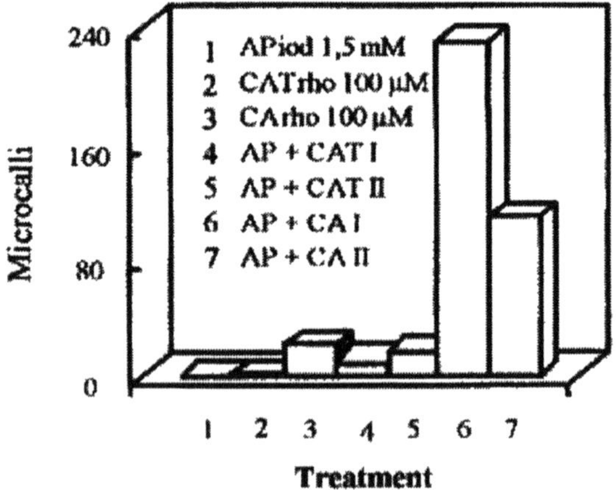

Figure 2 Microcalli arisen from parental protoplasts of *C. canephora* CV. Apoatã (AP) and *C. arabica* CVO's Catimor (CAT) and Catuaí Amarelo (CA) treated with the metabolic inhibitors rhodamine (ho) and iodoacetamide (iod), and which afterwards suffered (+) or not electrofusion. Microcalli were counted with a light microscope after two and a half months in culture.

4. Conclusions

Introgression of nematode resistance from *C. canephora* cv. Apoatã was attempted in diploid (DH_3) and tetraploid *C. arabica* (Catimor and Catuaí Amarelo) genotypes by protoplast fusion to evaluate the feasibility of inter-specific protoplast fusion as a tool in the parasexual improvement in *Coffea*. Apoatã and DH_3 cell suspensions showed a typical sigmoidal growth curve. Coincidentally, the lowest protoplast yields were observed for post-subculture intervals above the inflection point. High protoplast yields were achieved when two pectinases were combined to cellulase, at the highest concentrations tested. The twelve protoplast culture media tested did not allow any distinction between parental protoplasts Apoatã and DH_3, at microcalli growth stages. Alternatively, the metabolic inhibitor iodoacetamide and rhodamine, at the concentrations 2 and 0.9 mM, respectively, were effective in inhibiting microcalli formation from cultured parental protoplasts. The inhibitors allowed the accomplishment of 23 independent inter-specific electrofusion assessments, whose products led to the development of embryogenic differentiation.

5. References

Acuña R. & De Pena M. (1991), Plant regeneration from protoplasts of embryogenic cell suspensions of *Coffea arabıca* L. cv. Caturra. Plant Cell Rep., 10:345-348

Böttcher U., Aviv D., Galun E. (1989), Complementation between protoplasts treated with either of two metabolic inhibitors results in somatic-hybrid plants. Plant Sci., 63:67-77

Cordeiro A.T. (1999), Embriogênese somática indireta e fusão interespecífica de protoplastos em *Coffea*. Tese (Doutorado em Fisiologia Vegetal) – Universidade Federal de Viçosa, Viçosa-MG, Brazil

Fahleson J., Langercrantz V., Mouras A., Glimelius K. (1997), Characterization of somatic hybrids between *Brassica napus* and *Eruca sativa* using species-specific repetitive sequences and agenomic in situ hybridization. Plant Sci., 123:133-142

Fasuoli L.C. (1986), Genética e melhoramento do cafeeiro. In: Cultura do Cafeeiro – fatores que afetam a produtividade. A. B. Rena *et al.* (eds), Piracicaba, SP. Associação Brasileira Pesq. Potassa Fosfato. pp. 87-113

Gaikwad K., Kirti P.B., Sharma A., Prakash S., Chopra V.L. (1996), Cytogenetical and molecular investigations on somatic hybrids of *Sınapsıs alba* and *Brassıca juncea* and their backcross progeny. Plant Breeding, 115:480-483

Lima M.M.A., Gonçalves W. Fazuoli L.C., Tristão R.O. (1989), Estudo comparativo do ciclo de *Meloidogynbe incognita*, raça 3, em Mundo Novo e Apoatã. In: Congresso Brasileiro de Pesquisas Cafeeiras, 15, Maringá, 1989. Resumos. Rio de Janeiro, COTEC/DIPRO/IBC. pp. 128-129

Mazzafera P., Gonçalves W., Fernandes J.A R. (1989), Fenóis, peroxidase e polifenoloxidase na resistência do cafeeiro à *Meloidogyne incognita*. In: Congresso Brasileiro de Pesquisas Cafeeiras, 15, Maringá, 1989. Resumos. Rio de Janeiro, COTEC/DIPRO/IBC. pp. 4-6

Sihachakr D., Haicour R., Serraf I., Barrientos E., Herbreteau C., Ducreux G., Rossignol L., Souvannavong V. (1988), Electrofusion for the production of somatic hybrid plant of *Solanum melongena* L. and *Solanum khassianum* C.B. Clark. Plant Sci., 57:215-223

Söndahl M.R., Chapman M.S., Sharp W.R. (1980), Protoplast liberation, cell wall reconstitution, and callus proliferation in *Coffea arabica* L. callus tissues. Turrialba, 30:161-165

Spiral J., Pétiard V. (1991), Protoplast culture and regeneration in *Coffea* species. In: Colloque Scientifique International sur le Café, 14, San Francisco, 1991. ASIC (Paris). pp.383-390

Tahara M.,Yasuda T., Uchida N., Yamaguchi T. (1994), Formation of somatic embryos from protoplasts of *Coffea arabica* L. Hortscience, 29:172-174

Vasil V., Vasil I.K. (1986), Plant regeneration from friable embryogenic callus and cell suspension cultures of *Zea mays* L. J. Plant Physiol., 124:399-408

Waara S. & Glimelius K. (1995), The potential of somatic hybridization in crop breeding. Euphytica, 85:217-233

Zamarripa A., Ducos J.P., Bollon H., Dufour M., Pétiard V. (1991), Production d'embryons somatiques de caféier en milieu liquide: effets de la densité d'inoculation et du renouvellement du milieu. Café Cacao Thé, 25:233-244

Chapter 13

BIOCHEMICAL AND MOLECULAR STUDIES OF THE MAIN PROTEIN IN THE COFFEE ENDOSPERM

R. Acuña[a], R. Bassüner[c], V. Beilinson[c], H. Cortina[a], G. Cadena-Gómez[a], V. Montes[b], N. Nielsen[c]

[a]Centro Nacional de Investigaciones del Café, Chinchiná, Caldas, Colombia; [b]Departamento de Química, Universidad Nacional, Bogotá, Colombia; [c]United States Department of Agriculture, Agricultural Research Service and the Departments of Agronomy and Biochemistry, Purdue University, West Lafayette, IN 47907, USA

Running title: Coffee seed proteins

1. Introduction

Developing plant seeds accumulate and store carbohydrate, lipid, and protein that subsequently are used during germination as the source of energy, carbon, and nitrogen. A large number of storage protein genes have been cloned and characterized (Nielsen et. al., 1977). The expression patterns of seed proteins during seed development are highly regulated both developmentally (Golberg *et al.*, 1989; Perez-Grau and Golberg, 1989; Guerche *et. al.*, 1990) and metabolically (Wobus *et. al.*, 1995).

Coffee bean, the raw material from which all coffee products originate, is derived from the seeds of *Coffea* sp. Despite the immense economic importance of coffee bean, little is known about the biochemical and molecular properties of its seed proteins. Values for the protein content of coffee beans have been reported, and these are in the range of 8.7-12.2% (Macrae, 1985). Work has focused on the characterization of water-soluble proteins in different coffee varieties (Centi-Grossi *et. al.*, 1969; Amorim *et. al.*, 1975) or coffee beans processed by different methods

T. Sera et al. (eds.), Coffee Biotechnology and Quality, 157–169.

(Underwood and Deatherage, 1952). These reports described molecular mass profiles and protein contents. Protein isolates obtained from mature green Colombian *C. arabica* beans showed a reactive constituent part of coffee under conditions similar to the roasting process (Ludwig, 1995). More recently, a biochemical and molecular characterization and expression of the 11S-type storage protein from *C. arabica* endosperm was also reported (Rogers et al., 1999; Marraccini, 1999).
Attempts were made to identify legumin-like seed storage proteins of the 11S-size class (Nielsen *et. al.*, 1997) in coffee (Acuña *et. al.*, 1999). Two cDNAs that encode the proteins, a promoter for an 11S gene, and the sub-cellular localization of the protein in the seed endosperm are documented here.

2. Experimental

Coffee endosperms were harvested 150 and 220 days after flowering, and stored at -80°C until use. For immunocytochemistry, endosperms were harvested about 220 days after flowering and fixed with 3% glutaraldehyde in phosphate buffer. Extraction and fractionation of total seed protein was carried out using either immature or mature coffee seeds that had been frozen and stored at -80°C.

2.1. SEQUENTIAL FRACTIONATION OF COFFEE SEED PROTEINS:

Seed proteins were extracted sequentially from mature coffee endosperm in albumin extraction buffer, pH 4.5, (Sluyterman and Wijdens, 1970) and phosphate buffer (Globulin extraction buffer) and the remaining residue was extracted with 70% absolute ethanol and 10 mM DTT (Prolamine Fraction). Total coffee seed proteins were fractionated in a sucrose gradient (6 to 22%) and samples from each fraction, were analyzed by electrophoresis.

2.2. N-TERMINAL SEQUENCE ANALYSIS

Endosperm protein samples were electroblotted and the legumin-like coffee protein was cut from the stained blot membrane and processed in an Applied Biosystems Model A491 protein sequencer.

2.3. EXTRACTION AND PURIFICATION OF MRNA FROM IMMATURE COFFEE ENDOSPERM

Total RNA was isolated from frozen immature coffee endosperm using the method for recalcitrant plant tissues described by Schultz *et al.* (1994). To prove translatability of the isolated mRNA, *in vitro* mRNA cell-free translation was carried out in a reticulocyte lysate.

2.4. CDNA CLONING

Synthesis of full-length cDNA was made using the TimeSaver™ cDNA Synthesis Kit (Pharmacia). An *Eco*R I/*Not* I adapter was ligated to the ends of the resulting fragments. A sense oligonucleotide primer was synthesized on the basis of the N-terminal amino acid sequence determined for the acidic chain of the coffee legumin. A second antisense-degenerate oligonucleotide was synthesized based on N-terminal amino acid sequence determined for the basic chain. These primers were used to amplify nearly the entire acidic legumin chain. The PCR fragment was cloned into a pCRII vector and both strands were sequenced using an ALF automated DNA sequencer (Pharmacia). Based on the nucleotide sequence that was obtained, two additional oligonucleotide primers were synthesized and used in combination with a short specific primers designed on the basis of the EcoR I/Not I adapter sequence. These were used to amplify the basic chain cDNA and the N-terminal signal peptide, respectively. The overlapping sequences that were obtained were aligned to yield a full-length cDNA sequence.

2.5 DNA BLOT HYBRIDIZATION

Coffee genomic DNA was digested with various restriction enzymes and the restriction fragments were separated in agarose gels. A vacuum blotting system was used to transfer the DNA fragments from the gel to a membrane using an alkaline vacuum transfer protocol. An EcoR1 DNA fragment (~900 bp) that corresponded to the acidic chain of coffee legumin was purified by agarose gel electrophoresis and labelled and used for hybridization.

2.6. ELECTRON MICROSCOPY IMMUNOCYTOCHEMISTRY

Plant material was cut into small blocks of about 1-mm length and immediately fixed in 0.1M phosphate buffer, pH 7.3, containing 3% glutaraldehyde. Embedding was performed in LR White and in Epon resins. The sections were prehybridized in blocking solution, followed by hybridization with purified antibodies against coffee 11S proteins. Hybridization took place for several hours on a rocking platform. After washing in TBST, the grids were incubated with anti-rabbit IgG-gold conjugate diluted 15-fold into a blocking solution. This was followed by washings in TBST and water. The grids were incubated for two minutes in 1-% glutaraldehyde, followed by rinsing with water. To enhance contrast, the sections were treated in a LKB 2168 Ultra Stainer with a saturated solution of uranyl acetate, pH 4.0, for 20 min at 26°C, followed by incubation for two minutes at 20°C with 83mM lead acetate, pH 12. Finally, the thin sections were carbon coated in a vacuum evaporator (Ladd Research Industries). Inspections under the electron microscope (JEM 100 CX, Joel, Japan) were carried out at 80kV.

3. Results and Discussion

A legumin-like seed protein was purified from the endosperm of coffee (*C. arabica* L. cv. Colombia). In contrast to legumes, where efficient storage globulin extraction required buffered saline solutions well above the acidic pK_I of the globulins, coffee legumin was readily extracted with acidic aqueous buffers. Sub-units of coffee legumin had a M_r of about 55 kDa after one-dimensional SDS-PAGE electrophoresis in the absence of a reducing agent. In the presence of 2-mercaptoethanol, two polypeptides appeared that had apparent molecular masses of 33 and 24 kDa (Fig. 1).
This behaviour was typically found with legumin-like proteins. The mature sub-unit of 11S legumins contained acidic and basic chains that were linked via disulfide bonds, and these became separated from one another upon disulfide reduction. For comparison, an equivalent experiment was carried out with partially purified 11S legunmin from *Vicia faba* (lanes 3 and 8). The sedimentation of coffee endosperm proteins in sucrose density gradients was tested. The coffee legumin migrated like other 11S storage globulins in sucrose gradients (Fig. 2).
Messenger RNA was isolated from developing coffee seeds and used to prepare cDNA, which in turn served as a template to amplify legumin DNA by PCR. To deduce the sequence of primers to be used for PCR, the acidic and basic chains of the putative 11S proteins from coffee were purified from gels after SDS-PAGE and their N-terminal sequences were determined. These were used to deduce nucleotide sequences that could be used for PCR to generate clones that encode putative 11S coffee legumin-like molecules.

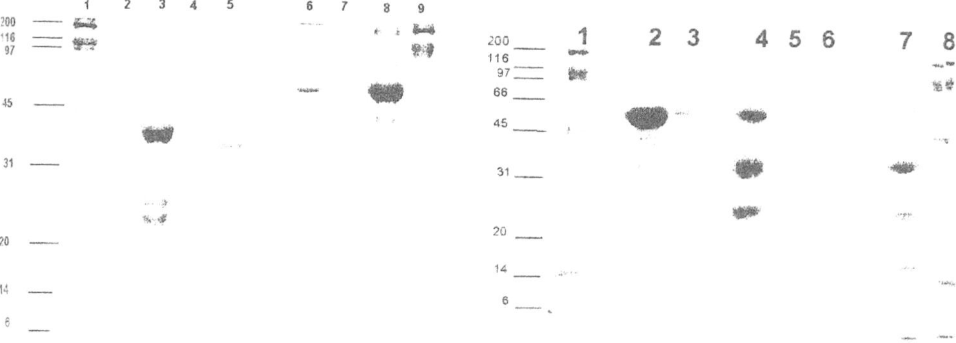

Fig. 1. A) Proteins extracted from mature and immature coffee endosperm: One-dimensional SDS-PAGE separation of seed proteins extracted in SDS buffer in presence (lanes 2-5) or absence (lanes 6-8) of 2-ME. Numbers in the right margin refer to molecular mass in kDa of protein standards. Lanes 1 and 9 size markers, lane 2 soybean conglycinin, lanes 3 and 8 total seed protein from faba bean, lanes 4 and 7 immature coffee seed proeins extracted, lanes 5 and 6 total seed proteins from coffee. B) Differential extraction of coffee seed proteins: One-dimensional electrophoretic separation of coffee seed proteins extracted sequentially with albumin, globulin, prolamin, and SDS buffers. Numbers in the right margin refer to the position and molecular mass in kDa of protein standards Lanes 1 ans 8 size markers, lanes 2 and 4 albumin fraction, lanes 3 and 5 globulin fraction, lane 6 prolamin fraction, lane 7 SDS fraction, lane 9 total coffee seed proteins extracted directly with SDS sample buffer. The samples in lanes 2 ans 3 contained no 2-ME, while those in lanes 4-9 contained 2-ME.

Two full-length cDNAs were generated from mRNA of developing seeds that were more than 98% homologous (Fig. 3). They had open reading frames of 1458 and 1467 bp. Each encoded legumin precursors consisted of 486 and 489 amino acids, respectively ((M_r= 54136 and 54818). The putative signal peptide at the N-terminal of the precursor was either 26 or 27 amino acids in length. Inspection of the deduced sequence revealed the presence of the N-G cleavage site that is conserve among prolegumins from diverse plant species (Nielsen *et. al.* 1997). Indeed, a proteolytic activity was isolated from immature coffee seeds that cleaved a recombinant prolegumin from *Vicia faba* (Acuña et. al. unpublished results).

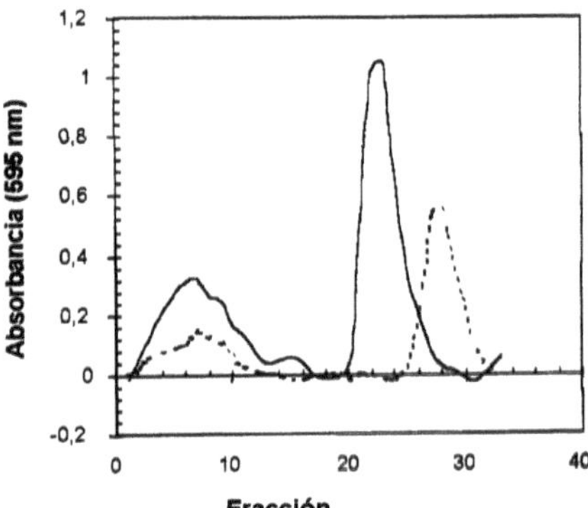

Fig 2. Sedimentation properties of proteins extracted from coffee endosperm.Proteins from the albumin fraction of coffee were separated in gradients of 6-22% sucrose at low (no added salt, dashed line) or high ionic strength (0.4 M KCl, solid line) buffer. Two prominent peak regions are observed with coffee samples around fractios 9 and 27 for sambples from gradients that contained no added salt. The latter peak shifted to a lower apparent density (fraction 23) in sucrose gradients that contained 0.4 M NaCl. Fraction 1 is from the top of the gradient and raction 33 is from the bottom.

The sequence signature assigned to the legumin family of globulins that recognized conserved regions within these molecules (ProSite database, Bairoch and Buchner 1994) were found in the coffee storage protein. The deduced amino acid sequence displayed highest similarity with rice (*Oriza sativa*) glutelins, followed by legumin B sub-units from faba bean (*Vicia faba*). The amino acid composition showed a low content of sulphur-containing amino acids and elevated levels of acidic and amidated amino acids.

```
aacacactacactctcctctgttgtcagagaatggctcactctcatatgatttctctttcct gtacgttctt   ttgttcctcggctg 90
aacacactacactctcctctgttgtcagagaatggctcactctcatatgatttctctttcct gtacgttctt---ttgttcctcggctg 87
                                 M  A  H  S  H  M  I  S  L  S S/L Y  V  L L/- L  F  L  G  C 20

tttggctcaactagggagacca agccaaggctc gggtaaaactcagtgc atattcagaagcttaatgcacaagaaccatccttcag 180
tttggctcaactagggagacca agccaaggctc gggtaaaactcagtgc atattcagaagcttaatgcacaagaaccatccttcag 177
 L  A  Q  L  G  R  P E/Q P  R  L G/R G  K  T  Q  C N/D I  Q  K  L  N  A  O  E  P  S  F  P 53

gttcccatcagaggctggtttaactgaattctgggattctaataatccagaatttgggtgcgctggtgtggaatttgagcgtaacactgt 270
gttcccatcagaggctggtttaactgaattctgggattctaataatccagaatttgggtgcgctggtgtggaatttgagcgtaacactgt 267
 F  P  S  E  A  G  L  T  E  F  W  D  S  N  N  P  E  F  G  C  A  G  V  E  F  E  R  N  T  V 83

ccaacctaagggccttcgtttgcctcattactctaacgtgcc aaattcgtctacgttgtcgaaggtaccggtgttcaaggcactgtgat 360
ccaacctaagggccttcgtttgcctcattactctaacgtgcc aaattcgtctacgttgtcgaaggtaccggtgttcaaggcactgtgat 357
 Q  P  K  G  L  R  L  P  H  Y  S  N  V  P  K  F  V  Y  V  V  E  G  T  G  V  Q  G  T  V  I 113

ccctggttgtgctgaaacatttgaatcccaggg gaatcatttt gggtggtcaggaacagccgggcaaagggcaagaagg--------- 441
ccctggttgtgctgaaacatttgaatcccaggg gaatcatttt gggtggtcaggaacagccgggcaaagggcaagaagg          447
 P  G  C  A  E  T  F  E  S  Q  G  E  S  F S/W G  G  Q  E  Q  P  G  K  G  Q  E  G -/Q-/E-/ 143

---ttccaaaggtggtcaggaagggc aaggcaaaggtttccagaccgccatcagaagctcagaaggttccaaaaaggagatgtccttat 528
   ttccaaaggtggtcaggaagggc aaggcaaaggtttccagaccgccatcagaagctcagaaggttccaaaaaggagatgtccttat 537
Q-/G S  K  G  G  Q  E  G Q/R R  Q  R  F  P  D  R  H  Q  K  L  R  R  F  Q  K  G  D  V  L  I 175

attgcttcctggtttcactcagtggacatataatgatggagatgttcc cttgtcactgtcgcacttcttgatgttgccaatgaggctaa 618
attgcttcctggtttcactcagtggacatataatgatggagatgttcc cttgtcactgtcgcacttcttgatgttgccaatgaggctaa 627
 L  L  P  G  F  T  Q  W  T  Y  N  D  G  D  V  P  L  V  T  V  A  L  L  D  V  A  N  E  A  N 205

tcagcttgatttgcagtccaggaaatttttcctagccgg aacccgcaacagggtggtggaaaggaaggccatcaaggccagcagcagca 708
tcagcttgatttgcagtccaggaaatttttcctagccgg aacccgcaacagggtggtggaaaggaaggccatcaaggccagcagcagca 717
 Q  L  D  L  Q  S  P  K  F  F  L  A  G  N  P  Q  Q  G  G  G  K  E  G  H  Q  G  Q  Q  Q  Q 235

gcatagaaacatcttctcaggatttgatgaccaacttttggc ga gctttcaatgttgacctcaaaataatacagaaattgaagggtcc 798
gcatagaaacatcttctcaggatttgatgaccaacttttggc ga gctttcaatgttgacctcaaaataatacagaaattgaagggtcc 807
 H  P  N  I  F  S  G  F  D  D  Q  L  L  A E/D A  F  N  V  D  L  K  I  I  Q  K  L  K  G  P 266

gaaagat aaaggggtagcacagtccgagctgaaaaacttcaactgttcctgcctgaatatagtgagcaagagcaacaaccccaacaaca 888
gaaagat aaaggggtagcacagtccgagctgaaaaacttcaactgttcctgcctgaatatagtgagcaagagcaacaaccccaacaaca 897
 K  D K/Q R  G  S  T  V  R  A  E  K  L  Q  L  F  L  P  E  Y  S  E  Q  E  Q  Q  P  Q  Q  Q 297

gcagg gcagcaacaaca ggtgttggaagaggatggagatccaa ggacttgaggaaactttgtgcacggtgaagcttagtgaaaacat 978
gcagg gcagcaacaaca ggtgttggaagaggatggagatccaa ggacttgaggaaactttgtgcacggtgaagcttagtgaaaacat 987
 Q G/E Q  Q  Q Q/H G  V  G  R  G  W  R  S  N  G  L  E  E  T  L  C  T  V  K  L  S  E  N  I 329

tggcctcccccaagaggctgatgtattcaatcctcgtgctggccgcattaccactgttaatagccaaaagattcctatcctcagcagcct 1068
tggcctcccccaagaggctgatgtattcaatcctcgtgctggccgcattaccactgttaatagccaaaagattcctatcctcagcagcct 1077
 G  L  P  Q  E  A  D  V  F  N  P  R  A  G  R  I  T  T  V  N  S  Q  K  I  P  I  L  S  S  L 359

ccaacttagtgcagaaagaggattcctctacagcaatgccattttgcaccacactggaatatcaatgcacatagtgccctgtatgtgat 1158
ccaacttagtgcagaaagaggattcctctacagcaatgccattttgcaccacactggaatatcaatgcacatagtgccctgtatgtgat 1167
 Q  L  S  A  E  P  G  F  L  Y  S  N  A  I  F  A  P  H  W  N  I  N  A  H  S  A  L  Y  V  I 389

tagaggaaatgcaagaattcaggtggtggatcacaaaggaaacaaagtttttgacgatgaagtaaaacagggtcagctaataattgtgcc 1248
tagaggaaatgcaagaattcaggtggtggatcacaaaggaaacaaagtttttgacgatgaagtaaaacagggtcagctaataattgtgcc 1257
 R  G  N  A  R  I  Q  V  V  D  H  K  G  N  K  V  F  D  D  E  V  K  Q  G  Q  L  I  I  V  P 419

acaatactttgctgtgatcaagaaagctggaaac aaggatttgagtacgttgcattcaagacgaacgacaatgccatgattaacccact 1338
acaatactttgctgtgatcaagaaagctggaaac aaggatttgagtacgttgcattcaagacgaacgacaatgccatgattaacccact 1347
 Q  Y  F  A  V  I  K  K  A  G  N E/Q G  F  E  Y  V  A  F  K  T  N  D  N  A  M  I  N  P  L 450

 gttggaagactttcggcatt cgagcaattcctgaggaagttttgaggagctctttccaaatttccagcgaggaagctgaggaattgaa 1428
 gttggaagactttcggcatt cgagcaattcctgaggaagttttgaggagctctttccaaatttccagcgaggaagctgaggaattgaa 1437
 V  G  R  L  S  A L/F R  A  I  P  E  E  V  L  P  S  S  F  Q  I  S  S  E  E  A  E  E  L  K 481

gtatggaagacaggaggctttgcttttgagtgagcagtctcagcagggaaaaag gaagttgcttgagctaattatgtaaaaataatcgt 1518
gtatggaagacaggaggctttgcttttgagtgagcagtctcagcagggaaaaag gaagttgcttgagctaattatgtaaaaataatcgt 1527
 Y  G  P  Q  E  A  L  L  L  S  E  Q  S  Q  Q  G  K  R  E  V  A  *

atattagtccatgcatggtctaccaactatatgtgtgaatctaattccaaaataaaatggtcaatggatgtaaagacatggcaatcctag 1608
atattagtccatgcatggtctaccaactatatgt                                                         1561

ccttactactggcgttgattgcgagaagtttgatgtttggtgaccatgagtcaataataaactatgataattaatgt              1685
```

Fig. 3. The cDNA nucleotide sequence for two coffee legumins and their deduced amino acid sequences. The N-terminal amino acid sequences for acidic and basic chains that were determines chemically are shown by solid horizontal lines above the sequences. Clone pCA4 includes bases 127-1023, pCA3 extends from 5′ linker to base 216, and clone pCA9 extends from base 208 to the 3′linker. The oligonucleotides denoted by the left and right facing open arrows permitted full coding regions of coffee legumin to be amplified. Differences between the nucleotide coding regions are indicated in reverse type, and the deduced amino acid sequences are provided above the nucleotide sequences. The single vertical arrow (↓) indicates the position where the signal peptide is cleaved from the prolegumin while the short, double vertical arrow (⇓) identifies where post-translational modification takes places to form the acidic and basic chains of mature subunits. Upper sequences corresponds to accession number U64443 and the lower sequence to AF054895.

The 11S storage protein genes from legumes contained a conserved regulatory sequence known as the legumin box (Bäumlein *et. al.*, 1986). Because the coffee storage proteins were deposited in endosperm tissue while legumes accumulate them in cotyledons, it was of interest to determine if the legumin box was also present in the promoter of the legumin gene from coffee.

```
-361  caagatgaatgtgtgtttgatttggggtttgattcatcaaaagccatcgtagcagataatgcaccttaccatgccattgc

-281  taaagtacaaaaatttcatgcaaatacaaacacaaaaggattgaacaatacatgtcagaaactcattgccaccaaggctt
                                  *tccatagc--atgcatgctgaagaatgtct*
-201  acacatcatccttggtgtaaagaagtgtcctcttcatagccatgcacaagactgagtagccaagtgtaaaatgaaaat

-121  tttgacgtgtcgattcctcatcttccaattacatgttataaaaggagccatttccaagctctaatcgccgcatcccctca

 -41  ccacaaaaacacactacactctcctctgttgtcagagaatg
```

Fig 4. Nucleotide sequence of the promoter for a coffee legumin gene. The first 400 bp of DNA upstream from the beginning of the coding region for a coffee legumin gene is shown. A putative legumin box is compared with a consensus sequence derived from legume genes. Homology is indicated for a nucleotide identified by reverse printing. Central to the legumin box is a degenerate CATGCAT element (underlined). As it is typical of corresponding genes from legumes, other CATGCAT elements are present in the upstream region. A TATA box element (double underlined) is also present.

Fig 4 presents the first 400 bp upstream from the beginning of the coding region for a coffee legumin gene. About 2.5 kb of nucleotide sequence for this promoter is reported in Accession No AF055300 in the GenBank database. Examination of a 5′-promoter region from a coffee legumin gene revealed a degenerated form of the legumin-box. The CATGCAT element, which is a central feature in the legumin box of legumes, is conserved in the coffee sequence. The putative legumin box is found in approximately the same location relative to the start codon as in the 11S genes of legumes. As in legume genes, CATGCAT elements besides the one in the legumin box were also present farther upstream in the coffee promoter.

Genomic DNA from *C. arabica* was digested with six different restriction endonucleases. After separation of the fragments by electrophoresis, single discrete fragments on DNA blots hybridized strongly to a cDNA probe for the acidic chain.

DNA from other species and commercially important cultivars that comprise the genus *Coffea* produced similar results (Fig. 5). These data suggested that the number and structure of the legumin genes were reasonably conserved among the various species of coffee.

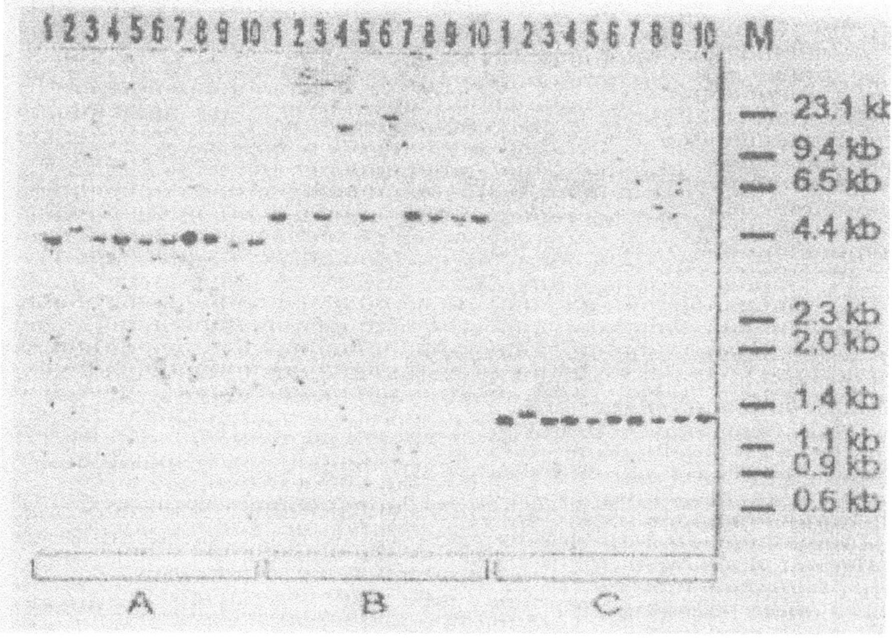

Fig. 5. Genomic DNA blot analysis of then different species and cultivars from the Coffea family. Genomic DNA blot analysis of ten differnt species and cultivars from Coffea family. Genomic DNA was digested with three restriction endonucleases (A Xba1, B Eco RV, C Eco R1). DNA fragments after this treatment were separated in a 0.8% agarose gel, transferred to a nitrocellulose membrane, and probed with the DNA insert from pCA4 at high stringency. The insert corresponds to the mature acidic chain and N-terminal of the basic chain. Species 1, Hibrido de Timor; species 2, C. eugenoides; species 3 C. liberica; pecies 4 C. kapakata; species 5, C. arabica cv. Caturra; species 6, C. canephora cv. Conilon; species 7, C. canephora cv. Robusta; species 8 C. arabica x C.. canephora; species 9, C. arabica; species 10, C. arabica cv. Colombia

Where polymorphisms were found, single rather than multiple bands were observed. Surprisingly, *C. eugenoides*, which has been considered as the ancestral parent of *C. arabica* (Orozco-Castillo *et al.*, 1996), did not produce a DNA fragment that hybridized to the coffee legumin probe. Studies revealed that *C. eugenoides* and *C. arabica* accessions shared 805 RAPD and organelle specific genome markers they developed. Perhaps the DNA fragments from this diploid species that contained the legumin gene were less than 1 kb and eluted from the gel. Nonetheless, the data did indicate that in *C. eugenoides*, the DNA encoding legumin and flanking regions was different in most of the others members of the coffee genus.

Fig 6A presents a low magnification electron micrograph of a typical coffee endosperm at 220 days after flowering. The coffee endosperm cells typically have relatively thick cell walls and an adjacent layer of cytoplasm that contained many electron-transparent vacuoles that were probably oil droplets. The central vacuole occupied the greatest proportion of each cell section. Both the electron-dense cytoplasmic layer that underlined the cell wall and the less electron-dense central vacuole became labelled with gold when the sections were reacted with antibody against coffee legumin (Fig 6B). Purified pre-immune serum did not hybridize significantly to the sections (data not shown).

Immunocytochemical studies revealed that some legumin was detected in the cytoplasm in mature coffee seeds, but that the majority of it was in large storage vacuoles that accounted for most of the cell volume Nonetheless, the electron-dense cytoplasm, particularly that around the periphery of the cell, also became heavily labelled with gold particles. The presence of signal peptides on the N-terminals of the proteins suggested that coffee legumin entered the secretory pathway. Consistent with this interpretation, the label found associated with the cytoplasm could be due to proteins synthesized in rough endoplasmic reticulum but not yet transported to the central vacuole. Interestingly, multiple smaller storage vacuoles (protein bodies) were not observed as in cotyledonary tissue from legumes or endosperm from oats (Adeli et al., 1984). Because the central storage vacuole accounts for such a large proportion of the total cellular volume at seed maturity, it contains the majority of label. Still, we were surprised by the intensity of gold label found in the electrondensecytoplasm adjacent to the cell wall, something we have not observed in legumes. Indeed, proteinaceous aggregates facing the cell wall of coffee endosperm cells have been reported by Dentan (1985). According to this report, the storage proteins accumulated from 94 to 143 days after pollination, primarily in smaller vacuoles in the vicinity of the cell wall. On days 171 and 202 after pollination, the accumulation of large central storage vacuole predominated.

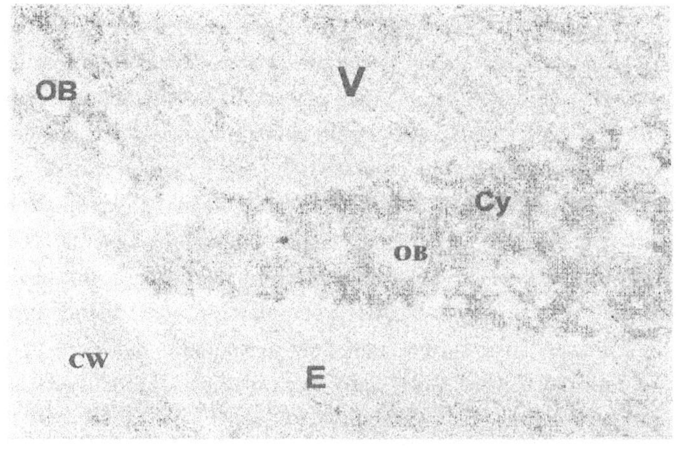

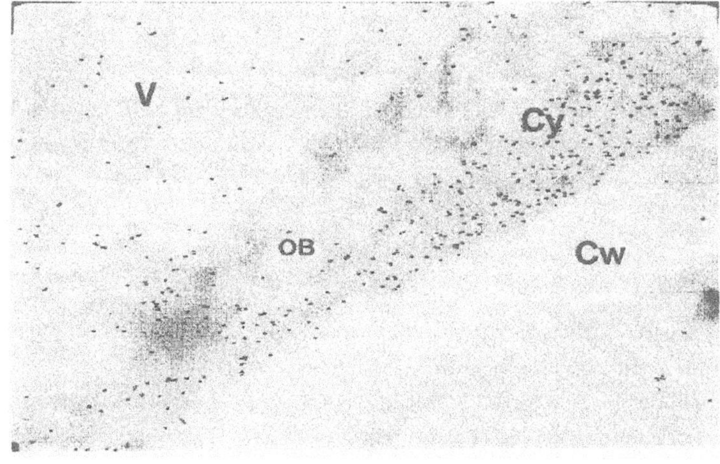

Fig 6. Immunolocalization of coffee legumin. A) Low magnification electron micrograph of a typical coffee endosperm cell at about 220 DAF. The cells contain a large central storage vacuola surrounded by cytoplasm. Contained within the cytoplasm are many small, electron-transparent vacuoles, presumably oil bodies. (B) High magnification electron micrograph showing immunolocalization of coffee legumin. Gold laceled antibody against coffee legumin were reacted with thin sections of coffee endosperm. CW: cell wall; Cyt: cytoplasm; V: Storage vacuole; OB: oil body.

4. Perspective

The coffee 11S protein, comprised in the green bean, are likely to participate in the formation of coffee aroma, either as intact protein, hydrolyzed or as free amino acids. Therefore, further studies about this protein could contribute understanding of coffee taste and aroma development. In other beans, such as cocoa and peanut, the size, sequence and chemical conformation of a peptide influence the aroma formed during heating and may control the formation of quantity of each compound. The coffee 11S legumin described here is the most abundant protein in the coffee endosperm and could contribute as an important precursor of such components formed during roasting. Furthermore, it would be interesting to know the existence of other coffee 11S genes within the Coffea species, giving rise to proteins with slightly different molecular masses or amino acid composition.

5. Acknowledgements

We thank Dr. David Asai, John Wilder, Mrs. Mary Bower, Dr. John Turek, Deborah Van Hahn, Phyllis Lockard and Deborah Sherman for their help. A grant from NCI and support to R. Acuña by COLCIENCIAS (Colombia) and National Coffee Growers of Colombia is gratefully acknowledged.

6. References

Acuña, R., Bassüner R., Beilinson V., Cortina H., Cadena-Gómez G., Montes V., Nielsen N. (1999), Coffee seeds contain 11S storage proteins. Physiol. Plantarum, 105:122-131

Adeli, K., Allan-Wojtas P., Altosaar I. (1984), Intra-cellular transport and posttranscriptional cleavage of oat globulin precursors. Plant Physiol., 76:16-20

Amorim, H.V., Josephson, R.V. (1975), Water soluble protein and nonprotein components of Brazilian green coffee beans. J. Food. Sci., 40:1179-1184

Bairoch A., Bucher P. (1994), PROSITE: Recent developments. Nucleic Acids Res., 22:3583-3589

Baümlein H., Wobus U., Pustell J. and Kafatos F. C. (1986), The legumin gene family: Structure of a B type gene of *Vicia faba* and a possible legumin gene specific regulatory element. Nucleic Acids Res., 14:2707-2720

Centi-Grossi M, Tassi-Mico C., Silano, V. (1969), Albumin fractionation of green seed varieties by acrylamide gel-electrophoresis. Phytochemistry, 8:1749-1751

Dentan E. (1985), Etude microscopique du developement et de la maturation du grain de café. ASIC 11°: 381-387

Goldberg, R.B., Barker S.J., Perez-Grau L. (1989), Regulation of gene expression during plant embryogenesis. Cell, 56: 149-160

Guerche P., Tire C., Grossi de Sa F., deClercq A., Van Montagu M., Krebbers E. (1990), Differential expression of the Arabidobpsis 2S albumin genes and the effect of increasing gene family size. Plant Cell, 2:469-478

Ludwig E., Raczek N., Kurzrock T. (1995), Contribution to composition and reactivity of coffee protein. ASIC, 16° Colloque, Kyoto, 359-365

Macrae, R. (1985), Nitrogenous components. In- Coffee, vol. 1:Chemistry. Clarke RJ and Macrae (eds.), Elsevier Applied Science Publishers, New York, pp. 115-152

Nielsen, N.C., Bassüner, R., Beaman, T. (1997), The biochemistry and cell biology of embryo storage proteins. Cellular and Molecular Biology of Plant Seed Development, In- Advances in Cellular and Molecular Biology of Plants, Vol. 4, Vasil, I. K. (ed.), Kluwer Academic Publishers, Dordrecht, pp 151-220

Orozco-Castillo C., Chalmers K.J., Powell W., Waugh R. (1996), RAPD and organelle specific PCR re-affirms taxonomic relationships within the gecin *Coffea*. Plant Cell Rep., 15:337-341

Perez-Grau L., Goldberg R.B. (1989), Soybean seed protein genes are regulated spatially during embryogenesis. Plant Cell, 1:1095-1109

Shultz, D.J. Craig, R , Cox-Foster, D.L., Mumma, R.O., Medford, J.I. (1994), RNA isolation from recalcitrant plant tissue. Plant Mol. Biol. Rep., 12:310-316

Sluyterman, L. A., Widjens J. (1970), An agarose mercurial column for the separation of mercaptopapain and non-mercaptopapain. Biochim. Biophys. Acta, 200:593-595

Underwood, G.E., Deatherage, F.E. (1952), Nitrogen compounds of coffee. Food Res., 17: 419-424

Wobus U., Borisjuk L., Panitz R., Manteuffel R., Baümleun H., Wohlfahrt T., Heim U., Weber H., Miséra S., Weschke W. (1995), Control of seed storage protein gene expression: New aspects on an old problem. J. Plant Physiol., 145:592-599

Chapter 14

IN VITRO EMBRYO CULTURE OF *COFFEA ARABICA*: THE INFLUENCE OF NAA AND BAP

M. Pasqual[1], J.M. Cavalcante-Alves[1], L.M.C.O. Andrade[1], A.B. Pereira[1], A.L.R.Maciel[1], R.D. de Castro[2]

1Laboratório de Cultura de Tecidos, 2Laboratorio de Sementes, Departamento de Agricultura da Universidade Federal de Lavras, 37200-000, Lavras – MG, Brazil

Running title: *Coffea arabica* embryo culture

1. Introduction

Inter-specific and inter-generic crossing offer plant breeders a way to increase genetic variability and transfer of desirable genes among species, especially from wild plants to cultivated ones (Gomathinayagam *et al.*, 1998). The use of hybridization between different species is frequently limited due to failures that can occur during pre- and post-fertilization, which can affect endosperm and/or embryo development leading to abortion or degeneration before maturation is reached (Mallikarjuna, 1999; *Sukno et al.*, 1999; Angra *et al.*, 1999). Hybrid embryos can be saved if they are removed before abortion and artificially cultivated in a nutritive medium (Asano and Imagawa, 1999). The embryo originated from a normal fecundation process can easily be separated and cultivated in aseptic conditions in an adequate culture medium. This will maintain them genetically stable, producing identical descendant to them. For the removal of the embryo, one need only to disinfect the external surface of the seed due to the fact that the embryo is located inside, in an sterile region of the seed. Therefore the *in vitro* contamination rate is very low in comparison to other cultures (Illg, 1985).
Low concentrations of auxin are known to induce normal embryo growth, whereas high concentrations have an inhibiting effect and can also induce the formation of callus (Raghavan and Srivastava, 1982). However, cytokinines have usually inhibited embryo

T. Sera et al. (eds.), Coffee Biotechnology and Quality, 171–178.

development (Raghavan and Torrey, 1964) even though Pinfield and Stobart (1972) have observed stimulus to Acer embryo development with the use of kinetine, which was also observed by Gomathinayagam *et al.* (1999) with the use of BAP to *Vigna* embryo culture. The aim of this study was to determine the best naftalenoacetic acid (NAA) and 6-benzilaminopurine (BAP) concentrations in *in vitro* embryo culture of *Coffea arabica* L.

2. Experimental

Embryos of *C. arabica* L. (cv. Catuaí Vermelho LCH 2077-2-5-44) were isolated from ripe fruits during the 'cherry' stage. The nutritive medium was Murashige and Skoog (1962), pH 5.9, plus 6 $mg.l^{-1}$ GA_3 and solidified with $7g.l^{-1}$ agar.

Seed parchments were removed and seeds were immersed in distilled water for 15 hours. After this period, seeds were plunged into 70% ethanol for one minute, followed by immersion in 2.5% sodium hypochloride for 30 minutes. Then the embryos were isolated and inoculated into the culture medium by placing one embryo per tube. The tubes were incubated at 26 ± 1°Cfor 60 days in a growth chamber with a photo period of 16 hours and light intensity of 16 _$M.m^{-2}.s^{-1}$. Evaluations were made for total number of sprouts, total number of leaves, sprouts lengths, fresh and dry weights. Treatments consisted of all possible combinations between concentrations of NAA (0.0; 0.01; 0.1 and 1.0 $mg.l^{-1}$) and BAP (0.0; 3.0; 6.0 and 9.0 $mg.l^{-1}$), resulting in a 4 x 4 factorial design using the totally randomized scheme with four replications.

For the statistical analysis, two methods of disintegrating the variation among the treatments were used. The first method corresponded to the deployment of the main factors and the interaction among them, typical of factorial assays. The second method made the use of the adjustment to response surface models (Box and Draper, 1987), when the differences among the treatments showed themselves significant by the F test. The backward test (Draper and Smith, 1981) was used for selecting multiple regression models. Once determined the proper model, it was tested the significance of the regression deviations to verify the applicability of the model.

3. Results and discussion

The summary of the analysis of variance for the evaluated parameters is shown in Table 1. This considered the variation among treatments according to the factorial scheme and variation due to the adjusted multiple regression model and regression deviation. It was observed that there was a significant interaction among the growing regulators used in relation to total number of sprouts, total number of leaves, sprouts larger than one cm, fresh and dry weights. It was also observed that the regression deviation was not significant for all variables analyzed, making clear that the models were satisfactorily adjusted.

Table 1. ANOVA for Shoot Number (SN), Shoots Larger than 1 cm (SL>1), Leaf Number (LN), Fresh Weight (FW) and Dry Weight (DS) versus NAA and BAP concentrations, after 60 days of *in vitro* culture

Source of		M. S.				
variation	D.F.	SN	SL>1	LN[1/]	FW	DW
Treatment	(15)	1,3098**	0,3091**	75,6338**	3027,7500**	80,9784
NAA	3	1,8018**	0,6563**	10,8965**	3158,3439**	96,3355
BAP	3	3,8671**	0,2998*	8,9978**	7126,3501**	170,8920*
NAA x BAP	9	0,2933	0,1965*	1,7723**	1618,0186*	45,8882
Regression	4	4,7187**	3,5211**	17,5756**	7607,1770**	
Desvio	11	0,0701	0,0797	0,4847	1362,5038	
Error	48	0,1748	0,0891	0,6240	741,2335	
C.V.(%)		23,94	23,85	30,42	47,43	

* and ** significant ANOVA at $P \leq 0.05$ or 0.01 repectively

3.1. NUMBER OF SPROUTS AND THOSE LARGER THAN ONE CENTIMETER

According to the adjusted regression surface model, the highest concentration of auxin, i.e. 1.00 mg.l-1 NAA, combined with 7.42 mg.L-1 BAP cytokinine promoted a greater number of sprouts, as estimated in the evaluated range. The expected number of sprouts for this combination would be 3.00 sprouts per explant (Fig. 1).

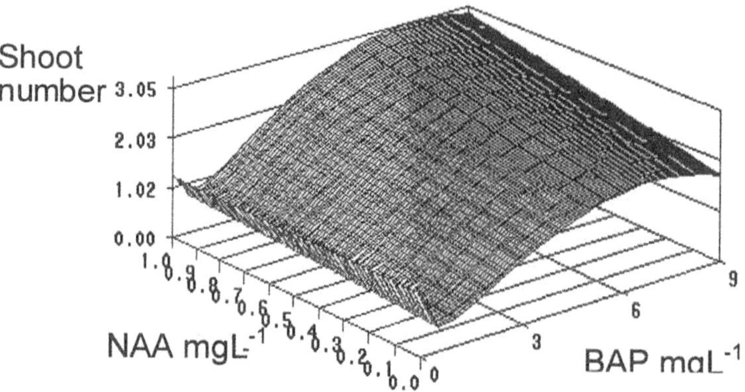

Figure 1. Response surface for shoot number versus NAA and BAP doses.

$$Y= 1.224745 + 1.378642\ X2 - 0.072996\ X22 - 1.989705\sqrt{X_2} + 0.379332\ X1\sqrt{X_2}$$

The maximum number of sprouts that were larger than one centimetre was obtained with the highest concentrations of both growth regulators (NAA 1.0 mg.l-1, BAP 9.0 mg.l-1) (Fig. 2), which appeared to interact with each other.

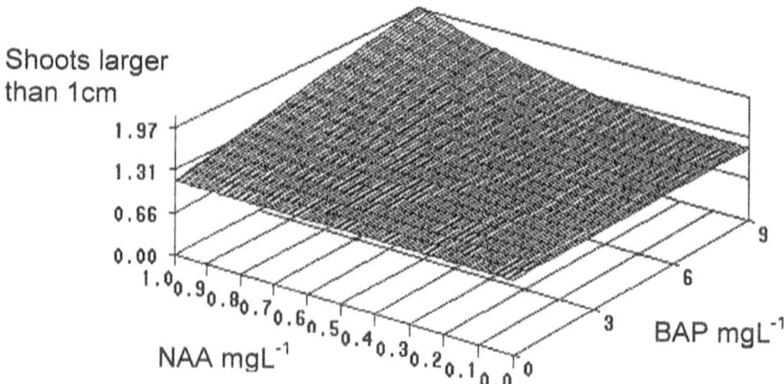

Figure 2. Response surface for shoot larger than 1 cm versus NAA and BAP doses ($Y= 1.147714 + 0.091410\ X_1^2X_2$)

3.2. TOTAL NUMBER OF LEAVES

The maximum number of leaves per explant was obtained with the highest auxine concentration, i.e. (1.0 mg.l-1 NAA) combined with 7.4 mg.l-1 BAP cytokinine (Fig. 3). The factorial analysis confirmed that there was interaction between the growth regulators as observed for the adjusted model (Table 1). According to the adjusted regression surface model, one can expect 12 as the total number of leaves per explant.

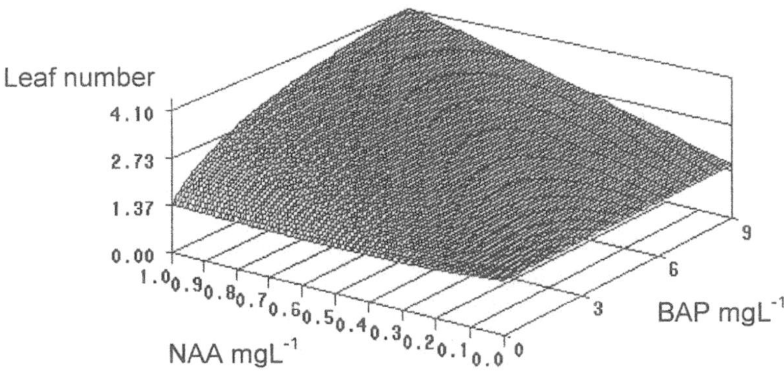

Figure 3. Response surface for leaf number versus NAA and BAP doses.
$Y = 1.5484 - 1.1504\ X_1 + 2.0890 \log(X_1 + 1)^2 + 1.6904 \log(X_1 + 1) \log(X_2+1)$

3.3. DRY AND FRESH WEIGHTS OF SPROUTS

As can be seen from the Fig. 4, intermediate concentrations of both growing regulators promoted greater responses of dry matter accumulation in the sprouts. The maximum response was obtained with a concentration of 0.53 mg.l-1 NAA and 6.00 mg.l-1 BAP, resulting in average dry weight of 148.53 mg.explant-1.

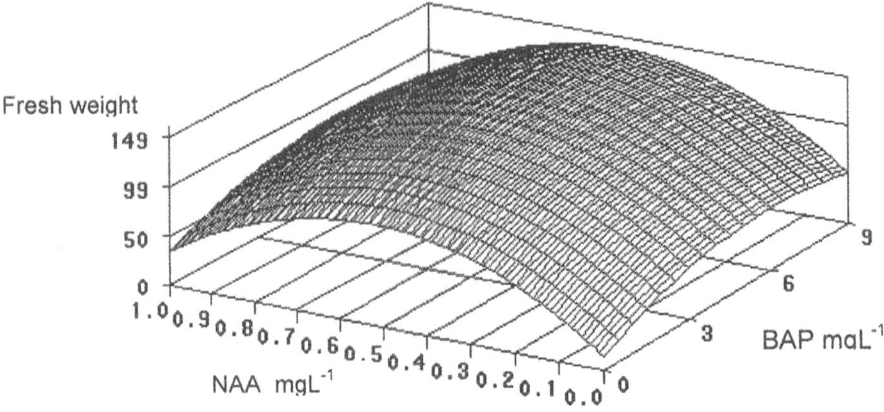

Figure 4. Response surface for fresh weight versus NAA and BAP doses.
$Y= 14.024855 + 322.720961 X_1 + 16.2252219 X_2 - 303.235632 X_1^2 - 1.352865 X_2^2$

The factorial analysis did not show interaction between the growth regulators in terms of dry matter accumulation (Table 1).

However, it could be observed in the adjusted multiple regression model that increasing concentrations of BAP had positive effects on the response of dry matter accumulation (Fig. 5).

Contrary to our observations, it has been observed that high concentrations of auxin inhibited embryo growth (George, 1994). In general, low concentrations of auxin favoured normal embryo growth, whereas high concentrations inhibited as well as promoted disorganized callus growth in embryo culture (Raghavan and Srivastava, 1982). Raghuramulu (1989) succeeded in the culture of embryos from C. arabica using 1.0 mg.l-1 of auxin and cytokinine, whereas higher or lower concentrations were not satisfactory. The author concluded that auxin concentration higher than 1.0 mg.l-1 became toxic, resulting in the death of the embryos. Effects of cytokinine have usually resulted in the inhibition of growth (Raghavan and Torrey 1964), despite the stimulus obtained with the use of kinetine (Pinfield and Stobart, 1972).

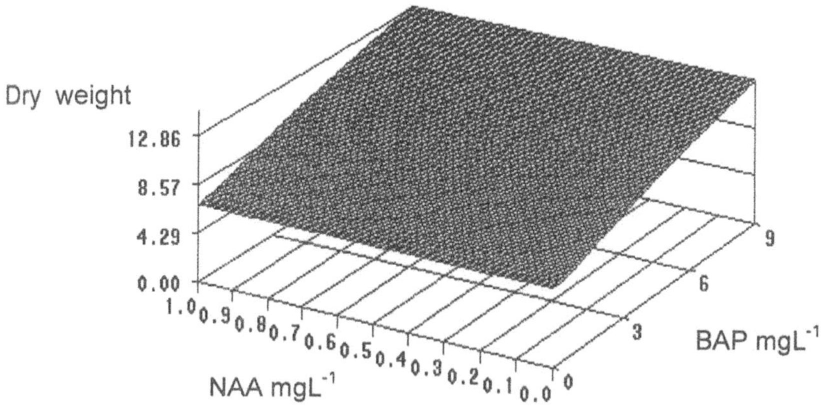

Figure 5. Response surface for dry weight versus NAA and BAP doses
$Y = 6.8095 + 1.1415 \log (X_2 + 1)^2$

4. Conclusions

The embryo culture technique is relatively simple and should represent a great potential in respect to the rescue of desirable C. arabica embryos, which would degenerate if not extracted from the developing fruits and seeds and cultivated in vitro. Although growing regulators are extensively used in embryo culture, it can be observed that their effects have been somewhat inconsistent and sometimes contradictory. In conclusion, considering that the effects of growth regulators are not nutritional, it is probable that the osmotic concentration and the mode of action of these substances are connected in some way with cell permeability and ion incorporation.

5. Acknowledgements

We acknowledge the financial support from CBP&D-Café and FAPEMIG.

6. References

Angra, D.C., Barbosa,M. M., Prestes, A.M., Fernandes, M.I.B.D. (1999), Embryo culture of back-crosses between hybrids of *Triticum aestivum* Thell. and *Agropyron elongatum* Host. & Beauv. Pesquisa Agropecuária Brasileira, 34, 209-215

Asano, Y., Imagawa, M. (1999) Hybrid seed formation among *Dioscorea opposita* Thunb. Cvs Nagaimo, Ichoimo, Tsukuneimo and *Dioscorea japonica* Thunb. J. Jap. Soc. Hort. Sci., 68, 591-597

Box, G.E.P., Draper, N. R. (1987), Empirical model-building and response surfaces. John Wiley & Sons, New York

Draper, N. R., Smith, H. (1981), Applied regression analysis. John Wiley & Sons, New York

George, E.F. (1993), Plant propagation by tissue culture. Part. 1 The Technology, 2nd edn., Edington: Exegetics

Gomathinayagam, P., Ram, S.G., Rathnaswamy, R., Rathnaswamy, N. M.(1999), Inter-specific hybridization between *Vigna unguiculata* (L.) Walp. and *V. vexillata* (L.) A. Rich. Through in vitro embryo culture. Euphytica, 102, 203-209

Illg, R.D. (1986), Metodologia de seleção in vitro para resistência a fatores causadores de estresse. In: Simpósio Nacional De Cultura De Tecidos Vegetais, 1, Brasília, Anais. Brasília, ABCTP/ EMBRAPA,. pp.45-47

Mallikarjuna, N. (1999), Ovule and embryo culture to obtain hybrids from inter-specific incompatible pollinations in chickpea. Euphytica, 110, 1-6

Murashige, T. and Skoog, F. (1962), A revised medium for rapid growth and bioassays with tobacco tissue cultures. Physiologia Plantarum, 15, 473-479

Pinfield, N. J., Stobar, A.K. (1972), Hormonal regulation of germination and early seedling development in Acer pseudoplatanus L. Planta, 104, 134-145

Raghavan, V., Srivastava, P.S. (1982), Embryo culture. In- Experimental Embryology of Vascular Plants, Johri, B.M. (ed.), Springer Verlag, Berlin, pp. 195-230

Raghavan, V. and Torrey, J.G. (1964), Inorganic nitrogen nutrition of the seedlings of the orchid, Cattleya. Am. J. Botany, 51, 264-274

Raghuramulu, Y. (1989), Anther and endosperm culture of Coffee. J. Res., 19(2), 71-81

Sukno, S., Ruso, J., Jan, C.C., Melero-Vara, J.M., Fernandez-Martinez, J.M. (1999), Inter-specific hybridization between sunflower and wild perennial *Helianthus* species via embryo rescue. Euphytica, 106:(1) 69-78

Chapter 15

DNA MARKERS FOR COFFEE TREE BREEDING

N.S. Sakiyama

Departamento de Fitotecnia, Universidade Federal de Viçosa, 36571-000 Viçosa - MG, Brazil

Running title : DNA markers

1. Introduction

Coffee cultivation in the world has benefited greatly from the successful breeding programmes, which have given the farmers productive cultivars adapted to specific cropping conditions. For example, presently in Brazil the improved arabica coffee cultivars (*Coffea arabica* L.) produces three to four times more than the cultivars used in the past. However, yield in this species seems to have reached a plateau, which is hard to overcome. Examples are the yield of improved Mundo Novo, Catuaí Vermelho and Catuaí Amarelo lines, which are still the most productive today. Therefore, one of the great challenges for the breeders is to increase the yield of the present cultivars. Another great challenge for the breeder is to transfer genes to improve coffee resistance/tolerance to disease and pests, since the arabica species is very susceptible to them. Although *C. arabica* has twice the number of chromosomes than the other species in the genus, some inter-specific hybridization has been successfully exploited. The world wide occurrence of coffee tree rust (*Hemileia vastatrix* Berk.et Br.) for example, and the damage it causes, have aroused the interest of the breeder in finding resistant arabica coffee cultivars by natural or artificial hybridization with *Coffea canephora* Pierre. Breeding by hybridization in coffee, as in other plants, requires identification of genetic variability for the characteristics to be improved, the choice of the most promising hybridization, and the application of methods, which favour selection of superior hybrids, lines or populations. In this context, DNA marker technologies have been proposed as useful tools for research programs. This technologies may be of great potential to help breeding, especially in coffee, because it is a perennial crop with a long juvenile period.

T. Sera et al. (eds.), Coffee Biotechnology and Quality, 179–185.

2. DNA markers

The term genetic marker is used to identify any morphological, physiological or molecular factor, which differentiates between two genotypes and is inherited by sexual reproduction. Genetic markers are usually phenotypic expressions of the differences between two individuals. However, there is a class among the molecular markers called DNA markers, which results from the direct manipulation of DNA molecules.

Several DNA marker technologies have been developed, all of which are based on the DNA polymorphism existing between genotypes. Depending on the technology used, the polymorphism among genotypes in a given locus occurs in two more common forms: a) in the form of DNA fragments present in one individual and absent in another; or b) in the form of different fragment lengths for the two individuals. These fragments are separated by electrophoresis and visualized as bands, using staining techniques or by hybridization with marked probes. The set of bands is called the electrophoresis standard.

The DNA markers may be qualified as dominant markers or co-dominant markers according to their inheritance type (Figs. 1 and 2). Dominant markers do not differentiate the heterozygote from the dominant homozygote individual, in a given studied locus. This generally occurs in markers with polymorphism based on the DNA fragment presence (dominant allele) or absence (recessive allele). The dominant markers are thus visualized in the form of the presence or absence of bands. With this type of marker, an F_2 population originating from self pollination of diploid heterozygote F_1 individuals segregates in a ratio of 3:1 (3 individuals with one band: 1 individual with no band). Co-dominant markers identify heterozygote from any homozygote individuals for the locus in question. This generally happens with the markers with polymorphism based on the difference of DNA fragment length. The co-dominant markers are thus visualized in the form of bands in different positions; each band corresponds to an alternative allele. With this marker, an F_2 population from self pollination of F_1 heterozygote diploid individuals segregates in the ratio of 1:2:1 (1 individual with only the first band: 2 individuals with both bands: 1 individual with only the second band). Exceptionally, a typically dominant marker becomes co-dominant when the heterozygote is identified by quantification of the DNA present in the band.

Parent 1	Parent 2	Hybrid	Parents
I		I	A (polymorphic band)
I	I	I	B (monomorphic band)

Figure 1: Electrophoretic pattern of a dominant marker; the hybrid is the same as parent 1 but different from parent 2. A and B represent 2 different loci.

Parent 1	Parent 2	Hybrid	Parents
I		I	A (polymorphic band)
	I	I	B (polymorphic band)

Figure 2: Electrophoretic pattern of a co-dominant marker; the hybrid is different from both parents. A and B represent alternative alleles of the same locus.

DNA markers are outstanding because they can be obtained in high numbers, because they have simple inheritance and consistent results, regardless of the cropping environmental condition of the plant, or of the type or age of the tissue sampled. These characteristics are even more relevant for coffee research, a perennial crop with a long juvenile period. Assessment by DNA markers may, for example, be carried out at initial plant development stages, or even in the seeds, giving identical results to those, which would be obtained at the adult phase. The coffee genotypes studied are often of different ages or are located in distinct environments, and even so, the results from these markers can be compared. They have been used in coffee to study the molecular pattern (DNA fingerprint) and linkage relationships.
Examples of DNA markers include RFLP (Restriction Fragment Length Polymorphism), STR (Short Tandem Repeats), SSR (Simple Sequence Repeats), RAPD (Randon Amplified Polymorphic DNA), SPAR (Single Primer Amplification Reaction), AFLP (Amplified Fragment Length Polymorphism), etc.

3. DNA fingerprint

DNA fingerprinting is the identification of the genotype using DNA markers. The molecular patterns enable differences and similarities to be identified at the DNA level. Fingerprinting has often been used in genetic diversity studies of species in germplasm banks and in nature. This additional data on genetic diversity helps the identification of new variants in the wild for introduction in germplasm banks, or of duplications, which can be eliminated. It also guides the formation of germplasm sub-samples, the so-called core collections, which should represent, with the minimum of repetition, the highest possible diversity present in a complete Collection. Core collections can be easily transferred to research centres to meet breeders' needs. Maintenance of different core collections, in various locations, is also an effective and economic strategy for preservation of the genetic diversity. DNA markers have been used in studies of genetic origin and diversity in coffee (Lashermes *et al.*, 1993; Orozco-Castillo *et al.*, 1994, 1996; Teixeria *et al.*, 1999a, b). Figure 3 shows the electrophoretic patterns obtained by RAPD markers in coffee.

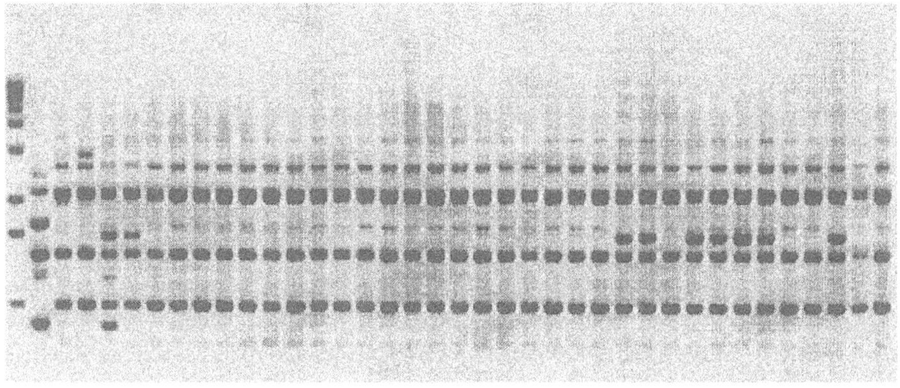

Figure 3: Electrophoretic pattern of coffee genotypes, obtained by RAPD markers.

Molecular patterns can be used as genetic descriptors to identify cultivars. The limitations of morphological markers used to describe and distinguish between cultivars, as laid out in the Brazilian Cultivar Protection Law (Law 9.456, April 25, 1997), can be avoided by the inclusion of DNA markers. RAPD markers were used in coffee to estimate the genetic distances among 16 commercial *C. arabica* lines, and, in spite of the small total variability, polymorphic loci were found even among closely related lines (Teiexeira *et al.*, 1999a,b).

The possible use of genetic divergence data assessed by DNA markers to predict F_1 hybrid performance has been investigated. Divergence may be valuable for predicting especially when measured among improved lines. When random markers are used, the maximum divergence observed for a pair of genotypes does not always correspond to the maximum high-parent heterosis value in the F_1 they generate. Therefore, prediction becomes more reliable if the markers used are associated with the characteristics to be improved in the hybrid and are complementary in the pair of parents.

DNA fingerprinting may be used in Marker Assisted Selection (MAS) in back-crosses to recover the recurrent genotype more efficiently. The introgression of interesting genes, such as those for resistance to disease and pests, is frequently carried out by the back-cross method. Recovering the recurrent genotype using this method is gradual, and several back crossed generations are needed. The MAS may reduce the number of back crosses needed to recover those individuals phenotypically more similar to the recurrent parent. If a molecular map is available, it is recommended that markers distributed uniformly along the genome be chosen. When dominant markers are used, those showing bands in the donor parent and showing absence in the recurrent parent should be chosen. At each back-cross generation, various alleles from the recurrent parent become fixed, and this may be observed by markers.

Other forms of using molecular patterns in coffee tree breeding are, for example, assessment of the cross pollination rate, confirmation of hybrids in artificial crosses, verification of the genetic purity in lines and clones, assessment of somaclonal variations, etc.

4. Linkage

Linkage is the phenomenon referring to a pattern of linked gene transmission. Linked genes are those localized in the same chromosome, which tend to remain united during sexual reproduction, that is, they do not segregate independently. Molecular markers can be studied by their linkages among themselves, or among them and the genes which control the characteristics of interest.

Three possibilities for using genetic linkage are exploited in the study of molecular markers: the molecular map, mapping genes which control the qualitative characteristics, and mapping chromosome regions responsible for quantitative characteristics (Quantitative Trait Loci – QTL).

The genetic map is a chromosome map, which informs the linear order of the genes along each chromosome and their distances measured by the probability of recombination. The maps are constructed to localize the genetic information physically. The relationship among the various characteristics of agronomic interest may be analyzed from the genetic map, for example, genetic linkage, pleiotropy or independent segregation. The construction of genetic map was difficult before molecular markers were developed, due to the need of a considerable large number of genes with simple inheritance and visible phenotypic segregation, and to the large number of crosses necessary to map each new gene. Mutant genes in crosses involving series of aneuploids or trisomics were frequently used to construct these maps. The molecular markers facilitate the development of genetic map in various ways: they behave as single mendelian factors, their allelic segregation is independent from the visible phenotypic segregation, they are stable and obtained in large numbers, and a map can be obtained using the segregating population from a single cross. A molecular map was constructed for coffee from a di-haploid *C. canephora* population (Paillard *et al.,* 1996). Molecular maps may be used in MAS, in mapping and cloning single genes and in mapping QTLs. The greatest impact of molecular map application in plant breeding will doubtless come from QTL, mapping for characteristics with agronomic interest.

Mapping the genes which control qualitative agronomically important characteristics in coffee trees may be useful to breeding programmes when used in MAS. This indirect selection can be used in programmes where pyramiding of disease resistant genes, which is difficult to obtain by conventional methods, is required. The indirect simultaneous selection of different characteristics is also possible using markers. DNA markers were identified for mapping self-incompatibility in *C. canephora* (Lashermes *et al.*, 1996) and resistance to *Colletotrichum kahawae* (Agwanda *et al.*, 1997) and *Hemileia vastatrix* (Tedesco *et al.*, 1999) *C. arabica*. The objective in the latter is to pyramid dominant coffee tree rust resistance genes in improved lines.

A quantitative characteristic is generally referred to as that which is conditioned by various genes with small similar effects. DNA markers, however, has shown that the quantitative characteristics may be conditioned by loci with different effects. The loci with detectable effects in the variation of a quantitative characteristic are called QTLs. QTLs can be identified by using a molecular map and they explain significant parts of the genetic variation of quantitative characteristics. An immediate application for QTLs mapping in breeding programs is the choice of parents with complementary QTLs, which have greater chances of displaying high-parent heterosis in the hybrids and favorable recombination in the segregant generations. Markers associated with QTLs may also be used in backcross programs where marker assisted selection is practiced. Strategies for germplasm assessment and use are new and important contributions from QTLs studies. Tanksley and McCouch (1997) proposed the strategy of searching gene in place of searching phenotype. The search for phenotypes of agronomic interest has been the traditional strategy for germplasm use in breeding. Germplasm can now also be assessed directly for the presence of useful genes by markers associated to QTLs, even though these loci do not result in observable phenotypic differences.

5. Conclusion

Genotype introduction in coffee breeding brought a significant progress in coffee cultivation in Brazil. The inter-specific hybridization has been exploited in breeding, as in research to obtain leaf rust and coffee berry borer resistant cultivars. The use of coffee germplasm could bring important benefits to breeding by the transfer of disease and pest resistance genes and of QTLs associated with these and other important agronomic characteristics, such as yield. In this context, DNA markers should play an important role in the modern coffee tree breeding programmes.

6. References

Agwanda, C.O.; Lashermes, P.; Trouslot, P.; Combes, M.C., Charrier, A. (1997), Identification of RAPD markers for resistance to coffee berry disease, *Colletotrichum kahawae*, in arabica coffee. Euphytica, 97: 241-248

Lashermes, P.; Cros,J.; Marmey, P., Charrier, A. (1993), Use of random amplified DNA markers to analyse variability and relationships of *Coffea* species. Genetic Resources Crop Evolution, 40: 91-99

Lashermes, P.; Couturon, E.; Moreau, N.; Paillard, M., Louarn, J. (1996), Inheritance and genetic mapping of self-incompatibility in *Coffea canephora* Pierre. Theor. Appl. Genet., 93: 458-462

Orozco-Castillo, C.; Chalmers, K.J.; Waugh, R. and Powell, W. (1994), Detection of genetic diversity and selective gene introgression in coffe using RAPD markers. Theo. Appl. Genetics, 87: 934-940

Orozco-Castillo, C.; Chalmers, K.J.; Powell, W. and Waugh, R. (1996), RAPD and organelle specific PCR re-affirms taxonomic relationships within the genus *Coffea*. Plant Cell Reports, 15: 337-341

Paillard, M., Lashermes, P., Pétiard, V. (1996), Construction of a molecular linkage map in coffee. Theor. Appl. Genet., 93: 41-47

Tanksley, S.D., Mc Couch, S.R. (1997), Seed banks and molecular maps: unlocking genetic potential from the wild. Science, 277: 1063-1066

Tedesco, N.S., Sakiyama, N.S., Zambolim, L., Teixeira, T.A., Pereira, A.A., Sakiyama, C.C.H. (1999), Mapeamento de gene de resistência a ferrugem-do-cafeeiro (*Hemileia vastatrix* Berk. et Br.) com marcador RAPD. Anais do III SIBAC, Londrina. in press

Teixeira, T.A., Zambolim, L., Sakiyama, N.S., Pereira, A.A., Silva, D.G. (1999a), Padrão molecular de clones de cafeeiro diferenciadores de *Hemileia vastatrix* Berk. et Br. Anais do III SIBAC, Londrina. in press

Teixeira, T.A., Sakiyama, N.S., Zambolim, L., Pereira, A.A., Sakiyama, C.C.H. (1999b), Caracterização de acessos de *Coffea* por marcadores RAPD. Anais do III SIBAC, Londrina. in press

Chapter 16

GENETIC POLYMORPHISM IN SPECIES AND HYBRIDS OF *COFFEA* REVEALED BY RAPD

P.M. Ruas[1], L.E. C. Diniz[1], C.F. Ruas[1], T. Sera[2]

[1]*Departamento de Biologia Geral, Universidade Estadual de Londrina, 86051-990 Londrina - PR, Brazil;* [2]*Instituto Agronômico do Paraná (IAPAR), 86001-970 Londrina-PR, Brazil*

Running title: Genetic polymorphism in coffee

1. Introduction

The genus *Coffea* (Rubiaceae) consists of approximately 80 species which are native from a wide region of tropical Africa, Madagascar and neighbouring island (Carvalho *et al.*, 1984). While botanists consider the coffee tree similar to any other plant of the Rubiaceae family, which reproduce seeds resembling beans, the international coffee trade is concerned with only two species, *Coffea arabica* and *C. canephora*. However, coffee beans can be produced by many other *Coffea* species. Therefore, the gene pool, useful for both *C. arabica* and *C. canephora* breeding, is made up of all *Coffea* species (Berthaud and Charrier, 1988).

The species *C. arabica* is produced mainly in Latin America and represents about 73% of the world production of coffee. *C. canephora*, on the other hand, is produced in Central and North Africa, corresponding to 80% of the production in that continent (Orozco-Castillo *et al.*, 1994). *C. canephora* is considered to produce a low quality beverage, however, it is an important source of resistance genes for several coffee diseases. Many of these genes have been transferred to *C. arabica* by artificial hybridization (Smith *et al.*, 1992).

T. Sera et al. (eds.), Coffee Biotechnology and Quality, 187–195.

Many efforts have been directed to explore and preserve coffee genetic resources. Various methods applied to describe variability present in *Coffea* species include morphological and agronomic characteristics, electrophoresis variants of allele distribution between and within populations, and genetic analysis in progenies of controlled crosses (Charrier and Berthaud, 1985). Isozyme pattern was useful to estimate genetic diversity among plant species such as rice (Peng *et al.*, 1988), *Chenopodium quinoa* (Wilson, 1988a, b, 1990), *Ananas* (DeWald *et al.*, 1992; Aradhya *et al.*, 1994), and coffee (Guedes and Rodrigues, 1974; Payne and Fair-Brothers, 1976; Berthou and Trouslot, 1977; Berthou *et al.*, 1980). However, electrophoresis analysis in accessions of *C. arabica* from Ethiopia and Quenia, with six enzymatic systems revealed low level of polymorphism (Louarn, 1978). The results contrasted with the high level of morphological variation observed in the same germplasm suggesting that isozymes were not appropriate for screening genetic diversity in *C. arabica* varieties and cultivars.

Molecular markers such as Random Amplified Polymorphic DNA (RAPD) and Restriction Fragment Length Polymorphism (RFLP) have been extensively and successfully used to study phylogenetic relationships in plant species (Gonzales and Ferrer, 1993; Heun *et al.*, 1994; Graham and McNicol, 1995), to study genetic differentiation among subspecies (Zhang *et al.*, 1992), and to access the inter- and intra-specific genetic diversity (Abo-elwafa *et al.*, 1995; Liu, 1996; M'Ribu and Hilu, 1994; Link *et al.*, 1995; Ruas *et al.*, 1995, 1999). The usefulness of RAPD technique to study the genus *Coffea* was demonstrated by Lashermes *et al.* (1993), Orozco-Castillo *et al.* (1994), and Lashermes *et al.* (1996).

Brazil is the largest coffee producer among more than 50 countries of Latin America, Africa, and Asia. Paraná state, located at South Brazil, is an important coffee producer. In this state, the Research Centre, Instituto Agronômico do Paraná (IAPAR) maintains a germplasm Collection with seven *Coffea* species, some of which have genes for resistance to drought, frost and diseases. In addition, the IAPAR Collection preserves 40 botanical varieties of *C. arabica*, and several populations of inter-specific hybrids. The phylogenetic relationships among the Coffea species and the origin of putative hybrids maintained in this Collection have not been clearly defined. We attempted to determine the phylogenetic relationship among the seven species (one species with two varieties) and six inter-specific hybrids of coffee using the RAPD analysis.

2. Experimental

2.1. PLANT MATERIALS

The species and hybrids used in this study were the part of germplasm Collection of the IAPAR, Londrina, Paraná, Brazil. The plant materials included: *C. arabica* var. *arabica*, *C. canephora* var. *robusta*, *C. canephora* var. *kouillou*, *C. dewevrei*, *C. eugenioides*, *C. kapakata*, *C. racemosa*, *C. stenophylla*, and six natural inter-specific hybrids.

2.2. DNA EXTRACTION, AMPLIFICATION AND GEL ELECTROPHORESIS

Genomic DNA were isolated from fresh leaf tissue following the CTAB method (Doyle and Doyle, 1987), except that CTAB was replaced by Mixed Alyl-trimethyl-ammonium bromide (MATAB, Sigma) in the extraction buffer. DNA concentration was estimated using a fluorometer DyNA Quant 200 (Höefer-Pharmacia), according to the manufacturer's instructions. Amplification reactions were carried out in a final volume of 15 μl containing 1X PCR buffer (10 mM Tris-HCl pH 8.3; 50 mM KCl); 1.5 mM $MgCl_2$; 0.1 mM each of dATP, dTTP, dCTP, and dGTP; 4 μM of 10 mer arbitrary primers (Operon Technologies); 1.8 U of *Taq* DNA polymerase (Klen *Taq*, AB Peptides), and 20 ng of template DNA. For each primer, a control reaction that lacked template DNA but contained all other reagents was included. Amplification was carried out using a PTC 100 (MJ Research) 60 well Thermal Cycler, programmed with 3 min at 94°C for initial DNA denaturation, followed by 48 cycles of 1 min at 94°C, 1 min 45 sec of 38°C, and 2 min at 72°C. The final cycler was followed by a 7 min extension at 72°C. Amplified DNA products were resolved in 1,4% agarose gels, stained with ethidium bromide and visualized under UV light. DNA markers were scored for the presence (1) and absence (0) of homologous DNA bands among species and hybrids. The data matrix was read by NTSYS-pc (Numerical Taxonomy and Multivariate Analysis for Personal Computers) V. 1.80. A dendrogram was constructed using an unweighted pair group method with arithmetic averages (UPGMA), based on Jaccard's similarity coefficient with the SAHN (Sequential Agglomerative, Hierarchical and Nested Clustering) routine.

3. Results and Discussion

A total of 150 RAPD markers were scored for the 37 primers tested among seven species, one with two varieties, and six inter-specific hybrids of *Coffea*. From the total markers obtained 97.8% were polymorphic. The genetic relationship among species and between species and hybrids was examined by cluster analysis (Fig. 2). The similarity matrix generated by RAPD data is shown in Table 1. The grouping association identified a level of similarity coefficient between species ranging from 0.58 (*C. eugenioides* and *C. racemosa*, *C. arabica* and *C. racemosa*) to 0.84 (*C. arabica* and *C. eugenioides*), revealing high level of genetic variation. The RAPD data showed that *C. arabica* has the closest association with C. *eugenioides* (84%), followed by *C. kapakata* (79%), *C. canephora* var. *robusta* (76%), *C. stenophylla* (73%), *C. dewevrei* (69%), *C. canephora kouillou* (68%), and *C. racemosa* (58%). These results are consistent with those of Lashermes *et al.* (1993) and Orozco-Castillo *et al.* (1994) who showed that *C. arabica* has low genetic relationship with *C. racemosa* and has more affinity with *C. eugenioides* than with *C. canephora*. Inter-specific F_1 hybrid (*C. arabica* vs. tetraploid *C. canephora*) showed a low but highly variable fertility (Owuor and Van Der Vossen, 1981).

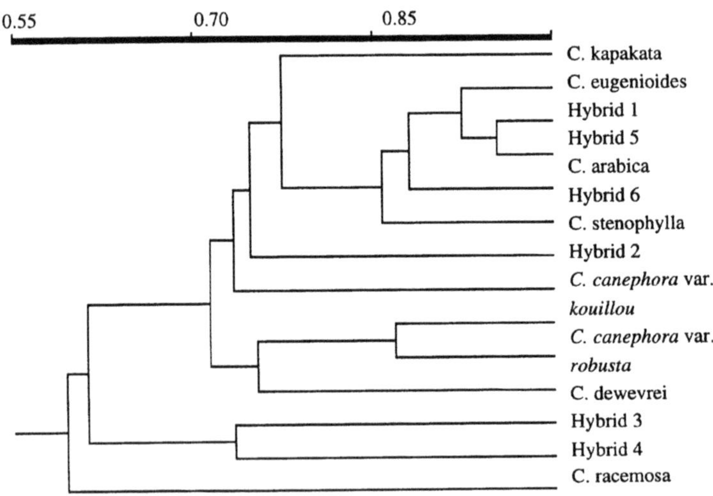

Figure 2. UPGMA dendrogram of genetic relationship of species and hybrids of *Coffea* based on RAPD markers.

Estimates of genetic distance for 3-8 enzyme systems (Berthou *et al.*, 1980) showed that the genetic distance between *C. canephora* and *C. eugenioides* are considerable. These data was also confirmed by Berthou *et al.* (1983) using electrophoresis of fragments of mitochondrial and chloroplast DNA cutting with restriction enzymes. In that investigation the authors observed that *C. arabica* and *C. eugenioides* have a high similarity and might have a common origin. Crossability data are also in agreement with the RAPD data. Charrier and Berthaud (1985) showed that *C. eugenioides* performs in general much better in crosses with all other diploid taxa than *C. canephora*. Raina *et al.* (1998), using Fluorescent *in situ* hybridization (FISH) in *C. eugenioides*, *C. congensis*, *C. canephora*, *C. liberica*, and *C. arabica* showed that the genomic DNA of *C. eugenioides* preferentially hybridized with the 22 chromosomes of *C. arabica*. The others 22 chromosomes of *C. arabica* hybridized with the genomic DNA of *C. congensis*. However, Lashermes *et al.* (1999) applying the RFLP and GISH techniques suggested that *C. eugenioides* and *C. canephora* were the putative progenitors of *C. arabica*. In this investigation, the close genetic similarity (0.84), revealed by RAPD markers between *C. arabica* and *C. eugenioides* (Table 1, Fig. 2) was, therefore, consistent with previous reports, suggesting that *C. eugenioides* was one of the ancestors of *C. arabica*. However, the participation of *C. canephora* is not clearly defined with RAPD data.

Table 1. Genetic similarity based on Jaccard's coefficient for 14 accessions of *Coffea* sp. obtained from 150 RAPD markers

No	Accessions	1	2	3	4	5	6	7	8	9	10	11	12	13	14
1	*C. kapakata*	1.00													
2	*C. can. kouillou*	0.70	1.00												
3	*C. can. robusta*	0.77	0.86	1.00											
4	*C. eugenioides*	0.77	0.69	0.70	1.00										
5	Hybrid 1	0.78	0.69	0.73	0.93	1.00									
6	*C. racemosa*	0.68	0.60	0.64	0.58	0.57	1.00								
7	Hybrid 2	0.74	0.66	0.74	0.71	0.76	0.67	1.00							
8	*C. arabica*	0.79	0.68	0.76	0.84	0.91	0.58	0.80	1.00						
9	Hybrid 3	0.68	0.62	0.68	0.72	0.73	0.54	0.68	0.75	1.00					
10	Hybrid 4	0.56	0.50	0.50	0.50	0.50	0.48	0.59	0.51	0.73	1.00				
11	*C. dewevrei*	0.72	0.75	0.75	0.72	0.72	0.62	0.68	0.69	0.66	0.52	1.00			
12	Hybrid 5	0.76	0.68	0.72	0.91	0.95	0.56	0.73	0.88	0.72	0.51	0.70	1.00		
13	*C. stenophylla*	0.67	0.70	0.71	0.79	0.76	0.62	0.68	0.73	0.68	0.51	0.72	0.76	1.00	
14	Hybrid 6	0.75	0.68	0.71	0.86	0.86	0.55	0.70	0.82	0.71	0.54	0.77	0.88	0.76	1.00

3.1. PUTATIVE PARENTS OF *COFFEA* HYBRIDS

The RAPD markers were also applied to deduce putative parents of six *Coffea* natural hybrids. These hybrids represent important sources of genes for resistance to several diseases of coffee.

3.1.1. Hybrid nos 1 and 5:

Clustered with a similarity coefficient of 0.95 these hybrids had a mean similarity of 0.91 and 0.90 with *C. eugenioides* and *C. arabica*, respectively (Table 1, Figure 2). Hybrids no 1 and no 5 were highly fertile tetraploid and they showed almost the same morphological patterns as *C. arabica*. The grouping of *C. arabica* and *C. eugenioides* (similarity of 0.84) based on molecular markers was consistent with taxonomy and pedigree information. Therefore, as suggested from morphological characters and from RAPD data, hybrids no 1 and no 5 could be originated from crosses between *C. eugenioides* and *C. arabica*, followed by successive backcrosses to the parental *arabica*.

3.1.2. Hybrid no 2:

It showed a similarity coefficient of 0.8 with *C. arabica* and of 0.67 with *C. racemosa* (Table 1). Since this hybrid was a sterile triploid, it might have originated from a diploid and a tetraploid parent. The phenotype of hybrid no 2 was similar to that of the tetraploid *C. arabica*. The leaves, however, were narrow as that of *C. racemosa*. As in *C. racemosa*, hybrid no 2 carried genes for resistance to drought, frost and leaf miner disease (*Perileucoptera coffeella*). The latter is considered as one of the most important disease in Brazilian coffee plantations. Therefore, on the basis of morphological and molecular data, *C. arabica* and *C. racemosa* were the putative parents of this hybrid.

3.1.3. Hybrid nos 3 and 4:

Clustered with a similarity coefficient of 0.73, hybrid no 4 was associated in 0.51 with *C. arabica* and in 0.52 with *C. dewevrei*. Hybrid no 3 on the contrary, associated in 0.75 with *C. arabica* and in 0.66 with *C. dewevrei*. Hybrid nos 3 and 4 were morphologically similar to both the species. The genetic association detected with RAPD markers (Table 1), between these hybrids and the diploid *C. dewevrei* and the tetraploid *C. arabica* suggested that F_1 hybrid plants (hybrid no 4), carrying gametes with specific chromosome combinations were back-crossed with the *C. arabica* parent and thus originated the hybrid no 3. Hybrid nos 3 and 4 had genes for resistance to frost, rust disease (*Hemileia vastatrix*), and nematodes (*Meloidogyne* sp.), which could be transferred do the cultivated species of *C. arabica.*

3.1.4. Hybrid no 6:

Hybrid no 6 was associated very close to *C. eugenioides* and *C. arabica*, followed by *C. dewevrei*, and *C. stenophylla* (Table 1, Fig. 2). It is known, however, that *C. stenophylla* is the female parent of hybrid no 6. Whereas artificial hybrids between this species and *C. arabica* have not yet been obtained. *C. stenophylla* constitutes an excellent source genes for resistance to the leaf miner. This species can easily cross with *C. dewevrei* (Carvalho, 1988), which also carries alleles for leaf miner. The morphological characteristics of hybrid no 6 was intermediate between *C. dewevrei* and *C. stenophylla*, except for the large leaves and the purple shoot which were similar to *C. dewevrei.* Although the results obtained with RAPD were not conclusive about the origin of hybrid no 6, the phenotypes of leaves and shoots suggested that *C. dewevrei* was the unknown parent of this hybrid. Therefore, hybrid no 6 could be used to transfer alleles from *C. stenophylla* to *C. arabica.*

4. Conclusions

It can be concluded that RAPD markers offer a reliable and effective means of assessing genetic variation of *Coffea* species and of deducing parents of unknown hybrids, which could be used to assist coffee-breeding programmes.

5. References

Abo-elwafa, A., Murai, K., Shimada, T. (1995), Intra and interspecies variation in *Lens* revealed by RAPD markers. *Theor. Appl. Genet.*, 90: 335-340

Aradhya, M.K., Zee, F., Manshardt, R.M. (1994), Isozyme variation in cultivated and wild pineapple. *Euphytica*, 79: 87-99

Berthaud, J., Charrier, A. (1988), Genetic Resources of *Coffea*. In- *Coffee Agronomy*, Vol. 4, R. J. Clarke & R. Macrae (eds.), Elsevier Applied Science Publishers Ltd., Amsterdam, Netherlands, pp. 1-42

Berthou, F., Trouslot, P. (1977), L'analyse du polymorphisme enzymatique dans le genre *Coffea*: adaptation d'une méthode d'électrophorèse en serie. Presented in 8th International Colloquium on the Chemistry of Coffee, ASIC, Paris, pp. 373-83

Berthou, F., Trouslot, P., Hamon, S., Vedel, F., Ouetier, F. (1980), Analyse en électrophorèse du polymorphisme biochimique des caféiers: variation enzymatique dans dix-huit populations sauvages: *C canephora, C. eugenioides* et *C. arabica. Café Cacao Thé*, 624: 313-326

Berthou, F., Mathieu, C., Vedel, F. (1983), Chloroplast and mitochondrial DNA variation as indicator of phylogenetic relationship in genus *Coffea* L. *Theor. Appl. Genet.*, 65: 77-84

Carvalho, A., Medina-Filho, H.P., Fazuoli, L.C. (1984), Evolução e melhoramento do cafeeiro. In- *Tópicos de Citogenética e Evolução de Plantas*, M.L.R. Aguiar-Perecin, P.S. Martins & G. Bandel (eds.), Sociedade Brasileira de Genética, Ribeirão Preto, S.P., Brazil. pp. 215-235

Carvalho, A. (1988), Principles and Practice of Coffee Plant Breeding for Productivity and Quality Factors: *Coffea arabica.* In- Coffee Agronomy, R. J. Clark & R. Macrae (eds.), Elsevier Applied Science Publishers Ltd., Amsterdam, Netherlands, pp. 129-165

Charrier, A., Berthaud, J. (1985), Botanical classification of coffee. In- *Coffee: Botany, Biochemistry and Production of Beans and Beverage*, M.M. Clifford & K. C. Wilson (eds.), Croom Hel Publ., Beckenham, UK, pp. 13-47

DeWald, M.G., Moore, G.A., Sherman, W.B. (1992), Isozymes in *Ananas* (pineapple): genetics and usefulness in taxonomy. *J. Amer. Soc. Hort. Sci.*, 117: 491-496

Doyle, J.J., Doyle, J.L. (1987), A Rapid DNA isolation procedure for small quantities of fresh leaf tissue. *Phytochemical Bulletin*, 19(1): 11-15

Gonzales, J.M., Ferrer, M. (1983), Random Amplified Polymorphic DNA analysis in *Hordeum* species. *Genome*, 36: 1029-1031

Graham, J., McNicol, R. J. (1995), An examination of the ability of RAPD markers to determine the relationships within and between *Rubus* species. *Theor. Appl. Genet.*, 90: 1128-1132

Guedes, M.E.M., Rodrigues Jr, C. J. (1974), Disc electrophoresis patterns of phenoloxidase from leaves of Coffee cultivars. *Portugaliae Acta Bioloica*, 13: 169-178

Heun, M., Murphy, J.P., Phillips, T.D. (1994), A comparison of RAPD and isozyme analysis for determining the genetic relationships among *Avena sterilis* L. accessions. *Theor. Appl. Genet.*, 87: 689-696

Lashermes, P., Cros, J., Marmey, P., Charrier, A. (1993), Use of random amplified DNA markers to analyze genetic variability and relationships of *Coffea* species. *Genetic Resources and Crop Evolution*, 40: 91-99

Lashermes, P., Trouslot, P., Anthony, F., Combes, M.C., Charrier, A. (1996), Genetic diversity for RAPD markers between cultivated and wild accessions of *Coffea arabica. Euphytica*, 87: 59-64

Lashermes, P., Combes, M.C., Roberts, J., Trouslot, P., D'Hont, A., Anthoni, F., Charrier, A. (1999), Molecular characterization and origin of the *Coffea arabica* L. genome. *Molec. Gen. Genet.*, 261: 259-266

Link, W., Dixkens, C., Singh, M., Schwall, M., Melchinger, A. E. (1995), Genetic diversity in European and Mediterranean faba bean germplasm revealed by RAPD. *Theor. Appl. Genet.*, 90: 27-32

Liu, C. J. (1996), Genetic diversity and relationships among *Lablab purpureum* genotypes evaluated using RAPD as markers. *Euphytica*, 90: 115-119

Louarn, J. (1978), Diversité compararée des descendances de *Coffea arabica* obtenues en autofécondation et en fécondation libre au Tonkoui. In- *Etude de la Structure et de la Variabilité Génétique des Caféiers*, A. Charrier (ed.), IFCC Bulletin, n. 14, pp. 75-78

M'Ribu, H. K., Hilu, K.W. (1994), Detection of inter-specific and intra-specific variation in *Panicum millets* through random amplified polymorphic DNA. *Theor. Appl. Genet.*, 88: 412-416

Orozco-Castillo, C., Chalmers, K.J., Powell, R.W. (1994), Detection of genetic diversity and selective gene introgression in coffee using RAPD markers. *Theor. Appl. Genet.*, 87: 934-940

Owuor, J.B.O., Van Der Vossen, H.A.M. (1981), Inter-specific hybridization between *Coffea arabica* L. and tetraploid *C. canephora* P. ex Fr. I. Fertility in F_1 hybrids and back-crosses to *C. arabica*. *Euphytica*, 30: 861-866

Payne, R. C., Fair-Brothers, D.E. (1976), Disc electrophoresis evidence for heterozygosity and phenotypic plasticity in selected lines of *Coffea arabica* L. *Bot. Gazette*, 137: 1-6

Peng, J.Y., Glaszman, J.C., Virmani, S. S. (1988), Heterosis and isozyme divergence in indica rice. Crops Sci., 28: 561-563

Raina, S.N., Mukai, Y., Yamamoto, M. (1998), *In situ* hybridization identifies the diploid progenitor species of *Coffea arabica* (Rubiaceae). *Theor. Appl. Genet.*, 97: 1204-1209

Ruas, P.M., Ruas, C F., Fairbanks, D. J., Andersen, W.R., Cabral, J.R.S. (1995), Genetic relationship among four varieties of pineapple, *Ananas comosus*, revealed by random amplified polymorphic DNA (RAPD) analysis. *Bras. J. Genet.*, 18: 413-416

Ruas, P.M., Bonifácio, A., Ruas, C.F., Fairbanks, D. J., Andersen, W.R. (1999), Genetic relationship among 19 accessions of six species of *Chenopodium* L., by Random Amplified Polymorphic DNA fragments (RAPD). *Euphytica*, 105: 25-32

Smith, N.J.H., Williams, J.R., Plucknett, D.L., Talbot, J.P. (1992), Tropical forests and their crops. Comstock Publishing Associates, Ithaca, NY., pp. 20-37

Wilson, H. D. (1988a), Quinua biosystematic I: free-living populations. *Econ. Botan.*, 42: 461-477

Wilson, H. D. (1988b), Quinua biosystematic II: free-living populations. *Econ. Botan.*, 42: 478-494

Wilson, H. D. (1990), Gene flow in squash species. *Bioscience*, 40: 449-455

Zhang, Qifa, Saghai Maroof, M. A., Lu, T.Y., Shen, B.Z. (1992), Genetic diversity and differentiation of indica and japonica rice detected by RFLP analysis. *Theor. Appl. Genet.*, 83: 495-499.

Chapter 17

GENETIC FINGERPRINTING OF COFFEE LEAF RUST DIFFERENTIALS WITH RAPD MARKERS

A. Santa Ram, H.L. Sreenath

Biotechnology Centre, Unit of Central Coffee Research Institute, Coffee Board, Mysore -570 006, India

Running title: Fingerprinting of coffee leaf rust

1. Introduction

Perennial plantation crops such as coffee have a long breeding and selection cycle. Hence, genetic analysis of such plants by conventional methods is a difficult task. Discerning segregating genes in F_2 generation in these crops is very difficult at the juvenile stage. DNA markers are reliable, free from environmental influence and can be assayed at any stage of plant growth with any type of tissue. Recent advances in the field of plant molecular genetics have resulted in the development of a series of DNA markers. Restriction fragment length polymorphism (RFLP), (Botstein *et al.*, 1980) random amplified polymorphic DNA (RAPD) (Williams *et al.*, 1990; Welsh and McClelland, 1990), amplified fragment length polymorphism (AFLP), and simple sequence repeats (SSRs), or microsatellites (Weber and May, 1989) are important among them.

RAPDs are relatively simple to generate and are ideally suited for a breeder's laboratory. In basic PCR, two selected primers of known nucleotide sequence are used to amplify the specific gene of interest (Mullis and Faloona, 1987). Thus, PCR is an effective tool in screening for the presence or, absence of a gene of interest in a variety

T. Sera et al. (eds.), Coffee Biotechnology and Quality, 197–208.

of target DNAs. In contrast, a simple primer of arbitrary nucleotide sequence is used in RAPD assay. Earlier work on RFLPs and RAPDs of coffee evaluated their use as indicators of phylogeny and genetic diversity (Berthou *et al.*, 1983; Orozco-Castillo *et al.*, 1994,1996; Lashermes *et al.*, 1995). Efforts were also devoted towards developing linkage maps of coffee using RFLP and RAPD data (Paillard *et al.*, 1996). Research work linking these markers to leaf rust resistance is lacking. However, a lot of back ground information is available on the genetics of rust resistance in coffee (Eskes, 1989). Several host types manifesting differential resistance to the pathogen races are identified.

In the genus *Coffea, C. arabica* is the only tetraploid species (2n=4x=44), manifesting an often-self-pollinating reproductive behaviour; remaining all other species are diploid (2n=2x=22) and self-incompatible. Arabica is commercially the most important species and is also the one susceptible to a variety of pests and diseases. Leaf rust caused by *Hemileia vastatrix* B. et Br. is of particular importance among the diseases of arabica coffee. Arabica coffee manifests differential race specific resistance against the rust fungus. In the gene bank of Central Coffee Research Institute (CCRI), there are 14 host types of *C. arabica,* including a wild Collection known as Rume Sudan which manifests non-specific (horizontal) resistance to the rust fungus, and two host types derived from *C. racemosa* Lour. and *C. congensis* Pierre. Present RAPD study was conducted on these plant materials to assess their polymorphism.

2. Experimental

2.1 MATERIALS OF THE STUDY

Fourteen host physiologic types of tetraploid *C. arabica* and the two diploid species, C. *racemosa* and C. *congensis* provided the material for the present study. Their genotypes, resistance groups and resistance is summarised in Table 1. These arabica host types studied could be classified into three distinct groups as follows: arabicas, arabicoids and hybrids.

Table 1. : Coffee leaf rust differentials used in the study.

Differentials	Resistance		
	Genes	Group	Intensity
Coffea arabica (tetraploid)			
Arabicas			
Bourbon	SH5	E	highly susceptible
Kents	SH2,5	D	highly susceptible
Geisha	SH1,5	C	moderately susceptible
S12 Kaffa	SH1,4,5	W	moderately susceptible
Cioccie	SH4,5	J	moderately susceptible
Agaro	SH4, 5	J	moderately susceptible
Rume Sudan	..	HR	highly resistant
Arabicoids			
S.288	SH3,5	G	moderately resistant
Hibrido-de-Timor	SH6,7,8,9	A	highly resistant
Hibrido-de-Timor	SH6	R	highly resistant
Hybrids			
Blue Mountain x Cioccie	SH2,4,5	Y	moderately resistant
Blue Mountain x S12 Kaffa	SH1,2,4,5	O	moderately resistant
Dilla & Alghe x S. 333	SH1,3,5	Z	moderately resistant
S. 333 x Dilla & Alghe	SH1,2,3,5	V	moderately resistant
Coffea racemosa (diploid)	..	F	highly susceptible
Coffea congensis (diploid)	..	B	moderately resistant

2.1.1. Arabicas:

These are tetraploid derived from wild arabica and are considered as pure arabica genotypes.

2.1.2. Arabicoids:

These are also tetraploid derived from the putative spontaneous hybridization of *C. arabica* with diploid species such as *C. liberica* (S. 288, S. 333), or *C. canephora* (HDT).

2.1.3. Hybrids:

These hosts are also tetraploid derived from artificial hybridization of arabica and the arabicoid, or between arabicas.

2.2. DNA ISOLATION

High molecular weight genomic DNA was isolated from the young leaves of field grown plants. About 10-15 g young coffee leaves were ground to a fine powder in liquid nitrogen in a pre-chilled mortar. The powder was scooped into four oakridge centrifuge tubes containing 10 ml of pre-warmed (60°C) extraction buffer (2% cetrimide (SRL) (w/v), 1.4 M NaCl, 20 mM EDTA, 100 mM Tris HCl, pH 8.0) containing 1% PVP and 0.2% β-mercaptoethanol, which were added to the buffer immediately before use. This extraction mixture was incubated in a water bath at 60°C for one hour with frequent stirring of the contents. To this, 0.6 volume of chloroform-isoamylalcohol (24:1) reagent was added and the contents of the tubes were thoroughly mixed. Then the tubes were spun at 10,000 rpm for about 15 min at 4^0C to separate the phases. The upper aqueous phase was transferred to fresh pre-cooled oakridge tubes without disturbing the inter-phase. To this, 0.5 volume of ice-cold isopropanol was added. At this stage, DNA precipitated as a whitish net work. DNA was pelleted by spinning the tubes at about 12,000-rpm for 15-20 min at 4°C. DNA pellet was washed with 10-mM ammonium acetate in 70% ethanol. The tubes were again spun at 10,000 rpm for 10 minutes and the washing solution was decanted. DNA pellet was dried under vacuum for about 30 min. Finally, the pellet was dissolved in 1000 µl TE buffer (10 mM Tris HCl, pH 8.0 –1 mM EDTA). These samples were stored at –76° C in an ultra freezer until further use.

2.3. PCR AMPLIFICATION- COMPONENTS AND CONDITIONS

Reaction mixture for PCR composed of 2.5 µl standard assay buffer (10X), 2 units of Taq DNA polymerase (Genei, India), 400 µM dNTPs, 25 ng of arbitrary primer (10-mer) (Operon, USA) and 75 ng of DNA sample. Total volume (25 µl) was made up with sterile distilled water. The sequences of the primers used in this study are presented in Table 2.

Table 2. Decamer primers used in the study

Primer	Nucleotide Sequence 5' 3'
OPA - 01	CAGGCCCTTC
OPB - 12	CCTTGACCGA
OPF - 04	GGTGATCAGG
OPF - 06	GGGAATTCGG
OPF - 07	CCGATATCCC
OPF - 15	CCAGTACTCC
OPAA - 01	AGACGGCTCC
OPAA - 12	GGACCTCTTG
OPAT - 01	CAGTGGTTCC
OPAT - 02	CAGGTCTAGG
OPAT - 11	CCAGATCTCC
OPAT - 18	CCAGCTGTGA

Amplification was carried out in a peltier thermal cycler (PTC 200 DNA Engine, MJ Research, USA) with the following programme: Step I- 94°C for 3 min (initial denaturation), step II- 45 cycles of 94°C for 30 sec (denaturation), 36°C for 1 min (Annealing), 72°C for 2 min (chain extension); step III- 72°C for 10 min (final chain extension).

2.4. ELECTROPHORESIS

PCR products were subjected to sub-marine electrophoresis (Biometra, Germany) at 10 v/cm in 1.2% agarose gel with 1X TBE as tray and gel buffers to resolve the RAPD bands. Either a 100 bp ladder or λDNA digest was used as molecular weight marker. The RAPD profiles were recorded by visual observation of ethidium bromide stained gels on an UV transilluminator, besides being photodocumented with a gel documentation system (Bio-Rad, USA).

2.5. ANALYSIS

RAPD band patterns were analysed using the unweighted pairgroup method of analysis (UPGMA) algorithm to understand the similarities and relationships of the genotypes studied and the results were presented in a dendrogram.

3. Results and Discussion

RAPD profiles generated on 16 host types of coffee with 12 primers revealed DNA polymorphism (Figs. 1-4).

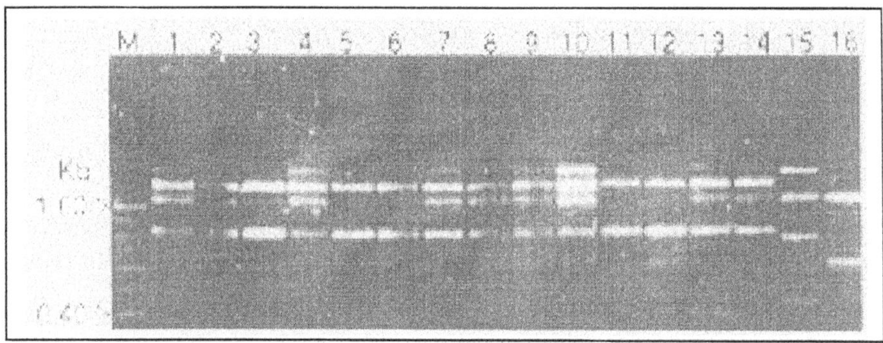

Figure 1 - RAPD profile of coffee leaf rust differentials generated with the primer OPB-12. Details of lanes. M = Marker (100 bp ladder), 1 = Bourbon, 2=Kents, 3=S.288, 4=S_{12} Kaffa, 5=Geisha, 6=Agaro, 7=Cioccie, 8=Rume Sudan, 9=HDT-A, 10=HDT-R,11=Blue Mountain x Cioccie, 12=S.333 x Dilla & Alghe, 13=Dilla & Alghe x S.333. 14=Blue Mountain x S_{12} Kaffa, 15= *C. racemosa*, 16= *C. congensis*.

A total of 132 RAPD loci were visualised from the PCR products of these primers. Of these, 85 were from *C. arabica* and 47 were from *C. congensis* and *C. racemosa.* In *C. arabica,* about 77% of RAPD bands were polymorphic. This high degree of polymorphism could be a reflection of the cryptic genetic variation among arabica varieties, which might have arisen as a means of local adaptations in the diverse geographic locations from where they were drawn. The primers OPA-01 and OPAT-18 distinguished most of the genotypes by producing characteristic product profiles (Figs. 2 and 3). Within each group genomic diversity was less obvious.

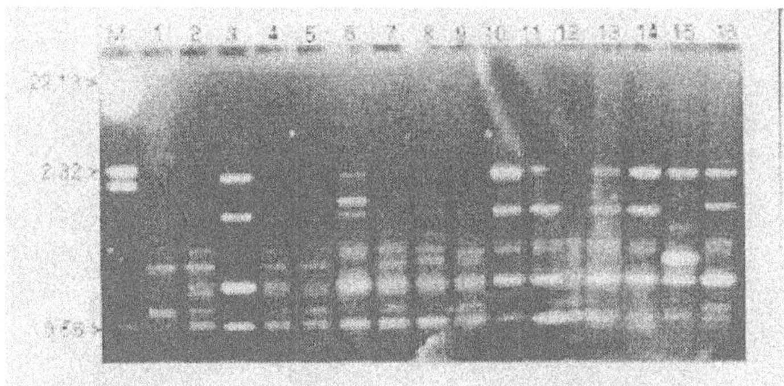

Figure 2 - RAPD profile generated with the primer OPA-01. Lane details same as in Fig. 1, except that marker is λ DNA Hind III digest. Note that most of the differentials can be distinguished using this primer and the two common fragments of *C. arabica* with *C. racemosa* and *C. congensis*.

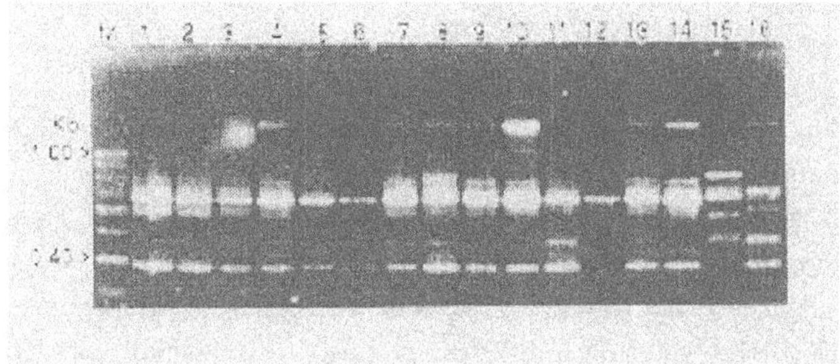

Figure 3 - RAPD profile generated with the primer OPAT-18. Lane details same as in Fig.1. Note the fragment of about 0.4 kb present in all differentials except S.333 x Dilla & Alghe (lane 12).

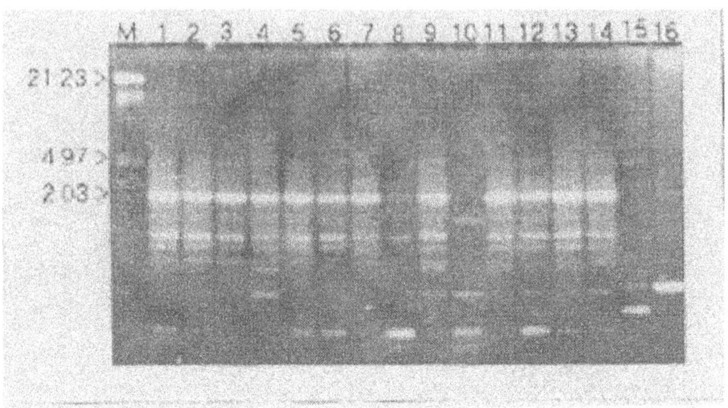

Figure 4 - RAPD profile generated with the primer OPF –06. Lane details same as in Fig.1 except that marker is λ-DNA Hind III digest.

An interesting observation in this study was the distinction between HDT genotypes. HDT-A genotype has been reported unique in carrying the rust resistance genes S_H6, 7, 8 and 9- all derived from *C. canephora* (Bettencourt *et al.*, 1992). Thus, it could be heterozygous at the two loci. Besides this heterozygosity, the two loci carrying different S_H genes make this genotype very distinct with durable resistance (Ram, 1995). HDT-A segregated into different groups, one of which was HDT-R. In the present study, it was observed that some of the primers could distinguish HDT-A and HDT-R by some RAPD bands (Fig. 5).

Besides the above distinctions, the primers also generated a number of common bands from all genotypes of arabica, arabicoid and the hybrids. These could represent constitutive DNA, which makes *C. arabica* a distinct species. Several of these bands could not be observed in the RAPD profiles of the diploid species, *C. racemosa* and *C. congensis* (Figs.1-4). Among the arabica descendants (arabica, arabicoid and the hybrids), there were some DNA fragments, which appeared to be constitutive but were absent in a few host types. This could be a possible indicator of the dynamic genomic differentiation (Fig. 3, 4).

The two host types of diploid species, *C. racemosa* and *C. congensis* could also be distinguished from all the arabica descendants, as well as within themselves by the RAPD bands. Possible homologies of DNA of these species with that of C. *arabica* was revealed by the common bands produced by some primers (Fig. 3). These have been indicative of the genomic distinction of *Coffea* and the possible inter-specific gene flow among the species of this genus (Orozco-Castillo *et al.*, 1994).

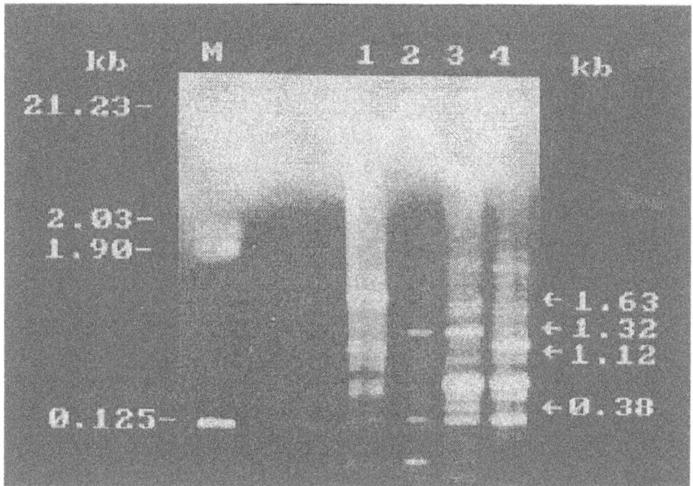

Figure 5 - RAPD bands of HDT-A and HDT-R. Lane details - M=Marker, λ DNA EcoRI Hind III digest; 3=HDT-A, 2 and 4=HDT-R. Bands in lanes 1 and 2 are generated by the primer OPF-6 and those in lanes 3 an generated by OPF-7.

From the dendrogram (Fig. 6), it can be inferred that the genotypes Bourbon and Kents were very closely related. The Ethiopian genotypes S_{12} Kaffa, Agaro, Geisha and Cioccie were similar and manifested a close similarity with Rume Sudan, Bourbon and Kents reflecting the large similarity of all arabica.

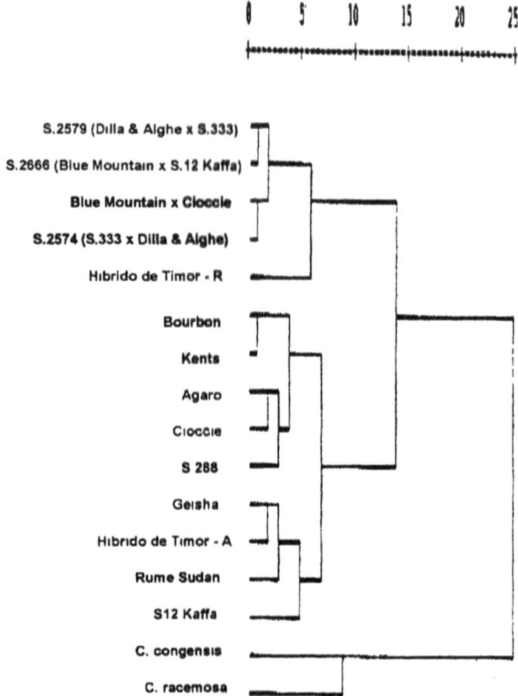

Figure 6 – Dendrogram showing the clustering of coffee leaf rust differentials.

Interestingly, the arabicoid S.288 and HDT-A also exhibited a close similarity with this group, indicating the probable involvement of Typicoid arabica in their evolution. The hybrids S.2574 (S.333 x Dilla and Alghe), S.2579 (Dilla & Alghe x S.333), S.2666 (Blue Mountain x S_{12} Kaffa) and Blue Mountain x Cioccie shared a closely similar banding pattern with HDT-R. Geographic origin of S.288 and S.333 (one of the parents of S.2574 and S.2579) could be traced back to the Doobla Estate in Chikmagalur District. Their putative derivation from Arabica-Liberica hybridization is reflected in the close similarity of their RAPD pattern. Relative distinctness of HDT-A and R genotypes is brought out in this analysis. In spite of clear distinction and 77% genetic polymorphism, arabica, arabicoid and hybrids manifested a close clustering of genotypes with genetic distances ranging from 0 to 5. The diploid species *C. racemosa* and *C. congensis* clustered distinctly from the tetraploid arabica.

From the foregoing discussion, it could be inferred that the genus *Coffea* was of monophyletic origin and drew support from the earlier work (Orozco-Castillo *et al.*, 1996). Also, it could be inferred that the genomic differentiation in *C. arabica* was in a dynamic flux resulting from possible reorganisation/reorientation of DNA at frequent intervals. Thus, there was adequate genetic diversity among the host differentials, which were drawn from geographically diverse locations, as well as among the land races from various provinces of Ethiopia, the home land of *C. arabica*.

4. Conclusions

Leaf rust is a major disease of coffee causing severe crop loss. Worldwide studies have revealed that resistance in the coffee plant and virulence in the rust pathogen were conditioned by a gene-for-gene host-pathogen relationship. Thus, specific genotypes of coffee manifest resistance to specific sets of rust races. There are 14 such differential hosts of *C. arabica* and two differential hosts of *C. racemosa and C. congensis* in the gene bank of CCRI. The present study was undertaken to fingerprint these differentials. The data indicated DNA polymorphism with clear differences among the differentials. Cluster analysis indicated a large degree of similarity among the genotypes of *C. arabica,* which fell into two major clusters.

5. References

Berthou F., Mathieu C., Vedel F. (1983), Chloroplast and mitochondrial DNA variation as indicator of phylogenetic relationships in the genus *Coffea* L. Theor. Appl. Genet., 65: 77-84

Bettencourt A.J., Lopes J., Palla S. (1992), Factores geneticos que a resistencia as racas de *Hemileia vastatrix* Berk. et Br. dos clones tipe dos gruipos 1, 2e 3 de derivados de Hibrido de Timor. Broteria Genetica, XIII (LXXX): 185-194

Botstein D., White R.L., Skolnick M., Davis R.W. (1980), Construction of a genetic map in man using restriction fragment length polymorphism. Amer J. Human Genet., 32: 314-331

Eskes A.B. (1989), Resistance. In - Coffee Rust: Epidemology, Resistance and Management, A. C. Kushalappa and A.B. Eskes (eds.) CRC Press, Boca Raton, pp. 171-291

Lashermes P., Combes M.C., Cros J., Trouslot P., Anthony F., Charrier A. (1995), Origin and genetic diversity of *Coffea arabica* L based on DNA molecular markers. In - Proc. XVI International Scientific Colloquium on Coffee (Kyoto), pp. 528-535

Mullis K.B., Faloona F.A. (1987), Specific synthesis of DNA *in vitro* via a polymerase catalysed chain reaction. Methods. Enzymol., 155: 335-350

Orozco-Castillo C., Chalmers K.J., Waugh R., Powell W. (1994), Detection of genetic diversity and selective gene introgression in coffee using RAPD markers Theor. Appl. Genet., 87: 934-940

Orozco-Castillo C., Chalmers K.J., Powell W., Waugh R. (1996), RAPD and organelle specific PCR re-affirms taxonomic relationships within the genus *Coffea*. Plant Cell Rep., 15: 337-341

Paillard M., Lashermes P., Petiard V. (1996), Construction of a molecular linkage map in coffee. Theor. Appl. Genet., 93: 41-47

Ram A.S. (1995), New dimensions in understanding inheritance of coffee rust resistance: A mendelian perspective. In- Proc. XVI International Scientific Colloquium on Coffee (Kyoto), pp. 548-553

Weber J.L., May P.E. (1989), Abundant class of human DNA polymorphisms which can be typed using polymerase chain reaction. Amer. J. Human Genet., 44: 388-396

Welsh J., McClelland M. (1990), Fingerprinting genomes using PCR with arbitrary primers. Nucleic Acids Res., 18: 7213-7218

Williams J.G.K., Kubelik A.R., Livak K.J., Rafalski J.A., Tingey S.V. (1990), DNA polymorphisms amplified by arbitrary primers are useful as genetic markers. Nucleic Acids Res., 18: 6531-6535

Chapter 18

COFFEE (*COFFEA* SP.) GENETIC TRANSFORMATION FOR INSECT RESISTANCE

M. Dufour [1], T. Leroy [2], C. Carasco-Lacombe [1], R. Philippe [1], C. Fenouillet [1]

[1]CIRAD-CP, B. P. 5035, 34032 Montpellier Cedex 1, France; [2]Centre de Recherches Nestlé, B. P. 9716, 37097 Tours Cedex 2, France

Running title: Coffee genetic transformation

1. Introduction

Coffee, as a woody species, has a long biological cycle and it takes between four and five years from seed to seed. Therefore, classical breeding programmes generally spread over 20 to 30 years. Unconventional breeding techniques would be of utmost importance for quick genetic progress. One of these techniques is genetic transformation. Using this process, it could be possible to insert selected traits in coffee without changing the whole genome. Genetic transformation is basically the introduction of foreign DNA into plant cells. Two main techniques are used for plant transformation: i) direct transformation, through biolistics (McCabe *et al.*, 1988), DNA uptake (Zhang and Wu, 1988), or protoplast electroporation (Fromm *et al.*, 1985), and, ii) indirect transformation using viruses or *Agrobacterium sp* (Bevan *et al.*, 1983).

Agrobacteria are natural soil-born bacteria that infect the majority of dicotyledons and a few monocotyledons. Their pathogenecity is due to the plasmid Ti-plasmid (tumour inducing) for *A. tumefaciens,* or Ri-plasmid (root inducing) for *A. rhizogenes*. Portions of the plasmid known as T-DNA are transferred into the nuclear genome of the plant. The attachment of the bacteria on the plant cells as well as the transfer itself is made possible by functions encoded by the *vir* region of the plasmid (Zupan and Zambryski, 1995). The transferred genes stably incorporate into the plant genome. The genes responsible for tumour formation can be removed (the plasmid is then called disarmed) and replaced with virtually any gene. These genes can, therefore, be transferred into plants. The efficiency of transformation is dependent on four factors:

T. Sera et al. (eds.), Coffee Biotechnology and Quality, 209–217.

a) Ability of *Agrobacterium* to efficiently transform cells in the target tissue (controlled by genetic and physiological factors of both the plant and *Agrobacterium*). The virulence of the *Agrobacterium* strain has to be tested first.
b) Efficient selection of all the transformed cells. Selectable and scorable (i.e. reporter) genes are needed.
c) Efficient regeneration from the transformed cells. It is the bottleneck in the process, especially for perennials. However in coffee, various regeneration pathways can be used like regeneration from protoplast, regeneration from leaf explants through organogenesis or somatic embryogenesis (Yasuda *et al.*, 1985; Zamarripa *et al.*, 1991; Berthouly and Michaux-Ferrière, 1996).
d) Stability of the incorporated DNA into the genome of the plant. In annual plants, this can be determined easily by selfing the plant in order to determine if the gene is inheritable. With perennials, Southern blot analysis must be used to identify different segments of the inserted T-DNA.

Genetic transformation was described for the first time in coffee by Barton *et al.* (1991) who worked on protoplast electroporation of *Coffea arabica*. However, no transgenic plant was regenerated from these experiments. Later on, Feng *et al.* (1992), using *C. arabica* cotyledons and somatic embryos integrated a foreign gene conferring resistance to Kanamycin (an antibiotic). Again, no plant was obtained. It was only in 1993 when transgenic plants were created by Spiral *et al.* using a co-culture of C. *canephora* somatic embryos with *Agrobacterium rhizogenes*. The regenerated plants were scored to have several DNA inserts. Unfortunately, most of these plants expressed a "hairy root" pattern. Since then, genetic transformation has been extended to other genotypes and other explant sources (Spiral and Pétiard, 1993; Sugiyama *et al.*, 1995). Leroy *et al.*, (1997) successfully used *A. tumefaciens*.

Protection against attacks by detrimental pests represents a major challenge for crop production in agriculture. This is especially true for coffee, for which a large percentage of the potential harvest yield of Brazil and East Africa is lost each year due to insects, in particular to the leaf miner (*Perileucoptera* sp.) (Guerreiro *et al.*, 1990). Since coffee is a perennial crop, it is not possible to use crop rotation as seen with annuals. Also, there are no breeding programmes using natural sources of resistance for that insect.

Presently, control of pest development is based on application of protective agro-chemicals. For the farmers, the products are expensive and their impact is often harmful for the environment. *Bacillus thuringiensis* (*B. t.*) is an entomocidal bacterium. For more than 35 years, crystalline δ-endotoxins, which accumulate during sporulation of this bacterium, have been available as an alternative to synthetic insecticides in controlling harvest losses. The crystalline inclusions containing the endotoxins are activated in the midgut of the insects and bind to specific high affinity receptors on cell membranes. Within minutes, the midgut cells are paralysed and disrupted. *B. t.* toxins are diverse and have been found active against many insects within the order of Lepidoptera,

Diptera and Coleoptera. The use of *B. t.* genes to transform plants is a feasible strategy to fight insects efficiently (Estruch *et al.*, 1997; Schuler *et al.*, 1998).
Several *B.t.* endotoxins have been tested against *Perileucoptera coffeella* Guérin Méneville and one, CryIA(c), was shown as very efficient (Guerreiro *et al.*, 1993, 1998). The gene corresponding to this toxin has been isolated, cloned, artificially synthesised, and included in a vector construction containing also a selectable and a marker genes. Several transgenic coffee plants actually resistant to the leaf miner have been obtained (Leroy *et al.*, 1999,2000). Here, we describe genetic transformation of four different coffee genotypes with the *cryIA(c)* gene and the verification of the gene expression through Western blot. The accuracy of the correlation with insect bioassays is also discussed.

2. Experimental

2.1. PLANT MATERIAL

Two *C. canephora* and two *C. arabica* genotypes were used (clones number 126 and 197, Catimor (8661-4), and Et29 x Ca5 F1 hybrid, respectively). They were propagated *in vitro* micro-cutting on MS-based medium (Murashige and Skoog, 1962) containing 4.44 μM BAP.

2.2.TRANSFORMATION PROTOCOL

LBA4404 disarmed strain of *A. tumefaciens* was used with the pBin19 plasmid (Bevan, 1984), engineered with three genes: *uidA*, coding for β-glucuronidase and containing an additional intron; the *csr1-1* gene conferring resistance to chlorsulfuron used as a selection agent; and a synthetic *cry1A(c)* gene (Sardana *et al.*, 1996), driven by the EF1α- promoter (Curie *et al.*, 1991). Leaf explants were cultured on specific semi-solid media and embryogenic cells and somatic embryos were obtained. Between 1,000 and 10,000 somatic embryos, or 10 to 50 grams embryogenic cells were used for transformation. These were soaked with bacteria ($O.D._{600nm}$: 0.3-0.5) for two hours and co-cultivated for three days in hormone-free MS medium. They were then cultivated in presence of cefotaxime (400 $mg.l^{-1}$) for 21 to 28 days, and then transferred to selective MS medium with cefotaxime (400 $mg.l^{-1}$) and chlorsulfuron (80 $\mu g.l^{-1}$). After three to six months, regeneration occurred and the somatic embryos germinated on MS medium supplemented with Morel vitamins, 1μM BAP and 0.03-M sucrose.

2.3. ROOTING AND HARDENING

Once the somatic embryos germinated and exhibited one, or two pairs of leaves, their base was sectioned with a scalpel and soaked for 12 hours in a sterile MS/2 medium containing IBA and NAA (50 $mg.l^{-1}$ each). They were then transferred to the greenhouse in peat pellets under high humidity conditions for at least three weeks. Relative humidity was then lowered progressively. Rooting occurred within a month.

2.4. HISTOCHEMICAL GUS ASSAY

β- glucuronidase was assayed by overnight incubation in the classical medium (Jefferson, 1987). In order to reduce endogenous activity of non-integrated *uidA* gene, 20% v/v methanol was added. (Kosugi *et al.*, 1990).

2.5. MOLECULAR ANALYSIS

Only the GUS positive plants were used for total DNA extraction. The protocol was described by Edwards *et al.* (1991) and modified by the addition of 100-mM sodium bisulphite. Southern blotting analysis was done as described by Leroy *et al.* (1999) and Spiral *et al.* (1999).

2.6. BIOASSAYS

Regenerated plants of different ages were put in presence of adult Tanzania leaf miners (*Leucoptera caffeina*) for 24 hours. Two weeks later, an overall score was affected to the plant and number of pups was recorded.

2.7. WESTERN BLOTS

Leaves were ground in liquid nitrogen and proteins were extracted from about one gram of fresh leaves in a standard buffer containing Tris –HCl pH8, 87% glycerol (57.5 $ml.l^{-1}$), EDTA, DTT (155 $mg.l^{-1}$), PMSF (175 $mg.l^{-1}$) and Triton 1$ml.l^{-1}$. The extracts were then concentrated using Millipore Ultrafree-4® (biomax-50 membranes) filtration units (30-50 min. centrifugation at 4500 rpm) and immediately checked for protein content (Bradford, 1976). The extracts were then aliquoted and denatured (Laemmli, 1970). A 10% SDS-Page gels were used and migration of the proteins was carried out for four hours at ca. 100 V. Gels were then transferred to nitro-cellulose membranes using the Bio-Rad transblot system in Towbin buffer (Towbin *et al.*, 1979) for one hour at 100 V.

Blocking occurred overnight in 0.5% non fat milk in TBS buffer supplemented with 0.1% (v/v) Tween-20. A polyclonal rabbit antiserum raised against the purified Cry1A(c) protein was used (1/2500 v/v) and detected by a goat anti-rabbit antiserum alkaline phosphatase (AP)-conjugate (Promega) at a dilution of 1/3300 (v/v).

3. Results and discussion

For the *C. canephora* genotypes, transformation efficiency was about 1%, which was approximately 1% of the explants submitted to the procedure produced embryogenic calli on selective medium. This result slightly increased in further experiments (Leroy *et al.*, 1999). Not every embryogenic callus gave rise to somatic embryo. About 30% of calli produced embryos, of which 80% germinated properly. More than 200 independent primary transformants were obtained using somatic embryos, or embryogenic calli for clone 126, and only 30 for clone 197. Three copies each of 120 primary tansformants of clone 126 were transferred to the greenhouse. The two arabica genotypes behaved differently. On selective medium, 0 to 5% of the primary Catimor explants produced directly somatic embryos, depending on the experiment. Less than 0.1% embryogenic calli appeared on the F_1 hybrid, 10% of which produced somatic embryos with a germination rate similar to that of *C. canephora*. Obviously, there was a wide genotype effect for response to the genetic transformation procedure. *C. canephora* is a diploid species in contrast with *C. arabica*, which is tetraploid. Such behaviour differences have already been noted for the aptitude to liquid medium somatic embryogenesis (Zamarripa *et al.*, 1991).

Chlorsulfuron is a herbicide and seems to be an adequate agent for selection of coffee transformed tissue. However, a certain rate of escape, up to 50% has been observed. Consequently, GUS assay was also used as a screen (Leroy *et al.*, 1999). Former experiments had shown that the antibiotic Kanamycin was not as efficient (Spiral *et al.*, 1993). Two types of analysis were carried out on the transformed plants growing in the greenhouse for at least two months. PCR and Southern blotting analysis established the presence of foreign genes. The quality of integration of the T-DNA and the number of copies integrated was documented by Southern analysis. Among 51 transformation events from *C. canephora*, the majority (69%) contained one copy of the foreign T-DNA. Up to five copies have been observed in other events (Leroy *et al.*, 1999). The presence of an expressed *cry*1A(c) gene was established by Western blotting (Fig. 1), performed on 21 independent transformation events (clone 126).

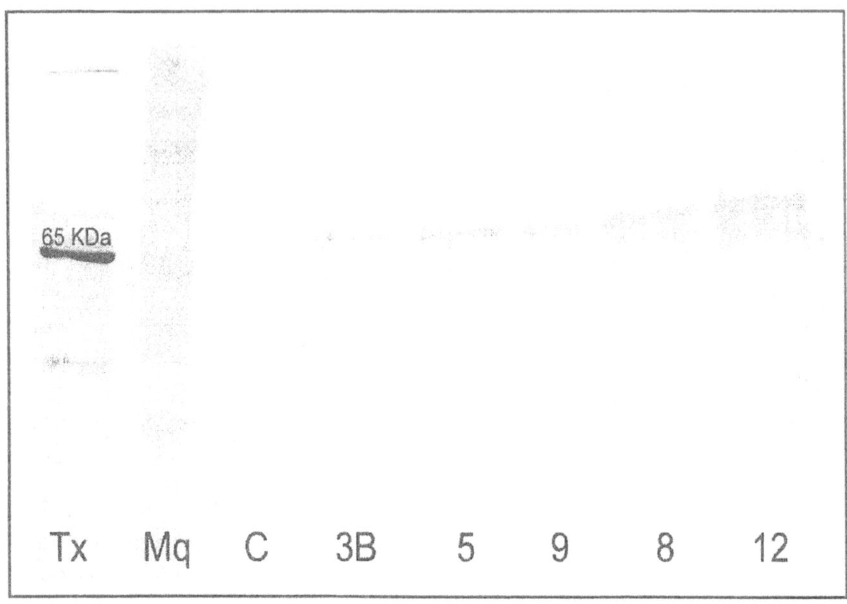

Figure 1 : Western blot membrane showing different plant profiles. C = control, non transgenic plant. Tx = purified CryIA(c) toxin.

Among these plants, the majority appeared rather resistant when submitted to the bioassay (Table1) with three plants being very resistant (note ≤ 1 and 0 pupa) and eight intermediate-resistant (1<note≤2 and number of pupae ≤ 2). All these plants had CryIA(c) protein activity detected by Western blotting., which confirmed the bioassay results. Five plants were susceptible; two of them were controls, two others did not have any CryIA(c) protein detected and one was Western positive. That plant was probably in a bad physiological state, which made it susceptible to the leaf miner, although the CryIA(c) protein was expressed. Also, the level of expression was perhaps low.

Besides these two categories, five plants were "intermediate-susceptible", their note being between 1.2-1.8 with a number of pupae from 3-9. Three were Western positive, but two did not have any CryIA(c) protein detected. The cryIA(c) gene was detected by PCR and Southern blot. These occurrences have already been discussed (Leroy *et al.*, 1999) and have been confirmed by recent Western blots. It could be likely that the toxin was expressed in a small amount (below 0.1% of total protein), rendering the plants somehow susceptible.

Table 1. Comparative results of Western blots and bioassays for 21 clone 126 independent transformation events.

Plant Number	Plant height (cm)	Bioassay note	Number of pupae	Western blot result[a]
1	17	0.9	0	+
14	37	0.9	2	+
10	26	1	0	+
20	33	1	0	+
15	29	1.1	0	+
11	24	1.1	0	+
5	47	1.1	2	+
18	48	1.2	7	+
6	35	1.4	2	+
3	23	1.5	0	+
16	35	1.6	3	+
8	29	1.7	0	+
12	33	1.7	2	+
7	34	1.8	3	+
4	45	2.5	19	+
19	48	1.2	5	-
9	50	1.6	9	-
Control A*	21	3	29	-
17	28	3	32	-
Control B*	17	3.9	11	-
2	39	3.8	29	-

[a]presence (+) or absence (-) of a band corresponding to CryIA (c) protein
*control A : non transformed plant ; control B : plant transformed with plasmid devoid of *cry* gene.

4. Conclusions

The transformation technique described here has been efficient on four different coffee genotypes although some differences have been observed in response according to genotype. A large number of plants was transferred to the greenhouse and does not show any phenotypic aberration. Integration of T-DNA in *C. canephora* clone 126 has been confirmed by Southern blot analysis and Western blot allowed to comment gene expression. It seems that expression varied according to the bioassays, and it would be interesting to try to quantify the amount of toxin through Western blot. Bioassays are routinely carried out at the laboratory. The next step is to perform them in the field under natural conditions. This will be achieved next year, within a field study carried out in

French Guyana and involving 1200 *C. canephora* clone 126 plants. The plants will be under study for five years. The purposes of this study will also include overall agronomic evaluation of the transgenic plants, including fruit production, transgenic pollen dissemination and impact of transgenic coffee on wild non-target insects. Prospects for future studies include widening of coffee genotype array, use of other selective genes such as PMI gene (phosphate-6-mannose isomerase), and identification of genes conferring resistance to other coffee diseases or pests. Quality or fruit ripening genes could also be identified for their use in future genetic transformation programmes.

5. References

Barton, C.R., Adams, T.L., Zarowitz, M. A. (1991), Stable transformation of foreign DNA into *Coffea arabica* plants. In Asic: 14ème colloque scientifique sur le café, San Francisco, pp. 460-464

Berthouly, M., Michaux-Ferrière, N. (1996), High frequency somatic embryogenesis in *Coffea canephora* (var. Robusta): induction conditions and histological evolution. Plant Cell, Tissue Organ Culture, 44: 169-176

Bevan M. (1984), Binary *Agrobacterium* vectors for plant transformation. Nucleic Acid Res. 12: 8711-8721

Bevan, M., Flavell, R., Chilton, M.-D. (1983), A chimaeric antibiotic resistance gene as a selectable marker for plant cell transformation. Nature, 304: 184-187

Bradford, M.D. (1976), A rapid and sensitive method for the quantitation of microgram quantities of protein utilizing the principle of protein dye binding. Anal. Biochem., 72: 248-254

Curie, C., Liboz, T., Bardet, C., Gander, E., Medale, C., Axelos, M., Lescure, B. (1991), Cis and trans-acting elements involved in the activation of *Arabidopsis thaliana* A1 gene encoding the translation elongation factors EF-1 alpha. Nucleic Acid Res., 19: 1305-1310

Edwards K., Johnston C., Thompson C. (1991), A simple method for PCR analysis. Nucleic Acid Res., 19: 1349

Estruch J.J., Carrozzi, N.B., Desai, N., Duck, N.B., Warren, G.W., Koziel, M.G. (1997), Transgenic plants : an emerging approach to pest control. Nature Biotechnol., 15: 137-141

Feng, A. V., Yang M.Z., Zheng X.S., Pan N. S., Chen Z. L. (1992), *Agrobacterium* mediated transformation of coffee (*Coffea arabica* L.), Chinese J Biotechnol., 8 (3) : 255-260

Fromm, M., Taylor, M., Walbot, V. (1985). Expression of genes transferred into monocot and dicot plant cells by electroporation.. Proc. Natl. Acad. Sci., USA, 82: 5824-5828

Guerreiro F. O., Penna M.F., Gonçalves W., Carvalho A. (1990), Melhoramento do cafeiiro: XLIII Selecao do cafeeiros resistentes ao bicho-miniero. Bragantia 49 (2): 291-304

Guerreiro, F. O., Denolf, P., Frutos, R., Peferoen, M., Eskes, A. B. (1993), L'identification de toxines de *Bacillus thuringiensis* actives contre *Perileucoptera coffeella*. Asic, 15ème colloque scientifique sur le café, Montpellier, France, pp. 329-335

Guerreiro, F. O., Denolf, P., Peferoen, M., Decazy, B., Eskes, A. B., Frutos, R. (1998), Susceptibility of the coffee leaf miner (*Perileucoptera spp*) to *Bacillus thuringiensis* δ endotoxins: a model for transgenic perennial crops resistant to endocarpic insects. Current Microbiol. 36: 175-179

Jefferson, R. (1987), Assaying chimeric genes in plants: the GUS gene fusion system. Plant Molec. Biol. Reporter 5 (4): 387-405

Kosugi, S., Ohashi, Y., Nakayama, K., Arai, Y. (1990), An improved assay for β-glucuronidase in transformed cells: methanol almost completely suppresses a putative endogenous β-glucuronidase activity. Plant Science 70: 133-140.

Laemmli, U.K. (1970), Cleavage of structural proteins during the assembly of the head of bacteriophage T4. Nature 227: 680-685

Leroy T., Royer M. Paillard M., Berthouly M., Spiral J., Tessereau S., Legavre T., Altosaar I. (1997), Introduction de gènes d'intérêt agronomique dans l'espèce *Coffea canephora* Pierre par transformation avec *Agrobacterium sp.* In Asic, 17ème colloque scientifique sur le café, Nairobi,Kenya, pp. 439-446

Leroy, T., Philippe, R., Royer, M., Frutos, R., Duris, R., Dufou,r M., Jourdan, I., Lacombe, C., Fenouillet, C. (1999), Genetically modified coffee trees for resistance to coffee leaf miner. Analysis of gene expression, resistance to insects and agronomic value. In Asic 18ème colloque scientifique sur le café, Helsinki, Finland

Leroy, T., Henry, A.-M ., Royer, M., Altosaar, I., Frutos,R., Duris, R., Philippe, R. (2000), Genetically modified coffee plants expressing the *Bacillus thuringiensis* cryIAc gene for resistance to leaf miner. Plant Cell Reports, 19: 382-389

McCabe, D., Swain, W., Martinelli, B., Christou, P. (1988), Stable transformation of soybean (*Glycine max*) by particle acceleration. Bio/Technology 6: 923-926

Murashige, T., Skoog, F. (1962), A revised medium for rapid growth and bioassays with tobacco tissue culture. Physiol. Plant., 15 : 473-497

Sardana, R., Dukiandjiev, S., Giband, M., Cheng, X.Y., Cowan, K., Sauder, C. and Altosaar, I. (1996), Construction and rapid testing of synthetic and modified toxin gene sequences CryIA (b&c) by expression in maize endosperm culture. Plant Cell Reports, 15 : 677-681

Schuler, T.H., Poppy, G. M., Kerry B. R, Denholm I. (1998), Insect resistant transgenic plants. Tib Tech., 16: 168-175

Spiral, J., Pétiard, V. (1993), Développement d'une méthode de transformation génétique appliquée à différentes espèces de caféiers et régénération de plantules transgéniques. In Asic, 15ème colloque scientifique sur le café, Montpellier, France, pp. 115-122

Spiral, J., Thierry, C., Paillard, M., Pétiard, V. (1993), Obtention de plantules de *Coffea canephora* Pierre (Robusta) transformées par *Agrobacterium rhizogenes*. C.R. Acad. Sci. Paris 316 (série III): 1-6

Spiral J., Leroy T., Paillard, M., Petiard V. (1999), Transgenic coffee (*Coffea* species). In-: Biotechnology in Agriculture and Forestry, Vol. 44, Y.P.S. Bajaj (ed.), Springer-Verlag, Berlin, pp 55-76

Sugiyama, M., Matsuoka, C., Takagi T. (1995), Transformation of coffee with *Agrobacterium rhizogenes*. In Asic, 16ème colloque scientifique sur le café, Kyoto, Japan, pp. 853-859

Towbin, H., Staehelin T., Gordon J. (1979), Electrophoretic transfer of proteins from polyacrylamide gels to nitocellulose sheets : Procedure and some applications. Proc. Natl. Acad. Sci., USA, 76: 4350-4354

Yasuda, T., Fujii, Y., Yamaguchi T. (1985), Embryogenic callus induction from *Coffea arabica* leaf explants by benzyladenine. Plant Cell Physiol., 26: 595-597

Zamarripa, A., Ducos, J.P., Tessereau, H., Bollon, H., Eskes, A., Pétiard, V. (1991), Production d'embryons somatiques de caféier en milieu liquide : effet de la densité d'inoculation et du renouvellement du milieu. Café Cacao Thé, XXXV (4): 233-244

Zhang, W. and Wu, R. (1988), Efficient regeneration of transgenic plants from rice protoplast and correctly regulated expression of the foreign gene in the plants. Theor. Appl. Gen., 76: 835-840

Zupan, J.R., Zambryski, P. (1995), Transfer of T-DNA from *Agrobacterium* to the plant cell. *Plant Physiol.* 107: 1041-1047

Chapter 19

DEVELOPMENT OF COFFEE TREES RESISTANT TO LEAF MINER

O. Guerreiro-Filho, H.P. Medina-Filho, L.C. Fazuoli, W. Gonçalves

Instituto Agronômico de Campinas, Centro de Café, CP 28; 13001-970, Campinas, SP, Brazil. E-mail: oliveiro@cec.iac.br

Running title: Trees resistant to leaf miner

1. Introduction

The production of coffee in Brazil reached 34 million bags in 1999 being 29,5 million from *Coffea arabica* cultivars. Since all such cultivars are susceptible to leaf miner (*Perileucoptera coffeella*), the main pest in Brazil, the yield and economic losses are considerable. It is under way for several years at the Instituto Agronômico de Campinas, a breeding program aiming at the development of high yielding, leaf miner resistant cultivars with standard arabica cup quality. The main steps of this program are discussed in this chapter.

2. Identification of sources of resistance

As *C. arabica* is the only tetraploid (2n = 44 chromosomes) species of *Coffea* the search for resistance to the insect had started among individuals of the species. It soon became evident that no cultivar, botanical variety, mutant or accession of *C. arabica* was resistant to the leaf miner. However, different levels of resistance were found among the diploid species (2n = 22 chromosomes) of *Coffea* such as *C. liberica*, *C. eugenioides*, *C. kapakata*, *C. salvatrix* and *C. racemosa*. The highest level of resistance was found in *C. stenophylla* where the larvae dies soon after the eggs hatch and start feeding in the leaves, eventually causing very small lesions (Inhering, 1912; Sein, 1942; Speer, 1949; Vicente-Chandler *et al.*, 1968; Medina-Filho *et al.*, 1977a; Guerreiro-Filho and Medina-Filho, 1987; Guerreiro-Filho *et al.*, 1987; Guerreiro-Filho *et al.*, 1991). Caffeine free species from Madagascar and Mascarene Islands such as *C. tetragona*, *C. bertrandi*, *C. millotii*, *C. resinosa*, *C. tsirananae*, *C. bonnieri*, *C. dolichophylla* and *C. farafanganensis* are also resistant (Guerreiro-Filho, 1994).

T. Sera et al. (eds.), Coffee Biotechnology and Quality, 219–227.

Despite this resistance, similarly to *C. stenophylla* (Carvalho and Monaco, 1967; Medina-Filho *et al.*, 1984), these species, are cross-incompatible with *C. arabica*. Purposely transfer of leaf miner resistance to *C. arabica* was started after the identification of two resistant segregantes (C1195-5-6-1 and C1195-5-6-2) of the first natural backcross to arabica of a triploid hybrid (33 chromosomes), of *C. racemosa* x *C. arabica*. In addition to the dominant gene(s) for leaf miner resistance, those two plants were drought tolerant, medium to early ripening and one of them was also resistant to rust. Although both were aneuploid, they had a quite favorable cytogenetic constitution of 44 and 45 chromosomes (Medina-Filho *et al.*, 1977b). This situation fostered the transfer of resistance to arabica cultivars, associating it with leaf rust (*Hemileia vastatrix)* resistance reinforced through crossings with Icatu selections (Guerreiro-Filho *et al.*, 1990).

3. Genetics, expression of resistance and the breeding methodology

Notwithstand the fact that in *C. stenophylla* the resistance to leaf miner is probably recessive (Guerreiro-Filho *et al.*, 1987), the analysis of a number of plants from over a dozen interspecific hybrids and backcrosses indicated that the resistance present in both *C. racemosa* and the diploid species is due to a major dominant gene with modifiers (Medina-Filho *et al.*, 1977a,b). Backcross progenies of C1195-5-6-1 and C1195-5-6-2 with Icatu were analyzed in laboratory conditions by Guerreiro-Filho (1989). The resistance of those lines originally present in *C. racemosa* and formerly identified in the field as plants with low levels of natural infestation was due to an inherent resistance of the leaf tissue and not due to a possible non-preference of the insect since they lay eggs and they hatch at the same rate in both resistant and susceptible plants, the difference between them being chiefly in the area of the subsequent lesions. Analysis of segregant progenies from selfings and backcrosses confirmed that the simple dominant nature of leaf miner resistance from *C. racemosa* derivatives C1195-5-6-1 and C1195-5-6-2 has probably an additional pair of complementary alleles, Lm_1 and Lm_2 (Guerreiro-Filho, 1994).

In field conditions, the intensity of infestation is not related to the size, thickness or ploidy levels of the leaves. It is however, dependent on the season and on the position of the leaves in the plant. The dry season and the top portion of the plants usually display an increased rate of infested leaves (Medina-Filho *et al.*, 1977a). The age of the leaves although not related with percentage of infestation in field conditions (Medina-Filho *et al.*, 1977a) it nevertheless does play an important role in the expression of the resistance as demonstrated in leaf discs by Guerreiro-Filho *et al.* (1999) in *C. racemosa* and its hybrids with *C. arabica* as the first two pairs of leaves in the branches are more resistant.

The resistance to leaf miner is dominant (Medina-Filho *et al.*, 1977a; Guerreiro-Filho *et al.*, 1999), originally present in a wild species (*C. racemosa*) which is diploid and with very poor agronomic performance in addition to an extremely unpleasant flavor of the roasted coffee beans. Provide its resistance should be transferred to a good cultivated tetraploid variety, the choice breeding strategy was the backcross method mixed with pedigree selection. In such procedure, along the generations (Table 1) plants were selected on the basis of agronomic performance, yield, resistance to leaf miner and rust. Among a multitude of other characters selected for and against, special attention was paid in discarding progenies with low rates of semi-empty fruits, of peaberry seeds and phenotypically angustifolia leaves at the nursery stage since such an array of phenotypes in coffee is strongly associated with abnormal meiotic behavior in the mother plant and the ensuing aneuploidy in the progenies (Cruz, 1972; Medina-Filho *et al.*, 1977b).

4. Screening Strategy

Screening for leaf miner resistance was formerly done solely in the field, under natural conditions during winter time in Campinas when the infestation is usually high (Medina-Filho *et al.*, 1977a, Guerreiro-Filho *et al.*, 1990). The selections were greatly expedited by the use of laboratory procedures for raising the insects (Katiyar and Ferrer, 1968) coupled with evaluation of damage in leaf discs (Figure 1D) maintained in controlled conditions (Guerreiro-Filho *et al.*, 1992; Guerreiro-Filho, 1994).

Presently, hundreds of six month old advanced progenies are firstly screened in the lath house where individual plants and entire progenies are discarded if heavily infested or showing large lesions, typical symptoms of susceptibility. Unaffected plants are then subjected to the lab tests and the resistant plants are transplanted to the field and evaluated in competitive trials for yield and several other agronomic attributes besides a subjective rating (1 to 10) for the intensity of natural infestation. Elite progenies are also checked for cup quality.

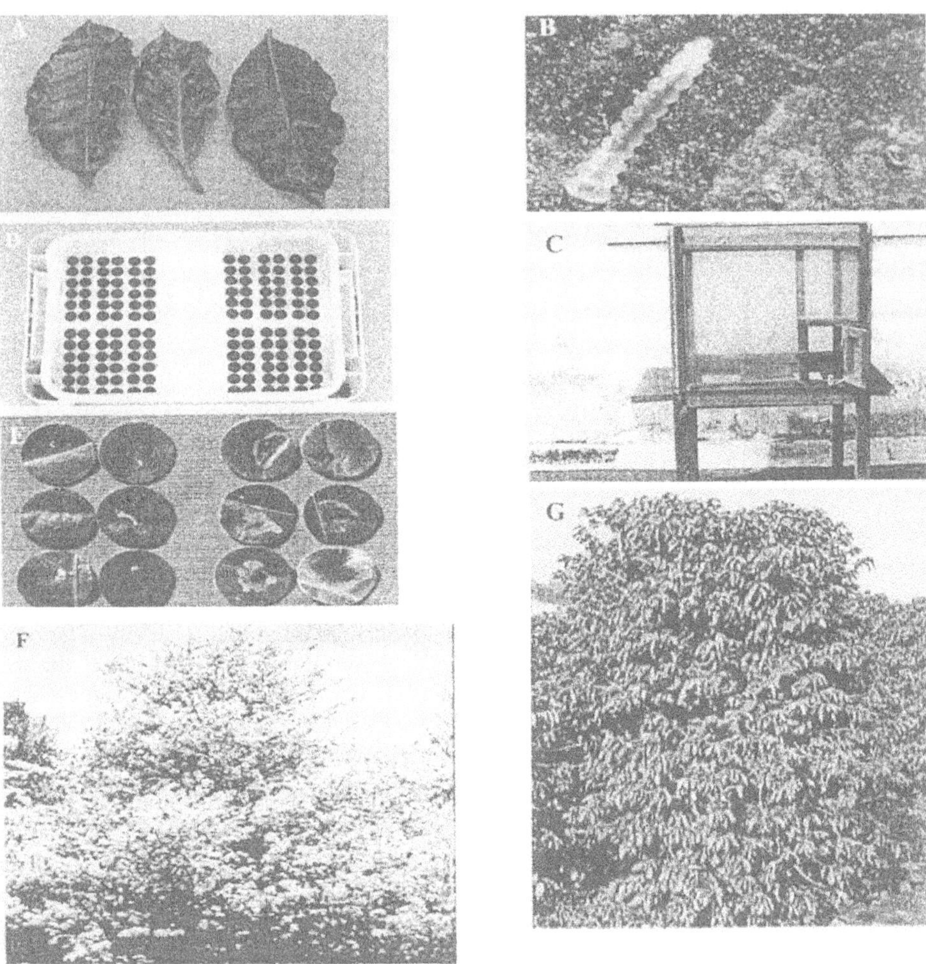

Figure 1. A) Typical large lesions of leaf miner in susceptible genotypes; B) Caterpillar and eggs of P. coffeella; C) Wooden frame covered with cotton material for raising the insects; D) Leaf discs in moist plastic box used for lab screenings; E) Leaf discs of resistant (R) and susceptible (S) genotypes after development of lesions; F) Blooming of *C. racemosa*, the donnor species of leaf miner resistance and G) Selected resistant plant of advanced backcross progeny.

Table 1. Development of some backcross (BC) and open-pollinated (OP) selected plants (*) and progenies (**) resistant (R) and susceptible (S) to leaf miner. IAC - # are lines of cv. Catuaí.

Generation	Population				
Parental	*C. racemosa* C1195 (R) x *C. arabica* cv Blue Mountain (S)				
Fl	C1195-5 (R) *(C. racemosa x C. arabica)*				
BC1	C1195-5-6 (R) *(C. arabica x* C1195-5)				
BC_2	C1195-5-6-1(R) *(C. arabica x* C1195-5-6)	C1195-5-6-2(R) *C. arabica x* C1195-5-6			
BC_3	H13684-7 (S) (IAC-81 x C1195-5-6-2)	H11877-4 (R) (IAC-24 x C1195-5-6-2)	H11877-11 (R) (IAC-24 x C1195-5-6-2)		
	H13684-7-H13684-7 [OP]	H11877-4-H11877-4 [OP]	H11877-11-H11877-11 [OP]	H11877-11-9(R) H11877-11 [OP]	H11877-11-18 (R) H11877-11 [OP]
BC_3F_2	H11421-17-23 (R) H11421-17 [OP]	H11536-3-49 (R) H11536-3 [OP]			
BC_3F_3	H11877-11-9-H11877-11-9 [OP]	H11877-11-18-H11877-11-18 [OP]	H11421-17-23-H11421-17-23 [OP]	H11536-3-49-H11536-3-49 [OP]	
	H13685-1* (R) (IAC-81 x H11421-11)	H13685-2 (S) (IAC-81 x H11421-11)	H13465-5 (S) (IAC-62 x H11877-)	H13465-7 (R) (IAC-62 x H11877-)	H14066-4 (R) (IAC-62 x H12114-1)
BC_4	H14066-11 (S) (IAC-62 x H12114-1)	H14060-7 (R) (IAC-62 x H12092-)	H14060-10 (S) (IAC-62 x H12092-)	H14136-5 (R) (IAC-100 x H11421-17)	H14104-4 (R) (IAC-72 x H12074-)
	H14066-12 (R) (IAC-62 x H12114-1)	H14066-13 (R) (IAC-62 x H12114-1)	H14940- (IAC-62 x H12037-1)	H14941- (IAC-62 x H11877-11)	H14945- (IAC-62 x H12037-1)
	H14949- (IAC-62 x H12092-5)	H14950- (IAC-62 x H11877-11)	H14961- (IAC-62 x H11421-17)	H14963- (IAC-62 x H11421-11)	
	H13685-1-** (IAC-81 x H11421-11) [OP]	H13685-2- (IAC-81 x H11421-11) [OP]	H13465-5- (IAC-62 x H11877-) [OP]	H13465-7- (IAC-62 x H11877-) [OP]	H14066-4- (IAC-62 x H12114-1) [OP]
BC_4F_2	H14066-11- (IAC-62 x H12114-1) [OP]	H14060-7- (IAC-62 x H12092-) [OP]	H14060-10- (IAC-62 x H12092-) [OP]	H14136-5- (IAC-100 x H11421-17) [OP]	H14104-4- (IAC-72 x H12074-) [OP]
	H14066-12- (IAC-62 x H12114-1) [OP]	H14066-13- (IAC-62 x H12114-1) [OP]			
BC_5	H14955- (IAC-62 x H13376-8)	H14926- (Pacas x H13685-1)	H14998- (IAC-62 x H14096-2)	H14954- (IAC-62 x H13685-1)	H14964- (IAC-62 x H13660-6)

5. Laboratory technique for raising insects and evaluating lesions in leaf discs

Massive production of insects is necessary for year round screening with an efficient infestation of leaves in laboratory conditions. For raising the insects it used the methodology described by Katiyar and Ferrer (1968) where adult insects are fed with 10% sucrose (Parra, 1985) for enhancing the number of eggs laid. Seedling of susceptible cultivars are kept overnight inside a wooden frame covered with cotton material, under 27 ± 2°C, 70 ± 10% relative humidity and photoperiod of 14 h (Figure 1C). Once infested the seedlings are removed and maintained at the same conditions until the insects became pupae. The leaves with the pupae are then detached from the plants and put back inside the wooden frame where the adults emerge. Tests of resistance are performed according to Guerreiro-Filho (1994). Detached leaves of 40 plants/day, are stuck on a base kept moist during exposure to the insects. Leaf discs containing eggs are then cut out with a cork borer and kept moist on top of a soggy sponge maintained inside a plastic box covered with a glass lid (Fig. 1D). After 15 days the lesions of the check discs from susceptible plants show large, roundish lesions (rating 4). Resistant plants rated 1 have small punctiform lesions while moderately resistant ones (rating 2) have threadlike lesions and moderately susceptible (rating 3) plants depict large irregular lesions. Only plants rated 1 and 2 are selected for further tests in the field.

6. Earliness

A conspicuous agronomic attribute of *C. racemosa* (Fig. 1F) besides leaf miner resistance and drought tolerance is its extreme earliness (Medina-Filho *et al.*, 1977a). Indeed, in Campinas, it is the earliest accession, amongst an extensive collection comprising several genera, a number of species and hundreds accessions of *C. arabica* from Ethiopia. In Campinas (23°S, 660m high) ripening in this species takes approximately only 100 days while the earliest arabica accessions and cv Bourbon takes at least twice as much time (Medina-Filho *et al.*, 1984). Earliness is an important characteristic in coffee provide the altitude and thus the coldness of the region where coffee is grown is directly related to both lateness and quality. Late cultivars such as the high yielding Catuaí can not be grown in cold regions since ripening and blooming coincide. The earliness characteristic of *C. racemosa*, can be observed, though not as intense, in its hybrids with *C. arabica* as well as in its backcross derivatives (Medina-Filho *et al.*, 1977b). The triploid F_1 hybrid *C. racemosa* X *C. arabica* is as early as some of the second backcross derivatives. Some advanced backcross selections are also quite early. Irrespective of a preliminar information that this characteristic seems to be recessive (Medina-Filho *et al.*, 1977b) it is certainly due to very few genetic factors, readily transmitted and observed in some advanced backcross generations (Guerreiro-Filho *et al.*, 1990) as shown also in Table 2.

7. Quality of the beverage

The *C. racemosa* coffee used to be locally consumed in south Mozambique, its native habitat. At ripening time the cherries give off a very unpleasant smell that persists and became even stronger after roasting and grinding the beans. Considering that chemical analyses of the aroma of arabica coffee have revealed more than 600 different elements subjected to complex interactions (Dart and Nursten, 1987), one would foresee that recovering the good aroma of arabica coffee after hybridizing it with such a poor quality species would be quite a difficult task. Fortunately this turned out to be not the case. Although F_1 hybrids yield an unpleasant beverage, it is however closer to arabica than to *C. racemosa* improving rapidly as the backcross generations progress (Carvalho *et al.*, 1990). Presently, advanced progenies of backcross generations have cup quality similar to standard cultivars. The same pattern of results have been previously obtained in the development of Icatu cultivars derived from *C. canephora* x *C. arabica* (Fazuoli, 1991).

Table 2. Cumulative yields of cherries for 5 years, resistance to rust (rated 1 to 4), leaf miner (rated 1 to 10) and earliness (rated 1 to 5) of selected plants of progeny trial in Campinas. MN and CV are standard, susceptible cultivars.

Trial	Progeny	Yield (Kg)	Rust resistance *H. vastatrix*	Leaf miner *P. coffeella*	Earliness
473	90/23	50	3	4	4
	MN 388-17	34	3	9	4
369	H13685	31	4	3	2
	CV 81	13	4	8	5
360	H13465	25	4	4	2
	MN 388-6	12	4	7	4
388	H14104	24	1	2	3
	MN 388-17	21	4	6	3

8. PRESENT STATUS OF THE BREEDING PROGRAM

BC_5 progenies selected for yield and leaf miner resistance have been assayed in field trials in a few Experimental Stations of IAC in São Paulo State. In general, they have good vigor (Fig. 3G), high percentage of normal flat beans and standard cup quality. Yield data and characteristics of some progenies are shown in Table 2. It can be seen that they outyield standard cultivars, are resistant to leaf miner and some are earlier. One of them is resistant to leaf rust. An additional generation of selfing will be necessary to pull out lines homozygous for Lm_1 and Lm_2 since those lines have an extra backcross to cv Catuaí Vermelho. The results of this program summarized herein indicate that considerable progress has been achieved in transferring to *C. arabica* cultivars the resistance to leaf miner found in *C. racemosa* thus stimulating the final selections for testing in yield trials at different regions of Brazil.

9. Aknowledgements

This work was supported by PNP&D Café/Embrapa and authors thank to CNPq for fellowship.

10. References

Carvalho A., Monaco L.C. (1967) Relações genéticas de espécies selecionadas de Coffea. Ciência e Cultura, São Paulo, (19)1:151-165

Carvalho A., Teixeira A.A., Fazuoli L.C, Guerreiro-Filho O. (1990), Qualidade da bebida de algumas espécies e populações derivadas de híbridos interespecíficos de *Coffea*. Bragantia, Campinas, 49(2):281-290

Cruz D.D. Aneuplóides em café. (1972), Aspectos morfológicos e citológicos na análise de duas progênies de café Mundo Novo (*Coffea arabica* L.). Piracicaba, ESALQ, USP, (PhD Thesis).

Dart S.K., Nursten H.E. (1987), Volatile Components. In: Coffee, Chemistry 1, R.J. Clarke and R. Macrae (eds.) New York, p.223-265

Fazuoli L.C. (1991), Metodologia, critérios e resultados de seleção em progênies do café Icatu com resistência a *Hemileia vastatrix*. Universidade Estadual de Campinas, Campinas (SP), 322p., (Doutorado – UNICAMP)

Guerreiro-Filho O. (1989), Avaliação da resistência de *Coffea spp* a *Perileucoptera coffeella* (Guérin-Meneville, 1842) (Lepidoptera-Lyonetiidae), Campinas, SP, 118p. Tese [mestrado].

Guerreiro-Filho O. (1994), Identification de gènes de résistance à *Perileucoptera coffeella* en vue de l'amélioration de *Coffea arabica* : Potentiel d'espèces diploïdes du genre *Coffea*; gènes de *Bacillus thuringiensis*. Thèse de Doctorat. ENSAM, Montpellier. 173 p.,

Guerreiro-Filho O., Medina-Filho H.P. (1987), Resistência genética à *Perileucoptera coffeella* em *Coffea salvatrix* e *Coffea brevipes*. Ciência e Cultura, 39(7):749-750

Guerreiro-Filho O., Gonçalves W., Mazzafera P., Carvalho A. (1987), Resistência de *C. stenophylla* ao bicho mineiro. In: Congresso Brasileiro De Pesquisas Cafeeiras, 14., Campinas, 1987. Resumos. Rio de Janeiro, IBC/GERCA, p.105-106.

Guerreiro-Filho O., Medina-Filho H.P., Gonçalves W., Carvalho A. (1990), Melhoramento do cafeeiro, XLIII. Seleção de cafeeiros resistentes ao bicho mineiro, Bragantia, Campinas (49)2:2911-304

Guerreiro-Filho O., Medina-Filho H.P., Carvalho A. (1991), Fontes de resistência ao bicho mineiro, *Perileucoptera coffeella* em *Coffea* spp. Bragantia, Campinas, 50(1):45-55

Guerreiro Filho O., Medina Filho H.P., Carvalho A. (1992), Método para seleção precoce de cafeeiros resistentes ao bicho mineiro *Perileucoptera coffeella*. Turrialba, 42 (3): 348-358

Guerreiro-Filho O., Silvarolla M.B., Eskes A.B. (1999), Expression and mode of inheritance of resistance to leaf miner . Euphytica, 105(1):7-15

Ihering R.V. (1912), Nossos cafezaes. Chácaras e Quintaes, São Paulo, 6(4):1-7

Katiyar K.P., Ferrer F. (1912), In: International Atomic Energy Agency, Viena, IAEA, 1968, p. 165-175.

Medina-Filho H.P., Carvalho A., Monaco L.C. (1977a), Melhoramento do cafeeiro. XXXVII - Observações sobre a resistência do cafeeiro ao bicho mineiro. Bragantia (36): 131-137

Medina-Filho H.P., Carvalho A., Medina D.M. (1977b), Germoplasma de *C. racemosa* e seu potencial no melhoramento do cafeeiro. Bragantia (36):XLIII-XLVI

Medina Filho H.P., Carvalho A., Sondahl M.R., Fazuoli L.C., Costa W.M. (1984) Coffee Breeding and related evolutionary aspects. Plant Breeding Reviews, vol. 2., p.157-193.

Parra J.R.P. (1985), Biologia comparada de *Perileucoptera coffeella* (Guérin - Méneville, 1842) (Lepidoptera - Lyonetiidae) visando seu zoneamento ecológico no Estado de São Paulo. Revista Brasileira de Entomologia, 29(1):45-76

Sein F. (1942), Estudio del minador de las hojas del café (*L. coffeella* Guérin). Ann. Rep. Fiscal Year 1940-1941. Univ. Puerto Rico Agricultural Exp. Sta. 7: 82-84

Speer M. (1949), Observações relativas à biologia do "bicho mineiro das folhas do cafeeiro" *Perileucoptera coffeella* (Guérin-Méneville) (Lepidoptera-Buccolatricidae). Arquivos do Instituto Biológico, 19(3):31-46

Vicente-Chandler J., Abruña F., Bosque-Lugo R., Silva S. (1968), Intensive coffee culture in Puerto Rico. Agricultural Experimental Station, Rio Piedras, 84p. (Bol. 211).

Chapter 20

THE ROLE OF BIOLOGICAL CONTROL IN AN INTEGRATED COFFEE BERRY BORER MANAGEMENT IN COLOMBIA

A.E. Bustillo

Disciplina de Entomología, Centro Nacional de Investigaciones de Café, Cenicafé, A. A. 2427, Manizales, Caldas, Colombia

Running title: Coffee berry borer management

1. Introduction

Coffee is the most important export crop of Colombia. Its importance grows if we consider the socio-economic aspect that supports many families and generates millions of direct and indirect jobs. The most important pest of this crop is the coffee berry borer, *Hypothenemus hampei* (Ferrari). This insect destroys the coffee berry and causes fall and weight loss of the harvested berry, but more important even is the loss in the quality of the beverage. The unique dependence on the use of chemical insecticides for its control is not recommended; therefore, it is suggested the implementation of an integrated management program based on the biological control to achieve a cost reduction in production, adding an ecological base, thus improving the quality of the coffee produced.

The coffee berry borer is considered as one of the most important pests to coffee all over the world (Le Pelley, 1968). It was introduced in the American continent from Africa to Brazil in 1923 and is distributed in almost all coffee growing countries, except Costa Rica and Panama, where so far, there is no information on its presence. The females of the borer are the ones initiating the attack upon penetrating the berry in the disk region of the crown of the fruit, thus reaching the endosperm and caving an egg

T. Sera et al. (eds.), Coffee Biotechnology and Quality, 229–237.

spawning chamber from which the larvae come out destroying portions of the seed and causing consequent weight loss.
Endosulfan is the most commonly used insecticide in the borer control, but it has been reported that borer has developed resistance to this product (Brun *et al.*, 1989). The threat to the biological equilibrium and the toxic risks in the coffee zones led countries such as Mexico, Guatemala, Salvador, Nicaragua, Honduras and Colombia to introduce African parasitoids in order to develop biological control programmes. The Colombian plan on biological control of borer began in 1988 when it was reported first. The National Federation of Coffee Growers began a research programme on the development of a mycoinsecticide based on *Beauveria bassiana* and the search of alternatives of insecticides of lesser toxic category.

2. Use of parasitoids

When the borer was brought from Central Africa, where it originated to Colombia, it did not bring along its natural enemies (which are considered to be normally present at the place of origin). These parasitoids were *Prorops nasuta* Waterston (Bethylidae) and *Heterospilus coffeicola* Schmeideknecht (Braconidae) both found by Heargraves in 1923 and 1924, respectively. Ticheler (1963) found *Cephalonomia stephanoderis* Betrem (Bethylidae) in 1960. A fourth eulophid endoparasitoid was found in Togo by Borbón (1989) which was described as *Phymastichus coffea* La Salle (La Salle, 1990). Among these parasitoids the bethylids (*C. stephanoderis* and *P. nasuta*) are so far the most promising candidates to be included in a coffee berry borer management programme, for being the ones that can be reared under laboratory conditions. Currently, research is being conducted for mass production of *P. coffea,* which was successfully introduced into Colombia in 1996 from Togo through quarantine in England. Field experiments are now underway to determine its behaviour and efficacy against the borer.
In Colombia, *H. hampei* is massively reared in laboratories using parchment coffee with the humidity content of 45%. The infested parchment coffees are kept in metal trays at 80% RH and 27°C. After a few days, the infested grains show feeding activity and 20 days later they have enough immature stage for parasitoid reproduction. The development of *C. stephanoderis* takes between 18-21 days at 25°C in dark. The parasitoid emergence is achieved by providing higher temperatures and light. In a similar manner *P. nasuta* is also produced. During 1993 it was possible to produce in Colombia close to three million parasitoids and in1994 the efficiency increased to produce close to six millions per month in three units of production. Till September

1998, over 1500 million parasitoids have been released on different regions of the country in borer infested farms not using insecticides for the purpose of insuring their establishment. This parasitoid has been established in all sites where it has been released. After five years of being released on the coffee crops of Nariño, the parasitoid remains active in this zone (Benavides *et al.*, 1994; Quintero *et al.*, 1998). Recent studies releasing large amounts of parasitoids in a ratio of at least one adult of *C. stephanoderis* per infested berry in commercial coffee farms have shown that it could be possible to reduce *H. hampei* populations. However, the costs to achieve this goal are still too high (Aristizábal *et al.*, 1997). Research is now being focused on the production of the coffee berry borer using artificial diets to lower the cost of parasitoid production (Portilla, 1999).

3. Use of entomopathogens

The entomopathogen fungus, *Beauveria bassiana (Bb)* is found naturally infecting *H. hampei* in almost all regions of Colombia where it appears. So far there has been a collection of 102 isolates of *Bb* from different places, out of which approximately half of them have shown activity against the borer.

A technique of bioassay for isolating the most pathogenic ones has been studied and developed (González *et al.*, 1993). The life cycle of *Bb* on the borer under laboratory conditions is completed in an average of 8.2 days from insect inoculation to spore production. This may, however, vary depending up on the isolate used and the laboratory conditions (mainly temperature). The importance of applying *Bb* to the insects to reactivate its pathogenicity has been established. When the fungus is cultured in artificial media for three or more generations the pathogenicity is considerably reduced and the average time to cause 50% mortality in the population is increased in comparison to the fungus activated on the borer (González *et al.*, 1993).

Studies with *Metarhizium anisopliae* (Bernal *et al.*, 1994) showed that this fungus could play a role in the control of *H. hampei* on the ground, upon infecting the borer population that emerged from the infested berries that previously fell to the ground. However, recent studies showed that *B. bassiana* has a greater persistence and efficacy than *M. anisopliae* under these circumstances (Bernal *et al.*, 1999).

4. Production of *Beauveria bassiana*

Two approaches have been investigated for the production of *Bb*, at industrial and artisan level. At the industrial level the technology has been transferred to the manufacturers for the production of the fungus (Morales *et al.*, 1991). Presently there are four private laboratories licensed by ICA, which provide the fungus prescribed for the control of the coffee berry borer. In 1992, five tons of fungi were used at a concentration of 1x 10^8 spores/g for experimental purposes. The *Bb* production for the coffee berry control was that of 60 tons in 1993 (Posada, 1993) and 100 tons for 1994. In the last four years (1995-1998) the production was estimated to be 200 tons per year with an average yield of 3 x 10^9 spores/g of commercial product.

A methodology was also studied for producing fungus at coffee grower level at the farm itself (Antía *et al.*, 1992). The average production of spores in the bottles was 5 x 10^{10} spores/100g of substrate at 25°C after a 24-day development period. Once the fungus completes its development it is ready to be used by the grower. The production of one bottle is enough to spray 100 trees at a dose of 5 x 10^8 spores/tree. Cenicafé has made available to growers a massive production unit of *B. bassiana* fungus in order to provide a pure inoculum for reproduction. The fungus produced in these units is of a high quality. Quality control test is performed to ensure: contaminant free, appropriate concentration, 100% viability, high pathogenicity on the borer in laboratory conditions (over 80%), and use of a strain of the fungus recently activated and of better field performance (Vélez *et al.*, 1997). The formulations of *B. bassiana* were evaluated under field conditions and in all cases the fungus was established in the borer population. *Bb* is effective only when the borer makes contact with the spores when attempting to penetrate the berry. If the insect is already inside the berry, it is difficult for the fungus to infect it.

The epizootiology of *Bb* was studied at a farm infested with the coffee berry borer in Ansermanuevo (Valle) at 1000 m.a.s.l., in a Colombia variety coffee crop planted at a density of 10,000 trees per hectare (Bustillo *et al.*, 1991). The infestation of the borer started in the centre part of the plantation during June 1990 when *Bb* was sprayed at a dosage of 1 x 10^8 spores/tree. After six months the borer was distributed all over the crop as well as the fungus. There was an evaluation of the borer infestation and fungus infection over 300 coffee branches taken between January and August of 1991. The sampling unit was a productive branch on which all the berries were counted as well as the infested ones and those affected by the fungus. The results indicated that a high proportion of the borer (75%) was infected by the fungus towards the end of the observation period, causing a high reduction of its population. Further studies on life table analysis to determine key mortality factors of the coffee berry borer in three different areas demonstrated that *B. bassiana* was the most important agent causing on average a constant mortality of 49% on the borer population (Ruiz, 1996). So far, the fungus has been found infecting *H. hampei* populations in all the Colombian coffee growing areas infested by the borer.

Table 1. Native enemies found attacking the coffee berry borer, *Hypothenemus hampei* and competitors in infested decaying berries in Colombia.

	BORER STAGE INFECTED
Hyphomycetes: Beauveria bassiana	Larva, pupa, adult
Beauveria brongniartii	Adult (laboratory).
Hirsutella eleutheratorum	Adult
Metarhizium anisopliae	Adult
Fusarium oxisporum	Adult
Paecilomyces lilacinus	Adult
Bacillaceae *: Bacillus* sp.	Larva
Enterobacteraceae *: Serratia* sp.-	Larva
Microsporidia : Near *Mattesia*	Adult
Braconidae : Near *Cryptoxilos* sp.	Adult
Formicidae *: Crematogaster* sp.	All stages
Pheidole sp.	All stages
Brachymyrmex sp.	All stages
Solenopsis sp.	All stages
Wasmannia sp.	All stages
¿Prenolepsis sp.?	All stages
¿Scoloposcelis sp.	Immature
INSECTS	
Formicidae	In berries of the tree and soil
Diptera	In berries of the tree
Coleoptera	In berries of the soil
FUNGI	In berries of the soil
BACTERIA	In berries of the soil
NEMATODES	In berries of the tree and soil

5. Native enemies of *Hypothenemus hampei*

In different coffee crops in Colombia infested by *H. hampei* several native enemies of the borer have been observed, especially at those places where insecticides was not applied. The Cenicafe's entomology discipline surveyed different coffee growing crops in Colombia and identified not only natural enemies of *H. hampei* but other organisms also competing for its niche (Table 1) (Bustillo *et al.*, 1998, Pérez *et al.*, 1996; Posada *et al.*, 1998).

The mortality factors of the borer associated with all organisms competing for a site, specially infested berries, are responsible for population reduction over time, which in turn lead to reduce control costs. The control practices used should keep the biological balance, that is, protect the beneficial fauna. Rainy seasons provide the moisture to the soil thus supplying the invasion of micro-organisms competing within the infested berries on the ground. The use of insecticides is counter indicated not only because they affect natural controllers of the borer but also because they eliminate beneficial fauna of other insects such as leaf miner, scales, mealy bugs, and defoliators, which do not become pests thanks to the control exercised by their natural enemies.

6. Integrated coffee berry borer management

Because the coffee berry borer is a pest introduced with no natural enemies in Colombia, the first strategy was to introduce parasitoids as well as entomopathogens in the coffee growing areas preserving and protecting the native beneficial fauna (Bustillo, 1993). A comprehensive management programme was then set up (Bustillo *et al.*, 1998), which included knowledge of the farm on blossoming seasons, infestation assessment, sanitary crop practices ("Re-Re"), control of *H. hampei* on the coffee processing area (Castro *et al.*, 1998), and biological components based on parasitoids such as: *C. stephanoderis*, *P. nasuta* (Orozco and Aristizábal, 1996), and entomopathogens such as: *B. bassiana* (Bustillo, 1998; Flórez *et al.*, 1997). The use of insecticides has been restricted to those cases in which the levels of infestation are higher than 5% and where the borer is found aggregated in small areas, in which case low toxicity products of low environmental impact are recommended (Categories III and IV) (Villalba *et al.*, 1995). The monitored plots of integrated pest management in coffee plantations have shown that this recommendations are feasible, so that the coffee grower may continue producing coffee, federation type, even with the presence of the coffee berry borer (Bustillo *et al.*, 1998).

7. Summary

The coffee berry borer, *H. hampei* is the main coffee pest and is practically found in all coffee growing regions all over the world, causing not only fruit loss but also harming the quality of the beverage. This insect was introduced in Colombia in 1988 without its natural enemies and spread in very high populations as reflected the damage caused. The research to control this pest was focused on the introduction and development of the highest possible number of biological control agents compatible in an integrated pest management programme. The *C. stephanoderis* and *P. nasuta* parasitoids were introduced and a new production technology allowing massive rearing of these species for field release was developed. Between 1994 and 1998, 1200 millions of *C. stephanoderis* and 300 millions of *P. nasuta* were released. Presently both the species are established in many coffee regions in Colombia and it is expected that very soon they would be introduced in 500,000 hectares of coffee plantations affected by the borer. A short-term goal has been the production and introduction of the coffee berry borer adult parasitoid, *Phymastychus coffea* in the Colombian coffee fields, which would complement the other two. In the future it is planned to introduce and colonize *H. coffeicola* from Uganda. The *Beauveria bassiana* fungus is also being used for the control of the borer. This programme has covered all the farms infested by the borer and now this fungus has been established in all affected regions.

The appropriate use of these biological agents and the protection of native beneficial fauna have become a useful tool along with control practices to make up an integrated management programme that keep the borer populations at levels not causing economical damage.

8. Conclusions

The integrated pest management approach is the best strategy for the control of the coffee berry borer, aiming to the biological control to be one of the main components. Beneficial agents have been introduced and developed in Colombia to be used in the control against this pest. The use of *C. stephanoderis* and *P. nasuta* parasitoids and of *B. bassiana* as well as the conservation of the native fauna in an integrated pest management programme is feasible as shown by research made in Colombia.

9. References

Antia O.P., Posada F.J., Bustillo A.E., González M.T. (1992), Producción en finca del hongo *Beauveria bassiana* para el control de la broca del café. Cenicafé, Avances Técnicos No. 182, p. 8

Aristizábal L.F., Baker P.S., Orozco J., Chaves B. (1997), Parasitismo de *Cephalonomia stephanoderis* Betrem sobre una población de *Hypothenemus hampei* (Ferrari) con niveles bajos de infestación en campo. Rev. Colombiana Ent., 23 (3-4): 157-164

Benavides P., Bustillo A.E., Montoya E.C. (1994), Avances sobre el uso del parasitoide *Cephalonomia stephanoderis* para el control de la broca del café, *Hypothenemus hampei*. Rev. Colombiana Ent., 20 (4): 247-253

Bernal M.G., Bustillo A.E., Posada F.J. (1994), Virulencia de aislamientos de *Metarhizium anisopliae* y su eficacia en campo sobre *Hypothenemus hampei.* Rev. Colombiana Ent. 20 (4): 229-233

Bernal M.G., Bustillo A.E., Cháves B., Benavides P. (1999), Efecto de *Beauveria bassiana* y *Metarhizium anisopliae* sobre poblaciones de *Hypothenemus hampei* (Coleoptera: Scolytidae) que emergen de frutos en el suelo. Rev. Colombiana Ent., 25 (1-2): 11-16

Borbón O. (1989), Bioecologie d'un ravageur des baies de caféier *Hypothenemus hampei* (Ferr.) (Coleoptera: Scolytidae) et des ses parasitoides au Togo. Phd Université Paul-Sabatier Toulouse Cedex

Brun L.O., Marcillaud C., Gaudichon V., Suckling D.M. (1989), Endosulfan resistance in *Hypothenemus hampei* (Coleoptera: Scolytidae) in New Caledonia. J. Econ. Entomol. 82 (5): 1312-1316

Bustillo A.E. (1993), El control biológico como un componente en un programa de manejo integrado de la broca del café, *Hypothenemus hampei*, en Colombia. Memorias XX Congreso de Socolen, Cali, Julio 13-16, 1993, pp.159-164

Bustillo A.E. (1998), Control of coffee berry borer, *Hypothenemus hampei*, by fungal pesticides in Colombia. *In*: Proc. Inter. Workshop, 9 – 13 December 1996, Berlin, Germany. Biotechnology for crop protection – it's potential for developing countries. Zentraistellefur Ernahrung und Landwirtschaft (ZEL)- Feldafing/Zschortau, pp. 219 –229

Bustillo A.E., Castillo H., Villalba D., Morales E., Vélez P.E. (1991), Evaluaciones de campo con el hongo *Beauveria bassiana* para el control de la broca del café, *Hypothenemus hampei* en Colombia. In: Colloque Scientifique International sur le café, 14. San Francisco. 14-19 Juillet, Paris, ASIC. pp. 679-686

Bustillo P.A.E., Cárdenas M.R., Villalba G.D., Benavides M.P., Orozco H.J., Posada F.F. (1998), Manejo integrado de la broca del café, *Hypothenemus hampei* (Ferrari) en Colombia. Chinchiná, Cenicafé, p. 134

Castro L., Benavides P., Bustillo A.E. (1998), Dispersión y mortalidad de *Hypothenemus hampei*, durante la recolección y beneficio del café. Manejo Integrado de Plagas (Costa Rica) 50: 19-28

Flórez E.; Bustillo A.E., Montoya E.C. (1997), Evaluación de equipos de aspersión para el control de *Hypothenemus hampei* con el hongo *Beauveria bassiana.* Revista Cenicafé (Colombia), 48 (2): 92-98

González M.T., Posada F.J., Bustillo A.E. (1993), Desarrollo de un bioensayo para evaluar la patogenicidad de *Beauveria bassiana* sobre *Hypothenemus hampei*. Revista Cenicafé 44(3): 93-102

La Salle J. (1990), A new genus and species of Tetrastichinae (Hymenoptera: Eulophidae) parasitic on the coffee berry borer, *Hypothenemus hampei* (Ferrari) (Coleoptera: Scolytidae), Bull. Ent. Res. 80: 7-10

Le Pelley R.H. (1968), Pests of coffee. Longmans, Green and Co. Ltd., London. p. 590

Morales E., Cruz F., Ocampo A., Rivera G., Morales B. (1991), Una aplicación de la biotecnología para el control de la broca del café. In: Colloque Scientifique International sur le café, 14. San Francisco. 14-19 Juillet, Paris, ASIC. pp. 521-526

Orozco J, Aristizábal L.F. (1996), Parasitoides de origen africano para el control de la broca del café. Avances Técnicos de Cenicafé No. 223. Chinchiná, enero de 1996

Pérez E.J., Posada F.J., González M.T. (1996), Patogenicidad de un aislamiento de *Fusarium* sp. encontrado infectando la broca del café, *Hypothenemus hampei*. Rev. Colombiana Ent., 22 (3): 177-181

Portilla M. (1999), Mass rearing technique for *Cephalonomia stephanoderis* (Hymenoptera: Bethylidae) on *Hypothenemus hampei* (Coleoptera: Scolytidae) developed using Cenibroca artificial diet. Rev. Colombiana Ent. 25 (1-2): 57-66

Posada F.J. (1993), Control biológico de la broca del café, *Hypothenemus hampei* (Ferrari) con hongos. Memorias XX Congreso Socolen, Cali Julio 13-16, 1993, 137-151

Posada F.J., Marin, P., Pérez, M. (1998), *Paecilomyces lilacinus*, enemigo natural de adultos de *Hypothenemus hampei*. Revista Cenicafé (Colombia), 49(1): 72 – 77

Quintero C., Bustillo A.E., Benavides, P., Cháves, B. (1998), Evidencias del establecimiento de *Cephalonomia stephanoderis* y *Prorops nasuta*, en cafetales del departamento de Nariño, Colombia. . Rev. Colombiana Ent., 24 (3-4): 141-147

Ruíz R. (1996), Efecto de la fenología del fruto del café sobre los parámetros de la tabla de vida de la broca del café, *Hypothenemus hampei*. Tesis Ing. Agr., Facultad de Ciencias Agropecuarias, Universidad de Caldas, Manizales, p. 87

Ticheler J.H.G. (1963), Estudio analítico de la epidemiología del escolítido de los granos de café, *Stephanoderes hampei* Ferr. en Costa de Marfil. (Traducción G. Quiceno). Revista Cenicafé 14 (4): 223-294

Vélez A.P.E., Posada F.F.J., Marín M.P., González G.M.T., Osorio V.E., Bustillo P.A.E. (1997), Técnicas para el control de calidad de formulaciones de hongos entomopatógenos. Boletín Técnico No 17. Cenicafé. Chinchiná, Caldas, Colombia. p. 37

Villalba D.A., Bustillo, A.E., Cháves, B. (1996), Evaluación de insecticidas para el control de la broca del café en Colombia. Revista Cenicafé (Colombia), 46 (3): 152-163

Chapter 21

CORRELATIONS BETWEEN EDAPHIC FACTORS AND *COFFEA ARABICA* FUNGAL PATHOGENS IN SOUTH PACIFIC

F. Pellegrin, F. Kohler, D. Nandris

IRD (ex-ORSTOM), Laboratory of Plant Pathology, BP 5045, 34032 Montpellier, France

Running title : Coffee pathosystem

1. Introduction

Epidemiological research on *Coffea arabica* (var. *Typica* and *Bourbon*) has been conducted in the Plant Pathology at the ORSTOM's New Caledonia Centre since 1991. Its aim has been to understand the functioning of the "pathosystem" which comprises coffee, its main pathogenic fungi (*Hemileia vastatrix*, *Colletotrichum gloeosporioides* and *Cercospora coffeicola*) and the environment, in order to model their inter-relationships. More specifically, the goal is to identify the environmental conditions which influence the dynamics of coffee diseases. Regular epidemiological surveys are carried out in arabica coffee plantations for this purpose. The biometric evaluations of these data take into account the spatial and temporal dimensions of the events occurring. The ultimate objective of this research is to use modelling to devise a decision-making tool, permitting a forecast of the risk of an epidemic in a defined environmental context (Nandris *et al.*, 1997).

2. Experimental

This research programme was mainly performed in New Caledonia, but also included a regional research dimension with the complementary investigations being carried out in Papua New Guinea (Coffee Research Institute, Kainantu, Eastern Highlands Province) and Vanuatu (Tanna Island). The experimental approach was based on monthly epidemiological observation carried out in 20 traditional coffee-growing plots using specific pathological survey methods (leaf-by-leaf inspection, etc.) and environmental characterisation procedures (weather station, soil analyses, etc.).
Management of the pathological and environmental data characterising each site was performed with the ORACLE database through a series of queries which generate

T. Sera et al. (eds.), Coffee Biotechnology and Quality, 239–243.

synoptic tables. The statistical interpretation of these data carried out with a software known as "Multivariate Analysis and Graphic Expression of Environmental Data" (Thioulouse et al., 1997). This software makes it possible to produce complex statistics which are highly relevant to the issues raised by such a spatio-temporal study:

- Three-way analysis of a "data cube" (variables x sites x observation dates). This was a three-stage approach: prior determination of an "inter-structure" which proposes an ordering of the sampling dates, establishment of a "compromise" describing common structures at various dates and, thirdly, characterisation of an "intra-structure" describing the divergences from these common compromise structures for each date.
- The costructure of two "data cubes" through analysis of co-inertia (Simier *et al.*, 1999). This approach simultaneously merges "interdate" analyses (temporal effects) and "intradate" analyses (spatial effects). It seeks the combinations of variables (or axes of co-inertia) which express the temporal co-variation and spatial co-structure between clusters of points representing data. The projection of these variables onto these axes defines planes determining the linkages between the two cubes.

3. Results and Discussion

In a given annual cycle, surveys confirmed the existence of a mosaic of highly diverse pathological situations (disease distributions, infection and mortality kinetics, etc.) within the experimental design. The three-way analysis produced a scatter diagramme (Fig. 1) on which the plots were located as determined by the vectors characterising either the pathology or the environment.

Analyses of co-inertia between cubes of pathological and environmental data have made it possible to describe the nature of the links between them. In particular, combinations of variables have been identified. These revealed significant trends (Fig. 2)

- Anthracnose emerged as being closely linked to edaphic factors such as high pH values, a good soil structure or low shade.
- High pH and shade values, and to a lesser extent, low altitude were the main factors influencing the behaviour of cercosporiosis.
- Rust differed from the two previous diseases, in that it was more likely to occur in sites featuring low soil pH values, low rainfall, quite high minimum temperatures, high shade and high altitude.
- This original correlation between environment and pathology was taken further by also considering results recorded over four successive survey years. In fact, the comparative analysis of the distribution of pathogens and epidemiological kinetics, on an annual basis, revealed that depending on the site, the pest situation may or may not remain similar from year to year.

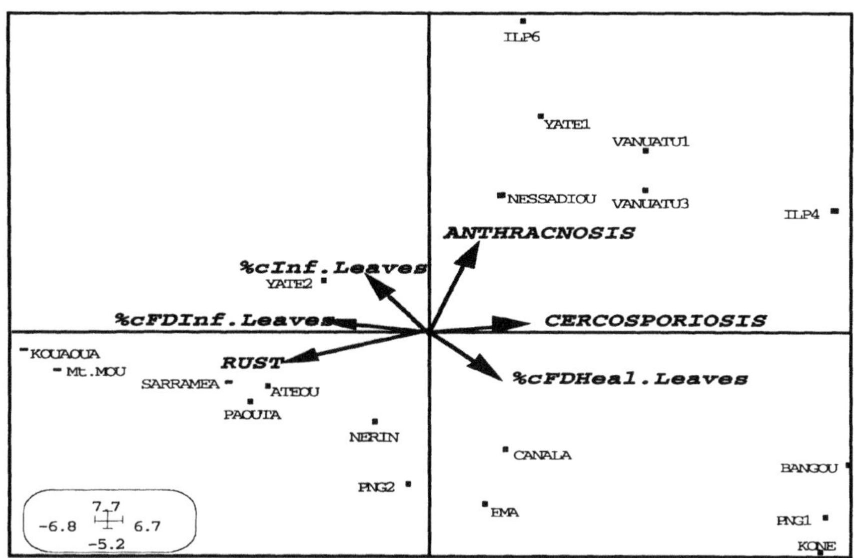

Scatter diagram of sites (New-Caledonia, Vanuatu and Papua New Guinea) in terms of disease parameters. {**RUST** = Sanitary Mark of the rust; **ANTHRACNOSIS** = Sanitary Mark of the anthracnose; **CERCOSPORIOSIS** = Sanitary Mark of the cercosporiosis; **%cInf.Leaves** = cumulative % of infected leaves; **%cFDInf.Leaves** = cumulative % of Fallen Infected Leaves; **%cFDHeal.Leaves** = cumulative % of Fallen Healthy Leaves}.

From an epidemiological point of view, almost 40% of these sites showed individual characteristics of "temporal stability", whereas the weather conditions were very varied. In such cases, taking into account previous co-inertia results, it was legitimate to think that the epidemiological diversity observed between sites in any given year was basically a reflection of edaphic diversity (e.g. pH, relief, fertility, etc.) because only these factors remained stable over the time-scale of the study. On the other hand, there were other sites which showed "instability", expressed from one year to the next in major differences in severity for the same pathogen. Such inter-annual variations supported the idea that disease diversity in these sites was dependent on climatic characteristics, which in New Caledonia varied greatly between 1992 and 1995.

The characterisation of sites where pathological determinism is highly different is used to detail the influence of soil and climate respectively on the disease. The major trends resulting from the above interpretations were used as a basis for modelling the risk of disease attributable either to rust or to anthracnose (cercosporiosis being insufficiently represented in situ to permit an analysis of this kind). Significant forecasts of the level of infestation of each plot at the end of the crop cycle were made, taking into account only the environmental parameters selected for each of the years studied. Although significant correlations have already been obtained, work is continuing to further improve this specific statistical approach.

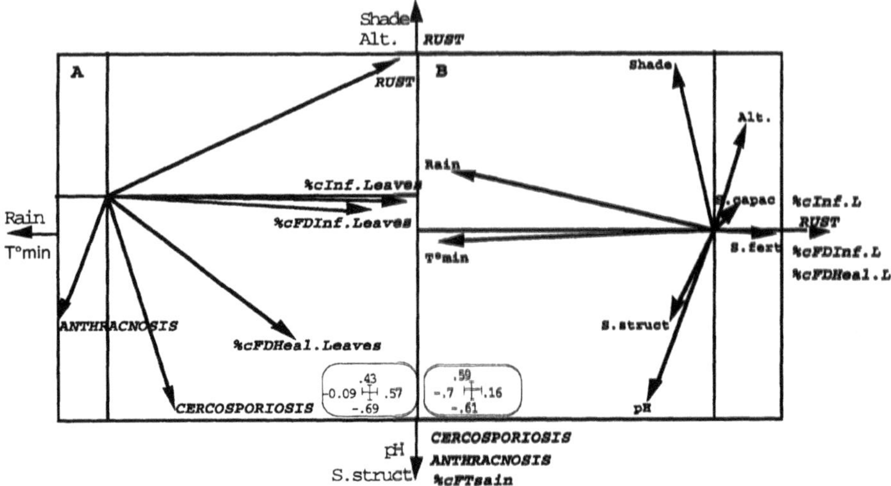

Co-structure of pathological data (left) and environmental data (right).
{**Rain** = cumulative rainfall over one month; **T° min.** = mean minimum temperatures; **S.fert** = soil fertility; **S.struct** = soil structure; **S.capac** = soil water capacity; **pH** = soil pH; **Alt** = altitude of site; **Shade** = amount of shade}

To clearly perceive the mechanisms studied, it is first necessary to evaluate the proportion of variability due to the genetic diversity of the populations of pathogens and to the possible host diversity:

- a biomolecular analysis of the *H. vastatrix* populations is going ahead in order to compare the various rust isolates. To detect possible genetic diversity between the 25 collected rust samples (20 from New Caledonia, 3 from Papua New Guinea, 2 from Vanuatu). Among the ITS, REP, ERIC and RAPD methodologies tried, only RAPD (with severe constraints for spore collection & DNA extraction) performed with OPA9 and OPC8 primers reveals a pattern polymorphism between the rust sources. In addition, after the sampling of spores from every tree in every survey plot, a rust race identification was done in Oeiras using the CIFC method (Rodrigues et al., 1975). Three different races (I, II, III) were identified, with race II definitely appearing as the most common one. In some plots, the coexistence of the II & I or II & III races was observed. Trials to link these results with the characteristics of the identified races and with the dynamics of the diseases are under way.
- an appraisal of *C. gloeosporioides* populations (isolated on the survey sites) was also carried out. Using biomolecular tools (RAPDs), 125 isolates collected in 6 surveyed plots (4 in New Caledonia, 1 in Vanuatu, 1 in Papua New Guinea) were analysed (Faugeron, 1996). A high level of genetic diversity was found in all populations and each isolate exhibited a unique RAPD haplotype. Genetic differentiation was found between three out of the six populations (Wright's index, Fst= 0.27), but no relationships could be evidenced between the genetic variability of the populations and other traits such as environmental or epidemiological characteristics of each plot. A high recombination rate could be hypothesised to

explain this situation. These evaluations of genome diversity are complemented by quantification of the pathogenicity.

- the intra-specific homogeneity of arabica plants is also being studied under a more general programme dealing with the genetic diversity of coffee in New Caledonia Indeed, the existence of rust resistant coffee hybrids (*canephora* x *arabica*) growing in some coffee plots in New Caledonia may be a source of genetic heterogeneity that must be assessed in terms of susceptibility of the trees. For each of the 15 plots in the epidemiological survey, biomolecular analyses (RAPD) of the genome of the surveyed coffee trees were performed in Montpellier using the methodology developed by Lashermes et al. (1996). On the one hand, among the studied trees, no *canephora* character was found, thus excluding any effect on their behaviour vs. diseases. On the other hand, a molecular polymorphism indicates the existence of two cultivated coffee types *typica* and *bourbon* as well as intraspecific hybrids, with respective ratios varying within the plots.

4. Conclusions

These results on the diversity of pathogens and of the host as well as the deletion of irrelevant information from data files (e.g. selection of trees and branches, and distinction between "stable" or "unstable" plots, etc.) are clarifying factors which, once included in the statistical model, would make it possible to increase the correlation coefficients between pathological and environmental characteristics and therefore the accuracy of disease risk forecasts. The concepts and methodologies developed under this programme are now being disseminated in the region and training programmes are being conducted for our partners in coffee disease work. These tools may be of interest to teams working on other crops for which pathogen incidence has become a serious source of concern.

5. References

Faugeron S. (1996), Structure génétique des populations de *Colletotrichum gloeoeporioides* responsable de l'anthracnose du caféier. Mémoire de DEA, USTL-ENSA, Montpellier, p. 21

Lashermes P., Trouslot O., Anthony F., Combes M.C., Charrier A., (1996), Genetic diversity for RAPD markers between cultivated and wild accessions of *Coffea arabica*, Euphytica 87: 59-64

Nandris D., Kohler F., Monimeau L., Pellegrin F. (1997), Approche intégrée du pathosystème/*Coffea arabica*. Proc. 17th International Conf. Coffee Sci., ASIC, Nairobi, Kenya, pp. 588-598

Rodrigues Jr. C.J., Bettencourt A.J., Rijo L. (1975), Races of the pathogen and resistance to coffee rust Ann. Rev. Phytopathol. 13: 49-70

Simier M., Blanc L., Pellegrin F., Nandris D. (1999), Approche simultanée de K couples de tableaux: application à l'étude des relations pathologie végétale. Rev. Stat. Appl., XLVII (1): 31-46

Thioulouse J., Chessel D., Doledec S., Olivier J.M., (1997), ADE-4 : a multivariate analysis and graphical display software, Stat. Comput., 7(1): 75-83

Chapter 22

PHYSIOLOGICAL STUDIES ON MYCORRHIZAL FUNGI PRODUCTION

A.A. De Araujo[1], M. de los A. Aquiahuatl Ramos[2], S. Roussos[1]

[1]*Laboratoire de Microbiologie de l'IRD, IFR-BAIM; ESIL Case 925, Universités de Provence et de Méditerranée, 163, Avenue de Luminy, F-13288 Marseille Cedex 9 France;* [2]*Departamento de Biotecnologia; UAM-I, Avenida Michoacan y la Purisima; Apartado Postal 55-535, Mexico DF C.P. 09340, Mexico*

Running Title: Technic for mycorrhizal fungi

1. Introduction

Micro-organisms are present in large numbers on and near the feeder roots of trees, and they play vital roles in numerous physiological processes. Pelmont (1993) stated that these dynamic processes are mediated by associations of micro-organisms participating in symbiotic root activities as nitrogen fixation and phosphorus mobilization. Mousain (1993) and Strullu (1991) confirmed that the major symbiotic associations on tree roots have been bacterial with *Rhizobium* or fungal in mycorrhizas. Mycorrhizas are defined as durable unions based in reciprocal exchanges between plant roots and fungi (Marx and Cordell, 1994). Each one optimizes its development due to this association. Mousain *et al.* (1998) opined that ectomycorrhizal fungus such as *Lactarius* Pers., *Pisolithus* Alb. & Schwein. and *Suillus* P. Karst. should be considered as fundamental micro-organisms for qualitative improvement of the trees and reforestation programs. De Araújo *et al.* (1008) studied the effect of culture media, initial pH, and salt concentration on the apical growth of four ectomycorrhizal fungi. However, not much information is available about ectomycorrhizal fungi biomass production and their metabolites. Because of this, a non-destructible technique was development to evaluate biomass production and

T. Sera et al. (eds.), Coffee Biotechnology and Quality, 245–253.

analyse metabolites as pigments, enzymes, sugar, and organic acids present in agar. Baar *et al.* (1997) determined mycelium biomass by dissolving agar media in hot water and filtering the solution before drying (100°C, 24 h) and determining dry weight as described by Oort (1981). As discussed by Jongbloed and Borst-Pauwels (1990), this method could cause loss of water-soluble compounds amounting to approximately 35% of biomass with little variation between isolates and no effect of age of cultures and composition of media on the amount of loss. Dry weight loss of biomass of *L. bicolor* as estimated by Jongbloed and Borst-Pauwels (1990) was approximately 19%. Gibson and Deacon (1990) described mycelium development of ectomycorrhizal fungi *in vitro* by radial growth and biomass production. Jongbloed and Borst-Pauwels (1990) hypothesized that radial growth reflected exploitation of resources, whereas biomass production was a measure of accumulation of carbon and nutrients.
Attempts were made to present some modifications introduced in the technique employed by Baar et al. (1997) and analysed by Jongbloed and Borst-Pauwels (1990) in order to evaluate glucose, pH, and produced biomass profiles after 5, 10 and 15 days of incubation at 25°C for *Suillus collinitus* and *Pisolithus tinctorius* strains in different culture media (PDA, BAF, MNM, MP, GM8, and MG). This non-destructible technique is being employed now to evaluate biomass production and analysis of metabolites such as pigments, enzymes, sugar, organic acids present in the agar.

2. Experimental

2.1. CULTURES AND MEDIA

Mycelia of mycorrhizal species were obtained from sporophores in *Pinus*. *Pisolithus tinctorius* (Pers.) Coker and Couch (PF 26) was isolated from Murcia region of Spain in 1991 and *Suillus collinitus* (Fr.) Kuntze (Sc 24) from La Grande-Motte (south-west region of France) in 1994. The mycelia of the strains were grown and maintained on Potato-Dextrose-Agar (PDA), pH 5.6. Six synthetic media as shown in Table 1 were used.

2.3. ANALYSIS

Colony diameters were measured at regular intervals up to 15 days. Results were expressed as means of diameter on three replicate plates. Average diameter was taken from tree replicates. Colony description was made in terms of its mycelium type, colour, margin aspect and characteristic features of the mycelium. Diffusive pigments presented in agar media were also described. Sugar consumption was determined according to Miller (1959).

Table 1. Media composition for ectomycorhizas nutritional studies.

Compounds	Culture media					
(g/l)	PDA	BAF	MNM	MP	GM8	MG
D-Glucose	20.0	30.0	10.0	10.0	31.25	10.0
Peptone	-	2.0	-	2.0	2.5	-
Yeast extract	-	0.2	-	-	-	-
Potatoes	200	-	-	-	-	-
Malt extract	-	-	3.0	-	-	-
$CaCl_2 . 2 H_20$	-	-	0.2	0.13	-	-
$Ca(N0_3)_2 . 4 H_20$	-	-	-	-	2.0	-
NaCl	-	-	0.025	-	-	-
KH_2PO_4	-	0.5	0.5	0.5	-	-
$(NH_4)_2HPO_4$	-	-	0.5	-	-	-
$MgSO_4 . 7 H_20$	-	0.5	0.15	-	-	-
Ammonium tartrate	-	-	-	-	1.0	-
$FeCl_3$	-	0.005	0.005	0.01	-	-
L-aspargin	-	-	-	-	1.0	-
Thiamine HCl	-	-	100 µg	-	-	-
Myo inositol	-	0.05	-	-	-	-
Oligoelement solution*	-	1.0 ml	-	-	-	-
Vitamin solution**	-	1.0 ml	-	1.0 ml	-	-
Agar	-	15.0	15.0	15.0	15.0	15.0
pH	5.6	6.0	6.0	7.5	5.0	6.5

*Oligoelement solution ($g.l^{-1}$): $CaCl_2$, 100; $MnSO_4$, 51; $ZnSO_4$ $7H_20$, 1

**Vitamin solution ($mg.l^{-1}$): Thiamine HCl, 500; Biotin, 10; Folic Acid, 100.

2.2. INOCULATION AND INCUBATION

A sterilized cellophane disc was placed on medium surface contained in Petri dishes (50-mm diameter). These were inoculated centrally with a mycelial block (3x3x2-mm) cut from the advancing margin of the 15 days old colony on PDA medium. The plates were wrapped in Parafilm and incubated at 25°C in dark.

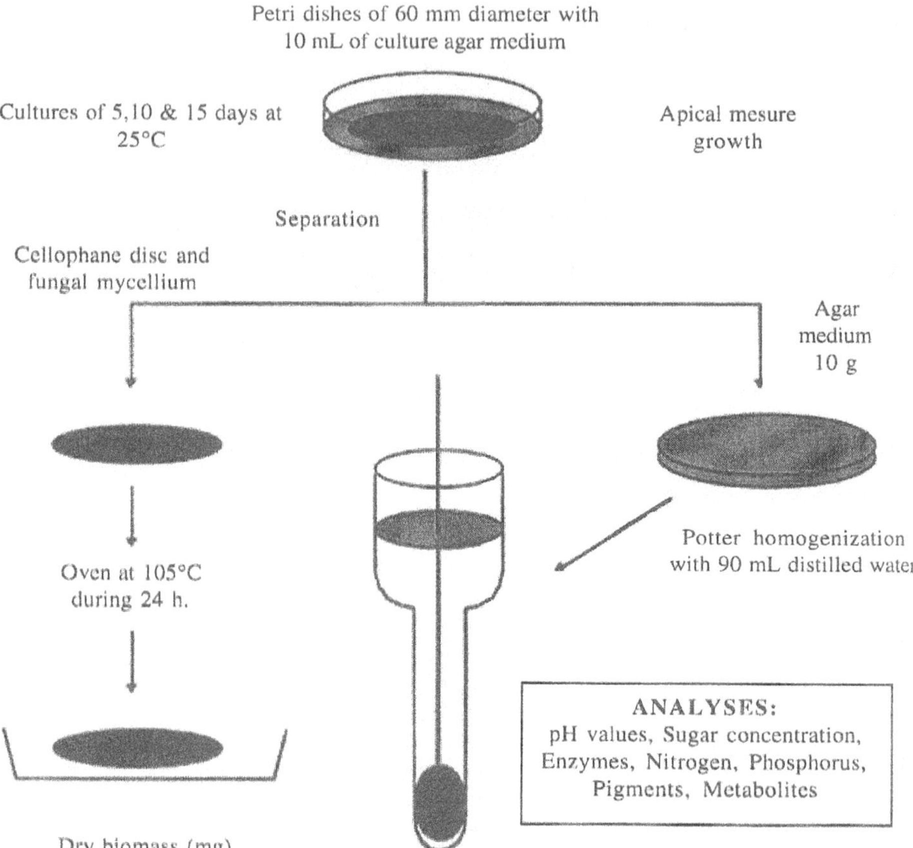

Figure 1. Scheme of fungi mycelia culture on agar surface containing a cellophane film and sampling treatment for biomass determination and metabolites produced analysis during colony development on solid surface.

2.3.1. Treatment of samples:

Figure 1 shows the treatment scheme of the samples. Agar medium (10 ml) was suspended with 90 ml of distilled water and homogenized with the Potter during 30 seconds at 4°C. Colonies growing on cellophane disc were used to measure biomass produced by drying the cellophane sheet containing the mycelium at 105°C during 24 hours. Controls were made with the cellophane sheet's weight determined similarly from un-inoculated Petri dishes. It represents an average of 15 samples.

3. Results and Discussion

Tables 2 and 3 show the consumption of glucose, biomass produced, colony diameter, and pH evolution for *S. collinitus* and *P. tinctorius* strains.

Table 2. Evolution of pH, biomass production and glucose used after 15 days incubation at 25°C for *Suillus collinitus* strain.

Culture Media	Glucose used (%)	Biomass production (g/l)	Colony diameter (mm)	pH
PDA	96	7.1	45	5.0
BAF	69	8.6	50	4.2
MNM	93	3.9	45	3.5
MP	27	1.3	24	5.8
GM8	0	1.6	18	4.8

Table 3. Evolution of pH, biomass production and glucose used after 15 days incubation at 25°C for *Pisolithus tinctorius* strain.

Culture Media	Glucose Used (%)	Biomass production(g/l)	Colony diameter(mm)	pH
PDA	52	5.4	23	5.8
BAF	41	7.6	37	4.3
MNM	77	4.5	43	3.4
MP	31	0.9	16	6.6
GM8	0	2.7	20	6.0
MG	24	1.4	33	5.7
MG	42	1.6	32	4.9

3.1. COMPARISON BETWEEN DIFFERENT CULTURE MEDIA FOR THE *S. COLLINITUS* GROWTH

Figure 2 shows different culture media, which were screened for maximum biomass production and sugar consumption. Evidently PDA medium was best for *Suillus collinitus* (Sc 24). After 15 days of growth, sugar consumption was 96% and biomass production was 7.1 g.l^{-1} (Table 2). BAF medium was favourable for biomass production, which suggested that strain Sc 24 needed yeast extract, oligo-elements and vitamins to grow. Growth patterns in MNM medium showed that this strain did not need vitamins to grow and that the decrease in the pH value after 15 days of incubation was due to the presence of PO_4^{3-} ions in its composition. Sugar consumption in this medium was 93% and nitrogen source was the limiting component.
Growth in GM8 medium showed that both the strains did not utilize glucose and little biomass was produced. It was, hence concluded that nitrate at a concentration of 2 g.l^{-1} had a negative effect for both strains. Similar results have been reported previously De Araújo *et al.* (1998).

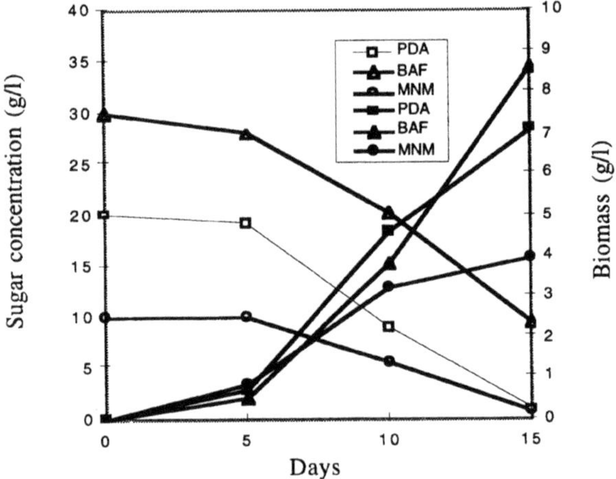

Figure 2. Glucose consumption (empty symbols) and biomass production (full symbols) for *Scuillus collinitus* grown on PDA, BAF, and MNM media at 25°C for 15 days.

3.1.1. Sugar consumption and biomass production profiles for Suillus collinitus

Sugar consumption in MG medium was 42% and very little biomass was produced. Mycelial growth was very poor due to absence of any nitrogen source. In MP medium only 27% sugar was consumed and almost no biomass was produced. Although GM8

medium contained nitrate in its composition, apparently Sc 24 for growth did not utilize this (Figure 2).
In view of these results, PDA media was selected for further studies. Torres and Honrubia (1991) studied the influence of culture media and pH on the growth of several strains of ectomycorrhizal fungi such as *S. collinitus*, *S. granulatus*, *Rhizopogon roselus*, *R. luteolus* and *Amanita muscaria* isolated from Murcia and Albacete (Spain). They cultivated the strains in different media such as MNM, PDA, MEA (2%), and the pH range was 5.5-7.5. For *S. collinitus*, after 60 days of incubation at 24°C on different culture media, they found that PDA was the best media and the colony presented 8.5-cm diameter.

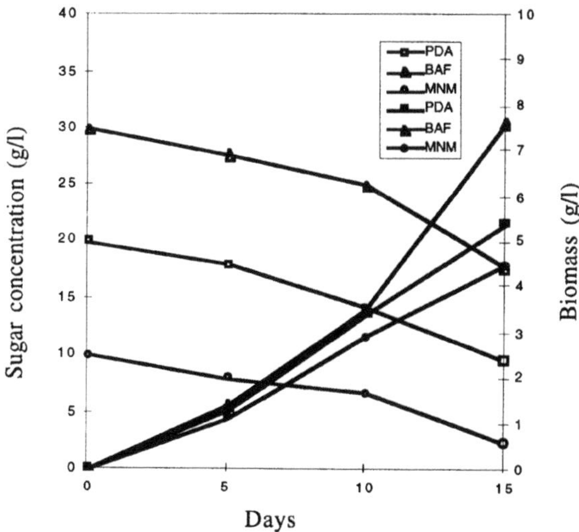

Figure 3: Glucose consumption (empty symbols) and biomass production (full symbols) for *Pisolithus tinctorius* grown on PDA, BAF, and MNM media at 25°C for 15 days.

3.2. COMPARISON BETWEEN DIFFERENT CULTURE MEDIA FOR THE *P. TINCTORIUS* GROWTH

As is evident from the Figure 3 and Table 3, better development was observed on PDA, BAF, and MNM media. The strain PF 26 did not utilize glucose at same level as the strain Sc 24. After 15 days of culture on PDA medium, 48% glucose was still present. Lower consumption of glucose by PF 26 was due to mainly its apical growth, which was slower in the medium. Perhaps nitrogen was limiting factor for this strain in these media. Apparently strain PF 26 strain tolerated the presence of nitrate, as it produced 2.7 $g.l^{-1}$ biomass in GM8 medium.

4. Summary and conclusions

A technique for studying the mycorrhizal fungi culture parameters on agar media was established. Two strains *viz. Suillus collinitus* and *Pisolithus tinctorius* were screened for biomass production and sugar consumption patterns in different media such as PDA, BAF, MNM, MP, GM8 and MG. Fungal biomass analysis was carried out by drying the biomass grown on cellophane sheet. Evolution of glucose, pH and produced biomass after 5, 10 and 15 days of incubation at 25°C. PDA medium appeared best. Data obtained for sugar consumption, pH values and fungal biomass production permitted to optimize culture conditions and study the physiology and metabolism of ectomycorrhizal fungi grown on solid agar media. One main advantage of this new technique was based in the fact that it was not destructible method.

5. Acknowledgements

The authors gratefully acknowledge Drs. D. Mousain and C. Plassard (Institut National de Recherche Agronomique, Montpellier, France) for providing the strains of ectomycorrhizal fungi. Álvaro A. De Araújo thanks the Brazilian government (CNPq) for a scholarship and Sfere (France). This investigation was supported by the European Union MYCOMED programme no AIR2 - CT 94-1149, and by IRD, Dept. DRV, France.

6. REFERENCES

Baar J., Comini B., Elferink M.O., Kuyper Th.W. (1997), Performance of four ectomycorrhizal fungi on organic and inorganic nitrogen sources. Mycol. Res., 101 (5): 523-529

De Araujo A.A., Hannibal L., Plassard C., Mousain D., Roussos S. (1998), Apical growth of the ectomycorrhizal fungi *Pisolithus, Suillus,* and *Lactarius* on solid culture media. Micol. Neotrop. Apl. 11: 23-34

Gibson F., Deacon J.W. (1990), Establishment of ectomycorrhizas in aseptic culture: effects of glucose, nitrogen and phosphorus in relation to succession. Mycol. Res., 94 (2): 166-172

Jongbloed R.H. Borst-Pauwels G.W.F.H. (1990), Effects of ammonium and pH on growth of some ectomycorrhizal fungi *in vitro*. Acta Bot. Neerl., 39, 349-358

Marx D.H., Cordell C.E. (1994), The use of specific ectomycorrhizas to improve artificial forestation practices. In- *Biotechnology of fungi for improving plant growth*. 2nd edn., J.M. Whipps and R.D. Lumsden (eds.), Cambridge University Press, Cambridge, pp.1-25

Miller G.L. (1959), Use of dinitrosalicylic acid reagent for determination of reducing sugars. Anal. Chem.,

31: 426-428

Mousain D. (1993), *Physiologie des arbres et arbustes en zones arides et semi-arides.* 1st ed. Editions John Libbey, Eurotext, Montrouge CEE Project, MYCOMED, Contract Number AIR 2-CT-94-1149

Mousain D., Plassard C., Roussos S., Brunel B., Karkouri K.E., Fréchet C.C., Laguette A.G., Bouckcim H., Gobert A., De Araujo A.A. (1998), Report on "Improvement of the quality of forest seedlings (*Pinus pinea, Pinus halepensis, Pinus nigra* subsp. *laricia, Quercus suber*) and Mediterranean reforestation using controlled Mycorrhizal infection (MYCOMED)", CEE Contract no AIR 2-CT-94-1149

Oort A.J.P. (1981), Nutritional requirements of *Lactarius* species, and cultural characteristics in relation to taxonomy. Verhandelingen. Koninklilke Nederlandse akademie van Wetenschappen, Afdeling Natuurwetenschappen, *76*: 1-87

Pelmont J. (1993), *Bacteries et Environnement.* 1st edn., Presses Universitaires de Grenoble, Grenoble

Strullu D.G. (1991), *Les mycorhizes des arbres et plantes cultivées*, 3rd edn., Lavoisier. Paris

Torres P., Honrubia M. (1991), Dinámica de crecimiento y caracterización de algunos hongos ectomicorrizicos en cultivo. Cryptogamie Mycol., 12 (3), 183-192

Chapter 23

CHARACTERIZATION OF *BEAUVERIA BASSIANA* AND *METARHIZIUM ANISOPLIAE* ISOLATES FOR POTENTIAL USE AGAINST THE COFFEE BERRY BORER

P.E. Vélez, M.T. González, A. Rivera, A.E. Bustillo, M.N., Estrada E.C. Montoya

Centro Nacional de Investigaciones de Café, Cenicafé, AA 2427, Manizales, Colombia

Running Title: Entomogenous fungi characterization

1. Introduction

The coffee berry borer (CBB), *Hypothenemus hampei* (Ferrari) (Coleoptera: Scolytidae), is the most serious pest of coffee in Colombia and many other countries. This pest originated in central Africa and was accidentally introduced to Brazil at the beginning of the century (Murphy and Moore, 1990). Since its arrival in Colombia in 1988, the Colombian Coffee Growers Federation and its research centre, CENICAFÉ, have promoted extensive research into alternative control strategies using an IPM system. Biological control with parasitoids and with entomopathogenic fungi (*Beauveria bassiana* and *Metharizium anisopliae*) are promising control methods, offering an environmentally benign alternative to chemical pesticides (Cadena, 1993). Coleoptera are known to be susceptible to *B. bassiana*. The fungus may have spread with the pest as the latter colonised new geographic areas or it may have "jumped" from other local insects (Bridge *et al.*, 1990). To exploit the potential application of *B. bassiana* and *M. anisopliae* against CBB in Colombia, we used morphological and biochemical techniques to characterize isolates in order to select the best strain to use as a mycoinsecticide against CBB.

T. Sera et al. (eds.), Coffee Biotechnology and Quality, 255–263.

2. Experimental

2.1. ISOLATES TESTED

A total of 73 isolates of Bb and 19 of Ma from the culture Collection held in the Entomology Department of the Coffee National Research Centre in Colombia were examined. Isolates were grown on Sabouraud Dextrose Agar (SDA) for 15 days at 25°C and passed once on comminuted cuticle from adult CBB (Vélez *et al.*, 1997) in order to maintain their infectivity for the target host.

2.2.VIABILITY

Viability was determined as percent germination in culture after 30 days of growth in SDA. Five aliquots of 5μl (1 x 10^6 spore/ml) were placed (each n = 10). Plates were incubated for 24-48h at 25°C. One drop of lactophenol blue was added to each aliquot in order to stop the germination process and microscopic examination at 40x of the germinated spore was carried out. The percent germination was computed on the basis of the number of spore producing germ tube per 100 counted in each aliquot. A germinated conidium was defined as one, which produced a germ-tube equal to or greater than the width of the conidium (Vélez *et al.*, 1997).

2.3. PATHOGENICITY

Sixty adult female CBB, divided into six groups of ten, were maintained individually in vials in a controlled environment chamber at 27 °C and 80-85 % RH. To determine pathogenicity adult CBB were submerged two minutes in a spore suspension of the fungi (1 x 10^7 spore/ml sterile distilled H_2O, 1 % Tween 80) with constant manual shaking. Pre-treated CBB were confined in a 4-ml glass vial with one coffee bean (45% moisture content) as the borer substrate, wetted filter paper and capped with a piece of sterile cotton and maintained at 23°C and 80% RH. Additional 60 adult female CBB without infection were included as control. Over 10 days, mortality and presence of sporulating fungus (symptoms and signs) on the insects were recorded using a stereoscopic microscope (González *et al.*, 1993; Vélez *et al.*, 1997). Lethal time fifty (LT_{50}) was measured by taking into account the time in which the inoculum of the fungus caused 50% of mortality on the insect population.

2.4. RATE OF MYCELIAL GROWTH

Mycelial growth was measured on colonies growing on SDA. Each plate was inoculated at the centre with 5μl of fungal spore suspension (1 x 10^6 spore/ml) and incubated at room temperature (25°C) for five days. Each treatment was replicated 10 times. After incubation, the diameter of the colony up to 30 d was measured. A mean diameter was calculated for each isolate.

2.5. SPORE SIZE AND PRODUCTION

Spore diameter was determined using a compound light microscope fitted with an ocular micrometer. For each isolate, mean diameter was calculated based on measurements of 100 spores. Spore production was quantified in all of the isolates by enumerating spores from 8, 15, 21 and 30 day old cultures grown on SDA. Spores from these cultures were suspended in 5-ml of 1% Tween 80, diluted (1/10), and spores were counted using a haemocytometer. Ten replications for treatment and six counts for replication were recorded (Vélez *et al.*, 1997).

2.6. ENZYME REACTION

Production and activity of extra-cellular enzymes were screened through the technique that involved incorporating particular substrates into modified agar growth media; degradation of the substrate was then detected after growth either by a colour change (pH indicator), a reduction in opacity, or liquefaction. The enzyme systems tested using this method were esterase, gelatinase, caseinase, and lipase. The nitrogen sources evaluated were urea, peptone and yeast extract and the carbon sources: citrate, starch, lactose, lanolin, N-acetyl glucosamine and glucosamine (Paterson and Bridge, 1994; Mugnai *et al.*, 1989). Six plates were set up for each treatment. Biochemical test results were analysed using the percent of plates that gave some change of colour. The time, at which substrate utilisation was visible by colour change, reduction in opacity or liquefaction was assessed for each isolate.
Isolates of Bb and Ma were classified according to pathogenicity against CBB, establishing 4 groups: group 1, pathogenicity <25%; group 2, > or = 25% and < 50%; group 3, > or = 50% and < 80%, and group 4, > or = 80% (Table 1).

2.7. STATISTICAL ANALYSIS

Variance analysis of the groups and Tukey test ($p=0.05$) were carried out. Likewise, a multivariate principal component analysis was carried out with a group of 17 isolates of Bb in which all the following variables were evaluated: pathogenicity against CBB, LT_{50}, germination rate, enzyme reaction, time of enzyme reaction, spore production, mycelial growth and spore size. A dendrogram and cluster statistic analysis were performed in order to select groups of isolates according to the evaluated variables, variance analysis of the clusters (p=0,05) and Tukey test (5%) were applied to the groups of isolates.

Table 1. Average, variation, minimum and maximum values of the variables tested in *Beauveria bassiana*.

GROUP	1		2		3		4	
VARIABLE	x	C.V.	x	C.V.	x	C.V.	x	C.V.
% pathogeniticy	14,09 d*	70,35	40,48 c	40,09	67,34 b	21,27	88,87 a	10,89
LT_{50} days	5,76 a	22,17	5,05 b	13,79	4,38 c	17,04	4,32 c	18,39
Spore production	12 b	128,89	11 bc	76,94	14 a	81,31	9 c	126,09
Mycelial growth rate	0,19 c	15,81	0,21 ab	55,01	0,20 bc	35,47	0,22 a	39,76
Spore size (μ)	2,33 a	18,01	2,28 a	18,87	2,27 a	14,83	2,36 a	19,23
% Germination	70,94 c	35,32	71,57 c	17,30	75,83 b	18,47	78,77 a	21,40
% Enzyme reaction	75 a	54,25	70 a	61,71	75 a	55,12	62 b	72,43
Time of enzyme reaction (days)	6 b	72,31	7 a	55,44	6 b	64,88	6 a	62,89

Means not followed by common letter are significantly different (P=0.05) by Tukey test.

3. Results and Discussion

3.1. ANALYSIS OF PATHOGENICITY

3.1.1. Beauveria bassiana(Bb):

In group 1, there were six isolates from Colombia, one from Thailand and one from Philippines. Among the most virulent isolates were isolates from different insect families, i. e. from Coleoptera (Curculionidae), Scolytidae (Scarabaeidae) and Homoptera. Two were from unknown origin and one was from a commercial formulation. Seventeen of these virulent isolates were obtained from Colombia, four from Antioquia and four from Risaralda, two of them had unknown origin and one another from other country. The variance analysis showed differences in Bb and Ma groups in the variable pathogenicity against CBB (Tukey test p=0.05) (Tables 1 and 2)
The results showed that the most pathogenic isolates of Bb (group 4) showed the highest germination rates and the lowest values for LT_{50} (Table 1). Groups 1 and 3 showed the highest values of enzyme reaction and group 3 showed the highest values of

spore production. The lowest values of enzyme reaction and spore production were found in group 4 isolates (Table 1).

The most pathogenic group of isolates of Bb (group 4) was composed of 24 isolates, 15 belonging to the Coleoptera. From this group, 11 isolates were isolated from the Scolytidae, three from the Curculionidae and one from the Scarabaeidae. In addition, this large group included five isolates from Lepidoptera, two from the Pyralidae, and one each from the Stenomidae, Geometridae and Cossidae. Regarding the geographical origin of Bb isolates, 17 came from Colombia, five from other countries, *viz.* from China, Thailand, Ecuador and Italy. In the Bb Colombian isolates of this group, there were four from Antioquia and four from Risaralda. This could be a re-isolation of the same fungal strain within a specific region with similar climatic conditions. Germination rates of the isolates of this group were between 75-92% with values higher than 80% in most of the isolates. Enzyme reactions showed averages between 41-75%, and most of them were between 50-60%.

Table 2. Average, variation, minimum and maximum values of the variables tested on *Metarhizium anisopliae.*

Group	3		4	
Variable	x	C.V.	x	C.V.
% Pathogenicity	55,47 b	15,07	88,73	8,97
% Germination	96,80 a	2,81	64,88 b	52,41
% Enzyme reaction	100,0 a	0	96,0	19,08
Time of enzyme reaction (days)	6 a	98,49	5,0 a	100,24

Means not followed by common letter are significantly different (P=0.05) by Tukey test.

Isolates of Bb with <25% pathogenicity for CBB (group 1) were composed of eight isolates (five from CBB). In this group, isolates from Lepidoptera and Homoptera were also found. They were isolated in Colombia (six), four in Antioquia (three of them from CBB) and one in Risaralda. Two were from Thailand and Philippines. The isolates of Bb from this group showed an average pathogenicity of 14%, with a minimum value of 5 and maximum of 25%. The average germination rate was 71% and ranged between 20-89%. The average enzyme reaction was 75% and ranged between 43-100% (Table 1).

With respect to the relationship between host and locality, the results suggested certain specificity for the CBB within the most pathogenic isolates, as 61% of the isolates came from Coleoptera and 43% of these came from CBB. This was in agreement with Prior (1992), bearing in mind that virulence might be higher when the isolates come from the same host insect. However, the CBB isolates (62.5%) in the least pathogenic group of isolates did not show any relationship with virulence against CBB. So, these results did not allow inferring specificity in those isolates. Another and contrasting possibility could that the pathogenicity of the group 1 isolates could be explained on co-evolution criteria, which suggests an equilibrium between host and pathogen over a long period (Prior, 1992). Thus the most virulent pathogen could be from hosts other than CBB (Hokkanen and Pimentel, 1984; Waage, 1990). In pathogenicity groups 2 and 3, it was observed that 30.3 and 62.5% of isolates, respectively came from CBB and they did not show a high virulence for CBB. Isolates from Lepidoptera and Homoptera showed higher levels of pathogenicity for CBB (group 4, Table 1). According to these criteria, it was suggested that the search for entomogenous fungi should be conducted in different places from that where the insect pest is found (Hokkanen and Pimentel, 1984; Waage, 1990; Prior, 1992; Hajek and St. Leger, 1994).

3.1.2. Metarhizium anisopliae (Ma):

Two pathogenicity groups were established for Ma. Most of the isolates killed more than 80% in the tests: nine from Colombia, three from Antioquia (Coleoptera) (Table 2). The most pathogenic Ma isolates (group 4) consisted of 18 isolates, nine from Coleoptera, including six from Scarabaeidae, one from Curculionidae and one from Scolytidae. The group also contained four isolates from Homoptera and one from Hymenoptera. Ma isolates came from Colombia (9), USA (1), Australia (3), Brazil (1), Belize (2), New Zealand (1) and one isolate of unknown origin. Among the Colombian isolates three from Antioquia Department (Coleoptera), one from Nariño, one from Tolima and two from Caquetá were remarkable. Since the majority of Ma isolates belonged to Coleoptera: Scarabaeidae, it could be assumed certain specificity toward insects of this family.

3.2. MULTIVARIATE ANALYSIS

Taking into account the variables tested in the Bb isolates, the multivariate analysis showed that the components that contributed most to the total variation were enzyme reaction, time of enzyme reaction, pathogenicity against CBB and spore production. Spore size, LT_{50}, mycelial growth and germination rate accounted for a lower degree of variation.

Fig. 1 shows the dendrogram separating the three groups of isolates in a descriptive way. For the cluster statistical analysis, the isolates were grouped so that those with pathogenicity against CBB between 53-80% were assigned to cluster 1; isolates with pathogenicity against CBB >80% in cluster 2, and those with pathogenicity against CBB< 53.3% in cluster 3.

Table 3. Average and variation of the new cluster analysis grouping of Beauveria bassiana isolates formed according to % pathogenicity to the coffee berry borer.

Varib.	Isola.	Pathogenicity (%)		LT_{50} (days)		Spore production (10^8)		Mycelial growth rate (cm)		Spore size (μ)		Germination (%)		Enzyme reaction (%)		Time (days)	
Group	Bb	x	CV	x	CV	x	CV	x	CV	x	CV	x	CV	x	CV	x	CV
1	9005, 9416, 9008, 9023, 9117, 9209	70.5 b	22.3	4.8 b	22.7	15 a	70.9	0.22 b	32.5	2.2 b	17.4	73.2 b	23.7	68 a	64.2	6 b	65.3
2	9116 9207, 9204, 9212, 9010, 9101	90.1 a	8.8	4.6 b	23.7	6 b	94.3	0.24 a	35.0	2.4 a	19.1	83.9 a	8.1	57 b	84.3	7 a	62.4
3	9120, 9108, 9305	25.7 c	69.4	6.1 a	18.3	17 a	102.3	0.19 c	13.7	2.2 b	16.1	83.7 a	83.7	74 a	55.8	6 c	61.1

Means not followed by common letter are significantly different (P=0.005) by Tukey test.

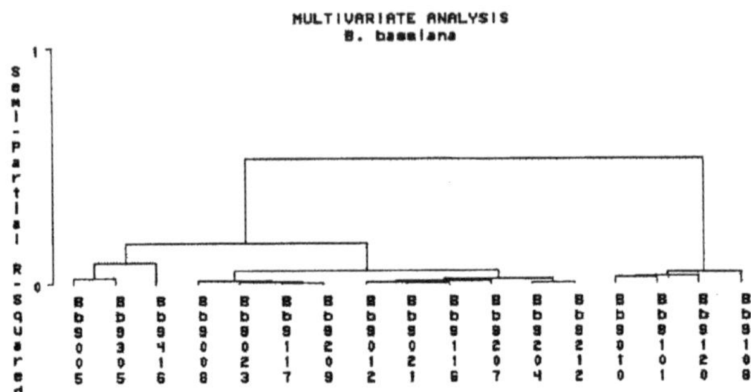

Figure 1. Dendrogram of the isolates of *B. bassiana*

The analysis showed three groups of them: six isolates in group 1; six isolates in group 2, and five isolates in group 3. The analysis of variance of the groups classified by cluster analysis showed differences between them. The Tukey test (5%) showed that the groups were statistically different in all the evaluated variables; however, the three groups were statistically separated in the variables CBB and mycelial growth (Table 3).

From the results, the variable percentage of pathogenicity to the CBB, used as classification criteria in the preliminary analysis, was one of the components, which had a major contribution to the total variation, according to multivariate analysis. Additionally this variable and the mycelial growth allowed the separation of different clusters in the statistical analysis (Table 3). It was notable that cluster 2, with the highest pathogenicities, had the lowest average of spore production and enzyme reaction. Additionally, that cluster showed the lowest averages of LT_{50} (Table 3). The highest averages of pathogenicity against CBB, mycelial growth, germination rate and time of enzyme reaction were showed by isolates from cluster 2. The pathogenicity against CBB and enzyme reaction did not match with the reports in which enzyme activity is related to the virulence against the pest (St. Leger *et al.*, 1988; Bidochka and Khachatourians, 1990). Likewise, previous experiments have shown no relationship between pathogenicity against CBB and spore production (Narváez, 1996) (Table 3).

Cluster 3 showed the lowest averages in pathogenicity against CBB and mycelial growth (Table 3), however, in this cluster the highest averages of LT_{50}, spore production and enzyme reaction were registered.

4. Conclusions

The characterization of the Collection isolates allowed selection of those variables that contributed most to the separation of isolates by groups, to identify the strains, to establish relationships between them according to their biology and locality, to define their host specificity and their potential to control the CCB, and other insect pests of economic importance in Colombia and other countries.

5. References

Bidochka M.J., Khachatourians G.G. (1990), Identification of *Beauveria bassiana* extra-cellular protease as a virulence factor in pathogenicity toward the migratory grasshopper, *Melanoplus sanguinipes*, J. Invertebrate Pathol., 56: 362 - 370

Bridge P.D., Abraham Y.J., Cornish M.C., Prior C., Moore D. (1990), The chemotaxonomy of *Beauveria bassiana* (Deuteromycotina: Hyphomycetes) isolates from the coffee berry borer *Hypothenemus hampei* (Coleoptera: Scolytidae), Mycopathologia, 111: 85-90

Cadena G.G. (1993), Políticas de la Federación Nacional de Cafeteros de Colombia para el control de la broca del café. In: Congr. de la Sociedad Colombiana de Entomología, 20. Cali (Colombia), Julio 13 – 16, p. 110

Gonzalez M.T, Posada, F.J., Bustillo, A.E. (1993), Desarrollo de un bioensayo para evaluar la patogenicidad de *Beauveria bassiana* sobre *Hypothenemus hampei*, Cenicafé, 44: 93-102

Hajek A.E., St. Leger R.J. (1994), Interactions between fungal pathogens and insect hosts, Ann. Rev. Entomol., 39: 293-322

Hokkanen H M.T., Pimentel D. (1984), New approach for selecting biological control agents, Can. Entomol., 116: 1109 - 1121

Mugnai L., Bridge P.D., Evans, H.C. (1989), A chemotaxonomic evaluation of the genus *Beauveria*, Mycol. Res., 92: 199 - 209

Murphey S., Moore D. (1990), Biological control of the coffee berry borer, *Hypothenemus hampei* (Ferrari) (Coleoptera, Scolytidae): previous programmes and possibilities for the future. Biocontrol News Inform., 11(2): 107-117

Narvez M del P. (1996), Estimación de la cantidad de esporas producidas por los aislamientos patogénicos de los hongos *Beauveria bassiana* y *Metarhizium anisopliae* cultivados en arroz y sobre la broca del café. Manizales (Colombia), Universidad Católica de Manizales. Facultad de Bacteriología, Tesis Bacterióloga y Laboratorista Clínica

Paterson R.R.M., Bridge P D. (1994), *Biochemical techniques for filamentous fungi*. IMI Technical Handbooks: No 1. CAB International. Surrey

Prior C. (1992), Discovery and characterization of fungal pathogens for locust and grasshopper control. In: Biological Control of Locusts and Grasshoppers. International Institute of Biological Control. Lomer, C. J., Prior, C. (eds.). CAB International, Melksham, UK, pp. 230 - 238

St. Leger R.J., Durrands P.K., Charnley A.K., Cooper R.M. (1988), Role of extra-cellular chymoelastase in the virulence of *Metarhizium anisopliae* for *Manduca sexta*, J. Invert. Pathol., 52: 285 - 292

Velez P.E., Posada F.J., Marin P., Gonzalez M.T., Osorio E., Bustillo A.E. (1997), Técnicas para el control de calidad de formulaciones de hongos entomopatógenos. Boletín Técnico Cenicafé No.17, p. 34

Waage J. (1990), Ecological theory and the selection of biological control agents. In: *Critical issues in biological control*. Mackauer, M., Ehler, L.E. and Roland, J. (eds.). Intercept, UK, pp. 135 - 157

Chapter 24

THE NEMATOPHAGOUS FUNGI HELPER BACTERIA (NHB): A NEW DIMENSION FOR THE BIOLOGICAL CONTROL OF ROOT KNOT NEMATODES BY TRAPPING FUNGI

R. Duponnois[1], J.L. Chotte[1], A.M. Bâ[2], S. Roussos[3]

[1] *Institut de Recherche pour le Développement (IRD), Laboratoire de Biopédologie. BP 1386, Dakar. Sénégal;* [2] *Institut Sénégalais de Recherche Agronomique (ISRA), BP 1386, Dakar. Sénégal,* [3] *IRD. Laboratoire de Mycologie et des FMS (LAMY-FMS). IFR-BAEM; Universités de Provence et de la Méditerranée, ESIL, Case 925; 163 Avenue de Luminy. F-13288, Marseille Cedex 9 France*

Running title: Helper bacteria of nematophagous fungi

1. Introduction

Many genera and species of plant parasitic nematodes are associated with coffee. These cause great financial losses to the coffee farmers. Among these pathogens, root-knot nematodes (*Meloidogyne* sp.) are the most abundant group (Table 1) and the most common species are *M. exigua*, *M. incognita* and *M. coffeicola*.

The symptoms of damage due to these pathogens are typical rounded galls on the root systems, white to yellowish brown becoming dark brown in ageing roots. The infested coffee plants show foliar chlorosis, leaf fall, general decline. Their growth is generally reduced and sometimes the plants died (Hutton et al., 1982; Lordello, 1984). The coffee plantations can be dramatically affected by the nematodes. For example, in Sao Paulo state (Brazil), they were destroyed by *M. incognita* with 5-year-old coffee plantations dying out (Lordello, 1984).

The control of nematodes is more difficult in a perennial crop than in annual or herbaceous crops. For example, the rotation schemes successfully used with annual crops are impractical with these long-term cultures. Moreover surviving roots of excised plants or old plants left in the field can provide nutrients for nematodes and consequently maintain the nematode population in the soil.

T. Sera et al. (eds.), Coffee Biotechnology and Quality, 265–276.

The control practices used by many farmers are generally based on:

(i) the production of seedlings without root-knot nematodes (i.e. disinfection of nursery soils),
(ii) the application of nematicides and/or the culture of resistant or tolerant cultivars.

Among these control measures, pesticide compounds are mostly applied to reduce nematode multiplication. However, they are expensive and toxic when used improperly. Thus, other control techniques such as biological control have been investigated. Besides various micro-organisms tested against *Meloidogyne* sp. such as arbuscular mycorrhizae (Hussey and Roncadori, 1982), eggs parasitic fungi (*Verticillium chlamydosporium*) (Kerry, 1990; Bourne *et al.*, 1994), rhizobacteria (Racke and Sikora, 1986) and fungal endophytes (Schuster *et al.*, 1995), research focused on nematophagous fungi (Cayrol, 1983; Pelagatti *et al.*, 1986; Duponnois *et al.*, 1996). The screening of efficient fungal strains against *Meloidogyne* sp. was based on the use of tests performed in axenic or controlled conditions (*in vitro* tests or glasshouse experiments with disinfected substrates). When these micro-organisms were transferred into the field, their antagonistic activity was often modulated (generally decreased) by the environmental conditions. In fact, the topics of this kind of research must integrated the requested qualities of the fungal strain but also its reply from interactions with the soil microbial community in the soil. In the laboratory of Biopedology (IRD, Dakar), we have developed a scientific programme based on:

(i) the (i) screening of efficient fungal strains for trapping juveniles of *Meloidogyne* sp.,
(ii) determination of potential effects of rhizobacteria on *in vitro* growth and nematode-trapping activity of *Arthrobotrys oligospora*, population development of plant parasitic nematodes
(iii) to develop a control method with the fungal strain incorporated into compost blocks. Tobacco and tomato plants, very susceptible to *Meloidogyne* sp., were used in these experiments.

2. Experimental

2.1. TEST OF THE POTENTIAL OF NEMATOPHAGOUS FUNGI AGAINST *MELOIDOGYNE* SP.

A Collection of nematophagous fungi isolated from several vegetable-producing areas in Senegal and Burkina Faso was maintained aseptically in dark at 25°C on the nutrient broth (8 $g.l^{-1}$) agar (20 $g.l^{-1}$) medium (Table 1). The trapping activity of each fungal strain was estimated using technique described by Duponnois *et al.*, (1996). Populations of different species of *Meloidogyne* sp. (*M. mayaguensis, M. javanica* and *M. incognita*) were reared on tomato (*Lycopersicon esculentum* Mill.), cv Roma roots. Two months after inoculation, the roots were harvested, cut into short pieces and placed in a mist chamber for one week for egg hatching (Seinhorst, 1950). Fungal agar plugs were taken from the margin of two week old colonies and transferred into Petri dishes filled with distilled water agar (20 $g.l^{-1}$).

Table 1. Nematophagous fungi used in the experiments

Identification	Code	Geographical origin	Author
Arthrobotrys oligospora	S 30	Burkina Faso	Sawadogo A. [(1)]
Arthrobotrys oligospora	S 31	Burkina Faso	Sawadogo A.
Arthrobotrys conoides	S 42	Burkina Faso	Sawadogo A.
Arthrobotrys sp.	BF 10	Burkina Faso	Sawadogo A.
Arthrobotrys sp.	BF 74	Burkina Faso	Sawadogo A.
Arthrobotrys sp.	BF 80	Burkina Faso	Sawadogo A.
Arthrobotrys sp.	SOSU 2	Burkina Faso	Sawadogo A.
Arthrobotrys sp.	ORS 18690 S2	Senegal	Duponnois R.
Arthrobotrys oligospora	ORS 18692 S5	Senegal	Duponnois R.
Arthrobotrys oligospora	ORS 18692 S7	Senegal	Duponnois R.

(1) INERA, Protection des Végétaux, BP 403, Bobo Dioulasso, Burkina Faso

One week later, 100 second-stage juveniles suspended in a water droplet were placed on these fungal cultures. After two days, the numbers of juveniles trapped by the fungus were counted under a dissecting microscope. Each combination *Meloidogyne* sp.-fungal isolate was replicated five times. The trapping rate (trapped juveniles/ total juveniles) was transformed by arcsin (sqrt) and treated with a one way analysis of variance, the mean values being compared with the Student's t-test at 0.05 probability level.
The ability of these fungal strains to control *Meloidogyne* sp. populations was examined through a glasshouse experiment. Solid fungus inocula were prepared in 0.5-dm^3 glass flasks containing 0.3-dm^3 compost. After autoclaving at 120°C for 40 min, the substrate was moistened to field capacity with liquid nutrient broth medium (8 $g.l^{-1}$), the jars closed with cotton wool and autoclaved a second time at 120°C for 20 min. After cooling, mycelial plugs samplings from each fungal culture on Petri dishes were put into flasks. The cultures were incubated for five weeks at 25°C.
The fungal inoculum was mixed to an autoclaved (140°C, 40 min) sandy soil (pH H_2O 7.1; fine silt 0.6%; coarse silt 1.4%; fine sand 61.6%) at the rate 1:100 (v/v). This mixture was distributed in 60-cm^3 polythene cells (the control treatment received the same quantity of compost but without fungus). Then a one-week old tomato seedling was transferred into each cell. After one week, the tomato plants were inoculated with 5-

ml suspension of 100 7-day-old second stage juveniles of *M. mayaguensis* (water alone for the control). The cells were placed in a glasshouse under natural climatic conditions (temperatures between 20-35°C, about 15 hours light per day). The treatments were watered daily and arranged in a randomized complete block design with 14 replicates. One month after the nematode inoculation, the plants were uprooted and the roots washed. Shoots were dried at 65°C for one week and weighed. Galls induced by *M. mayaguensis* were counted. Then roots were cut into 1-2 cm pieces and placed in a mist chamber for two weeks to recover hatched juveniles (Seinhorst, 1950). Roots were then oven-dried and weighed. The data were treated with one way analysis of variance and the mean values were compared with the Student's "t" test ($P < 0.05$). For nematodes, data were previously transformed by log (x+1).

2.2. STUDY OF THE POTENTIAL EFFECTS OF RHIZOBACTERIA ON THE ANTAGONISTIC ACTIVITY OF *ARTHROBOTRYS OLIGOSPORA* AGAINST *MELOIDOGYNE* SP.

Study of the potential effects of rhizobacteria on the antagonistic activity of *Arthrobotrys oligospora* against *Meloidogyne* sp. was carried out as described by Duponnois and Bâ (1998). The authors have demonstrated that a large nematophagous fungus population was associated with the presence of bacteria belonging to the group of fluorescent *Pseudomonas*. Attempts were made to test the effect of these bacteria on the fungal growth (a strain of *A. oligospora* T41), on its predatory activity and to evaluate the bacterial influence on the fungal control of the multiplication of *Meloidogyne* sp. population.

Bacterial strains were cultured in Petri dishes on 0.3% TSB agar medium at 25°C for two days. The bacterial cultures were suspended in 5-ml of sterile magnesium sulphate solution ($MgSO_4$, 0.1 M). The control treatment was prepared in the same way from a Petri dish containing the same agar medium but without bacteria. One set of experiment was performed by direct liquid contact between the mycelium and the bacteria. Fungal plugs sampled as described above were dipped in the bacterial suspensions or in the control solution for 1-2 min and transferred into empty Petri dishes (Duponnois and Garbaye, 1990). Another set of experiment was made with no liquid contact. Two-compartment dishes were used. The fungal plugs were laid on the dry bottom of the dish for one compartment while the other was filled by 0.3% TSB agar medium inoculated by the bacterial strains (uninoculated for control treatment). Since the wall separating the two compartments did not touch the lid of the dish, the gas diffusion from one side to the other was permitted.

In both the experiments, two dishes, each with three mycelial plugs, were prepared for each treatment and incubated at 25°C for four days. Observation and numeration were carried out through the lid with a stereomicroscope and the mean radial growth in two perpendicular directions was calculated. The data were statistically compared to the control treatments without bacteria with the Student "t" test ($P<0.05$).

The fungal plugs, dipped in the bacterial suspensions or control treatment were transferred to Petri dishes filled with distilled water agar (20 $g.l^{-1}$). There were five

replicates per treatment. The Petri dishes were incubated at 25° C in the dark. Two weeks later, 100 7-day-old-second stage J2 of *M. mayaguensis* suspended in 100 μl sterile distilled water were placed on the fungal cultures. Populations of *M. mayaguensis* were reared on tomato (*L. esculentum* Mill cv. Roma). Two months after inoculation, roots were cut into 2-3 cm pieces and placed in a mist chamber for one week to allow nematode eggs to hatch and collect the juveniles (J2) (Seinhorst, 1950). After 48 hours, the numbers of the juveniles trapped by the fungus were counted. The data were statistically treated as described above.
Tobacco seedlings (*Nicotiana tabacum* L. var. Paraguay x Claro) were grown in 60 ml polythene pots filled with an autoclaved sandy soil (140°C, 40 min) containing (%) clay 3.9, silt 2.9, sand 92.2, carbon 3.7, nitrogen 0.45 (pH-H_2O 8.3) and inoculated with 1-mg dry weight of fungal biomass and/or 5-ml of each bacterial suspension (about 10^{12} colony forming unit- $cfu.ml^1$). The fungal strain was grown in one litre flask filled with 0.5 litre of 0.3% TSB medium for two weeks at 25°C. The fungal suspension was then filtered and the mycelium was collected, washed three times in $MgSO_4$ 0.1-M solution and finally re-suspended in $MgSO_4$ 0.1 M solution. The bacterial isolates were cultured in 3 $g.l^{-1}$ liquid Difco tryptic soy broth in glass flasks under agitation for eight days at 25°C, centrifuged (2400 g, 10 min) and the pellet was re-suspended in $MgSO_4$ 0.1 M solution. The control treatments were performed by injecting 1-ml of $MgSO_4$ 0.1 M solution without either fungus or bacteria in the soil. There were ten replicates per treatment. After two-month culture, 10 tobacco plants from each treatment were transferred in 10-litre pots filled with the same soil as above but non-autoclaved. The pots were placed in a glasshouse under natural conditions (temperature between 20-35°C, about 15 hours light per day). After two-month culture, the plants were harvested and the roots were gently washed. The soil from each pot was mixed, a 250-g sub-sample was taken and the nematodes were extracted by the Seinhorst's (1962) elutriation technique. The oven-dried weight of shoot (one week at 65°C) was measured. Each root system was then cut into 2-3 cm pieces and placed in a mist chamber for two weeks in order to recover nematodes (Seinhorst, 1950). The nematodes were counted under a stereomicroscope (magnification x 150). The means of treatment vs. control (not inoculated) were compared with Student's 't' test at 0.05 probability level. For the nematode populations, data were transformed by log (x + 1) prior to analysis.

2. 3. DEVELOPMENT OF A CONTROL METHOD WITH THE FUNGAL STRAIN INCORPORATED INTO COMPOST BLOCKS

The fungal inoculum (strain ORS 18692S7) was prepared using the compost as described above. It was diluted with the same compost (without fungus) at the concentration 1:100 (v/v). This substrate was used to make small blocks (4 x 4 x 4 cm^3) with a mechanical apparatus (F.A.O. patent). Each block received a tomato seedling, which was cultured during three weeks in a glasshouse. The blocks with the seedlings were then transferred to the plots (2.5 x 2.5 m; 25 plants per plot) separated from one another by 2 m.

The treatments were arranged in a randomised complete block design with 10 replicates. The control treatment consisted of blocks without fungus. This experiment was conducted from May to July (maximal temperature <35°C) on the same soil as that used in the section 2.1. The height, mortality and shoot and root biomass were determined after two months. Every month, from transplanting to the end of the experiment, one tomato plant was uprooted from each plot. The nematodes were extracted from the roots as described above. A 250-g sample of the soil surrounding the plant was sampled in each plot and the nematodes were extracted (Seinhorst, 1962). The data were treated with a one way analysis of variance and the mean values were compared with the Student's 't' test ($P < 0.05$). For nematodes, data were previously transformed by log (x+1).

3. Results and Discussion

The juveniles of *M. mayaguensis* were trapped by all the fungal strains (Table 2).

Table 2. Predatory activity (expressed as % of trapped *Meloidogyne* spp. juveniles) of the fungal isolates against 3 species of *Meloidogyne* in axenic conditions.

Fungal isolates	M. mayaguensis	*M. incognita*	M. javanica
ORS 18690 S2	11 c [1]	0	0
ORS 18692 S5	26 b	3 b	0
ORS 18692 S7	74 a	0	0
S 30	78 a	65 a	0
S 31	82 a	70 a	20
S 42	82 a	60 a	0
BF 10	10 c	4 b	0
BF 74	9 c	2 b	0
BF 80	14 c	2 b	0
SOSU 2	8 c	16 b	0

data in the same column followed by the same letter did not significantly differ according to the one way analysis of variance ($P < 0.05$)

The higher rates were recorded with ORS 18692 S7, S 30, S 31 and S 42. The higher trapping activities against *M. incognita* were observed with S 30, S 31 and S 42. On the opposite, *M. javanica* juveniles were only affected by the fungal strain S 31.
The growth of tomato plants was significantly increased when the strains ORS 18690 S2, ORS 18692 S7 and S42 were inoculated. The fungal isolate S31 stimulated the root development. The numbers of juveniles of *M. mayaguensis* per plant were significantly lower in the fungal treatments than in the control. The same effect was observed with the gall indexes excepted with ORS 18692 S7 (Table 3).

Table 3. Effect of the fungal isolates on the growth of tomato plants infested with 100 juveniles of *M. mayaguensis* per plant and on the development of the nematode.

Fungal strains	Shoot biomass (mg dry weight)	Root biomass (mg dry weight)	Number of galls per plant	Number of juveniles per plant
Control	23.1 b [(1)]	101 b	21.3 a	7046 a
ORS 18690 S2	31.3 a	194 a	15.4 b	4008 b
ORS 18692 S7	31.3 a	184 a	18.3 ab	2112 b
S 31	22.3 b	174 a	13.6 b	3611 b
S 42	31.9 a	163 a	14.4 b	2470 b
BF 10	21.7 b	109 b	12.7 b	2527 b
SOSU 2	22.9 b	140 ab	12.4 b	2132 b

*: significantly different from the control according to Student's t-test (P<0.05).

**: significantly different from the control according to the Student's t-test (P<0.01).

data in the same column followed by the same letter did not significantly differ according to the one way analysis of variance (P < 0.05)

The fungal growth was increased by seven bacterial isolates (S22, S51, S73, G10, G33, G36 and SG9 when bacteria were tested for direct tropic effect (liquid contact) (Table 4). Only three bacterial isolates improved the fungal growth when the micro-organisms confronted by gaseous way. Seven bacterial isolates (S51, S109, G36, G93, G95, SG8 and SG9) enhanced the predatory activity of the fungus T 41 on *M. mayaguensis* (Table 4).

The two main plant-parasitic nematodes genus identified in the tobacco root systems and in the soil were *Meloidogyne* sp. and *Rotylenchulus reniformis* (Table 5). *A. oligospora* T41 did not affect *Meloidogyne* sp. Only two bacterial isolates (G93 and SG19), inoculated without T41, significantly inhibited the multiplication of the root-knot nematodes. However, when the dual inoculation (T41 + bacterial strain) was performed, the number of juveniles per plant decreased significantly in the treatments with G10, G36 and G95. The multiplication of *R. reniformis* was inhibited in the bacterial treatment S73 (Table 5). When the fungus was added, an inhibition of the nematode development was recorded in all the treatments, with the exception of treatment SG18 + T41.

Table 4. *In vitro* effect of fluorescent ***Pseudomonas*** on the radial growth (direct and gazeous confrontation) and predacious activity of ***Arthrobotrys*** sp. T41. *: significantly different from the control according to Student's t-test (P<0.05). **: significantly different from the control according to the Student's t-test (P<0.01)

Bacterial treatment	Direct confrontation Radial growth (mm)	Gazeous confrontation Radial Growth (mm)	Predacious activity (%) (trapped J2s/total J2s)
Control	63,5	13,8	45,2
S22	69,9**	16,8	56,9
S51	69,2**	15,5	60,0
S73	67,0*	19,0**	55,4
S109	65,8	16,5	78,6
G10	66,2*	16,1	39,4
G12	63,8	16,3	46,9
G33	67,0**	18,0*	37,4
G36	69,4**	17,9*	59,0
G93	65,1	16,0	75,0
G95	61,9	17,3	61,3
SG1	65,4	17,4	58,3
SG8	66,2	17,3	72,9
SG9	66,5*	17,1	81,7
SG12	63,8	16,3	37,3
SG18	62,7	14,1	35,4
SG19	61,9	14,1	41,9

*: significantly different from the control according to Student's test (P<0.05).
** : significantly different from the control according to Student's test (P<0.01).

Table 5. Effect of the bacterial strains on the number of juveniles of *Meloidogyne* spp. and *Rotylenchulus reniformis* per tocacco plants inoculated or not with the nematophagous fungus *Arthrobotrys* sp. T41.

Strains of	*Meloidogyne* spp.		*Rotylenchulus reniformis*	
Bacteria	Bacteria alone	Bacteria + T41	Bacteria alone	Bacteria + T41
Controls[1]	98944	90080	79452	41950 *
S22	186820	75890	54200 *	20780 *
S51	33410	35650	118240	ND
S73	23674	194411	35790 *	43270 *
S109	71720	104180	70040	29840 *
G10	67010	23000 *	75280	31320 *
G12	48343	49655	98630	21350 *
G33	57660	74562	65700	41137 *
G36	75370	12740 *	80790	43530 *
G93	14806 *	84212	43526 *	39925 *
G95	68740	7250 *	86131	47310 *
SG1	63700	89990	73160	26010 *
SG8	147072	101551	59050 *	29670 *
SG9	59640	146770	43730 *	28670 *
SG12	48138	57910	69826	37400 *
SG18	35140	50140	51890 *	75160
SG19	7640 *	103643	42710 *	44070 *

[1] Without bacteria
* : significantly different from the control (not inoculated treatment) according to the Student's t test ($P < 0.05$). ND : Not Determined.

At the end of the field experiment (two months), the fungal treatment was responsible for a significant increase in the height, shoot and root biomass of the tomato plants and a significant decrease of the mortality (Table 6).

Table 6. Effect of *Arthrobotrys oligospora* ORS 18692 S7 on height, mortality, average shoot and root biomass of tomato plants in the field experiment after 2 month culture.

Parameters	Compost block without fungus	Compost blocks with fungus
Height (cm)*	23,2 b	31,8 a
Mortality (%)	36,8 a	16,4 b
Shoot biomass (g dry weight per plant)	7,1 b	12,6 a
Root biomass (g dry eight per plant)	0,7 b	1,8 a

*: for each parameter, data in the same lign followed by the same letter did not significantly differ according to the one way analysis of variance (P<0.05)

Moreover the number of juveniles of *Meloidogyne* sp. per gram of root biomass and per dm^3 of soil was significantly greater in the control than in the treatment with ORS 18692 S7 (Fig. 1).

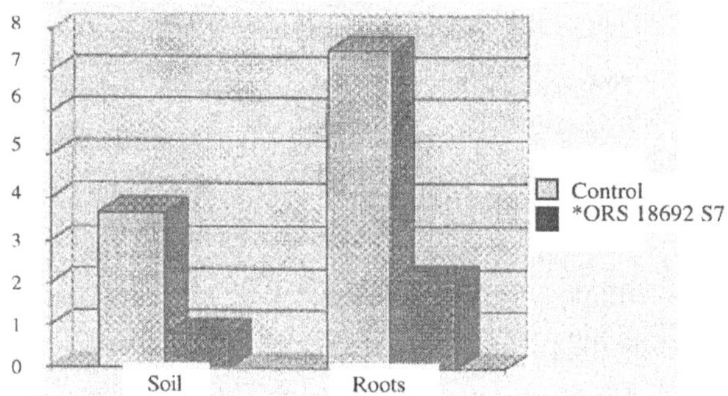

Figure 1. Effects of the nematophagous fungus ORS 18692 S7 on the multiplication of *Meloidogyne* spp. juveniles in tomato roots and in the soil. For each parameter (root and soil), the columns indexed with the same letter are not significantly different ($P < 0.05$). The data of the nematode population in the soil are expressed as x 100, in the roots, the numbers are calculated per g of dry weight of root biomass and divided by 10^5.

The different fungal isolates of *Arthrobotrys* from Senegal and Burkina Faso were able to trap the juveniles of *M. mayaguensis* and *M. incognita*. Thus, these could be good candidates for biological control in coffee plantations as *M. incognita* has occurred for many years in separate or mixed populations with *M. exigua, M. coffeicola*. Moreover, the fungal isolates can act against *M. mayaguensis*, which has caused serious damages and has spread throughout vegetable-producing areas in West Africa. It is now present under various agro-climatic conditions (Mateille *et al.*, 1996).
In the glasshouse experiment, the fungi showed a great influence on the growth of tomato plants in spite of a low rate of inoculation (1:100, v/v). Compared to the

concentrations commonly used in the controlled ectomycorrhization (1:10, v/v) (Duponnois and Garbaye, 1991), these results suggested that the nematophagous fungi have a great capacity to colonize the soil and consequently trapped the juveniles before infecting the roots. The decrease of the number of galls per plant in the fungal treatments showed this physiological advantage. Moreover, these fungi can also use phytoparasitic nematodes as an energy source.
However, all these experiments have been performed in controlled conditions and during a short period (about one month). It is well known that the efficacy of these fungi strongly interact with environmental factors such moisture, pH, temperature and organic matter (Mosse, 1972). Recent researches have shown that the biological activities of soil micro-organisms can be enhanced by some rhizosphere bacteria such as fluorescent *Pseudomonas* (Duponnois *et al.*, 1993; Mateille and Duponnois, 1996). Present findings focus on the great interactions between this group of bacteria and the nematophagous fungus. These bacterial isolates have been termed as Nematophagous Helper Bacteria (NHB) (Duponnois *et al.*, 1998). From a practical point of view, it is generally assessed that the production of spores and mycelial biomass by nematophagous fungi is very limited (Cayrol, 1988). Use of NHB can decrease the quantities of fungal inoculum added to the soil because of their beneficial effect on the saprophytic growth and the predacious activities of the fungal strains. These NHB could also enlarge the effect of this practice on other pathogenic micro-organisms such as *Rotylenchulus reniformis*, which are also widespread plant parasitic nematodes. As we have demonstrated the compost blocks could be a good culture support for the biological control against nematodes with nematophagous fungi. It is simple to add together the fungus and a selected bacterial strain inside the block. This new concept of the biological control could integrate the properties of the fungus and those from the bacteria (plant growth promoting Rhizobacteria, etc.) and their beneficial interactions. However, the mechanisms involved between these two types of micro-organisms needs to be elucidated in order to increase the efficiency of this dual inoculation.

4. Summary

Present studies on the biological control with nematophagous fungi against the root-knot nematodes showed that the fungi could act against *M. incognita* and *M. mayaguensis*, the most infective species in West Africa but also frequently detected through coffee plantations. They can also control the development of *Meloidogyne* populations on tomato plants. A culture method was developed using compost blocks in which the fungus was inoculated. In the field experiment, the growth of tomato plants increased while the nematode development decreased. However, all these experiments have been performed during a short period. It was found that the fungus could be associated with bacterial isolates, called NHB, which stimulated the fungal activity against *Meloidogyne*. This, however, also involved a depressive effect on other plant parasitic nematodes such as *R. reniformis*.

5. Reference

Bourne J.M., Kerry B.R. De Leij, F.A.A.M. (1994), Methods for the study of *Verticillium chlamydosporum* in the rhizosphere. J. Nematol., 26: 587-591

Cayrol J.C. (1983), Lutte biologique contre les *Meloidogyne* au moyen *d'Arthrobotrys irregularis*. Revue Nématol., 6: 265-273

Cayrol J.C. (1988), Lutte biologique contre les Meloidogyne au moyen d'Arthrobotrys irregularis. Bull. OEPP, 18: 73-75

Duponnois R. Garbaye J. (1990), Some mechanisms involved in growth stimulation of ectomycorrhizal fungi by bacteria. Can. J. Botany, 68: 2148-2152

Duponnois R. Garbaye J. (1991), Mycorrhization helper bacteria associated with the Douglas fir-*Laccaria laccata* symbiosis : effects *in vitro* and in glasshouse conditions. Ann. Sci. Forest., 48: 239-251

Duponnois R., Garbaye J., Bouchard D., Churin J.L. (1993), The fungus-specificity of mycorrhization helper bacteria (MHBs) used as an alternative to soil fumigation for ectomycorrhizal inoculation of bare-root Douglas-fir planting stocks with *Laccaria laccata*. Plant Soil, 157: 257-262

Duponnois R., Sene V., Sawadogo A., Fargette M., Mateille T. (1996), Effects of different species and strains of the nematophagous fungus *Arthrobotrys* sp. from West Africa on *Meloidogyne* species. Technology transfer in biological control: from research to practice. IOBC/OILB. Montpellier.

Duponnois R., Bâ A.M. (1998), Influence of the microbial community of a sahel soil on the interactions between *Meloidogyne javanica* and *Pasteuria penetrans*. Nematologica, 44: 331-343

Hussey R., Roncadori R.W. (1982), Vesicular-arbuscular mycorrhizae may limit nematode activity and improve plant growth. Plant Disease, 66: 9-14

Hutton D.G., Eason-Heath S.A.E., Coates-Beckford P.L. (1982), Control of *Meloidogyne incognita* affecting coffee. In: Proc. III Res. Conf. Root-Knot Nematodes Meloidogyne sp., January 11-15, Region I, pp. 109-113

Kerry B. R. (1990), An assessment of progress toward microbial control of plant-parasitic nematodes. J. Nematol., 22: 621-631

Lordello L.G.E. (1984), Nematoides das plantas cultivadas. 8th Ed. Livraria Nobel S.A. p. 314

Mateille T., Duponnois R. (1996), Organismes auxiliaires de la lutte contre les nématodes phytoparasites. Applications aux cultures maraîchères. Séminaire national sur la protection des cultures et des stocks dans une agriculture régénératrice. Thiès, Sénégal,

Mosse B. (1972), The influence of soil type and Endogone strain on the growth of mycorrhizal plants in phosphate deficient soils. Revue Ecol. Biol. Soil, 10: 529-537

Pelagatti O., Nencetti V., Caroppo S. (1986), Utilizzazione del formulato R350 a based i *Arthrobotrys irregularis* nel controllo di *Meloidogyne incognita*. Redia, 89: 276-283

Racke J., Sikora R.A. (1986), Influence of rhizobacteria on *Globodera pallida* early root infection and *Erwinia carotovora* tuber soft rot of potato. Revue Nematol., 9: 305-309

Schuster R.P., Amin N., Sikora R.A. (1995), Utilization of fungal endophytes for the biological control of *Radopholus similis* on banana. Nematologica, 41: 342

Seinhorst J.W. (1950), De betekenis van de toestand van de grond voor het optreden van aanstasting door het stengelaaltje Ditylenchus dipsaci Kühn Filipjev.- Tijdschrift Plantenziekten, 56: 292-349

Seinhorst J.W. (1962), Modifications of the elutriation method for extracting nematodes from soil. Nematologica, 8: 117-128

Chapter 25

USE OF SOLID STATE FERMENTATION FOR THE PRODUCTION OF FUNGAL BIOPESTICIDES SPORES FOR INSECT CONTROL

S. Roussos[1], C. Bagnis[1], O. Besnard[2], C. Sigoillot[1], R. Duponnois[3],C. Augur[1]

[1]*Laboratoire de Microbiologie IRD, IFR-BAIM; Universités de Provence et de la Méditerranée, Case 925, ESIL, 163 Avenue de Luminy, 13288 Marseille Cedex 9, France;* [2]*Société Mycos; Parc Scientifique Agropolis II, 34397 Montpellier Cedex 5, France;* [3]*Laboratoire de Biopédologie IRD, BP 1386, Dakar, Senegal*

Running title: Fungal biopesticides production in SSF

1. Introduction

Numerous viruses, bacteria, nematodes and insects cause significant losses to various crops, including coffee in tropical regions (Table 1). Biological control of coffee pest consists in using natural enemies of coffee pests in order to control their population and proliferation (Dufour *et al.*, 1999; Bustillo, 1999; Muller *et al.*, 1999). There are over 400 species of fungi that can attack and kill specifically nematodes and insect pests (Duponnois *et al.*, 1996; Jenkins, 1995; Schoeller and Rubner, 1994; Segers *et al.*, 1999). The concept of using fungal pathogens to control insects is by no means new (Daoust *et al.*, 1982; Fargues *et al.*, 1979; Ferron, 1967). However, in the last 20 years research stimulated mainly by the resistance of insects to chemical pesticides and by enhanced public awareness of the environment has brought closer the possibility of exploiting such organisms commercially (Hokkanen and Lynch, 1995; Leij, 1992). Recent successes in exploiting nematophagous and entomophagous fungi as biopesticides demonstrated that there could be a high potential in using such micro-organisms to protect coffee plants from known pests (Bustillo, 1999; Dufour *et al.*, 1999; Villain *et al.*, 1999).

T. Sera et al. (eds.), Coffee Biotechnology and Quality, 277–286.

Table 1. Number of diseases affecting some tropical cultures.

Plant	Virus	Bacteria	Fungi	Insects	Nematodes
Pineapple	2	3	3	4	3
Banana	4	2	5	1	1
Cacao	1	-	4	1	-
Coffee	-	-	2	3	2
Sugarcane	2	3	4	3	1
Cotton	2	1	4	3	1
Corn	2	1	3	12	-

A biopesticide is defined as an active biological material with specific action against insects, nematodes or pathogenic micro-organisms (Table 2). Important characteristics of biopesticides include (1) specificity of action, (2) biodegradability, and (3) non-toxicity to pollinic insects (Lopez-Llorca, 1992; Feng et al., 1994).

Table 2. Examples of entomopathogens fungi used in biological control.

Insects		Entomopatogenic fungi
Order	Genera and species	
Ortoptera	*Schistocerca gregaria*	*Metarhizium anisopliae*
Dermaptera	*Forficula spp.*	*Entomophtora formicula*
Eteroptera	*Scotinophora*	*Beauveria bassiana*
Omoptera	*Trialeurodes*	*Aschersonia aleuridis*
Tisanoptera	*Thrips tabaci*	*Verticillium lecanii*
Coleoptera	*Hypothenemus hampei*	*Beauveria bassiana*

In the following sections, we intend to discuss about a few of insects antagonistic fungi used as entomopathogens in biological control. Life cycle of filamentous fungi includes the following five steps: (1) dormancy of the spore with a long time conservation, (2) germination of the spore, (3) apical mycelium growth, (4) conidiogenesis, and (5) conidiospore production (Roussos, 1985). In order to use filamentous fungi as biopesticides it is very important to select strains with high ability to produce conidiospores, commonly named spores (Rivillas Osorio *et al.*, 1999). For large scale spore production on a solid medium, there are different techniques: (1) agar culture media (e.g. potato dextrose agar- PDA) in Erlenmeyer flask, Roux bottle or in a Disc Fermenter, (2) natural substrates such as wheat bran, sugar beet, etc., (3) natural supports (sugarcane bagasse, wheat bran, vermiculite, polyurethane foam, etc) complemented by a nutritive solution containing carbon and nitrogen sources as well as

minerals (Raimbault and Roussos, 1985; Raimbault *et al.*, 1989; Jenkins, 1995; Agosin *et al.*, 1997). Fungal growth and sporulation is commonly carried out using plastic bags containing solid substrates, in column bioreactors (Jenkins and Goettel, 1997) or in a Zymotis which is a bioreactor for pilot and industrial scale spore production (Montero *et al.*, 1989; Roussos *et al.*, 1993).
No worthwhile commercial exploitation of solid state fermentation (SSF) has been specifically elaborated in spite of intensive research and development efforts throughout the world (Lonsane *et al.*, 1985). One of the major reasons for this situation is the lack of efficient techniques to develop active spore inoculum, required in large quantity in SSF system. The form, age and ratio of inoculum are of critical importance in SSF system relying on larger inoculum ratio to control contamination (Lonsane *et al.*, 1992). The spores are usually preferred over vegetative or mycelium in SSF system due to ease in mixing of the inoculum with autoclaved moist solids (Pandey, 1994). The productivity of the system is also influenced by the age of the inoculum. Moreover, viability of the spores is of great significance. The development of a large-scale inoculum has been specified as one of the areas which poses problems in scale-up of submerged fermentation (SmF) processes (Joshi and Pandey, 1999). In fact, this will be more problematic in SSF system due to involvement of lower water activity, complex medium constituents, use of water insoluble polymeric substrates and high heterogeneity (Viniegra-Gonzalez, 1997). Work was, therefore, undertaken to develop efficient strategies for large-scale inoculum development on agar media and also in the SSF system involving the use of bagasse as solid support (Roussos *et al.*, 1991). *Trichoderma harzianum* was selected for the studies due to its significant industrial importance in the production of cellulases (Deschamps *et al.*, 1985), biopesticides (Elad *et al.*1993), antibiotics (Okuda *et al.*, 1982), protein enrichment of cassava flour (Muindi and Hanssen, 1981) and flavour compounds (Sarhy-Bagnon *et al.*, 1997).

2. Conidiospore production on agar medium

Erlenmeyer flasks and disc fermenter have been used for the production of conidiosopres on agar medium (Roussos et al., 1991), which contained (g/l): cassava flour 40; KH_2PO_4 2; $(NH_4)_2SO_4$ 4; urea 1; $CaCl_2$ 1; Agar 15; distilled water 1 l. The pH was adjusted to 5.6 using 2-N HCl. The medium in desired quantity was added to the bioreactors for autoclaving at 110°C for 30 min. Growth and sporulation of the cultures were allowed to take place at ambient temperature (28±1°C) without any pH control. However, the disc fermenter was aerated at the rate of 40 l of humidified and sterile air/h during for seven days (Table 3). Viability of conidiospores was studied using the medium of Douglas *et al.* (1979) as per the methodology of Roussos (1985). The conidiospores were counted using a Malassez's haemocytometer.

Table 3. Sporulation of five strains of filamentous fung igrown on agar enriched with various substrates in Earlenmeyer flasks at 25°C

Filamentous fungi	Spore production* (Number g^{-1} C source)	Carbon source	C/N ratio
Trichoderma harzanum	14.1×10^9	Cassava flour	14
Aspergillus niger	5.2×10^9	Cassava flour	24
Beauveria bassiana	7.2×10^9	Cassava flour	24
Aspergillus terreus	9.9×10^9	Molasses	14
Metarhizium flavoviride	6.5×10^9	Wheat bran + bagasse	ND
Penicillium roque fortii	7.2×10^9	Molasses	2

* Number of co per gram of carbon source

The extent of conidiospore formation per cm^2 of culture surface area in disc fermenter was similar to that on agar media in Erlenmeyer flasks (Table 4). Conidiospores from disc fermenter were sufficient for inoculation of 100 kg moist cassava flour medium for protein enrichment of cassava or 100 kg moist bagasse + wheat bran medium for enzymes production. Calculations are based on rate of inoculation of these media 10^{10} spores/kg moist solid medium with standardized values (Raimbault and Alazard, 1980). Larger inoculum would, however, be required for larger pilot or village level plant of 5-8 tons (wet wt.) per day capacity. Using disc fermenter, the inoculum for such plants can be produced but will require using 20 fermenters working in tandem. However, such a strategy is highly unthinkable to put into practice and it would be cost-intensive with respect to both capital and operating expenses. It is also not practical as the use of agar medium at such large scale is highly laborious and agar itself is expensive. It is, therefore, necessary to search for alternative strategy, simpler and less expensive, for large-scale production of conidiospores.

Table 4: Spore production by *Trichoderma harzianum* in various bioreactors

Bioreactor	Conidiospore production	
	Per gram of cassava flour	Per area (cm^2) or volume unit
Erlenmeyer flask (20 ml)	1.1×10^{10}	* 1.7×10^8
Disk fermentor	9.3×10^9	* 2.2×10^8
FMS Column (18 g)	5.0×10^{10}	** 8.8×10^8
Zymotis (21 kg load)	5.0×10^{10}	** 7.7×10^8

* Results are expressed per area (cm2) or ** per volume unit (cm3) of SSF medium

3. Conidiospore production by solid state fermentation system

The use of inert solid support to absorb liquid medium in solid state fermentation (SSF) system has been pioneered by Raimbault et al. (1989) in order to facilitate selective and homogenous development of mycelia as well as the study of the physiology and growth of fungi. Two different types of bioreactors for SSF systems have been developed and include column fermenter (Raimbault and Alazard, 1980), agitated reactor (Durand *et al.*, 1997), and static reactor such as Zymotis (Roussos *et al.*, 1993). These bioreactors offer excellent potential for conidiospores production (De Araujo *et al.*, 1997).

The use of bagasse as support to absorb starch containing liquid medium in column fermenter gave nearly equal conidiospores production when compared to that on agar medium in flasks as well as disc fermenters (Roussos *et al.*, 1991). The inclusion of feather meal in the medium (10.75 g in 100 g mixture of dry bagasse and starch) resulted in a five fold increase in conidiospores production (Montero *et al.*, 1989). The moist solid medium used in column fermenter and Zymotis contained (g): bagasse 80, cassava flour 20, $(NH_4)_2SO_4$ 0.3, urea 1.3, KH_2PO_4 2.5, $CaCl_2$ 2, birds feathers meal 10.7 and tap water 100 ml. The ingredients were mixed thoroughly and the moist medium was filled in cloth sacks (6 kg, wet weight) for autoclaving at 121°C for 15 min. After cooling to about 30°C, it was mixed thoroughly with the liquid inoculum obtained from disc fermenter so as to provide $3x10^7$ spores/g cassava flour initially present in the autoclaved medium. The final moisture content of the medium was 75%. The inoculated medium was used to fill column fermenters (18g/column). The Zymotis compartments were charged to occupy 5 and 10 cm length with 50-cm height as well as 10-cm length with 30-cm height. In another case, Zymotis was charged with 21-kg moist medium and the medium in this case contained a mixture of sugarcane bagasse and cassava flour at the ratio of 80:20. The medium used was autoclaved at 110°C for 90 min.

The column fermenter assembly and the design of Zymotis have been described elsewhere with operating procedures (Raimbault and Alazard, 1980; Prebois et al., 1985; Roussos *et al.*, 1993). In both cases, fermentation was carried out at 29°C for 6 days with aeration by humidified air at a rate of 4 l/h/column and 300 l/h/kg dry solids in Zymotis. At the end of the fermentation, fermented solids were removed from the bioreactors and the conidiospores were harvested from 10-g material in 0.01% Tween 80 solution as per the methodology described for conidiospores production on agar medium. Productivity of the conidiospores in Zymotis operated in four different substrate load conditions was equal to that from column fermenter (Table 4). In fact, conidiospore production in Zymotis per gram cassava flour was 5 times higher than that on agar medium in flasks. This constituted a tremendous success in development of large-scale inocula. Conidiospores formed in 21-kg moist medium in Zymotis were sufficient to inoculate 5-tons of cassava flour or bagasse and wheat bran media in SSF processes. The maximum working capacity of Zymotis is 42 kg moist medium which

provides sufficient inoculum for 10-tons of wet cassava flour or bagasse and wheat bran media.
Productivity per gram cassava flour has been higher in Zymotis with 21-kg moist medium load as compared to other loads, though productivity in terms of conidiospores/cm^3 of culture surface area is comparatively lower. This is because of the use of 20:80 ratio of cassava flour and sugarcane bagasse in the former as compared to 30:70 ratio in the latter case. Consequently, the effective surface area contained more of the inert solid support. It may be possible to achieve better productivity by using higher cassava flour concentration.

4. Studies on conidiospores preservation and viability

Conidiospore suspensions from the disc fermenter are highly dilute due to the need of large volumes of liquid used to recover conidiospores completely from agar surfaces with an efficiency of recovery of 98%. Unless used immediately, preservation at 4°C becomes essential. Viability of conidiospores after such preservation for 1, 26 and 53 days was 97.3, 84.4 and 83.2%, respectively. Vacuum concentration of the suspension at 40°C to reduce bulk volume, however, resulted in merely 9.4% viability. Reduction in viability was more drastic when temperatures of 50 and 60°C were used along with vacuum concentration. Conidiospores of *T. harzianum* are very sensitive to temperature (Roussos *et al.*, 1989). The strain grows at an optimum temperature of 29°C and does not grow at 35°C. It is interesting to note that the viability was 38.8% when bagasse were added to the suspension before vacuum concentration at 40°C. Bagasse probably absorbed the conidiospores and imparted protection during vacuum concentration. Conidiospores, when used for production of cellulases in column fermenter and Zymotis, performed equally well and produced 19 IU of APF and 200 of ACMC activity per gram substrate dry matter.
It is emphasised that the use of SSF system for the production of the conidiospores could overcome problems associated with the dilute nature of the conidiospores suspensions from disc fermenter. For example, the need for recovery of conidiospores could be avoided and the spores containing fermented mass can be used directly as is done in a number of food fermentation (Lonsane *et al.*, 1992). Alternatively, the fermented bagasse containing *T. harzianum* spores was dried at 20°C to a moisture content of 8-12%, without any appreciable loss of spore viability for use at a later date (Table 5).

In this respect, Zymotis offers advantages as it is possible to dry the fermented solids *in situ* by passing dried hot air through the loop used during fermentation to supply humid air. The inoculum grown on wheat bran and other solid substrates has been stored after drying up to six months without appreciable loss of viability of spores.

Table 5. Kinetics of spore production by *T. harzianum* grown on a mixture of sugar beet and sugarcane bagasse (75/25). Spores viability before and after air drying at 20° C.

Incubation at 29° C (days)	Substrate moisture (%)		Spores produced (x $10^8 g^{-1}$ DM)		Viable spore (x 10	
	Fresh	Dry	Before drying	After drying	Before drying	After Drying
4	80.1	3.4	38.7	35.0	5.0	13.1
5	86.6	3.9	45.8	48.9	21.5	40.2
6	81.4	3.3	54.0	36.4	41.2	40.2
7	85.9	2.4	47.9	35.5	50.0	45.4

The data indicate high potential and many advantages in producing large-scale conidiospores (inoculum) in Zymotis for pilot and large-scale SSF systems. Conidiospores, thus, produced could also be used as such in SmF processes or if necessary, after recovery in sterile water. The dried fermented solids with its high concentration of conidiospores of *T. harzianum* could also be used directly as biopesticide.

5. Summary

It is estimated that over 400 species of fungi can attack and kill specifically nematodes and insect pests. In the past two decades, much attention has been paid on research and commercial development of biopesticides to control insects. This is stimulated by the increased resistance of insect's pests to chemical insecticides and by a greater public awareness of the environmental impacts of agro-chemicals. Recent success in using nematophagous and entomophagous fungi as biopesticides demonstrated their high potential to control coffee pests. Studies on the fungal sporulation physiology have allowed defining optimal conditions for the mass production of conidiospores of filamentous fungi such as *Beauveria bassiana, Metarhizium flavoviride, Paecilomyces fumosoroseus,* and *Trichoderma harzianum.* The pilot plant scale production of conidiospores can be achieved using solid state fermentation

6. References

Agosin E., Cotoras M., Muñoz G., San Martin R., Volpe D. (1997), Comparative properties of *Trichoderma harzianum* spore produced under solid state and submerged culture conditions. In - Advances in solid state fermentation, Roussos, S., Lonsane, B.K., Raimbault, M. and Viniegra-Gonzalez, G. (eds.), Kluwer Acad. Publ., Dordrecht, pp. 463-473

Bustillo A. (1999), El papel del control biológico en un programa de manejo integrado de la broca del café, *Hypothenemus hampei* en Colombia. Proc. III SIBAC, Londrina 24-28 May 1999, Brazil

Daoust R.A., Ward M.G., Roberts D.W. (1982), Effects of formulation on the virulence of *Metarhizium anisopliae* conidia against mosquito larvae. J. Invert. Pathol., 40: 228-236

De Araujo A.A., Lepilleur C., Delcourt S., Colavitti P., Roussos S. (1997), Laboratory scale bioreactors for study of fungal physiology and metabolism in solid state fermentation system. In- Advances in solid state fermentation, Roussos, S., Lonsane, B.K., Raimbault, M. and Viniegraz-Gonzalez, G. (eds.), Kluwer Acad. Publ., Dordrecht, pp. 93-111

Deschamps F., Giuliano C., Asther M., Huet M-C., Roussos S. (1985), Cellulases production by *Trichoderma harzianum* in static and mixed solid state fermentation reactors under non-aseptic conditions. Biotechnol. Bioeng., 27: 1385-1388

Douglas K.A., Hoking A.D. and Pitt J.I. (1979), Dichloran rose bengal medium for enumeration and isolation of moulds from foods. Appl. Environ. Microbiol., 37: 959-964

Dufour B., Barrera J.F., Decazy B. (1999), La Broca de los frutos del cafeto: La lucha biológica como solución. In- Desafios de la caficultura en Centroamérica. Bertrand B and Rapidel B. (eds.), IICA.PROMECAFE: CIRAD: IRD: CCCR.FRANCIA, San José, Costa Rica, pp. 293-325

Duponnois R., Mateille T., Sene V., Sawadogo A., Fargette M. (1996), Effect of different West African species and strains of *Arthrobotrys* nematophagus fungi on meloidogyne species. Entomophaga, 41: 475-483

Durand A., Renaud R., Maratray J., Almanza S. (1997), The INRA-Dijon reactors: designs and applications. In- Advances in solid state fermentation, Roussos, S., Lonsane, B.K., Raimbault, M. and Viniegra-Gonzalez, G. (Eds.), Kluwer Acad. Publ., Dordrecht, pp. 71-92

Elad Y., Kirshner B., Gotlib Y. (1993). Attempts to control *Botrytis cinerea* on roses by pre- and postharvest treatments with biological and chemical agents. Crop Protection, 12: 69-73

Fargues J., Robert P.H., Reisinger O. (1979), Formulation des productions de masse de l'hyphomycète entomopathogène *Beauveria bassiana* en vue des applications phytosanitaires. Ann. Zool. Ecol. Anim., 11: 247-257

Feng M.G., Poprawski T.J., Khachatourians G.G. (1994), Production, formulation and application of the entomopathogenic fungus *Beauveria bassiana* for insect control: current status. Biocontrol Sci. Technol., 4: 3-34

Ferron P. (1967), Etude en laboratoire des conditions écologiques favorisant le développement de la mycose à *Beauveria tenella* du ver blanc. Entomophaga, 12: 257-293

Hokkanen H.M.T., Lynch J.M. (1995), Biological control: Benefits and Risks. Plant and Microbial Biotechnology Research Series 4, University Press, Cambridge, p. 304

Jenkins N.E. (1995), Studies on mass production and field efficacy of *Metarhizium flavoviride* for Biological control locusts and grasshoppers. PhD Thesis, Cranfield University

Jenkins N.E., Goettel M.S. (1997), Methods for mass-production of microbial control agents of grasshoppers and locusts. Mem. Entomol. Soc. Canada, 171: 37-48

Joshi V.K., Pandey A. (1999), Biotechnology food fermentation In- Biotechnology: Food Fermentation, Joshi, V.K. and Pandey, Ashok. (eds.), Educational Publishers & Distributors, New Delhi Vol. 1, pp. 1-24

Leij de F.A.A.M. (1992), The significance of ecology in the development of *Verticillium chlamydosporium* as a biological control agent against root-knot nematodes (*Meloidogyne* sp.). PhD Thesis, University of Wageningen, The Netherlands

Lonsane B.K., Ghildyal N.P., Budiatman S., Ramakrishna S.V. (1985), Engineering aspects of solid state fermentation. *Enzyme Microb. Technol.*, 7: 258-265

Lonsane B.K., Saucedo-Castañeda G., Raimbault M., Roussos S., Viniegra-Gonzalez G., Ghildyal N. P., Ramakrishna M., Krishnaiah M.M. (1992), Scale-up strategies for solid state fermentation systems. *Process Biochem.* 27: 259-273

Lopez-Llorca L.V. (1992), Los hongos parásitos de invertebrados y su potencial como agentes de control biológico. Rev. Iberoamer. Micolo., 9: 17-22

Montero F.A., Roussos S., Olmos A. (1989), Producción de biopesticidas a nivel planta piloto para el uso en cultivo de café. In- I Seminario Internacional de Biotecnología en la Indústria Cafetalera. Roussos, S., Licona, F.R. and Gutierrez, R.M. (eds.), Memorias I SIBAC, Xalapa, Veracruz, pp. 199-202

Muindi P.J., Hanssen J.F. (1981), Protein enrichment of cassava root meal by *Trichoderma harzianum* for animal feed. J. Sci. Food Agric. 32: 655-661

Müller R., Berry D., Bieysse D. (1999), La anthracnosis de los frutos: un grave peligro para la caficultura Centroamericana. In- Desafios de la caficultura en Centroamérica. Bertrand, B and Rapidel, B. (eds.), IICA.PROMECAFE: CIRAD: IRD: CCCR.FRANCIA, San José, Costa Rica, pp. 261-291.

Okuda T., Fujiwara A., Fujiwara M. (1982), Correlation between species of *Trichoderma* and production patterns of isonitrile antibiotics. Agric. Biol. Chem., 46: 1811-1822

Pandey Ashok (1994), Solid state fermentation- an overview. In- Solid State Fermentation, Pandey, A. (ed.), Wiley Eastern Publisher, New Delhi, pp. 3-10

Prebois J.P., Raimbault M., Roussos S. (1985), Biofermenteur statique pour la culture de champignons filamenteux en milieu solide. *Brevet Français* N° 85.17.934

Raimbault M., Alazard D. (1980), Culture method to study fungal growth in solid fermentation. Eur. J. Appl. Microbiol. Biotechnol., 9: 199-209

Raimbault M., Roussos S. (1985), Procédé de production de spores de champignons filamenteux. Brevet Français n° 85.08555. Brevet Européen N° EP 0 223 809 B1

Raimbault M., Roussos S., Oriol E., Viniegra G., Gutierrez M., Barrios-Gonzalez J. (1989), Procédé de culture de microorganismes sur milieu solide constitué d'un support solide, absorbant, compressible et non fermentable. *Brevet Français* N° 89. 06558

Rivillas Osorio C.A., Leguizamon Caycedo, J.E., Gil Vallejo, L.F. (1999), Recomendaciones para el manejo de la roya del cafeto en Colombia. Cenicafé Boletin Técnico, 19: 7-36

Rombach, M.C., Aguda R.M., Roberts D.W. (1988), Production of *Beauveria bassiana* in different liquid media and subsequent conidiation of dry mycelium. Entomophaga, 33: 315-324

Roussos S. (1985), Croissance de *Trichoderma harzianum* par fermentation en milieu solide: Physiologie, sporulation et production de cellulases. *Thèse d'Etat*, Université de Provence, Marseille

Roussos S., Aquiahuatl M-A. Brizuela M-A., Olmos A., Rodriguez W., Viniegra-Gonzalez G. (1989), Production, conservation y viability of filamentous fungi inoculum for solid substrate fermentation. Micol. Neotrop. Apl., 2, 3-17

Roussos S., Olmos A., Raimbault M., Saucedo-Castaneda G., Lonsane B. K. (1991), Strategies for large scale inoculum development for solid state fermentation system: Conidiospores of *Trichoderma harzianum*. Biotechnol. Tech., 5: 415-420

Roussos S., Raimbault M., Prebois J-P., Lonsane B.K. (1993), Zymotis, A large scale solid state fermenter: Design and evaluation. Applied Biochem. Biotechnol., 42: 37-52

Sarhy-Bagnon V., Lozano P., Pioch D., Roussos S. (1997), Coconut-like aroma production by *Trichoderma harzianum* in solid state fermentation. In-Advances in solid state fermentation, Roussos, S., Lonsane, B.K., Raimbault, M. and Viniegra-Gonzalez, G. (eds.), Kluwer Acad. Publ., Dordrecht, pp. 379- 391

Schoeller M., Rubner A. (1994), Predacious activity of the nematode-destroying fungus *Arthrobotrys oligospora* in dependence of the medium composition. Microbiol. Res., 149: 145-149

Segers R., Butt T.M., Carder J.H., Keen J.N., Kerry B.R., Peberdy J.F. (1999), The subtilisins of fungal pathogens of insects, nematodes and plants: distribution and variation. Mycol. Res., 103: 395-402

Villain L., Anzueto F., Hernandez A., Sarah J.L. (1999), Los nematodos parasitos del cafeto. In-Desafios de la caficultura en Centroamérica. Bertrand, B. and Rapidel, B. (eds.), IICA.PROMECAFE: CIRAD: IRD: CCCR.FRANCIA, San José, Costa Rica, pp. 327-367

Viniegra-Gonzalez G. (1997), Solid state fermentation: Definition, characteristics, limitations and monitoring. In-Advances in solid state fermentation, Roussos, S., Lonsane, B.K., Raimbault, M. and Viniegra-Gonzalez, G. (eds.), Kluwer Acad. Publ., Dordrecht, pp. 5-22

Chapter 26

AMYLASE AND PROTEASE INHIBITORS AS ALTERNATIVE AGAINST HERBIVOROUS INSECT

A. Valencia Jiménez

Departamento de Química, Facultad de Ciencias Exactas y Naturales. Universidad de Caldas, Manizales-Colombia

Running Title: Coffee resistance to insect herbivores

1. Introduction

Insects are no doubt the highest economically important phytosanitary problem for many cultivated plants; since they attacks different organs from the early development stages to the steps next to harvesting,thus causing severe losses in production as well as the product quality. Integrated pest management strategy (IPM), which aims at reducing the damage and insect population level is a useful strategy for this (Bustillo 2000). The control measures that have been considered are the chemical control, use of parasitoids, entomopathogens, cultural control and the development of plants resistant to plague.
There are several alternatives for the development of varieties resistant to a given pest. Firstly, the source of resistance to a specific bug needs to be identified by germplasm studies of the species. The insect biology knowledge and the influence of biotic and abiotic factors on its biology, specially, feeding habits, spawning, movement, growth and fertility parameters establish the detailed knowledge of the intrinsic nature of the plague, through an exploration of the metabolic and physiological routes that may be the target of inhibition (Fig.1). Likewise, the digestive apparatus of insects, specially its digestive enzymes such as protease and amylase, become one of the most susceptible points in their physiology. This metabolic route is the principal energy supply way of the insect.
The problem of the insects deserves consideringresearch leading to explore other ways such as the study of mechanisms resistant to the attack and more specifically the use of anti-metabolic compounds of natural occurrence. Some of these compounds correspond to the digestive enzyme inhibitors of the insect, such as protease and amylase inhibitors, which have an enormous potential to be included within the genetic improvement programmes through biotechnology.

T. Sera et al. (eds.), Coffee Biotechnology and Quality, 287–296.

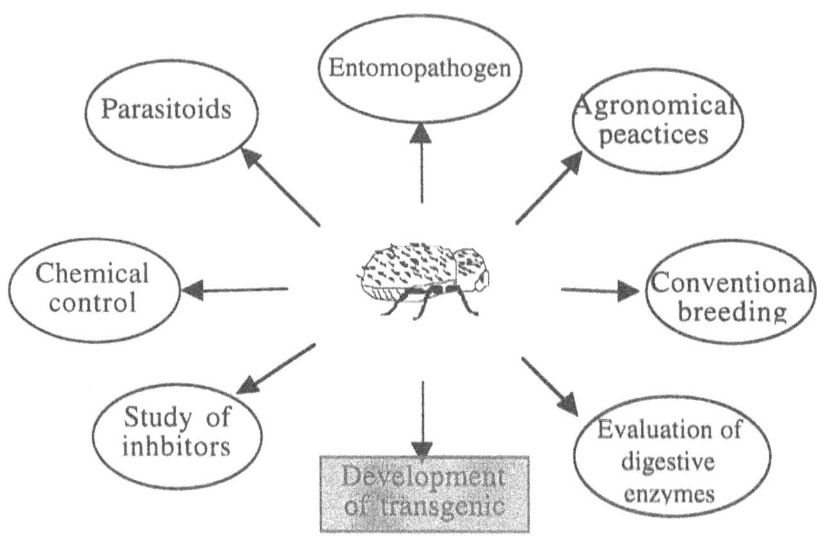

Figure 1. Integrated pest management strategy.

The development of transgenic plants resistant to insects could contribute significantly to a reduction in the use of agro-chemicals, which actually do not provide a total protection. An indiscriminate use of these chemicals could lead to the growth of insect resistance as a consequence of the high selection pressure. This methodology conveys a reduction of costs in the plague management, and elimination of toxic waste in food products, health conservation and environmental protection.

1. 1. BACKGROUND

For several years, research has been oriented to the assessment of regulation and biosynthesis of specific compounds associated with plant defence mechanisms to the attack of insect and pathogens. These compounds were considered as secondary metabolic waste of plants, with unknown functions. However, it is known today that many of these compounds act alone or together playing a role in the resistance of plants to the attack of insects and pathogens (Ryan, 1990). These metabolites include lectins (Chrispeels and Raikhel, 1991; Huesing *et al.*, 1991), wax (Eigenbrode *et al.*, 1991), phenols (Summers and Felton, 1994), amino acids (Rosenthal and Dahlman, 1991; Rosenthal *et al.*, 1989), sugars (Juvik *et al.*, 1994; Liedl *et al.*, 1995), amylase inhibitors (Huesing *et al.*, 1991; Shade *et al.*, 1994) and proteinase inhibitors (Burgess *et al.*, 1994; Orr *et al.*, 1994;

Wolfson and Murdock, 1995; Ryan 1973,1990; Boulter *et al.*, 1990). These chemical compounds have provided several systems for the study of inducible defence mechanisms in plants. In many plants, they can be synthesized as an answer to the plague or pathogens attack, by generating intra- or inter-cellular signals that activate genes codifying for their synthesis (Ryan, 1990).

Most of these inhibitors have been found in high concentrations in seeds and tubers of some plants, especially grains and legumes (Ryan, 1990; Green and Ryan, 1972) and have been isolated from mung bean (Baumgartner *et al.*, 1976; Kapur *et al.*, 1989), rye and wheat (Chang and Tsen, 1981; Gatehouse *et al.*, 1986; Deponte *et al.*, 1976), guandúl (Godbole *et al.*, 1994b), potato (Eddy *et al.*, 1980; Ishikawa *et al.*, 1994; Orr *et al*, 1994), cowpea (Gatehouse *et al.*, 1980; Hilder *et al.*, 1989), amaranths (Rodriguez *et al.*,1993), tobacco (Geoffroy *et al.*, 1990), tomato (Walker-Simons and Ryan, 1977), pea (Domoney *et al.*, 1993), chickpea (Mulimani *et al.*, 1994) and corn (Blanco-Labra and Iturbe-Chiñas, 1981) among others.

The proteinase and amylase inhibitors are produced in the vegetable tissue in different proportions. Generally, they are proteins with molecular weights under 50Kda. In some cases the molecular weight could be below 20 KDa (Boisen and Djurtoft, 1981; Chang and Tsen, 1981; Gatehouse *et al.*, 1980; Geoffroy *et al.*, 1990; Godbole *et al.*, 1994a; Rodriguez *et al.*, 1993; Domoney *et al.*, 1993; Blanco-Labra and Iturbe-Chiñas, 1981; Weselake *et al.*, 1983; Garcia-Carreño *et al.*, 1993), appearing in the form of dimers and tetramers. Many inhibitors are the by-products of multigenic families and it is not common to find isoinhibitors expressing different specificity towards proteases (Ryan, 1990).

2. Physiological role of amylase inhibitors

The physiological role of amylase inhibitors in plants is unknown (Gatehouse *et al.*, 1986). There are evidences indicating that it could act as a proteic reserve in seeds. Nevertheless, the inhibitors do not generally affect endogenous amylases, except in rare cases. The specificity of amylase and protease inhibitors of enzymes is well known. The H-28 amylase inhibitor showed inhibitory activity very specific for insect amylases, and no inhibitory activity for vegetable, animal or fungi source amylases, including one of human saliva. It is also known that there are many examples where amylase and proteinase inhibitors, specially those isolated for cereals, have exhibited inhibition to the amylases from insect guts (Burgess *et al.*, 1994; Johnston *et al.*, 1993, Christeller and Shaw, 1989; Chagolla *et al.*, 1994; Grant *et al.*, 1995; Shade *et al*, 1994; Weselake *et al.*, 1983), which in turn has led to propose, that digestive enzyme inhibitors play a protective role against predator insects.

During an insect attack to a plant organ, a highly volatile compound production is unchained, such as the jasmonate and metyljasmonate, which are airborne to other organs inside and outside the plant. Once there, they start a series of biochemical reactions, leading to the synthesis of compound providing resistance to the insect attack. These compounds of acquired systemic resistance (ASR), helped by signal molecules, such as jasmonates and methyljasmonates, include alkaloids and very specially protease inhibitors (Enyedi *et al.*, 1992).

3. Effect of inhibitors on the digestive physiology of insects

When inhibitors are present in high concentrations in food, they can significantly alter several digestive and physiological processes, thus interfering with normal growth and development of insects. Amylase and protease inhibitors are not a direct problem for humans, since foods containing a high concentration of such are subject to cooking processes at high temperatures, which lead to inactivating them (Ryan, 1990). Several physiological effects are manifested in the insect as a consequence of inhibitors intake. These include:

- Moulting inhibition (Gatehouse *et al.*, 1992)
- Pancreas hypertrophy (Liener and Kakade, 1980)
- Antinutritional effect (Broadway, 1989; Houseman *et al*, 1989)
- Fertility reduction (Spats and Harris, 1984)
- Digestive enzyme hyperproduction (Ryan, 1990)
- Proteolysis inhibition (Ryan, 1990)

A better knowledge of the regulation of the digestive physiology of insects and of the actual effects of amylase and protease inhibitors on these processes will allow knowing the potential of such inhibitors on the systems and natural defence mechanisms of plants.

4. Assessment of biological activity of inhibitors

In general, there are several ways that can be used to assess the biological activity against herbivore insects. The most direct way measures the capacity of an inhibitor to inhibit *in vitro* the enzymatic activity coming from the intestine of a target organism (Hatakeyama *et al.*, 1992; Kang and Fuchs, 1973). The procedure involves extracting enzymes from the middle intestine of the insect, its incubation with the inhibitor and the analysis of the remaining activity (Valencia *et al.*, 1994). This test, leads to determining the type of enzyme present in the insect intestine (Dionysius *et al.*, 1993; Houseman and Thie, 1993), and establishes the grade of link between the enzyme and the inhibitor. However, this approach will not always indicate the insect's physiological response to the treatment (Broadway, 1995a).

Another way to evidence the response of one organism to inhibitor treatments consists of incorporating the inhibitors to artificial diets and evaluate growth, development and survival of the insect (Broadway, 1995b; Burgess *et al.*, 1994; Johnston *et al.*, 1993; Orr *et al.*, 1994; Gatehouse *et al.*, 1986; Steffens *et al.*, 1978; Lipke *et al.*, 1954). The response of the insect to the presence of inhibitors could also be assessed at plant level (Broadway, 1995c; Green and Ryan, 1972; Hilder *et al.*, 1987; Shade *et al.*, 1994; Schmidt, 1994).

5. Genetic transformation

The first transformation of a plant with a gene for protease inhibitor was developed by Hilder *et al.* (1987). The gene for the trypsine inhibitor from cowpea (*Vigna unguiculata*) (CpTI) was transgenically inserted in plants of tobacco, which showed higher resistance to *Heliothis virescens* attack, than plants without the presence of the gene codifying for the inhibitor (Gatehouse *et al.*, 1992).

Table 1. Transgenic plants expressing amylase and/or protease inhibitors.

Plant	Type of inhibitor expressed	Reference
Nicotiana tabacum	Cowpea Trypsin Inhibitor (CpTI)	Hilder *et al.*, 1987
	Tomato Inhibitor I	Gatehouse *et al.*, 1992
	Tomato Inhibitor II	Johnson *et al* , 1989
	Tomato Inhibitor II	Gatehouse *et al.*, 1992
Pisum sativum	α-amylase Inhibitor (AI-Pv)	Shade, 1994
Medicago sativa	Proteinase Inhibitor	Thomas *et al.*, 1994
Gossypium hirsutum	Cowpea Trypsin Inhibitor (CpTI)	Vidhu *et al.*, 1997

This protein was considered as a valuable source to be transferred to other vegetable species, via genetic engineering for several reasons. It is an effective anti-metabolite against a large number of insects-plagues of Coleoptera, Orthoptera and Lepidoptera orders with no detrimental effects on mammals and is, besides, a small protein with an 80 KDa molecular weight (Gatehouse *et al.*, 1992).

The genetic transformation of plants through the expression of digestive enzyme inhibitors of insects has been used by other research groups in economically important crops, thus showing the potential of such as mechanisms of defence to the attack of herbivore insects. (Table 1). These included pea (Schmidt, 1994; Shade *et al.*, 1994), alfalfa (Thomas *et al.*, 1994), tobacco (Hilder *et al.*, 1987) and tomato (Graham *et al.*, 1986). The expression of these proteins in transgenic plants has demonstrated the durability of ongoing generations of the vegetable specie, such as the case of transgenic pea (Table 2) thus, showing the genetic stability of the transformed plants.

Table 2. Development of the pea borer (*Bruchus pisorum*) in T_5 seeds of six plants of transgenic pea (F_{10})*.

Plants	No. of seeds		No. of infected seeds		No. of seed infested and adult emerged	
	T_5	Control	T_5	Control	T_5	Control
1	50	38	27	26	0	23
2	45	44	42	35	0	32
3	50	42	30	35	0	30
4	48	46	35	30	0	25
5	40	35	26	27	0	21
6	46	39	29	28	0	26
Total	279	244	189	181	0	157

*Schroeder *et al., 1995*

The genetic manipulation of plants, as part of an improvement programme may have a significant contribution in the production of plants having resistant to insects by transferring only the selected gene and shortening production time for new varieties. This method offers many advantages on the exclusive use of agro-chemicals (Gatehouse *et al.*, 1992) such as 1) provides long protection, 2) insects are affected in their most sensitive stage, 3) the protection is independent from factors such as climate, 4) protects vegetable tissue of difficult access to insecticides, 5) the effects are restricted only to the plant, and 6) does not affect environment, it is biodegradable and non-toxic to man and animal.
Keeping in mind the specificity of these inhibitors on the digestive enzymes of the insect (Richardson, 1977), they are considered excellent markers to be included in genetic improvement programs, leading to find their expression in vegetable tissue susceptible to the attack of plagues or pathogens (Gatehouse *et al.*, 1992).

6. Conclusions and Futuristic approach

Plants produce diverse proteins capable of inhibiting a determined enzyme, which grants resistance to the attack of herbivore insects. The potential of inhibitors of amylases and protease in insects, as an alternative to the integrated management of several insects has great significance. For this, it is necessary to purify and characterize the digestive enzymes present in the insect intestine, as well as of the inhibitors or these enzymes. Since the results of *in vitro* enzymatic tests do not always correspond to those found in field tests with transgenic material, it is necessary to include these inhibitors on diets for the growth of insect in such way that the information generated throughout the process allows selecting digestive enzyme inhibitors that may be utilized in future in the production of transgenic plants. It would generate a better knowledge of the action mechanisms of these compounds and the degree of interaction with digestive enzymes being studied.

7. References

Baumgarnert B., Chrispeels, J.M. (1976), Partial characterization of a protease inhibitor, which inhibits the major endopeptidase, present in the cotyledons of mung beans. Plant Physiology, 58: 1-6

Blanco-Labra A., Iturbe-Chinas, F.A. (1981), Purification and Characterization of an α-amylase inhibitors from maize (*Zea maize*). J. Food Biochem., 5: 1-17

Boisen S., Djurtoft, R. (1981), Trypsin inhibitor from Rye endosperm: Purificationand properties. Cereal Chem., 58 (3): 194-198

Boulter D., Gatehouse, J.A., Gatehouse, A.M.R., Hilder A.V. (1990), Genetic engineering of plants for insect resistance. Endeavour, New Series, 14: 85-190

Broadway M.R. (1989), Characterization and ecological implications of midgut proteolytic activity in larval *Pieris rapae* and *Trichoplusia ni*. J. Chem. Ecol., 15: 2101-2113

Broadway M.R. (1995a), Dietary proteinase inhibitors alter complement of midgut proteases. From Entomological Society of America Conf. Recent advances in the nutritional physiology of herbivorous insects. (Mimeografiado), p.29

Broadway M.R. (1995b), Potential use of natural products as resistance factors against herbivorous insects. Symp. Cornell University, (Mimeografiado), p.3

Broadway M.R. (1995c), Are insects resistant to plant proteinase inhibitors? J. Insect Physiol., 41(2), 107-116

Burgess E.P.J., Main C.A., Stevens P.S., Christeller J.T., Gatehouse A.M.R., Laing W.A. (1994), Effects of protease inhibitor concentration and combination on the survival, growth and gut enzyme activities of the black field cricket, *Teleogryllus commodus*. J. Insect Physiol., 40 (9), 803-811

Bustillo A.E. (2000) The role of biological control in our integrated coffee berry borer management in Colombia. In Proc. III SIBAC, 24-28 Mai 1999, Londrina, Brazil. Coffee Biotechnology and Quality

Chagolla L.A., Blanco L.A., Patty A., Sanchez R., Pongor S. (1994), A novel _-amylase inhibitor from amaranth (*Amaranthus hypocondriacus*) seeds. J. Biol. Chem., 269 (38), 23675-23680

Chang C.R., Tsen C. C. (1981), Isolation of trypsin inhibitors from Rye, Triticale, and Wheat samples. Cereal Chem., 58 (3), 207-210

Chrispeels J.M., Raikhel V.N. (1991), Lectins, lectin genes and their role in plant defence. Plant Cell, 3, 1-9

Christeller J.T, Shaw B.D. (1989), The interaction of a range of serine proteinase inhibitors with bovine trypsin and *Costelytra zealandica* trypsin. Insect Biochem., 19 (3), 233-241

Deponte R., Parlamenti R., Petrucci, T., Silano V., Tomasi M. (1976), Albumin-amylase inhibitor families from wheat flour. Cereal Chem., 53, 805-820

Dionysius A.D., Hoek S.K., Milne, M.J., Slattery, L.S. (1993), Trypsin-like enzyme from sand crab (*Portunus pelagicus*) purification and characterization. J. Food Sci., 58 (4), 780-784

Domoney C., Welham T., Sidebottom C. (1993), Purification and characterization of pisum seed trypsin inhibitors. J. Exp. Botany, 44 (261), 701-709

Eddy L.J., Derr E. J., Hass G.M. (1980), Chymotrypsin inhibitor from potatoes interaction with target enzymes. Phytochemistry, 19: 757-761

Eigenbrode S.D., Stoner, K.A., Shelton A.M., Kain W.C. (1991), Characteristics of glossy leaf waxes associated with resistance to diamondback moth (Lepidoptera: Plutellidae) in *Brassica oleracea*. J. Eco. Entomol., 84: 1609-1618

Enyedi A.J., Yalpani, N., Silverman, P., Raskin, I. (1992), Signal molecules in systemic plant resistance to pathogens and pests. Cell, 70: 879-886

Garcia-Carreno F.L., Dimes L.E., Haard F.N. (1993), Substrate gel electrophoresis for composition and molecular weight of proteinases or proteinaceus proteinase inhibitors. Anal. Biochem., 214: 65-69

Gatehouse A.M.R., Hilder V.A., Boulter D. (1992), *Plant genetic manipulationfor cropprotection*. Wallingford (England) CAB International, (Biotechnology in Agriculture). No 7.

Gatehouse A.M.R, Gatehouse A.J., Boulter D. (1980), Isolation and characterization of trypsin inhibitors from cowpea (*Vigna unguiculata*). Phytochemistry, 19: 751-756

Gatehouse A.M.R., Fenton, A.K., Jepson, I., Pavey, J.D. (1986), The effects of amylase inhibitors on insect storage pests: inhibition of alpha *amylase in vitro* and effects on development *in vivo*. J. Sci. Food Agric., 37, 27-734

Geoffroy P., Legrand M., Fritig, B. (1990), Isolation and characterization of a proteinaceous inhibitor of microbial proteinases induced during the hypersensitive reaction of tobacco to tobacco mosaic virus. Mole. Plant-Microbe Interactions, 3 (5), 327-333

Godbole, A. S., Krishna, G. T., Bhatia, R. C. (1994a), Purification and characterization of protease inhibitors from pigeon pea (*Cajanus cajan* (L.) Millsp) seeds. J. Sci. Food Agric., 64, 87-93

Godbole A.S., Krishna, G.T., Bhatia, R.C. (1994b), Changes in protease inhibitory activity from pigeon pea (*Cajanus cajan* (L.) Millsp) during seed development and germination. J. Sci. Food Agric., 66, 497-501

Graham S.J., Hall G., Pearce G., Ryan A.C. (1986), Regulation of Synthesis of proteinase inhibitors I and II mRNAs in leaves of wounded tomato plants. Planta, 169, 399-405

Grant G., Edwards, E.J. Pusztai, A. (1995), Alpha amylase inhibitor levels in seeds generally available in Europe. J. Sci. Food Agric., 67, 235-238

Green R.T. Ryan, C.A. (1972), Wound-induced proteinase inhibitor in plant leaves: A possible defence mechanism against insects. Science, 175, 776-777

Hatakeyama T., Kohzaki, H. Yamasaki, N. (1992), A microassay for proteases using succinyl casein as a substrate. Anal. Biochem., 204, 181-184

Hilder A.V., Gatehouse, A.M.R., Sheerman E. S., Barker F. R. Boulter D. (1987), A Novel mechanism of insect resistance engineered into tobacco. Nature, 330, 160-163

Hilder V.A., Barker, F.R., Samour, A. R., Gatehouse, A. M. R., Gatehouse, J. A. Boulter, D. (1989), Protein and cDNA sequences of Bowman-Birk protease inhibitors from the cowpea (*Vigna unguiculata* Walp.). Plant Mole. Biol., 13, 701-710

Houseman J.G. Thie, N.M.R. (1993), Difference in digestive proteolysis in the stored maize beetles: *Sitophylus zeamais* (Coleoptera: Curculionidae) and *Prostephanus truncatus* (Coleoptera: Bostrichidae). J. Eco. Entomol., 86 (4), 1049-1054

Houseman J.G., Downe, A.E.R. Philogene B.J.R. (1989), Partial characterization of proteinase activity in the larval midgut of the European corn-borer *Ostrinia nubilalis* Hubner (Lepidoptera: Pyralidae). Can. J. Zool., 67 (4), 864-868

Huesing E.J , Shade E.R., Chrispeels J. M., Murdock, L.L. (1991), Alpha amylase inhibitor, not phytohemagglutinin explains resistance of common bean seeds to cowpea weevil. Plant Physiol., 96, 993-996

Ishikawa A., Ohta S., Matsuoka K., Hattori T., Nakamura, K.A (1994), family of potato genes that encode kunitz-type proteinase inhibitors: Structural comparisons and differential expression. Plant Cell Physiol., 35 (2), 303-312

Johnson R., Narvaez A.J. An G., Ryan C.A. (1989), Expression of proteinase inhibitors I and II in transgenic tobacco plants: effects on natural defence against Manduca sexta larvae. Proc. Natl. Acad. Sci. (USA), 86: 9871-9875

Johnston K.A., Gatehouse J.A., Anstee, J.H. (1993), Effects of soybean protease inhibitors on the growth and development of larval *Helicoverpa armigera*. J. Insect Physiol., 39 (8), 657-664

Juvik J.A., Shapiro J.A., Young T.E., Mutschler M.A. (1994), Acylglucoses from wild tomatoes alter behaviour and reduce growth and survival of *Helicoverpa zea* and *Spodoptera exigua* (Lepidoptera: Noctuidae) J. Eco. Entomol., 87, 482-492

Kang, S., Fuchs, M.S. (1973), An improvementin the hummel chymotrypsinassay. Anal. Biochem., 54: 262-265

Kapur R., Tan-Wilson L. A., Wilson, A.K. (1989), Isolation and partial characterization of a subtilisin inhibitor from the mung bean (*Vigna radiata*). Plant Physiol., 91: 106-112

Liener I.E., Kakade, M.L. (1980), Protease inhibitors. In- Toxic constituents of plant foodstuffs, Liener, I. E. (ed), Academic Press, New York, pp. 7-71

Lipke H., Fraenkel G.S., Liener, I.E. (1954), Effect of soybean inhibitors on growth of *Tribolium confusum*. J. Agric. Food Chem., 2, 410-415

Liedl B.E., Lawson D.M., White K.K, Shapiro J.A., Cohen D.E., Carson W.E., Trumble J.T., Mutschler M.A. (1995), Acylsugars of wild tomato *Lycopersicum pennellii* alters settling and reduces oviposition of *Bemisia argentifolli* (Homoptera: Aleyroedidae). J. Eco. Entomol., 88, 742-748

Mulimani, V.H , Rudrappa, G., Supriya, D. (1994), Alfa-Amylase Inhibitors in chickpea (*Cicer arietinum L*). J. Sci Food Agric., 64, 413-415

Orr, L.G., Strickland, A.J., Walsh, T.A. (1994), Inhibition of *Diabrotica* larval growth by a multicystatin from potato tubers. J. Insect Physiol., 40 (10), 893-900

Richardson M. (1977), The proteinase inhibitors of plant and micro-organisms. Phytochemistry, 16, 59-169

RodriguezS V., Segura, N.M., Chagolla, L.A., Verver, A., Cortina, V., Martinez, G.N., Blanco, L.A. (1993), Purification, characterization, and complete amino acid sequence of a trypsin inhibitor from Amaranth (*Amaranthus hypocondriacus*) seeds. Plant Physiol., 103, 1407-1412

Rosenthal G. A. and Dahlman, D. L. (1991), Incorporation of L-canavanine into proteins and the expression of its antimetabolic effects. J. Agric. Food Chem., 39, 987-990

Rosenthal G.A., Reichhart J.M., Hoffmann, J.A. (1989), L-canavanine incorporation into vitellogenin and macromolecular conformation. J. Biol. Chem., 264: 13693-13696

Ryan C. A. (1973), Proteolytic enzymes and their inhibitors in plants. Ann. Rev. Plant Physiol., 24, 173-196

Ryan C. A. (1990), Protease inhibitors in plants: Genes for improving defences against insects and pathogens. Ann. Rev. Phytopathol., 28, 425-449

Schroeder E.H., Gollasch S., Moore A., Tabe L.A., Craig S., Hardie D.C., Chrispeels M.J., Spencer D., Higgins, T. J. (1995), Bean α- amylase inhibitor confers resistance to the pea weevil (*Bruchus pisorum*) in transgenic peas (*Pisum satıvum* L.). Plant Physiol., 107, 1233-1239

Shade E.R., Schroeder E.H., Pueyo J.J., Tabe M.L., Murdock L.L., Higgins V.T.J., Chrispeels J.M. (1994), Transgenic pea seeds expressing the alpha-amylase inhibitor of the common bean are resistant to bruchid beetles. Biotechnology, 12: 793-796

Schmidt K. (1994), Genetic engineering yields first pest-resistant seeds. Science, 265: 739

Spats G.E., Harris, R.L. (1984), Reduction of fecundity, egg hatch and survival in adult horn flies fed protease inhibitors. Southwest Entomol., 4: 399-413

Steffens R., Fox R.F., Kassell B. (1978), Effect of trypsin inhibitors on growth and metamorphosis of corn borer larvae *Ostrinıa nubılalıs* (Hubner). J. Agric. Food Chem., 26 (1): 170-174

Summers C.B., Felton G.W. (1994), Prooxidant effects of phenolic acids on the generalist herbivore *Helıcoverpa zea* (Lepidoptera: Noctuidae): Potential mode of action for phenolic compounds in plant anti-herbivore chemistry. Insect Biochem. Mole. Biol., 24: 943-953

Thomas J.C., Wasmann C.C., Echt C., Dunn R.L., Bohnert H.J. McCoy T.J. (1994), Introduction and expression of an insect proteinase inhibitor in alfalfa (*Medicago satıva* L.). Plant Cell Reports, 14: 31-36

Valencia J.A., Ruiz S.L., Gonzalez M.T., Riano N.M., Posada F.J. (1994), Efecto de inhibidores de proteinasas sobre la actividad tripsina y quimotripsina de *Hypothenemus hampei* (Ferrari). Cenicafé (Colombia), 45 (2), 51-59

Vidhu A.S., Nath P.A., Sane P.V. (1997), Development of insect resistant transgenic plants using plant genes: Expression, of cowpea trypsin inhibitor in transgenic tobacco plants. Current Sci., 72 (10), 741-747

Walker-Simons M., Ryan A.C. (1977), Immunological identification of proteinase inhibitors I and II isolated tomato leaf vacuoles. Plant Physiol., 60, 61-63

Weselake J.R., Mac Gregor, W.A., Hill, D.R., Duckworth W.H. (1983), Purification and characteristics of an endogenous _-amylase inhibitor from barley kernels. Plant Physiol., 73, 1008-1012

Wolfson J.L., Murdock L.L. (1995), Potential use of protease inhibitors for host plant resistance: a Test case. Environ. Entomol., 24 (1), 52-57

Chapter 27

PROPERTIES OF AMYLASES OF COFFEE BERRIES BORER, *HYPOTHENEMUS HAMPEI* (FERRARI) - COLEOPTERA: SCOLYTIDAE

C.P. Martínez Díaz[1], A. Valencia Jiménez[2], M.T. González García[1]A.E. Bustillo

[1]*Disciplina de Entomología, Centro Nacional de Investigaciones de Café Cenicafe, Chinchiná, Caldas, Colombia;* [2]*Departamento de Química, Facultad de Ciencias Exactas y Naturales, Universidad de Caldas, Manizales, Colombia*

Running title: Properties of *H. hampei* α amylases

1. Introduction

In the studies directed to combat and protect crops from insects attacking plants, it is important to have a better knowledge of its biochemistry, specially on the most important enzymes involved in digestive processes so that by knowing these enzymes a strategy for allowing blockade or interruption of their function could be designed, that is, to produce an inhibition addressed specifically against this enzyme (Blanco *et al.*, 1996).

The digestive metabolism of insects resembles that of vertebrates. They have enzymes adapted to their diet. If they live on a diet particularly rich in one substance, they would usually produce abundant enzymes for their digestion (i.e., amylases for diets rich in carbohydrates). However, if the diet is highly restricted, the synthesis of the corresponding enzymes would be limited (Applebaum, 1985). Most of the food required for the nutrition of insects consists of polymers such as starch, cellulose, hemicellulose and proteins. The digestive processes of these polymers occurs in three stages: during the onset occurs the decrease of the molecular weight of polymeric compounds through the action of polymeric hydrolases (amylase, cellulase, hemicellulase and endoprotease); in the intermediate stage, the oligomers resulting are hydrolyzed by oligomeric hydrolases, which act on partially degraded fragments. The products of this phase are dimers or small oligomers such as maltose, cellobiose and dipeptides. In the final stage of digestion, the dimers breakdown into monomers by hydrolases such as maltase, cellobiase dipeptidase, tripeptidase, amminopeptidase and carboxypeptidase, which are normally located in the epithelium membrane of the intestine (Terra, 1990). It is very

T. Sera et al. (eds.), Coffee Biotechnology and Quality, 297–306.

likely that the digestive carbohydrase of insects hydrolyse poly, oligo and disaccharide to their corresponding monosaccharide in order to prepare them for their absorption the same way as the carbohydrase of vertebrates. In the latter, the polymeric hydrolases are secreted mainly by salivary glands and the epithelium of the middle intestine. Two categories of carbohydrases are acknowledged, which are commonly labelled as α-amylases and β-amylases. The α-amylases (1,4- α-D-glucan glucanohydrolases) is characterized for attacking internal glycosidic bonds of starch or glycogen. It is considered that the attack (α-1,4) yields random breakdowns that generate a mixture of dextrins of variable chain lengths, although some evidences point out that starch hydrolysis follows a certain order (Bohinski, 1983).
The amylases of insects play an important role in the digestion of the starch. The Coleoptera order is extremely large; consequently it is difficult to stereotype the distribution of digestive enzymes of these insects. The digestion of starch by insect amylases has been demonstrated in *S. granarius* (Silano *et al.*, 1975; Baker, 1983); *Tribolium castaneum* (Krishna and Saxena, 1962); *Trogoderma* sp. (Krishna, 1955); *Rhyzopertha dominica* (Baker, 1991); *Sitophillus zeamais* (Baker, 1983; Sandoval, 1991); *Sitophillus orizae* (L) (Baker, 1987; Yetter *et al.*, 1979); *Callosobruchus maculatus* (Campos *et al.*, 1989); *Zabrotes subfasciatus* (Lemos *et al.*, 1990) and *Tenebrio molitor* (Applebaum, 1964). A better understanding of the enzymes involved in the use of food by the insect and of their main properties would allow a better understanding of the role played in the resistance mechanisms of the insect in front of potential inhibitors.

2. Experimental

2.1. BIOLOGICAL MATERIAL

The adult borers were obtained from the rearing unit of parasitoids of Cenicafé (Chinchiná-Caldas). They were raised in dry pergamino coffee with a humidity of 45% and maintained at 27°C and relative humidity between 65-75%.

2.2. EXTRACTION SOLUTIONS

The amylases were extracted by homogenizing 1-g of adult borers with 5-ml each of the following solutions: NaCl 1%, H_2O, Succinate buffer 50 mM pH 4.0 and 6.0; citrate buffer 50 mM pH 3.0 and buffer Tris- HCl 50 mM pH 8.0. The homogenized was centrifuged at 10,000 x g for 30 min and the resulting supernatant was stored at –20°C as an enzyme source. All the extractions were done at 4°C.

2.3. ASSESSMENT OF AMYLASE ACTIVITY

Amylase activity was determined according to Hopkins and Bird method (1954). An aliquot of the enzyme solution (50μl) was mixed with 10mM NaCl, 20mM $CaCl_2$ and 500μl of 0.125% of soluble starch solution (Sigma Chemical Co.) in 50mM of citrate buffer (pH 5.0) and incubated for 15 min at 30°C. The reaction was stopped by adding 5-ml of an iodine solution (I: 0.5% and KI: 5 %). The samples were read at 580 nm in a spectrophotometer (UNICAM UV2). The activity was determined using five repetitions.

2.4. EFFECT OF PH

The optimum pH of amylases in *H. hampei* was determined in 50 mM quantity of different buffers such as citrate buffer, pH 3.0 and 5.0, succinate buffer pH 4.0 and 6.0, phosphate buffer pH 7.0, Tris-HCl buffer pH 8.0 and borate buffer pH 9.0. For all solutions, 10mM of NaCl and 20 mM of $CaCl_2$ was also added.

2.5. THERMAL STABILITY

The thermal stability of amylases in the presence of NaCl and $CaCl_2$ was determined by pre-incubating the enzymatic extract in activity buffer for five minutes at 25, 30, 40, 50, 60 and 70°C. Five repetitions were made of each treatment.

2.6. ZYMOGRAM ACTIVITY

The amylolytic activity was detected in zymograms *in situ* (Campos *et al.*, 1989). The enzymatic extract samples were analyzed in native conditions on polyacrylamide gels (PAGE) at 15%, containing co-polymerized starch at a concentration of 0.25% using a vertical Miniprotean II Bio- Rad electrophoresis chamber. The running time was three hours at 100 volts and 4°C. The gel was softly rinsed with deionized water and placed in citrate buffer, pH 5.0, containing 10mM NaCl and 20 mM $CaCl_2$ for two hours at 35°C. The activity bands were detected by staining of gel for 30 minutes with an iodine solution (I: 0.5% and KI: 5 %).

3. Results and Discussion

3.1. EXTRACTION SOLUTIONS

Fig.1 shows the effect of the different extracting solutions used on the relative activity of the enzyme. The maximum activity came about with the 1% NaCl. The activity of the amylase in insect such as *S. zeamais* and *S. granarius* increased significantly when the enzyme was pre-incubated with NaCl (Baker, 1983), which corroborated the data obtained on *H. hampei*. The effect of NaCl on the activation of the enzyme is a feature of many mammal and bacteria amylases (Robyt and Whelan, 1968), as well as of many insects (Baker, 1983; Doane, 1969; Hori, 1971; Terra *et al.*, 1977). However, the amylase present in *C. chinensis* was inhibited by chlorine (Podoler and Applebaum, 1971). In extraction solutions assessed, the relative activity of the amylase was higher than 60%.

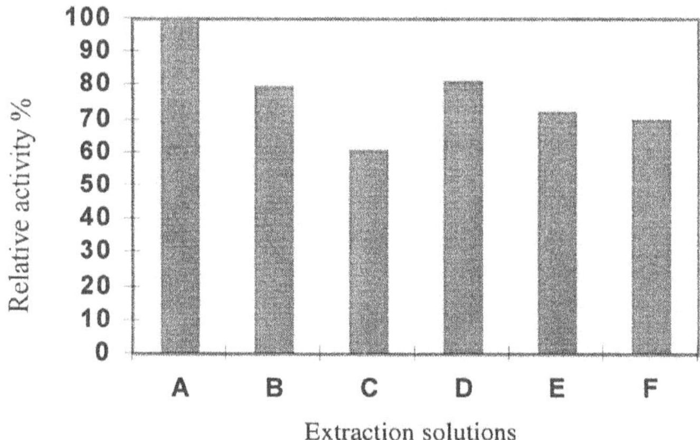

Figure 1. Effect of six Extraction solutions on starch hydrolysis by *Hypothenemus hampei* α-amylases. A, NaCl 1%. B, H_2O. C, Citrate buffer pH 3.0. D, Succinate buffer pH 4.0. E, Succinate buffer pH 6.0 F, Tris-HCl buffer pH 8.0. Each of the solutions contained 10 mM of NaCl and 20mM of $CaCl_2$.

3.2.EFFECT OF PH

Insect amylases has generally been found very active in ranges of slightly acid to neutral pH (Baker, 1983). In *H. hampei* the results showed a higher peak of activity at pH 5.0 (Fig.2.), which was present between 4.6-5.8 range for several insects of Coleóptera order. For example, *Callosobruchus chinensis* had an optimal pH of 5.2-5.4 (Podoler and Applebaum, 1971), pH 4.6-5.2 for *Tribolium castaneum* (Applebaum and Konijn, 1965); pH 5.8 for *Tenebrio molitor* (L) (Buonocore *et al.*, 1976), pH 6.5-7.0 for *Costelytra zealandica* (Biggs and McGregor, 1996) and pH 4.5-5.5 for *Sitophilus zeamais* and *Sitophilus granarius* (Baker, 1983).

The optimal pH for the amylolytic activity in *H. hampei* determined in the presence of Cl^- and Ca^{++} showed that the enzyme was very active in range of pH 4.0-7.0. All these properties coincided with those typical for α-amylases (Ming *et al.*, 1992).

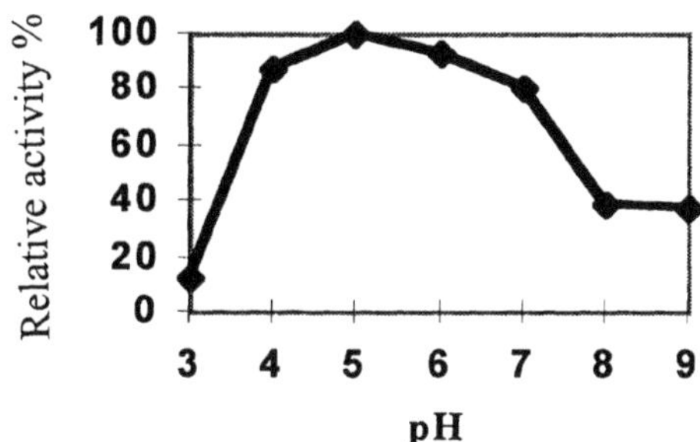

Figure 2. Effect of pH on starch hydrolysis at 30°C by *Hypothenemus hampei* α-amylases.

3.3. THERMAL STABILITY

The optimal activity of the enzyme was observed at 40°C with a slight decrease at 50°C, which decreased further with increase in temperature possibly due to the denaturization of the enzyme (Fig.3).

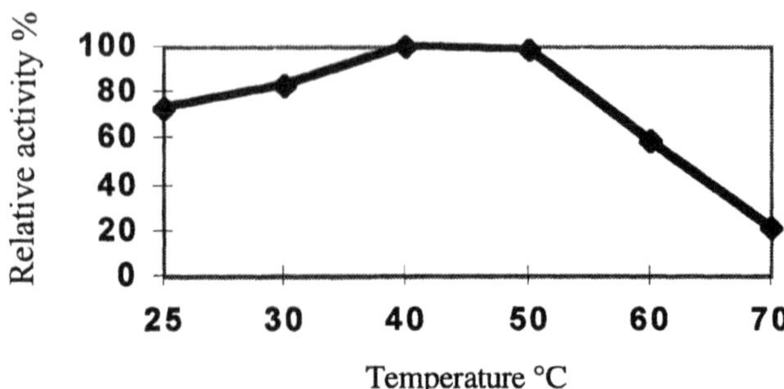

Figure 3. Effect of temperature on amylase activity from *Hypothenemus hampei.*

In general, the amylase from *H. hampei* showed good thermal stability in a range between 30-55°C, which was attributed to the present of Ca^{++} in activity buffer, due to the fact that amylases are metallo-enzymes, which need Ca^{++} for the catalytic activity (Robyt and Whelan, 1968). The incubation of *H. hampei* amylase for more than 5 minutes at 70°C conveys the almost total lost of enzymatic activity. Similar results was found for amylases of *S. zeamais* and *S. granarius* (Baker, 1983). The enzymes have different activation and conversion energy; therefore, a change in temperature can cause a shift in one of them (Lee and Anstee, 1995).

3.4. ZYMOGRAM OF ACTIVITY

The isoenzymatic analysis carried out for the extraction of *H. hampei* adults showed three bands of amylase activity; one of which represented a higher enzymatic activity percentage (Fig.4). This indicated that it was possible to purify and characterize this enzyme aiming to develop strategies allowing evaluation of possible inhibitors of utility in the improvement of coffee plant resistant to *H. hampei.* Similar results were reported for *S. zeamais*, which showed two isoenzymatic forms of amylases in polyacrylamide gels of 7.5%. The presence of a high number of small activity bands close to the main bands could be the result of devices (Doane, 1967). Two of these bands were also found in the extract of intestines, which showed that this band corresponded to intestinal amylases (image not shown). Terra *et al.* 1985 showed that the activity of α-amylases in *T. molitor* was associated with the digestive tract (80% of the midgut and 20% with foregut). However, Silva and Terra (1994) demonstrated that the activity of these enzymes started when the insects felt hunger and their food in take increased. In spite of the studies with homogenized of insects, larvae or the intestines themselves, the knowledge of special distribution of digestion in insects is limited to the knowledge of the zones were it occurs (Vazquez, 1997), thus remaining the need for isolating and characterizing digestive enzymes of salivary glands, belly and luminary spaces inside and out the peritrophic membrane coating the food in the midgut, as well as for identifying dimming mechanisms of pH that can affect the digestion.

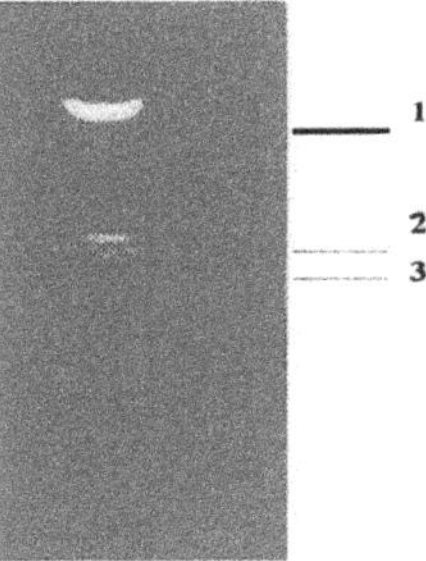

Figure 4. Zymogram of amylase activity from *Hypothenemus hampei*, in native electrophoresis gel at 15%. After the electrophoresis the gel was incubated in Citrate buffer 50 mM pH 5.0, containing NaCl 10mM and $CaCl_2$ 20mM for 120 minutes at 35°C. Finally, the gel was staining with iodine solution. The clear zones indicate enzymatic activity band on dark blue background.

As reported by Ishimoto and Kitamura (1988) α-amylases inhibitors were responsible for the protection of *P. vulgaris* to the attack from *Callosobruchus chinensis*. A high toxicity of the inhibitor may cause death of larvae when they feed from artificial diets containing this inhibitor.

4. Summary and Conclusions

Amylase is an active enzyme in the digestive tract of insects. The knowledge of the amylolytic activity of the borer could be a useful tool in the selection of inhibitors for these enzymes that may be used for genetic improvement of coffee. Studies were made to select the suitable extraction medium (H_2O, NaCl and buffer solutions) and effect of pH (3.0 to 8.0) and temperature (25-70°C) on amylolytic activity. The results showed that the best extraction solution for amylase was 1% NaCl and optimal pH and

temperature was 5.0 and 40°C, respectively. The zymogram of *H. hampei* showed several bands of activity, two of which were enzymes.
The knowledge generated by this work could be important for it allows us learning about some of the main properties of amylases of the coffee berry borer. With this information, it could be possible to evaluate different potential inhibitors with specific activity against α-amylases from *H. hampei*.

5. Acknowledgements

The authors thank Colciencias, Federacion Nacional de Cafeteros de Colombia and the Universidad de Caldas-Manizales for the financial support for this work.

6. References

Applebaum, S.W. (1964), The action pattern and physiological role of *Tenebrio* larval amylase. J. Insect Physiol., 10: 897-906

Applebaum, S.W., Konijn, A.M. (1965), The utilization of starch by larvae of the flour beetle, *Tribolium castaneum*. J. Nutr., 85: 275-282

Applebaum, S.W. (1985), Biochemistry of digestion. In- Comprehensive and Insect Physiology, Biochemistry and Pharmacology, G. A. Kerkut, L.I. Gilbert (eds.), Oxford Pergamon Press, Oxford, Vol. 7. pp. 279-311

Baker, J.E. (1983), Properties of amylases from midgut of larvae of *Sitophilus zeamais* and *Sitophilus oryzae*. Insect Biochem., 13: 421-428

Baker, J.E. (1987), Purification of isoamylases from the rice weevil, *Sitophilus orizae* (Coleoptera: Curculionidae), by high-performance liquid chromatography and their interaction whit partially purified amylase inhibitors from wheat. Insect Biochem., 17 (1): 37-44

Baker, J.E. (1991), Purification and partial characterization of ∝-amylase allosymes from the leeser grain borer, *Rhyzopertha dominica*. Insect Biochem., 21 (3): 303-311

Bigos, D.R., McGregor, P.G (1996), Gut pH and amylase and protease activity in larvae of the New Zealand grass grub *Costelytra zealandica*; Coleoptera: Scarabaeidae as a basis for selecting inhibitors. Insect Biochem. Mole. Biol., 26 (1): 69-75

Blanco L.A., Martinez G.A., Sandoval C.A., Delano, F.J. (1996), Purification and characterization of a digestive cathepsin D proteinase isolated from *Tribolium castaneum* (Herbst). Insect Biochem. Mole. Biol., 26 (1): 95-100

Bohinski, R.C. (1983), Modern concepts in biochemistry. 4th edn., Allen and Bacon

Bunocore, V., Poerio, E., Silano, V., Tomasi, M. (1976), Physical and catalytic properties of ∝-amylase from *Tenebrio molitor* L. Larvae. Biochem. J., 153: 621-625

Campos, F.A.P., Xavier-Filho, J., Silva, C.P., Ary, M.B. (1989), Resolution and partial characterization of proteinases and ∝-amylases from midgut of larvae of the bruchid beetle *Callosobruchus maculatus* (F). Comp. Biochem. Physiol., 92B: 51-57

Doane, W.W. (1967), Quantitation of amylases in *Drosophila* separated by acrylamide gel electrophoresis. J. Exp. Zool., 164: 363-378

Doane, W.W. (1969), Amylase variants in *Drosophila melanogaster*: linkage studies and characterization of enzyme extracts. J. Exp. Zool., 171: 321-342

Hopkins, R.H., Bird, R. (1954), Action of alpha amylase on amylose. Biochemistry, 56: 86-96

Hori, K. (1971), Physiological conditions in the midgut in relation to starch digestion and salivary amylase of the bug *Lygus disponsi*. J. Insect Physiol., 17: 1153-1167

Ishimoto, M., Kitamura, K. (1988), Identification of the growth inhibitor on azuki bean weevil in kidney bean (*Phaseolus vulgaris* L.). Jap. J. Breed., 38: 367-370

Krishna, S.S. (1955), Physiology of digestion in *Trogoderma* larva. J. Zool. Soc. India, 7: 170-176

Krishna, S.S., Saxena, N.K. (1962), Digestion and absorption on food in *Tribolium castaneum* (Herbst). Physiol. Zool., 35: 66-78

Lee, M.J., Anstee, J.H. (1995), Endoproteases from the midgut of larval *Spodoptera littoralis* include a chymotrypsin-like enzyme with an extended binding site. Insect Biochem. Mole. Biol., 25 : 49-61

Lemos, F.J.A., Campos, F.A. P., Silva, C.P., Xavier-Filho, I. (1990), Proteinases and amylases of larval midgut of *Zabrotes subfasciatus* (Boh.) (Coleoptera: Bruchidae) reared in Cowpea (*Vigna ungiculata* (L.) Walp.) seeds. Entomol. Exp. Appl., 56: 219-227

Ming, S.C., Guohua, F., Zen, K.C., Richardson, M., Rodriguez, V S., Reeck, R.R., Kramer, J. K. (1992),∝ - amylases from three species of stored grain Coleoptera and their inhibition by wheat and corn proteinaceous inhibitors. Insect Biochem. Mole. Biol., 22 (3): 261-268

Podoler, H., Applebaum, S.W. (1971), The α amylase of the beetle *Callosobruchus chinensis*: properties. Biochem. J., 121: 321-325

Robyt, J.F., Whelan, W.J. (1968), The α amylases. In- Starch and its Derivatives, J. A. Radley (ed.), Chapman & Hall, London, pp. 430-476

Sandoval, C.M.L. (1991), Purificación y caracterización de enzimas larvales de cuatro insectos que atacan al maíz durante su almacenamiento. Tesis Licenciatura, Universidad de Guanajuato, Irapuato, México

Silano, V., Furia, M., Gianfreda, E., Marci, A., Palescandolo, R., Rab, A., Scardi, V., Stella, E., Valfre, F. (1975), Inhibition of amylases from different origins by albumins from the wheat kernel. Acta Biochim. Biophys., 39: 170–178

Silva, C.P., Terra, W.R. (1994), Digestive and absorptive sites along the midgut of the cotton seed sucker bug *Dysdercus peruvianus* (Hemiptera:Pyrrhocoridae). Insect Biochem. Mole. Biol., 24: 493-505

Terra, W.R., Ferreira, C., De Bianchi, A.G. (1977), Action pattern, kinetical properties and electrophoretical studies on an alpha-amylase present in midgut homogenates from *Rhynchosciara americana* (Diptera) larvae. Comp. Biochem. Physiol., 56B: 201-209

Terra, W.R., Ferreira, C., Bastos, F. (1985), Phylogenetic considerations of insect digestion. Disaccharidases and spatial organization of digestion in the *Tenebrio molitor* larvae. Insect Biochem., 15: 443-449

Terra, W.R. (1990), Evolution of digestive systems of insects. Ann. Rev. Entomol., 35: 181-200

Vazquez, A.M. (1997), Anatomical, enzymatic, and microbiological studies on the digestive system of *Prostephanus truncatus* (Hom). PhD Thesis, University of Leicester, England

Yetter, M.A., Saundres, R. M., Boles, H.P. (1979), α amylase inhibitors from wheat kernels as factors in resistance to postharvest insects. Cereal Chem., 56: 243-244

Chapter 28

NEW DEVELOPMENTS IN MASS PRODUCTION OF PARASITOIDS CEPHALONOMIA STEPHANODERIS (HYMENOPTERA: BETHYLIDAE) ON HYPOTHENEMUS HAMPEI (COLEOPTERA: SCOLYTIDAE) REARED USING ARTIFICIAL DIET

A. Villacorta, S.M. Torrecillas

Instituto Agronômico do Paraná-IAPAR, Rod. Celso Garcia Cid. Km 375, CEP 86001-970, Londrina-PR, Brazil

Running title: Mass production of parasitoids

1. Introduction

The coffee berry borer (CBB) *Hypothenemus hampei* (Ferrari) (Coleoptera: Scolytidae) is reared for the mass rearing of the parasitoid *Cephalonomia stephanoderis* Betrem (Hymenoptera: Bethylidae) by means of two systems, a) in fruits naturally infested by its host (Barrera, 1994), and/or b) in parchment coffee artificially infested with the host. One of the problems with both methods is their high cost of production. The rearing of CBB upon an artificial diet could be a solution to this problem. The first artificial diet for CBB was developed by Villacorta (1985). Commercial diets have bean developed for the mass rearing of *C. stephanoderis* (Villacorta and Barrera 1993, 1996 and Portilla 1999). Conceptual model and methodology could be a useful mean for rearing CBB and its parasitoid, *C. stephanoderis*. Based in recent developments and experiences in several countries of Latin America, it is intended to set a pilot plant for the mass production of *C. stephanoderis* in Brazil,

T. Sera et al. (eds.), Coffee Biotechnology and Quality, 307–312.

2. Artificial diet for coffee berry borer

Basically all the published commercial diets for CBB contains sugar, yeast, casein, coffee beans meal, agar, ethanol, antibiotics and antioxidant. In Brazil, an improved diet as described by Villacorta and Barrera (1996) is being used. It consists of two parts, a) industrial agar 12g, coffee beans meal, 100g; casein 20g; sugar 14g, and b) torula yeast, 20g; Nipagin, 1g; potassium sorbate, 0.8g; ethanol, 10 ml; formaldehyde (37%) 2 ml and 650 ml of tap water.

The cost of the ingredients to produce one litre of diet for CBB varies from country to country, because some ingredients have to be imported. One important ingredient that increases the cost is the bacteriological agar. In place of this, industrial agar could be used by which one can have up to 40% reduction in cost of one litre of diet. Similarly yeast should be obtained from the local market.

More research is needed on the nutritional requirements of the CBB, especially on understanding of the relationship of the CBB with micro-organisms. Villacorta (1989) described its relationship with *Trichoderma* sp. It was observed that the insect did not reproduce in diets composed of a mixture of coffee bean powder, agar, water and antibiotics (Villacorta and Barrera, 1996). Rojas et al. (1999) described *Fusarium solani* as a symbiotic micro-organism associated with CBB and also the effects of antibiotics used in the diet on the fecundity of the CBB.

3. Infestation of the diet

The individual CBB used to commence the rearing process using the artificial diet is collected from coffee berries infested in the field. Upon dissection of the fruit, pre-pupae and pupae of the CBB should be selected along with active beetle females, especially those that appear eager to fly (which normally indicates that they have mated). These should be surface sterilized (Villacorta and Barrera, 1993). It is recommended to use nine adult females to one male to ensure fertilization.

4. Production of the CBB

The CBB diet is distributed amongst flat bottom glass tubes (1.7 x 7-cm) to a depth of 3-cm each or into rectangular plastic boxes of 11.5 x 11.5 x 3.5-cm to a depth of 2.5-cm using a grille squeezed into the diet while still hot to produce pellets. The tubes/boxes are sealed with cotton wool. The infested tubes/ boxes are maintained at 26±2°C, 60±5% R.H. in dark. The infested material is opened after 80-90 days. The diet pellet drops out by tapping lightly on the base of the tube/box. The total contents of the tube or the plastic boxes are placed on the top of a device consisting of sieves of two sizes: 27 and 67 mesh/inch. Different life of the CBB, fine dust and eggs are separated by gentle mechanical agitation of the diet pellets. After surface sterilization, the active adults are used to infest new diet. After 85-90 days, the population reaches between 36,500-43,500 borers per litre of diet.

5. Rearing C. stephanoderis

After 85-90 days, the diet contains about 16-29% pre-pupae and pupae (out of total number of stages of CBB). These stages of the CBB are preferred by *C. stephanoderis* for oviposition. This is considered to be the ideal time for using the diet to rear *C. stephanoderis*, as at later stage, the proportion of the adult borers markedly increases. Two methods are proposed for parasitoid rearing, a) the exposure of pellets containing the beetles to the parasitoids, and b) the exposure *in vitro* involving the use of pre-pupae and pupae of the borer separated from the diet during sieving.

6. Rearing in infested diet pellets

The diet pellets are placed in round plastic boxes with diameter of 22-cm and depth of 10-cm with a hole cut out in the lid and covered and sealed with gauze. Two parasitoids are used to parasitize each pellet. The material is placed in a rearing room or chamber under controlled conditions (27±2°C and 75±5% R.H.). From 17-day onwards, the new generation of parasitoids begins to emerge.

7. Exposure *in vitro*

By sieving, it is possible to separate 50% of the CBB pre-pupae and pupae from the diet pellets. The surface of a piece of corrugated cardboard (10.5 x 9.5-cm) is impregnated with arabic gum diluted with water, followed by applying a coating of finely ground coffee seeds. The selected pupae, pre-pupae, larvae and eggs of the CBB are placed at the end of the corrugations. The prepared cardboard is then placed in a plastic box (11 x 10 x 5-cm) with a hole cut out in the lid (5 x 5-cm), then covered and sealed with gauze. In each box, 25-30 parasitoids are placed and the boxes are then kept in a rearing room (27±2°C and 75±5% R.H.). After 10 days, the formation of cocoons can be observed grouped at one or other ends of the corrugations, because of the maternal behaviour of the parasitoids.

8. Discussion and Conclusions

The techniques described (Fig. 1) are designed to maximize the utilization of the biological material produced from the diet. It could be possible to reduce the cost of the diet for CBB by using industrial agar and locally produced yeast. Research is required into the behaviour of the CBB when raised in artificial diet to answer questions concerning the nutritional effect of the diet upon the fecundity and fitness of the CBB. Research is also necessary into developing improved methods of automation of the preparation (mixing, filling containers, etc.) which could reduce significantly the cost of labour and lead the way towards economic mass rearing of the parasitoid *C. stephanoderis*. This would allow the use of this technique in an integrated pest management for CBB based on innundative releases of the parasitoids.

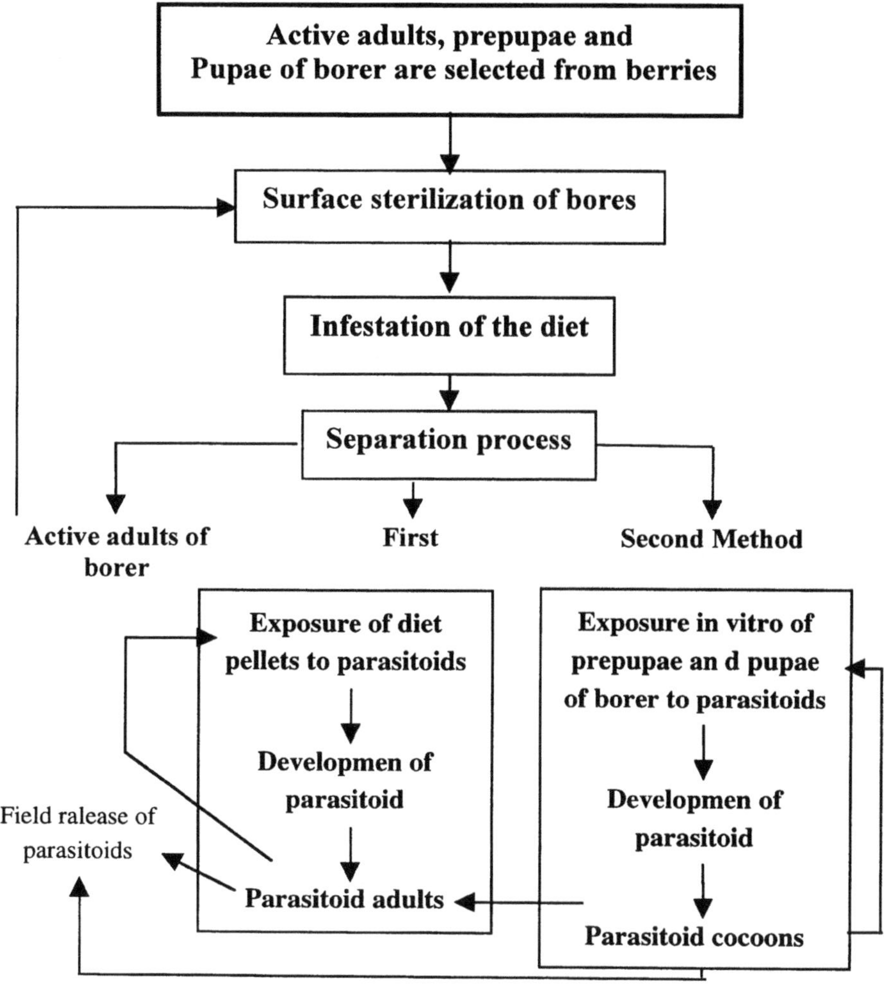

Figure 1. The rearing of *Cephalonomia* sp. Using CBB reared on the artificial diet.

9. Summary

For rearing the coffee berry borer *Hypothenemus hampei* (Ferrari) and its parasitoid *Cephalonomia stephanoderis* Betrem, use of artificial diets offers significant help in reducing the cost of production of the coffee berry borer (CBB). Studies revealed that after 85-90 days after infestation of the artificial diet, cylindrical pellets of the diet contained all live stages of the CBB, and dust from insect activity on the diet also contained all developmental stages of the CBB. In order to maximize the use of this biological material to rear *C. stephanoderis,* two production systems are proposed. The first method uses diet pellets simulation coffee fruits attacked by CBB, and the second uses the pupal and pre-pupal found in the dust from sieved pellets, which are then exposed *in vitro* to the parasitoids. This rearing technique offers an alternative to future industrial production of the parasitoid with high quality control.

10. References

Benavides G.M., Portilla M. (1990), Uso del café pergamino para la cria de *Hypothenemus hampei* y de su parasitoide *Cephalonomia stephanoderis* en Colombia. Cenicafe, 41: 114-116

Portilla M., Bustillo A. (1995), Nuevas investigaciones en la cria masiva de *Hypothenemus hampei* y de sus parasitoides *Cephalonomia stephanoderis* y *Prorops nasuta.* Rev. Colomb. Entomol., 21(1): 25-33

Portilla M. (1999), Mass production of *Cephalonomia stephanoderis* on *Hypothenemus hampei* reared using artificial diet. PhD Thesis, University of London, England

Rojas M.G., Morales-Ramos J.A., Harrington T.C. (1999), Association between *Hypothenemus hampei* (Coleoptera: Scolytidae) and *Fusarium solani* (Moniliales: Tuberculariacea). Ann. Entomol. Soc. Am., 92(1): 98-100

Villacorta A. (1985), Dieta meridica para criação de sucessivas gerações de *Hypothenemus hampei* (Ferrari, 1867)(Coleoptera: Scolytidae). An. Soc. Entomol. Brasil.,14: 315-319

Villacorta A (1989), Aspectos nutricionales de la cria de la broca del café (Coleoptera: Scolytidae). III Taller regional de broca, Antigua, Guatemala, IICA-PROMECAFE, pp. 181-186

Villacorta A., Barrera J.F. (1993), Nova dieta meridica para criação de *Hypothenemus hampei* (Ferrari)(Coleoptera: Scolytidae). An. Soc. Entomol. Brasil., 22: 405-409

Villacorta A., Barrera J.F. (1996), Techniques for mass rearing of the parasitoid *Cephalonomia stephanoderis* (Hymenoptera: Betylidae) on *Hypothenemus hampei* (Coleoptera: Scolytidae) using an artificial diet. Vedalia, 3: 45-48

Chapter 29

PRODUCTION COST OF HYBRID SEEDS OF COFFEE OBTAINED MANUALLY

S Fadelli[1], T. Sera[2]

[1]*Universidade Estadual de Londrina (UEL), PIBIC/CNPq/IAPAR;*
[2]*Instituto Agronômico do Paraná (IAPAR), Londrina, Brazil*

Running title: Hybrid seeds of coffee

1. Introduction

In spite of the high productivity potential of the coffee cultivars recommended for the Paraná State in Brazil, which could be as high as 75 bags of 60 kg clean coffee/ha/year under good cultivation conditions (including years without production) (Sera, 1995), the average productivity of the State (14 bags/ha/year, according to Demarchi, 1998) is lower than a rational and competitive coffee crop. To improve productivity with lower cost/bag, one of the alternatives is the formation of commercial F_1 hybrid seedling, taking advantage of the heterosis usually observed in hybridization. In this way, the profitability, mainly in small and medium properties that represent the great majority of the coffee culture area of the State could be better.
Several reports have described the heterosis in *Coffea arabica* with average up to 30% hybrid F_1 cultivars (Srinivasan and Vishveshavara, 1978; Walyaro, 1983; Ameha and Belachew, 1985; Neto *et al.*, 1993; Bertrand *et al.*, 1997). If the use of F_1 hybrid seed is not economical, it could be possible to cultivate the F_2 hybrid plants commercially, taking sufficient advantage of about half of the heterosis observed in the F_1 hybrid (Sera and Alves, 1999). The commercial seedling of F_1 hybrid can still be used for the genetic improvement in other traits resistance to rust, which could provide a greater economy of time with the reduction of the number of necessary selection cycles to obtain new lineage type cultivars and allow the breeders to test several combinations of lineage,

T. Sera et al. (eds.), Coffee Biotechnology and Quality, 313–319.

seedling the most promising as cultivars in a smallest period (Berthouly, 1997; Sera and Alves, 1999).

2. Experimental

By using manual hybridization a single cross was used among the Mundo Novo (lineage IAC 376-4), Catuaí Vermelho (lineage IAC2077-2-5-81) and Icatu Precoce (lineage IAC 3282) cultivars as female parents, and IAPAR 59 and Tupi cultivars as male parents. The emasculation and the pollination were accomplished during two flowering periods of 1998. The emasculation was accomplished three to four days before flowers opening. Because of climatic conditions on the coffee plant physiology, branches of the half of the skirt beside sun exhibition in the morning were preferred. Pollens were collected at the end of the day before the opening of the flowers when they were in the apprenticeship of developed floral bud. Pollination happened in the hottest period of the day (starting from the 10.00am until the end of the afternoon). The accomplishment and cleaning of the crossings to eliminate fruits and flowers coming from the previous and posterior of flowering period (August to December) were done four to five days before the opening of the flowers.

When the fruits of the crossings matured, they were pricked, shelled, demucilated and sowed in the nursery. The production cost of manually obtained F_1 hybrid seeds in coffee were estimated through the following two equations:

a) Determination of flowers to be emasculated (E) to obtain one Kg of seeds:

$$E = S \div 2 \div (1 - (b \div 2)) \div (1 - (c \div 2)) \div (1 - d) \qquad (1)$$

Where:

S = Seeds/Kg (5.000 of mesh 16 cultivar, 4.500 of mesh 17 cultivar and 4.000 of mesh 18 cultivar); b = rate of "moca" fruits; c = rate of empty fruits; d = rate of flowers that don't bear fruits.

b) Determination of cost/kg of hybrid seeds:

Spent time in hours for emasculation, pollination and cleaning was divided by hours effectively worked/man/day (adopting six hours/day/man). We obtained the number of necessary days for manual production of one kg of F_1 hybrid seed. Multiplying the value of one day for the number of necessary days to produce each kg of hybrid seed plus material expenses, it would be possible to estimate the cost (C) of hybrid seed in US$/kg:

$$C = \{ [(E + x) + (E + y) + z + p] + 6 \} \times t \times f \times k + s \qquad (2)$$

Where:

E= amount of flowers to emasculate; x= emasculation speed (flowers/hour); y= pollination speed (flowers/hour), z= time spent in hours for elimination of flower buds and out-crossed and self-pollinated fruits; p= time spent in hours to fruits harvest, shell and demucilating and drying; t= medium value of a day in US$/day; f= number of lost seeds by other factors as break branch, drill attack, environmental stress (physiologic fall of fruits), etc.; k= rate factor that considers the extra worked hours, and s= cost of the used materials (paper sacks, twine, labels, tongs and plastic box)/kg of seed.

3. Results and Discussion

Table 1 shows average values in the emasculation, pollination and cleaning (elimination of flowers originated from previous and posterior flowering periods in relation to the flowering period of controlled hybridization), harvest and processing of the seeds for each flowering period in each cultivar used as female progenitor. These values confirmed the expectation of obtaining larger revenue in the emasculation in the largest coffee plant flowering period, except for Catuaí cultivar that showed similar results in two flowering periods different in sizes. For the pollination, there was not any difference between them, being its revenue determined mainly by the climate of the day and the pollen amount of male progenitor's flower not having, seemingly, relationship between size of the flowering period and its revenue.

Table 1. Emasculation revenue, pollination, cleaning of the crossings and processing of the seeds, classification in size of the flowering period and useful days for emasculation in two flowering periods of the coffee plant in 1998.

Treatment	Catuaí		Mundo Novo		Icatu Precoce	
Flowering period	1°	2°	1°	2°	1°	2°
Emasculation [1]	252	266	311	238	414	219
Pollination[1]	630	399	574	327	596	394
Cleaning[2]	-	14	50	18	39	9
Collecting and processing of the seeds[3]	150	150	150	150	150	150
Useful days/male emasculation.	3	2	3	2	3	2
Classification /flowering period[4]	G	P/M	G	P/M	G	P/M

[1] flowers/hour; [2] minutes; [3] seed minutes/kg; [4] G = big, P/M = small the average

Table 2. Fruit set percentage and "mocas" fruits for six coffee hybrids in two flowering periods of the coffee plant in 1998.

Coffee hybrids	Fruit set percentage			"mocas" fruits percentage		
Flowering period	1°	2°	Average	1°	2°	Average
Mundo Novo x IAPAR 59	57.7	55.7	56.7	22.9	31.8	27.4
Mundo Novo x Tupi	40.7	65.9	53.3	17.8	10.3	14.1
Catuaí x IAPAR 59	76.7	60.2	68.45	27.8	22.0	24.9
Catuaí x Tupi	66.2	50.0	58.1	20.6	12.0	16.3
Icatu Precoce x IAPAR 59	42.1	70.1	56.1	30.9	0.0	15.5
Icatu Precoce X Tupi	-	57.9	57.9	-	12.1	12.1

In the two appraised flowering periods, it was observed that the largest fruit set percentage was of the hybrids that used Catuaí as feminine parental, followed by hybrids using Mundo Novo and Icatu Precoce, having small difference using different pollinators, in the same cultivar (Table 2). The differences in fertilization rate of the flowers in the crossings inside of a same cultivar were consequences of the conditions in which the crossings were achieved (mainly in the pollination) in each flowering period. These included sun exposure face, position of the branch in the plant in which the crossing was accomplished, humidity and temperature of the day when the pollination took place. There was similarity in the fertilized flowers percentage for the two pollinator cultivars in Mundo Novo and Icatu Precoce, while in Catuaí the largest fruit set percentage was so when IAPAR 59 was used as pollinator. The "moca" fruits percentage was largest in the crossings with largest fruit set percentage probably because of nutrition problem or by insufficiency of pollination.

Table 3. Estimation of the cost/kg of hybrid seed produced manually for six hybrids in two flowering periods of the coffee plant in 1998.

Hybrid	Cost / kg (US$)[1]		Percentage of increase of the cost in the second flowering period[2] ( )[3]
	First flowering period[2]	Second flowering period[2] ( )[3]	
Mundo Novo x IAPAR 59	29.58	36.44 (43.31)	18.8 (31.7)
Mundo Novo x Tupi	31.74	31.29 (37.14)	-1.4 (14.5)
Catuaí x IAPAR59	29.58	29.58 (35.09)	0.0 (15.7)
Catuaí x Tupi	30.95	30.72 (36.46)	-0.7 (15.1)
Icatu Precoce x IAPAR 59	28.16	28.44 (33.71)	1.0 (16.5)
Icatu Precoce x Tupi	-	35.29 (41.94)	-
Average	30.00	31.96 (37.89)	6.1 (20.8)

[1] US$1.00 = R$1.75; [2] k = 1.20; [3] k = 1.00

On average, the first flowering period presented the smallest cost/kg of hybrid seed (US$30.00) due to the largest emasculation and pollination speed and a larger flowering period than the second one. In the second flowering period, the cost was 6.6% superior than the first flowering period (US$31.96), although that difference arose to 20.8%. The cost differences among hybrids seemed inconsistent and not significant.
In relation to the components of the cost of manual production of hybrid seeds (Table 4), the emasculation was the strongest component (52.56%) due the necessity in the retreat of the flower buds and fruits of another flowering periods and the necessity of overtime work because of frequent flowering during the week-ends.

Table 4. Components of the production cost/kg of manually produced hybrid seeds of Icatu Precoce x IAPAR 59 hybrid in a flowering period classified as big in the 1998.

Component	Cost (US$)[1]	Cost percentage
Emasculation [2]	14.80	52.56
Pollination	7.27	25.83
Cleaning of the branches	1.43	5.07
Collecting and processing of the seeds	2.38	8.46
Used materials	2.27	8.08
Total	28.16	100.00

[1] US$1.00 = R$1.75; [2] including overtimes (US$4.31 and 15.32 of the cost Kg of hybrid seed).

The manual operation was responsible for 84.68% of the cost/kg of hybrid seed indicating the need of new studies to improve and to make possible the use of another techniques to emasculate such as the genetic male sterility that could represent a decrease of 52.56% of the cost and the chemical emasculation and flowering periods concentration (through irrigation) as alternative to decrease the cost/kg.

4. Conclusions

It can be concluded that the cost/kg of manually produced hybrid seed could be around US$30.00 in the flowering periods classified as major (second and third flowering period of the coffee plant), where better revenues and smallest cost/kg of manually produced hybrid seed were obtained. With regard to the production components, new studies are necessary, mainly on the utilization of genes for male sterility, chemical and fiscal emasculation, concentration of flowering periods and control of flowering by pruning to obtain hybrid seed accessible to the coffee farmers.

5. Acknowledgements

This research was supported by Consórcio de P & D Café/Funcafé.

6. References

Ameha M., Belachew B. (1985), Heterosis goes yield in crosses of indigenous coffee selected it goes yield and resistance to coffee berry disease. Acta Horticul., 158: 247-351

Berthouly M. (1997), Biotecnologías y técnicas de reproducción de materiales promissores en *Coffea arabica*. Paper presented at 18° Simposio Latinoamericano de Cafeicultura, 16-19 of September, San José, Costa Rica

Bertrand B., Aguilar G., Santacreo R., Anthony F., Etienne H., Eskes A.B., Charrier, A. (1997), Comportement d'hybrides F1 de *Coffea arabica* pour la vigueur, la production et la fetilité en Amérique Centrale Paper presented at 17ème Colloque Scientifique International sur le Café, 20 -25 juillet, Nairobi, Kenya

Demarchi M. (1998), Café. Acompanhamento da situação agropecuária do Paraná, 24(1): 13-26

Neto K.A., Ferreira J.B.D. (1980), Vigor de híbrido em cruzamentos de *Coffea arabica*. Paper presented at 8th Congresso Brasileiro de Pesquisas Cafeeiras, 25-28 November, Campos do Jordão, São Paulo, Brazil

Neto K.A., Miguel A.E., Queiroz A.R. (1993), Estudo de híbridos de *Coffea arabica*–catimor versus catuaí, catindú versus catuaí e outros. Paper presented at 19th Congresso Brasileiro de Pesquisas Cafeeiras, 23-26 de Novembro, Três Pontas, Minas Gerais, Brazil

Sera T. (1995), Cultivares na nova cafeicultura para o Paraná. Paper presented at 2th Encontro Paranaense de Café, 6-7 de outubro, Santo Antônio da Platina, Paraná, Brasil

Sera T., Alves S.J. (1999), Melhoramento genético de plantas perenes. In: *Melhoramento genético de plantas*, Destro, D.; Montálvan, R. (org.), Editora Universidade Estadual de Londrina, Londrina, Paraná, Brazil

Srinivasan C.S., Vishvashwara S. (1978), Heterosis and stability goes yield in arabica coffee. Indian J. Gen. Plant Breed., 38(3): 416-420

Walyaro D.J.A. (1983), Consideration in breeding goes improved yield on quality Arabica Coffee (Arabic Coffea L.). PhD Thesis, Agricultural University, Wageningen, The Netherlands

Chapter 30

ELECTRIC CONDUCTIVITY OF EXUDATES OF GREEN COFFEE AND ITS RELATIONSHIP WITH THE QUALITY OF THE BEVERAGE

C.E.C. Prete[1], T. Sera[2], I.C.B. Fonseca[1]

[1] *Departamento de Agronomia, Universidade Estadual de Londrina (UEL), caixa postal 6001, 86051-990, Londrina - PR, Brazil.* [2] *Área de Melhoramento e Genética (AMG), Instituto Agronômico do Paraná (IAPAR), Londrina - PR, Brazil.*

Running title: Quality evaluation

1. Introduction

Coffee production economics depends on the effort of controlling costs, increasing productivity and improving beverage quality. Current proceedings of commercial evaluation of coffee quality are based on empirical and subjective parameters because they depend on personal sensorial judgement (vision, taste, flavour, for example) and personal skills acquired with many years of experience. Thus, usual procedures complemented with the adoption of physical and chemical methods would be more real and objective for the determination of the quality of coffee.
The work of Amorim (1978) relating biochemical and histochemical aspect of coffee bean with deterioration of it's quality, contributed a great deal to raise the hypothesis that cell membrane loses permeability and structure contributes to coffee's deterioration. Clifford (1985) described that the worst kinds of coffee, in terms of beverage's quality, would have less soluble proteins, less free amino acids, more chlorogenic acid, less hydrolysable phenols, less ascorbic acid, less carbohydrates and free of oily acids together with the reduction of lipids content. These results indicated that oxidative reactions occurred during coffee bean's deterioration process, suggesting intense peroxidation of lipids. Among the various organic compounds, polyphenol oxidase (PPO) has been termed as an important one, which affects the quality of coffee

T. Sera et al. (eds.), Coffee Biotechnology and Quality, 321–338.

beverage (Amorim and Silva, 1968). *In vivo* PPO is found adhering with the membranes and it can be activated only when liberated from the membranes. The activated PPO oxidizes the chlorogenic acids, turning them into quinones. Interestingly, formation of quinone results inhibition of PPO activity (Amorim, 1978). Any environment factor that alters membrane's structure (inset attack, micro-organism infection, physiological alterations and mechanical damages) causes coffee bean deterioration. Once the cell membrane is broken, there is a larger contact with the enzymes and the chemical compounds that are present intra- and extra-cellular in the bean. This causes chemical reactions that modify the original green coffee bean chemical compounds, which in turn affects the organoleptical characters from the infusion prepared with this coffee. This sequence suggests that the lower quality of coffee beans with low activity of PPO is related to cell membrane damage, and the reduction of PPO as a consequence of these damages.

Although coffee bean beverage quality and taste are decisive in establishing its price, the causes of product's quality variation are now being cleared. The discovery of 2, 4-6 tricloroanisole (TCA) present in coffee samples from the "Rio" variety, which suffered the action of *Aspergillus* sp., was related to the lower beverage quality (Amorim and Mello, 1991). The process of deterioration seems complex with inter-dependent mechanism and no plain theory could give definite explanation and offer accurate description of the process. Obviously, the degeneration of membrane cells and loss of control of permeability are among the first events that characterize deterioration.

Tests based on the membrane's loss of integrity were developed, analyzed and utilized by Mathews and Bradnock (1968), Abdul-Baki and Anderson (1972), McDonald Jr. and Wilson (1979), Yaklichi *et al* (1979), Steere *et al.* (1981), Powell (1986). In these studies, the seeds were immersed in water and during the process of inhibition, according to the degree of integrity of their membranes, exuded cytoplasmic solutes in the liquid environment. The solutes, with the electrolytic properties, had electric charges that could be measured with a conductivity meter. These seeds with lower vigour liberated high amount of electrolytes in the solution, resulting in a higher index of electric conductivity, or in higher concentration of potassium ions, mainly. Attempts were made to confirm the hypothesis that there could be a relationship between the test of electric conductivity of the coffee beans and its quality. This could provide to develop a faster, easier and safer test method with lower cost that can complement the evaluation of coffee bean quality, facilitating the coffee beverage quality attainment.

2. Experimental

2.1. METHODOLOGY AND ANALYSIS

2.1.1. Electric conductivity:
The methodology used was the one proposed by Loeffler *et al.* (1988), which consisted of using four samples of 50 beans in each parcel, without selecting defective (black ones, brocades, unripe or sour) or heavy beans (accuracy of 0.1 g). These were immersed distilled water and stored at 25°C. After a period of 3.5 h, the conductivity of the solution was determined by a Digimed CD-20 apparatus. The results were expressed in $\mu S.cm^{-1}.g^{-1}$.

2.1.2. Potassium lixiviation:
After reading electric conductivity the solution (without the grains) was transferred into glasses and a sample was taken to determine the amount of lixiviated potassium. Potassium analysis was done in a photometer (Digimed NK-2002) without previous digestion and expressed as ppm/g. Bean moisture was analyzed by heating at 105±3°C for 24 hours (Brazil, 1976).

2.2. DETERMINATION OF BEVERAGE QUALITY

The beverage quality determination was done through the "sensorial test" conducted at the UNICAFÉ labs of coffee classification and quality; one from FEMECAP - Campinas-SP and one another from BMF - São Paulo/SP.

2.2.1. Sensorial evaluation or "cup test":
From every sample, 100g coffee beans were taken and were roasted until the grains had a chocolate-like colour. After roasting, the coffee was ground in special grinders with medium granulation. An infusion was prepared putting 10 g of powder for each 100 ml of boiling water. From each sample, five cups were prepared, which were tasted by experts in the different classification labs. The results of the tasting of each cup were turned into numerical expressions (Table 1) as proposed by Garrutti and Conagin (1961) and Miya *et al.* (1973, 1974).

Table 1. Organoleptical characteristics and degree of the coffee standard beverages.

Beverage classification	Organoleptical characteristics	Points
Strictly Soft	Beverage of very soft and sweet taste	24
Softish	Beverage of soft prevailing sweet taste	18
Soft	Beverage of soft, although with a slight astringent taste	13
Hard	Beverage with a astringent, rough taste	11
"Riada"	Beverage with a slight taste of iodoform or phenolic acid	7
"Rio"	Beverage with a hard and impleasant taste, reminding iodoform of phenolic acid	1

2.2.2. Classification by defectives beans:

At the *Bolsa de Mercadorias e Futuros-SP* coffee classification Lab, every sample was tested on a black board paper as described by Official Brazilian Classification Table (Teixeira *et al*, 1974).

2.3. EFFECTS OF DEFECTIVE BEANS ON THE ELECTRIC CONDUCTIVITY OF THE EXUDATES

Samples of coffee Mundo Novo (1991/92 harvest) were obtained from the Department of Agriculture of ESALQ-Piracicaba, and were separated for the following defects: unripe beans, brocades beans, sour beans, black beans and black unripe beans or "stinker". Four repetitions of 10 beans for each kind of defect were separated in order to determine: a) beans weight in a precision scale of 0.001g, b) electric conductivity, c) Potassium lixiviation, and d) moisture. The experiment was made in a totally random design with five repetitions.

2.4. EFFECTS OF DRYING TEMPERATURES OF COFFEE BEANS HARVESTED AT THE MATURATION STAGES CHERRY AND UNRIPE OVER THE ELECTRIC CONDUCTIVITY OF BEANS

In a Mundo Novo LCP-388-17 belonging to the Department of Agriculture of ESALQ in Piracicaba, a mixture of green beans and coffee cherries was harvested on May 27,

1991. From this mixture the cherry fruits were manually separated from the green ones and from each stage of maturation, 16 plots of 600 g each were composed (distributed at the interior of the tray). Every four plots of cherry fruits and unripe or green ones were put to dry simultaneously at different temperatures, a) natural drying at 20°C, and b) artificial drying at 30, 45 and 60°C. Initial moisture of the cherry and unripe fruits were 69.7 and 71.5%, respectively. When the fruits achieved moisture as 12%, the weight of each plot was determined as the final and compared with the initial weight. The parcels were stored in a lab environment for three weeks. After this period, each parcel was peeled in a machine in order to obtain samples of "Pinhalense" trademark.
From each sample, the regular beans, unripe bean, black-unripe beans, black and black unripe beans were separated and weighed. Then the percentage of beans classified in each coloration was calculated. From each sample, four sub-samples of 50 seeds were drawn, which were weighed and immersed in water as described above and resubmitted to electrical conductivity analysis using a conductivity meter. In the exudate of each sample, the amount of potassium was determined as described above. The experiment was done using a totally random design using a factorial scheme (stages of maturation) X 4 (Temperatures of dryness) with four replications.

2.5. EFFECT OF SITES AND HARVEST METHODS ON THE ELECTRIC CONDUCTIVITY OF THE COFFEE BEANS EXUDATES

Fruits of coffee were harvested from ten sites at three maturation stages: a) cherry maturation stage followed by pulp removal, b) fruits of coffee at the dry maturation stage and harvested over the cloth, constituting a mixture, and c) coffee fruits at the dry maturation stage picked from the ground (so-called coffee from the ground or swept from the ground. Each portion was constituted by approximately 12 kg of coffee with husk. The samples of coffee fruits were collected between June and September 1991. The ten sites of collection are characterized below (Table 2).
After the harvest, the fruits were dried until they reached 12% of moisture and remained stored in jute bags until November 1991 in an appropriate environment at the seeds room of D.A.H./ESALQ, Piracicaba. During November 1991, the different lots were again divided into five samples of two liters each, peeled in machine (Pinhalense type) to take off the sample. Each sample was the subdivided to three sub-samples of 100 g, which were packed into plastic bags, forming three groups with 150 samples. Each group was sent to different sites (Londrina, Campinas, and São Paulo) for sensorial test.

Table 2. Characteristics of the studied sites.

Sites of colection	Latitude	Longitude	Altitude
Piracicaba - SP	22° 45'S	47° 40'W	540 m
Patrocínio - MG	18° 57'S	47° 00'W	933 m
Campinas - SP	22° 53'S	47° 06'W	562 m
Alfenas - MG	21° 28'S	45° 58'W	889 m
Machado - MG	21° 40'S	45° 55'W	873 m
Mococa - SP	21° 28'S	47° 01'W	665 m
Pindorama - SP	21° 13'S	48° 56'W	562 m
Garça - SP	22° 14'S	49° 36'W	663 m
Londrina - PR	23° 23'S	51° 11'W	566 m
Maringá - PR	23° 26'S	52° 02'W	570 m

From each sample, four other sub-samples of 50 seeds were taken, which ones after were weighed and immersed in water (as described previously) for electrical conductivity analysis. The results were submitted to statistic analysis. Correlations between the parameters and confidence intervals between electrical conductivity and beverage quality were analyzed.

2.6. EVALUATION OF COFFEE CV. ICATU PROGENIES BY ELECTRIC CONDUCTIVITY OF EXUDATE

The best F_3, F_4, F_5 and F_6 "Icatu" progenies from two back-crosses of Icatu experiments were appraised. The six better plants from each one of the 23 better progenies by productivity and other agronomic characteristics were chosen. IAPAR-59 and Bourbon Vermelho were used as reference cultivars. Cherry fruits were collected on the same date for every plant from all progenies and reference cultivars during the first half of June 1993. One litre of collected fruits was shelled and dried in the sun without the husk and later tested by the electric conductivity method. Electric conductivity of the exudate determination was accomplished according to the methodology developed by Prete (1992) expressing the results in $\mu S.cm^{-1}.g^{-1}$. The progenies evaluation by the electric conductivity of the exudate was made from variance analysis complemented by the average values from Tukey test for casual outlining with 20 Icatu progenies and two reference cultivars.

2.7. ELECTRIC CONDUCTIVITY OF EXUDATES OF COFFEE FRUITS IN DIFFERENT MATURATION STAGES

The experiment was on July 7, 1993. Coffee fruits of Catuaí Vermelho cultivar were harvested in a cloth for five maturation stages: green, sugarcane-green, cherry (mature), dry-raisin and dry fruits with four replications of the two litres of fruits in each maturation stages. Fruits were placed for drying on sieves with periodic revolving until they reached close to 11% humidity. Then the electric conductivity of the exudate was tested (Prete, 1992). Results were subjected to the analysis of variance and Tukey test ($P \leq 0.05$) at the level of 5% significance.

3. Results and discussion

3.1. EFFECT OF THE DEFECTIVE BEANS ON THE EXUDATES ELECTRIC CONDUCTIVITY

Beans having similar size showed different medium weight and the regular beans (without damage) showed highest weight. Defective beans showed following order, from the heaviest to the lightest: unripe beans, sour, brocades, blacks and unripe blacks. The values of electric conductivity, potassium lixiviation and water level were significantly low for the regular beans in comparison to the damaged ones and among the damaged beans, the black-unripe ones showed the highest values. The results as shown in the Table 3 showed the following increasing order of ion lixiviation and electric conductivity: regular or normal beans, brocades similar to the unripe beans sour, blacks and black-unriped beans. The moisture level of the beans followed the increasing order as regular or normal beans, brocades, unripe beans, sour ones similar to the black ones and black-unripe ones.

Illy *et al* ., (1982) carried out an analysis about the reflective characteristics of the defective beans aiming to eliminate them from the coffee lost through electronic colour selection. Besides reflective studies, the characteristics of damaged bean surface and cells through electronic microscopy were also performed. They reported that cellular disorder increased from unripe beans to sour beans and from those to black ones.

Table 3. Values of average weight of bean, electric conductivity potassium lixiviation and water level of regular beans and defectives beans of coffee cv. "Mundo Novo", from Piracicaba, State of São Paulo, Brazil.

Treatment	Weight of individual bean (g)	Electric, Conductivity µS.cm^{-1}.g^{-1})	Potássium Lixiviated (ppm/g)	Moisture after imbibition (%)
Normal	0,1320 A	40,28 A	15,56 A	38,49 A
Brocades	0,1086 CD	93,94 B	38,73 B	46,97 B
Unripe	0,1210 B	108,94 B	35,63 B	57,78 C
Sour	0,1134 BC	201,47 C	68,72 C	65,77 D
Black	0,1000 D	258,14 D	87,48 D	69,43 D
Black Unripe	0,0800 E	358,77 E	119,42 E	74,53 E
Values of "F"	51,49**	123,54**	141,97**	148,09**
C.V.%	5,12	13,49	12,42	2,99
DMS 5%	0,01	46,66	14,47	2,93

(Averages followed by different letters, at the colunn differ themselves (P<0,05) through Tukey Test) (** It indicates P<0,01).

The cell disorder allowed water to penetrate and to spread as easy as the more damaged the cellular tissues were. Results shown in Table 3 are in similar lines. Regular beans moved from 12% to 38.49% moisture after 3.5h of immersion, while the unripe beans, sour beans, blacks and black-unriped ones showed from 12% to 57.78, 65.77, 69.43 and 74.53% moisture, respectively. Disintegration of cellular membranes that were evaluated by the electric conductivity tests and potassium lixiviation followed the same order of seriousness of coffee imperfections.

3.2. DRY TEMPERATURE EFFECTS ON THE ELECTRIC CONDUCTIVITY OF COFFEE GRAINS EXUDATE HARVESTED AT DIFFERENT MATURATION STAGES

Table 4 shows that drying of harvested cherry stage fruits has resulted in lower damage of regular beans. The cherry fruits naturally dried in the shade showed the best quality beverage, while artificial drying at a 60°C temperature changed beverage quality from the "hard standard" to the "hard acid standard". Fruits harvested at the green stage depreciated coffee's quality evaluated by the number of damages and beverages quality.

Table 4. Value in weight percentage of normal or regular beans, unripes and dark-unripe, blacks and black unripe number of defectives and beverage quality of coffee samples harvested at the green stage and cherry maturation stage submitted to different drying temperatures.

Maturation stage	Drying temp.	Normal	Unripe	Dark Unripe	Black	Black Unripe	Defectives Number	Beverage Quality
		%						
Green	20°C	7,5	92,5	-	-	-	480	Hard (Unripe)
	30°C	7,7	86,0	6,3	-	-	540	Hard (Unripe/Acid)
	45°C	6,9	56,7	18,4	6,3	11,7	705	Hard(Unripe/Acid)
	60°C	-	3,1	-	-	96,9	1800	Not to be drunk
Cherry	20°C	100	-	-	-	-	-	Softish
	30°C	100	-	-	-	-	-	Hard
	45°C	100	-	-	-	-	-	Hard
	60°C	100	-	-	-	-	-	Hard (Ácid)

(Averages followed by different letters, at the column differ themselves (P<0,05) through Tukey Test) (** It indicates P<0,01).

The results showed that as drying temperature increased the beverage quality depreciation increased too. When the beans were dried at 30°C, presence of black beans and black-unripe beans was not seen, which at 45°C was 10.7 and 6.3% for black beans and black-unripe beans, respectively (Table 4). When the green fruits were dried at 60°C, the beans showed 96.9% damages for the black-unripe beans and 3.1% damages for the unripe beans. It was shown that the number of damages got to the number of 1800 and beverage from these beans was not fit for drinking at all.

Results recorded in Table 5 showed that the electric conductivity values and potassium lixiviation were significant higher for the beans resulted from fruits that were harvested at the green maturation stage in a comparison to the cherries at all drying temperatures. As the drying temperature increased, the resulting beans showed higher electric conductivity of the exudate and higher potassium lixiviation. This behaviour was more true for the unripe harvested fruits, while the harvested fruits at the cherry stage did not differ in electric conductivity when dried at 45°C, but their beverage quality, electric conductivity and potassium lixiviation changed significantly when dried at 60°C.

Table 5. Values of exudate's electric conductivity and potassium lixiviation of coffee samples harvested at the green maturation stage and cherry stage submitted to different drying temperatures.

Drying	Electrical Conductivity ($\mu S.cm^{-1}.g^{-1}$)		Potassium Lixiviated (ppm/g)	
Temperature	Cherry	Green	Cherry	Green
20°C	36,19 Aa	52,60 Ba	14,67 Aa	21,11 Ba
30°C	44,78 Aab	103,85 Bb	18,30 Aa	42,49 Bb
45°C	58,47 Aab	170,69 Bc	26,44 Ab	65,28 Bc
60°C	83,95 Ac	304,44 Bd	36,73 Ac	113,29 Bd
X	55,85	157,89	24,03	60,54
DMS 5% (Maturation)	13,33		5,19	
DMS 5% (Temperature)	17,54		6,84	

(Averages followed by different letters, at the column differ themselves ($P<0,05$) through Tukey Test) (** It indicates $P<0,01$).

3.3. EFFECT OF LOCAL AND HARVEST METHODS ON THE ELECTRIC CONDUCTIVITY OF COFFEE BEANS

The analysis of electric conductivity data variation and beverage quality of coffee samples, harvested from different sites under different harvest methods revealed significant "F" values. The variation analysis revealed "F" values that were significant among beverages quality taste and for all other interactions that were analyzed.

Table 6. Values or scores of coffee samples beverage quality of coffee samples from different sites and different methods of harvest and processing.

Sits	Sensorial Testing Lab	Harvest and Processing Method		
		Pulp	Mixed	Swept
1. Patrocínio	1	12,36 A	10,20 B	9,72 B
	2	13.20 A	11.80 AB	11.60 B
	3	15.60 A	11.00 B	11.00 B
2. Mococa	1	14.60 A	10.84 B	10.68 B
	2	13.60 A	11.80 A	11.00 B
	3	11.00 A	9.63 A	9.08 B
3. Machado	1	15.20 A	11.00 B	10.56 B
	2	11.40 A	11.00 A	11.00 A
	3	13.80 A	11.00 B	10.28 B
4. Campinas	1	15.00 A	10.84 B	8.60 B
	2	12.80 A	11.00 B	11.00 B
	3	13.80 A	11.00 B	11.00 B
5. Garça	1	13.60 A	7.80 B	7.80 B
	2	14.60 A	11.40 B	11.80 B
	3	11.40 A	11.00 A	10.20 A
6. Pindorama	1	12.83 A	9.08 AB	9.23 B
	2	14.20 A	12.40 AB	10.20 B
	3	10.20 A	10.04 AB	7.08 B
7. Piracicaba	1	14.80 A	8.76 B	7.48 B
	2	12.60 A	11.80 A	8.60 B
	3	11.00 A	11.00 A	7.48 B
8. Maringá	1	13.32 A	9.72 B	5.08 C
	2	13.80 A	11.00 B	7.40 C
	3	11.40 A	9.63 A	9.08 B
9. Londrina	1	14.03 A	10.68 B	4.84 C
	2	11.80 A	11.40 A	6.60 B
	3	11.00 A	9.40 A	5.80 B
10. Alfenas	1	15.63 A	11.40 B	2.76 C
	2	11.40 A	11.00 A	1.00 B
	3	12.40 A	10.20 A	5.80 B

*1. FEMECAP - Campinas - SP; 2. BMF - São Paulo – SP; 3. UNICAFÉ - Londrina – PR (Averages followed by different letters, at the column differ themselves ($P<0,05$) through Tukey Test) (** It indicates $P<0,01$).

As can be seen from the Tables 6 and 7, from every harvest site, the coffee harvested at the cherry stage, followed by pulping, showed superior quality of beverage. The beverage quality, among the mixed coffee and the swept coffee, showed differences according to harvest site and sensorial quality test site ("cup test").

Table 7. Number of defectives coffee samples from different production sites ,and different harvest methods.

Prodution Sites	Harvest Methods		
	Pulp	Mixed	Swept
1. Patrocínio	8 A	86 B	83 B
2. Mococa	34 A	208 C	166 B
3. Machado	21 A	100 B	112 C
4. Campinas	34 A	152 B	360 C
5. Garça	19 A	46 C	34 B
6. Pindorama	41 A	113 C	79 B
7. Piracicaba	20 A	297 C	224 B
8. Maringá	70 A	77 A	82 A
9. Londrina	16 A	240 C	90 B
10. Alfenas	23 A	78 B	115 C

(Averages followed by different letters, at the column differ themselves ($P<0,05$) through Tukey Test) (** It indicates $P<0,01$).

Data recorded in Table 8 confirmed the superiority of pulped coffee independent of harvest site. Mixed coffee and swept coffee showed variable behaviour according to the harvest site, explaining the significant interaction that was found.

For each local of beverage quality evaluation, the data of electric conductivity and beverage standard were statistically related with each other, which revealed that there was an inverse relation between beverage quality and electric conductivity (Fig. 1).

Table 8. Values of electrical conductivity ($\mu S.cm^{-1}.g^{-1}$) of coffee samples from different sites and different methods of harvest and processing.

Prodution		Harvest Methods	
Sites	Pulp	Mixed	Swept
1. Patrocínio	98,37 A	88,46 A	107,84 A
2. Mococa	103,30 A	125,66 B	134,03 B
3. Machado	80,96 A	118,15 B	109,21 B
4. Campinas	123,19 A	137,27 A	163,30 B
5. Garça	117,31 A	132,89 B	123,78 B
6. Pindorama	96,43 A	89,73 A	109,37 A
7. Piracicaba	108,08 A	168,00 C	136,81 B
8. Maringá	133,07 A	192,13 C	158,04 B
9. Londrina	105,25 A	161,46 B	143,83 B
10. Alfenas	112,93 A	124,18 A	173,58 B

(Averages followed by different letters, at the column differ themselves ($P<0,05$) through Tukey Test) (** It indicates $P<0,01$).

3.4. EVALUATION OF COFFEE CV. ICATU PROGENIES BY ELECTRIC CONDUCTIVITY OF EXUDATE

The environment influences all coffee attributes, including beverage aroma and flavour and its limits are certain by the genotype. Carvalho (1982) described the origin of coffee plants from Icatu germplasm that means calm or serenity in *Tupi* Indian language (Monaco *et al.*, 1974). It was obtained from the initial crossing between *C. arabica* cv. Bourbon Vermelho and *C. canephora* cv. Robusta in 1950 and from two back-crosses from *C. arabica* cv. Mundo Novo with subsequent selections practiced in the F_2, F_3 and F_4 generations by free pollinations from the best plants. Therefore, Icatu is a germplasm that has about 87.5% of *C. arabica* genes and 12.5% of *C. canephora* genes. In spite of having *C. canephora* genes, Icatu has beverage quality similar to *C. arabica.* It has been stressed that the decisive factors of beverage quality should be dominant in relation to *C. canephora* ones (Fazuoli *et al.*, 1977).

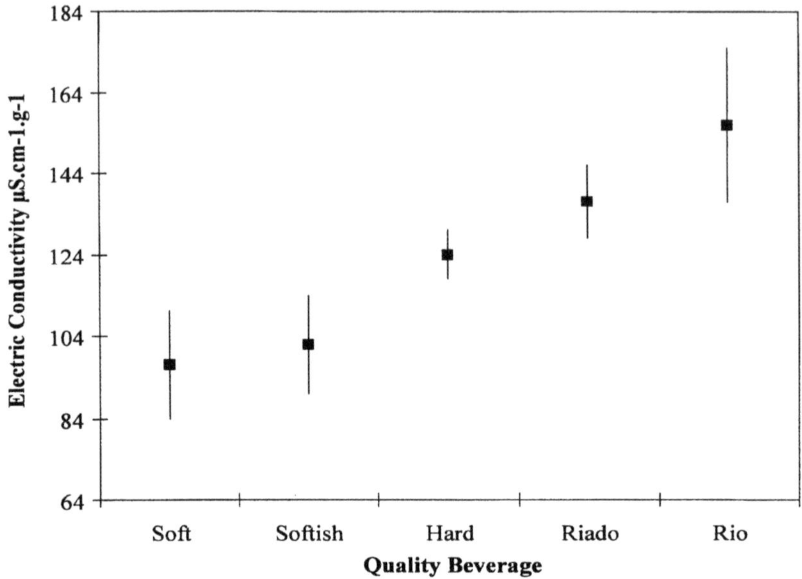

Figure 1. Relation between quality beverage standard soft ; softish ; hard ;"Riado"; and "Rio" evaluated in Campinas, State of São Paulo, Brazil, and trust gaps of electric conductivity (μS.cm^{-1}.g^{-1}) values of 150 coffee samples.

The results of the statistical analysis of the data are shown in Table 9. The majority of the progenies didn't differ to each other and to the references cultivars. Icatu lineage no 20 presented smaller electric conductivity value. Icatu lineage no 2 presented the largest conductivity value.

3.5. COFFEE GRAINS ELECTRIC CONDUCTIVITY EXUDATE AT DIFFERENT MATURATION STAGES

Freire and Miguel (1985), working with coffees in several maturation stages as green, sugarcane-green, cherry, dry-resin and dry fruits demonstrated that the best fruit quality was during the cherry stage.
Harvesting before this stage results high percentage of green fruits and cause damage in the type and beverage, which could be as high as 20% of losses in relation to the final revenue. Teixeira (1984) observed that harvesting of coffee at green maturation stage

resulted worse quality toast with consequent inferiority in the beverage quality. This also results smaller grains weight and size

Table 9. Medium electric conductivity ($\mu S.cm^{-1}.g^{-1}$) of the exudates of coffe Icatu progenies and Red Bourbon and IAPAR-59 cultivars.

Progeny	Conductivity
ICATU 20	64,37 b
ICATU 10	70,27 ab
ICATU 5	77,76 ab
ICATU AMARELO 6	81,19 ab
ICATU 1	83,58 ab
ICATU 8	91,39 ab
ICATU 11	92,01 ab
ICATU 15	95,12 ab
ICATU 13	97,39 ab
ICATU VERMELHO 24	97,77 ab
ICATU 14	108,27 ab
ICATU VERMELHO19	109,54 ab
ICATU 23	112,29 ab
ICATU AMARELO 30	112,58 ab
ICATU VERMELHO 7	115,42 ab
ICATU TARDIO AMARELO 4	115,98 ab
ICATU AMARELO 28	122,14 ab
ICATU VERMELHO 17	122,16 ab
ICATU VERMELHO 22	124,33 ab
ICATU 2	128,75 a
BOURBON VERMELHO 9	74,72 ab
IAPAR-59 29	87,07 ab

Averages followed by the same letter don't differ to each other in the level of 5% of significance on Tukey Test. DMS = 62,125 ($\mu S.cm^{-1}.g^{-1}$); CV = 25,86%

The undesirable biochemical transformations that happen in the grain of green fruit before and after the crop and that carries the formation of an inferior beverage are mainly of enzymatic nature, involving the polyphenol oxidase, glycosidase, lipase and protease. Some of these biochemical transformations degrade cellular wall and membranes and can change the collaboration of the grain and of the silver film, which can alter the beverage quality (Amorim and Teixeira, 1975).

Pimenta (1995) demonstrated through physical-chemistry analyses that among fruits in cherry stage, dry-resin and dry fruits the differences of the titrable acidity test, soluble solids, total sugars, phenolic compounds, polyphenol oxidase and beverage quality didn't show significant differences (Table 10).

Table 10. Medium electric conductivity of beans exudates ($\mu S.cm^{-1}.g^{-1}$) of harvest coffee fruits in different maturation stages.

Maturation Stages				
Green	Sugarcane green	Dry-raisin	Cherry	Dry
149,94 a[(1)]	105,73 b	79,60 c	76,63 c	76,04 c

(1) Averages followed by the same letter don't differ to each other in the level of 5% of significance by Tukey test; DMS = 21,17 (μS.cm-1.g-1); CV = 9,95%

4. Conclusion

Electric conductivity suffers a notorious effect from the defective coffee beans (black-unripe beans, black, sour, unripe and bored) because this sequence reports to the order of degradation importance of membrane system. The increment of harvested fruits at the unripe maturation stage depreciates coffee's quality, which could be much more if the fruits are to be submitted to drying at temperatures higher than 45°C and evaluated by the electric conductivity test. The pulp of fruits rendered better quality coffee (better beverage, lower number of defectives and lower electric conductivity) at all the ten harvest sites. There is an inverse relation between beverage standard and electric conductivity, that means, the better the beverage quality, the lower the values of electric conductivity of coffee beans exudates are.

5. References

Abdul-Baki A.A., Anderson J.D. (1972), Physiological and biochemical deterioration of seeds. p.283-315. *In* T. T. Kozlowski (ed.) Seed Biology. Acad. Press, New York,

Amorim H.V. (1978), Aspectos bioquímicos e histoquímicos do grão de café verde relacionados com a deterioração da qualidade. Piracicaba, SP, 85p. (Livre-Docência Escola Superior de Agricultura "Luiz de Queiroz").

Amorim H.V., Mello M. (1992). Significance of Enzymes in Nonalcoholic Coffee Beverage. p.189-209 *In* P. F. Fox (ed.) Food Enzymology. Elsevier , London.

Amorim H.V., Silva D.M. (1968), Relationship between the polyphenol oxidase activity of coffee beans and the quality of the beverage. Nature. 219: 381-382.

Amorim H.V., Teixeira A.A. (1975), Transformações bioquímicas, químicas e físicas dos grãos de café verde e a qualidade da bebida. In Congresso Brasileiro de Pesquisas Cafeeiras, Curitiba, PR. p.21

Brasil, Ministério da Agricultura (1976), Regras para análise de sementes. Dept°. Nac. Veg. DISEM. 188p.

Clifford M.N. (1985), Chemical and Physical Aspects of Green Coffee and Coffee Products. p.305-359. *In* M. N. Clifford e K. C. Wilson (ed.) Coffee, Botany, Brochemsitry and Production of Beans and Beverage. Croom helm, London.

Carvalho A. (1982), Melhoramento do cafeeiro: Cruzamento entre C. arabica e C. canephora. p.363-368. *In* Dixieme Colloque Scientifique Internacional sur le café. Salvador, BA.

Fazuoli L.C., Carvalho A., Monaco L.C., Teixeira A.A. (1977). Qualidade da bebida do café Icatu. Bragantia 36:165-72.

Freire A.C.F., Miguel A.C. (1985), Rendimento e qualidade do café colhido nos diversos estádios de maturação em Varginha-MG. In Congresso Brasileiro de Pesquisas Cafeeiras, 12. Caxambu, MG. p.210-214

Garruti R.S., Conagin A. (1961), Escala de valores para avaliação da qualidade de bebida de café. Bragantia 20:557-562.

Illy E., Brumen G., Mastropasqua L., Maughan W. (1982), Study on the characteristics and the industrial sorting of defective beans in green coffee lots. p.99-128. *In* Coloquio Científico Internacional sobre o Café, 10. Salvador, BA.

Loeffler T.M., Tekrony D.M., Egli D.B. (1988), The bulk conductivity test as an indicator of soybean seed quality. Journal of Seed Technology 12:37-53.

Mathews S., Bradnock W.T. (1968). Relationship between seed exudation and field emergence in peas and french beans. Hort. Res. 8:89-93.

Miya E.E., Garrut R.S., Chaib M.A., Angelucci E., Figuereido I., Shirose I. (1973). Defeitos do café e qualidade da bebida. Col. Int. Tecn. Alim. 5:417-432.

McDonald Jr.,M.B., Wilson D.O. (1979), An assessment of the standadization and the ability of the ASA-610 to rapidly predict potential soybean germination. Journal of Seed Techology 4:01-11.

Monaco L.C., Carvalho A., Fazuoli L.C. (1974). Melhoramento do cafeeiro Germoplasma do café Icatu e seu potencial no melhoramento. p.103. *In* Congresso de Pesquisas Cafeeiras, 2. Poços de Caldas, MG.

Pimenta C.J. (1995), Qualidade do café (Coffea arabica L.) originado de frutos colhidos em quatro estádios de maturação (Dissertação de Mestrado Escola Superior de Agricultura de Lavras).

Powell A.A. (1986), Cell membranes and seed leachate conductivity in relation to the quality of seed for sowing. Journal of Seed Technology 10:81-100.

Steere W.C., Leveengood W.C., Bondie J.M. (1981). An eletronic for evaluating seed germinaion and vigor. Seed Science and Technology 9:567-576.

Teixeira A.A., Pereira L.S.P., Pinto J.C.A. (1974), Classificação de café. Noções Gerais. IBC/GERCA - RJ, 88P.

Teixeira A.A. (1984), Observações sobre várias características do café colhido verde e maduro. In Congres Brasileiro de Pesquisas Cafeeiras, 11. Londrina, PR. p.227-228

Yaklich R.W., Kulik M.M., Anderson J.D. (1979), Evaluation of vigor tests in soybean seeds: relationship of ATP, conductibity and radioactive tracer multiple interia laboratory tests to field performance. Crop Science 19:806-810.

Chapter 31

RECENT DEVELOPMENTS IN BRAZILIAN COFFEE QUALITY: NEW PROCESSING SYSTEMS, BEVERAGE CHARACTERISTICS AND CONSUMER PREFERENCES

J.G. Cortez [1], H.C. Menezez [2]

[1] *Divisão de Desenvolvimento Rural, Ministério da Agricultura e do Abastecimento, 13087-000 Campinas (SP) Brazil Email: cortezcafe@hotmail.com* [2] *Departamento de Tecnologia de Alimentos, Faculdade de Engenharia de Alimentos, CP 6121, 13083-970 Campinas (SP) Brazil.*

Running title: Brazilian Coffee Quality

1. Introduction

Coffee cultivation in Brazil showed remarkable transformations during the sixties, due to the introduction of new varieties developed by the Genetic Section of the Agronomic Institute of Campinas (IAC), to the generalized use of mineral fertilization, to the renovation of old plantations in some traditional regions and to the financial and technical support given by the Brazilian Coffee Institute (IBC). Although the greatest attention was given to plant productivity, the recommendations with respect to quality led to better harvesting practices, to the separation of fermented beans in contact with the soil, and to the use of a wet-processing system in regions showing beverage problems (Silva & Cortez, 1998).
The introduction of coffee cultivation in new areas, due to aspects of geographical expansion or to avoid the frost areas, brought into view some new concepts on the quality of Brazilian coffees, especially on those produced in regions with dry and cold winters (harvesting and processing seasons), like in the Cerrado Region (Triângulo Mineiro / Alto Paranaíba). Under these conditions, the phenological cycles are slower and more regular, harvesting at the ideal stage of maturation is favored, fermentative processes are not stimulated during the drying of the beans and the sensory characteristics of the product are more indicated for modern systems of consumption, like the "expresso" preparation.
Fermentative processes have been defined as the prolonged action of microorganisms, such as bacteria, fungi and yeast from the genus *Klebsiella*, *Aspergillus* and *Fusarium*

T. Sera et al. (eds.), Coffee Biotechnology and Quality, 339–346.

and *Saccharomyces spp*. The prevalence of action was determined by the mucilagem composition (length of the pectinolytic chain, degree of esterification) and the pectinesterase / pectin-lyase production of each microrganism (Jones & Jones, 1984). When the climatic or technological conditions are favorable for microbial development, a breakdown of the mucilage is promoted with the concomitant production of alcohols and acids. If the fermentative process is interrupted just after the consumption of the mucilaginous layer, the endosperm is not attacked and the quality of the beverage is preserved; this is observed when the coffee beans are withdrawn from the fermentation tanks (wet method of processing) or when the humidity of the seed is carried to the point where there is no microbial development (coffee drying). Otherwise, when the fermentative process is not interrupted, the cellular layers are degraded, several chemical compounds are destroyed and the beverage is altered, with the occurrence of the Rio flavor, earthy or fermented tastes and aromas and the formation of defective beans like black and "stinker" beans (Dentan, 1989; Dentan, 1991).

This greater knowledge of the fermentative processes that coffee beans are subjected to, during maturation, harvesting and the processing phases, led to the development of a new processing method, called "semi-wet method". Under this system, which can be employed together with the use of strip harvesting (mechanical or manual) and the separation of the three maturation phases of the beans (immature, ripe and overripe cherries); after the dehulling (exocarp) of the mature beans, they are disposed into the drying places in thin layers and carefully dried to the point where there is any fermentative process. This also enables the separation of immature beans, with a strong metallic taste, due to the temporary relation between the isomers of the chlorogenic acids (Menezes, 1990) and the decrease of fermented or defective beans, leading to a great improvement in coffee quality, even in regions of irregular maturation and/or with Rio taste.

The general concepts of Brazilian coffee quality, under these new correlations of climate, applied technology and sensory characteristics, have led to the establishment of "brands" or origin certificates, sustaining marketing promotions and attending the diversification of products, beverage preparations and consumer preferences. Included in this context are a widening of the "gourmet" market, a greater variety on the Brazilian market and the diffusion of robusta coffees, produced in the States of Espirito Santo, Souther Bahia and Rondonia.

In this paper, the results of physical and sensory analyses of samples obtained from several regions, production and processing systems are presented, with some comments about possible forms of consumption, with a view to a wider integration between producers, traders, roasters and consumers in the coffee agribusiness.

2. Material and methods

The samples analyzed were obtained from several agricultural systems, according to the following models:

- cultivated species : *Coffea arabica* and *Coffea canephora*
- climatic conditions of the region (especially during harvesting and processing)
- cultural practices (spacing, irrigation, organic programming)
- harvesting and processing (wet, semi-wet processing, drying)
- destination (preference for consumption, exportation).

The physical and sensory analyses followed the instructions of the Official Classification by Types and Beverages adopted in Brazil (MIC/IBC, s.d.), according to the descriptions of aspect, drying, roast (graduations from excellent, good and regular to bad) and beverages (Soft, Hard, Ryoy and Rio). To these, were added the sensory descriptions from the Test Unit of the International Coffee Organization (Feria-Morales, 1989), according to the presence and intensities of taste (sweet, sour, acid, salty and bitter), mouthfeel (body and astringency) and flavors/aromas (chemical, fermented, chocolate-like, cereal, etc).

3. Results and discussion

Table 1 shows some sensory results from samples obtained from the State of Parana, in the region known as "Norte Velho" (Ribeirão Claro, Tomasina), as compared to those from a dry-method of processing in Londrina and another sample from Apucarana (altitude of 900 m).

Table 1 - Physical and sensory characteristics of arabica coffee samples from the State of Parana.

origin	processing	aspect	beverage	acidity	body	aroma
Londrina	dry	good	Rio	regular	high	medicinal
Apucarana	semi-wet	good	Soft	high	high	pleasant
Ribeirão Claro	semi-wet	good	Soft	regular	high	pleasant
Tomasina	semi-wet	excellent	Soft	regular	high	pleasant

The wet conditions during winter, without a hydric deficit during the harvesting period, allow for the occurrence of prolonged fermentative processes and the Rio taste. Thus, the semi-wet system is more adequate for high-density plantations (more than 5,000

plants per hectare) and for irregular maturation, allowing the expression of desirable sensory characteristics (body, acidity) and avoiding the development of the Rio flavor. The use of a slow drying, with a short interval during the last hours of drying ("resting period") and just before storage, results in higher scores for the aspect of the beans. The Apucarana region shows a less-humid winter (due to the altitude), in addition to slower maturation, resulting in better scores for the beverage.
In the State of São Paulo, three distinct coffee regions can be found: *Araraquarense* (Catanduva, São José do Rio Preto), ***Média Paulista*** (Garça, Marília) and *Alta Mogiana* (Franca, Pedregulho), with one other region (*Alta Sorocabana:* Pirajuí, Avaré) which has the same climatic conditions as Ribeirão Claro and Londrina. The first two regions are characterized by plantations at 500-600 m, with high average temperatures during the year and high hydric deficit during the winter, which induces a shorter phenologycal cycle between flowering periods and bean maturation and the occurrence of a Hard beverage, with high scores for astringency. In the Alta Mogiana region, the plantations are located at high altitudes (900 - 1.000 m), with low temperatures and high hydric deficit during the harvesting period, which leads to the production of a Soft beverage, with high body scores and chocolate-like aroma. Table 2 shows some physical and sensory aspects of samples collected in these regions.

Table 2 - Physical and sensory aspects of arabica coffee from the State of São Paulo.

Origin	processing	aspect	beverage	body	astringency	aroma
Avaré	dry	good	Rio	high	low	medicinal
Catanduva	dry	regular	Hard	regular	high	metallic
Marília	dry	regular	Hard	regular	high	metallic
Franca	dry	regular	Soft	high	low	cocoa

Due to the ample extensions of land suitable for coffee cultivation and the diversity of climatic conditions, the State of Minas Gerais shows four distinct coffee regions: *Sudeste* (Ouro Fino, Alfenas, Varginha), with high to medium altitudes, amplitude of rainy seasons (high to medium) and beverage variations (Soft to Rio) especially due to the proximity to water reservoirs; *Triângulo Mineiro / Alto Paranaíba*, with the use of *irrigation* where the temperatures and hydric deficit are very high (Araguari, Monte Carmelo) and *without irrigation* in areas at higher altitudes (São Gotardo, Serra do Salitre), with beverages varying from Hard to Soft, and *Zona da Mata / Alto Rio Doce* (Caratinga, Manhuaçú / Capelinha, Novo Cruzeiro) where the combination of high temperatures and high air humidity allow for the occurrence of the Rio taste (dry method) or Soft beverage (wet or semi-wet methods).

Table 3 - Sensory characteristics of samples from the State of Minas Gerais.

Origin	cultivation	processing	beverage	body	acidity	market
Ouro Fino	normal	semi-wet	Soft	medium	high	gourmet
Alfenas	normal	dry	Rio	medium	medium	Rio taste
Alfenas	normal	semi-wet	Soft	medium	high	gourmet
Araguari	irrigation	dry	Hard	high	medium	expresso
São Gotardo	normal	dry	Soft	high	medium	expresso
Manhuaçú	normal	dry	Rio	medium	medium	Rio taste
Manhuaçú	normal	wet / s-w	Soft	high	high	gourmet
C. Paranaíba	organic	dry	Soft	high	medium	gourmet

The coffee beans supply a great diversity of markets, from those who appreciate the Rio taste (Saudi Arabia, Greece, Turkey) to the gourmet and "expresso" markets (USA, Europe, Japan), also including the production of organic coffee (Carmo do Paranaíba), without the use of pesticides and herbicides and showing great advantages with respect to aspect (increase in the size of the beans), production (decrease in two year variations in production) and beverage (one or two points above the average grades for body and aroma). The results of samples from the State of Minas Gerais are shown in Table 3.

The State of Espirito Santo shows two distinct coffee regions, characterized by the altitude of cultivation and the species of coffee: *arabica cultivation*, at altitudes above 600 m and average annual temperatures below 23 °C (Santa Teresa, Domingos Martins) and *robusta cultivation*, especially the conillon variety, at altitudes below 600 m (São Gabriel da Palha, Colatina). Due to the irregular rainfall distribution, the use of irrigation is required in robusta areas, to assure the necessary availability of water during bean development (Summer). Some of the transition areas between regions (Pancas, Conceição do Castelo) are suitable for both robusta and arabica cultivations, leading to some different sensory characteristics for each species. A classification of types and beverages for robustas is now being suggested, the terms including one for roasting (graduation varying from excellent - almost entirely uniform color - to bad - heterogeneous color) and one for beverage (excellent - neutral flavor and little acidity - to regular - strong taste of robusta). Table 4 shows the sensorial analysis of arabica and robusta coffees from samples obtained in the State of Espirito Santo.

Table 4. Sensory classification of coffees from the State of Espirito Santo.

Origin	specie	processing	roast	beverage	acidity	body	market
S.Teresa	arabica	dry	regular	Rio	medium	low	Rio taste
S.Teresa	arabica	wet / s.w.	good	Soft	high	medium	gourmet
S.G.Palha	robusta	dry	bad	regular	low	low	instant
S.G.Palha	robusta	wet / s.w.	good	excellent	low	high	exportation
Pancas	robusta	dry	good	good	low	high	exportation

The State of Bahia is currently one of the regions showing great expansion of its plantations, besides an efficient renovation of traditional arabica areas. The accentuated expansion of the use of robusta coffees on the internal market, together with problems with the cultivation of cocoa in Southern Bahia, has led to the substitution of cocoa plantations with robusta, especially in Ilhéus, Camacã and Canavieiras. At higher altitudes, the coffee farmers are looking for new forms of technology, such as the use of irrigation and drying. This is the case in Vitoria da Conquista Plateau (Barra do Choça), where the use of irrigation during the summer (high rates of bean development) and drying at lower altitudes (caatinga) has led to excellent results with respect to the appearance of the beans and better beverage scores, giving rise to coffee known as "Café da Caatinga". The more northern plantations, such as those located in Mucugê, in the Chapada da Diamantina, are also trying new forms of marketing, combining the use of hand-picked harvesting with semi-wet processing, giving better characteristics to the beverage. This can be used as an item of the propaganda for the region, together with its natural beauty. Another area showing great development of its coffee plantations is the area called Western Bahia (Barreiras). The system of cultivation is by irrigation and mechanization is extended to harvesting and drying, with self-propelled harvesting machines and mechanical dryers. New projects for coffee cultivation by irrigation and high levels of mechanization are under study in other areas, opening new frontiers for coffee in Bahia. Table 5 shows some physical and sensory results of the coffee from Bahia.

Table 5 - Physical and sensory characteristics of arabica and robusta coffees from Bahia.

Origin	specie	cultivation	harvesting	processing	drying	beverage
Canavieiras	robusta	normal	normal	semi-wet	mechanic	excellent
B. Choça	arabica	irrigation	hand picking	semi-wet	caatinga	Soft
Mucugê	arabica	normal	hand picking	semi-wet	mechanic	Soft
Barreiras	arabica	irrigation	mechanical	semi-wet	mechanic	Hard

New plantations of Robusta coffee in the State of Rondonia have been developed, along with some studies on the introduction of other robusta varieties, showing high potential with respect to productivity and sensory characteristics. Table 6 shows some proprieties of these new varieties, compared with a sample of the conillon variety.

Table 6 – Sensory characteristics of different varieties of robustas from the State of Rondonia.

Cultivar	processing	aspect	roasting	aromas	body	acidity
Robusta 2258-1	semi-wet	good	fairly good	low (robusta)	fairly good	low
Robusta Laurenti	semi-wet	good	fairly good	low (robusta)	fairly good	low
Robusta 1641	semi-wet	good	fairly good	low (robusta)	fairly good	low
Robusta Conillon	semi-wet	fairly good	fairly good	low (robusta)	regular	regular

4. Conclusions

Brazilian coffee is showing substantial changes with respect to quality aspects. Developing from the old concept of a prevalence of the Hard taste, metallic aromas and an occurrence of the Rio taste, the diffusion and invigoration of the binomial productivity – quality can now be seen, especially amongst the coffee producers. New processing systems, alongside agricultural practices such as irrigation, have result in the availability of batches of coffee on the market, with less defects, a more homogeneous appearance, better results during roasting and a greater acceptance by the consumers.

The use of cultural practices, such as irrigation, pest and disease control and a better quantification and fractionation of nutrients ensure lower production costs, higher yields and a general strengthening of the agricultural activity. A better knowledge of the climatic conditions in each region leads to a more adequate choice of the species to cultivate, the correct population of coffee plants per area, the assessment of the potential production, the selection of one system of harvesting, the separation of defective beans, the type of drying, the sensory characteristics of the beverage and an adequation to one or another system of roasting and consumption.

Many of these results come from a better understanding of the fermentation processes occurring during maturation, harvesting, processing and storage of the coffee beans. It can be seen that some regions are more prone to fermentation processes than others: This is especially true of plantations near great reservoirs or warm-humid conditions during maturation and processing. However the separation of defective beans and the rapid dehydration of the beans in suitable drying places, as used in the semi-wet process, hinders the action of endogenous fungi and bacteria, and creates a better environment for coffee quality preservation. These results can be achieved by the use of bioproducts or by the pulverization of biostatics, like quaternary ammonium products (Fegatex). The potential use of biotechnology in the formation of aromas is also a possibility.

5. Acknowledgments

We are tankful to Mr. José Luís B. Toledo (Sâo Paulo Cereal and Commodities Exchange) and Mr. Alaerte T. Barbosa (Coffee Training Center – Brazilian Roasters Association) for helping in the samples analysis. We are also grateful to Mr. Lourenço Del Guerra and Mr. Marco Antonio Mendonça (Pinhalense S/A Máquinas Agrícolas) for providing the coffee samples from the States of Espírito Santo and Bahia and Dr. Wilson Veneziano (EMBRAPA – Vilhena) for the robusta samples from the State of Rondonia

6. References

Dentan E. (1989), Étude microscopique de quelques types de cafés défectueux : grains noirs, blanchâtres, cireux et "ardidos". Paper presented at 13 th Colloque Scientifique Internationale sur le Café, Paipa, p. 283-301.

Dentan E. (1991), Étude microscopique de quelques types de cafés défectueux. II: Grains à goût d'herbe, de terre, de moisi; grains puants, endommagés par des insectes. Paper presented at 14 th Colloque Scientifique Internationale sur le Café, San Francisco, p. 293-311.

Feria-Morales A. (1989), The research and test unit of the International Coffee Organization scientific activities focusing on the quality of coffee. Paper presented at 13th Colloque Scientifique Internationale sur le Café, Paipa, p. 159-180.

Jones K.L., Jones S.E. (1984), Fermentations involved in the production of cocoa, coffee and tea. *In*: Progress in industrial microbiology. *Vol. 19:* Modern applications of traditional biotechnologies. Bushell, M.E. (Ed.), Elsevier Ed., New York, p. 411-455

Menezes H.C. (1990), Variação de monoisômeros e diisômeros do ácido cafeoilquínico com a maturação do café. Campinas, 120 p. Tese (Doutor em Tecnologia de Alimentos, UNICAMP).

Ministerio da industria e comercio / Instituto Brasileiro do café (s.d.) *Classificação do Café: noções gerais.* Rio de Janeiro: Setor de Publicação Gráfica, IBC/GERCA, 117 p.

Silva L.F., Cortez J.G. (1998), A qualidade do café no Brasil – Histórico e perspectivas. *Cadernos de C & T da Embrapa*, vol. 15 (1): 63-76.

Chapter 32

ORGANOLEPTIC PROPERTIES OF ESPRESSO COFFEE AS INFLUENCED BY COFFEE BOTANICAL VARIETY

M. Petracco

Illycaffe' S.p.A., via Flavia, 110 - Trieste, Italy

Running title: Espresso coffee quality

1. Introduction

Espresso is a way to enjoy a cup of coffee that is gaining large popularity world-wide. Its roots are to be searched for in the Italian culture of foods and beverages, which developed a typical life-style, linked to coffee drinking (Illy and Illy, 1990). Its main marks are a) extemporaneous preparation on express order, b) brewing by a specific method using high water pressure, and c) rapid extraction admitting into the cup just the best material. The resulting beverage is very peculiar from a physical and chemical perspective too (Petracco 1989). However, its main characters are of sensory nature. All human senses, with exception of hearing, are involved in appreciation of an Espresso cup:

- -vision evaluates foam's aspect, examining its colour and its consistency and persistence,
- -touch assesses the beverage mouthfeel, or "body", a property linked with density and viscosity,
- -taste judges the bitter/acidic balance and the presence of a sweet caramelic after-taste,
- -olfaction appreciates both fragrances, by direct inhaling of the vapours arising from the cup, and flavour or nasal perception of the volatile substances evolving in the mouth.

Espresso extraction method produces in the cup a high concentration of sensorial active substances. Special care must, therefore, be devoted to prevent and to eliminate defects from the raw material. Moreover, in order to offer the customers the finest sensory quality, attention should be devoted to several variables, tending to affect the sensory

T. Sera et al. (eds.), Coffee Biotechnology and Quality, 347–353.

balance appreciated by consumers. Common knowledge, partly validated in selected environments by experimental data (Guyot *et al.*, 1996) claims that factors such as growing region and altitude as well as climatic situation and processing technique can have an effect on overall coffee quality, mainly in the meaning of increased acidity and aroma. Unfortunately, little is known about their influence on Espresso beverage. Besides, genetic parameters were long neglected to the benefit of local traditional cultivars and, with some exception (Moschetto et al., 1995; Moreno et al., 1995; Roche, 1995), coffee improvement programmes relying on breeding used to focus mainly on agronomic traits such as increased yield and resistance to diseases. Espresso quality may significantly be affected by the genetic variability brought by diverse characteristics of coffee plants grown together in the same geographical conditions and under the same farm practice.

2. Experimental

About 20,000 individually identified *Coffea arabica* plants were sown and grown in a premium coffee producing region, located in Brazil at about 21° South at 1040 m altitude, where the average temperature varies between 20-26.4°C (Sondahl and Bragin, 1991). A possible clustering criterion to describe this coffee population is to group them according to the shared botanical variety origin such as Caturra acronym for CA, Catimor- CR, Red Catuai- CTR, Yellow Catuai- CTY, Icatu- IC. Among well-appreciated conventional cultivars, two hybrid varieties derived from inter-specific crosses were chosen: Catimor and Icatu. The reason to do so was to explore cup quality of plants carrying some traits of the Robusta heritage, which makes them very interesting to all coffee technologists because of some attractive agronomic traits that have been introduced by the hybridisation procedure.
It is commonly accepted that the name Robusta itself has been imparted to the most cultivated variety of *C. canephora* because of its resistance in humid tropical environments. It shows marked resistance against the leaf rust (*Hemileia vastatrix*) (Osorio Garcia, 1990) and against root nematodes (*Meloidogyne* species) (Anzueto *et al.*, 1995). Coffee breeders have been trying since long to cross robusta with tetraploid arabica. A spontaneous hybrid with arabica phenotype was found in Timor hybrid. Further crossing with the Caturra variety led to a rust-resistant progenies that got various names in different countries: Catimor in Brazil and Central America, Colombia in Colombia and Ruiru 11 in Kenya. A distinct successful trial was performed in Brazil, where a tetraploid robusta plant was produced by colchicine-induced chromosome duplication. Its crossing with arabica variety Mundo Novo produced several resistant and high-yielding lines that were called Icatu.
Since no robusta character was detected during preliminary cup tasting for Espresso coffee, we were highly interested in selecting superior beverage producing lines of these agronomically promising varieties. The farming practices utilised were the typical ones of local tradition: full sunlight, semi-dense plot of 1666 plants/ha, with some quality-oriented operations like adequate fertilisation, regular pesticide spraying, weed control,

handpicking harvesting, de-pulping, cherry processing by fermentation. Besides agronomic and chemical evaluations, the individual adult plants were assessed for "Espresso quality". Harvest was accomplished plant-by-plant by careful handpicking to avoid the presence of unripe, overripe or damaged cherries. The crop of each individual plant of about 3 kg of cherries was de-pulped and the resulting mucilaginous parchment beans were fermented in individual buckets with the help of a pectolytic enzyme (Isozym) to standardise the fermentation time among all samples. After sun drying and de-hulling, the samples were shipped to the Illycaffé corporate laboratory in Trieste (Italy), where the organoleptic analysis was performed.

The green coffee samples were prepared according to the Illycaffè' lab procedures established for evaluating commercial samples (Illy and Viani, 1995): a) visible defects (black, broken, unripe beans) were eliminated manually, b) each sample was roasted to a typical Italian roast, using 80 g per sample under controlled conditions in lab scale roasters (Probat), and then ground to an appropriate particle size in professional grinders (Faema), c) each roasted and ground sample was evaluated in three different cups, according to three key preparation techniques:

Espresso- the classic espresso cup prepared under standardised and thoroughly controlled conditions samples (Illy and Viani, 1995). The resulting beverage is a solution of sugars, acids and caffeine, in which a distinct phase made of lipid micronic droplets is dispersed as an emulsion. Overall concentration can be as high as 60 g/l.

Infusion- a brewing method widely used in Northern Europe and US. Boiling water is poured on coarsely ground coffee powder and allowed to rest for a given period before filtering away the spent-grounds from the remaining clear liquid coffee. The concentration of the beverage is low (below 20 g/l) and only the soluble substances pass into the cup, imparting it an aromatic pattern typical of a filtered coffee beverage.

Diluted espresso- where a little aliquot of espresso is taken and diluted with hot water up to the same total solids content as of the infusion. This way the high concentration of regular espresso does not hinder any longer the evaluation of some weaker aroma nuances, and the difference between the solution aromatic pattern and the emulsion one can be determined.

Twelve samples per day were submitted to a cup tasting panel formed by at least three trained professional experts and were examined in blind absolute tests where some complex variables, e.g. overall merit, were determined by comparison with a mental paradigm present in each assessor's memory by previous experience. The scale used encompassed scores ranging from -3 to 0 in the negative rank and from 0 to +3 in the positive one, defining "good for espresso" only those samples that score from +1 to +3.

3. Results and Discussion

Two data sets pertaining to two consecutive crops 1990 and 1991 were available. A total of 585 plants were evaluated in 1990 and 225 more in 1991, giving a total pool of more than 800 samples. As can be seen from Fig. 1 on overall merit distribution of these sensory analyses, just a few samples (less than 10%), namely those scoring ≤-2 were penalised for objectionable taste (stinking, harsh, immature). This result might be related to the "wet" processing method, which allows eliminating by floatation some improper beans, so enhancing quality (Illy and Viani, 1995). Unfortunately, only as little as 20% of samples were reported as "good for espresso" as defined above (merit ≥ +1).

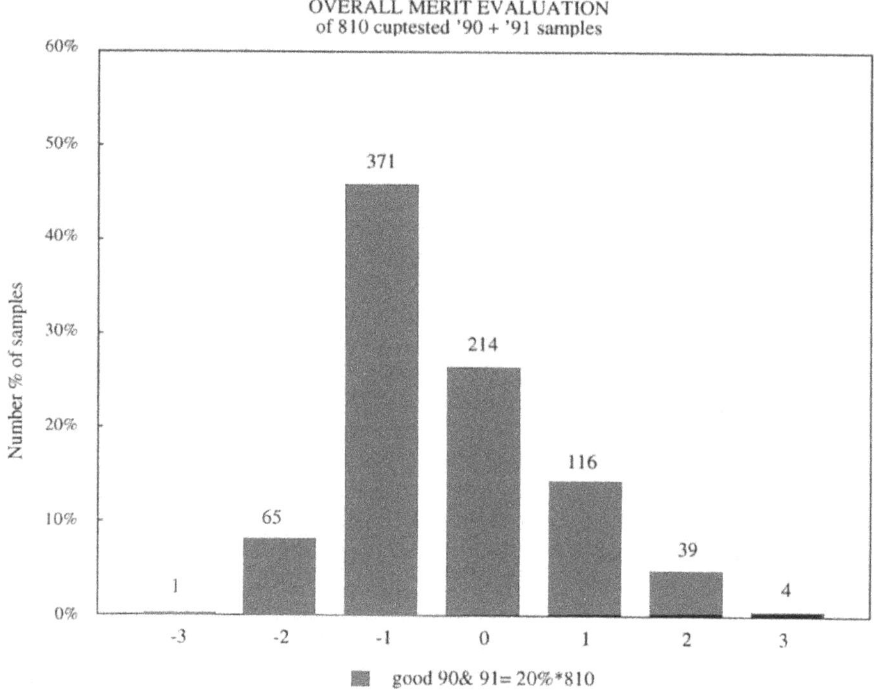

Figure 1. Overal merit evaluation of 810 cuptested "90+91" samples.

With the objective of finding a relationship between coffee genotype and "Espresso quality", we organised the 1990 data as displayed in Fig.2. All the examined coffee varieties exhibited irregular sensory characteristic with presence of both good and bad samples. However an evident "Espresso quality" trend emerged when considering the percentage of "acceptable for espresso" samples (merit ≥ 0):

Caturra	CA	45 %
Catimor	CR	29 %
Red Catuai	CTR	35 %
Yellow Catuai	CTY	31 %
Icatu	IC	46 %

This tendency was confirmed when examining the 1991 data (where Caturra figures had to be removed due to the too little number of samples analysed):

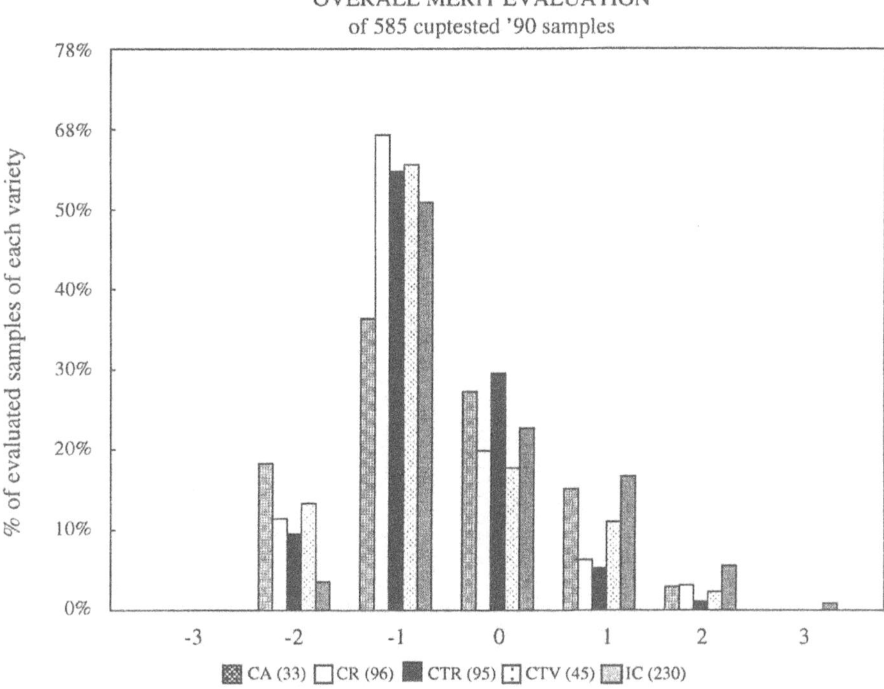

Figure 2. Overal merit evaluation of 585 cuptested "90" samples.

All coffee varieties showed fluctuation from year to year appraisal, which could be explained by a general quality improvement in 1991 samples. When comparing directly the 1990 + 1991 combined data of the two varieties, Catimor and Icatu, the quality gap was particularly striking (Fig.3).

Catimor	CR	41 %
Red Catuai	CTR	49 %
Yellow Catuai	CTY	64 %
Icatu	IC	63 %

In the range of "acceptable for ESPRESSO" (merit ≥ 0) Icatu scored an outstanding 51% compared to Catimor's 31%. In a more strict classification criterion, namely the quality class called "good for Espresso" (merit ≥ +1), the performance of Icatu was more than double of Catimor samples: 24% vs. 10%. The presence of 12% of bad Catimor samples (merit ≤ -2), compared to the 4% of bad Icatu samples did not appear so important when considering that some defective or ill-fated samples might be present due to process accidents.

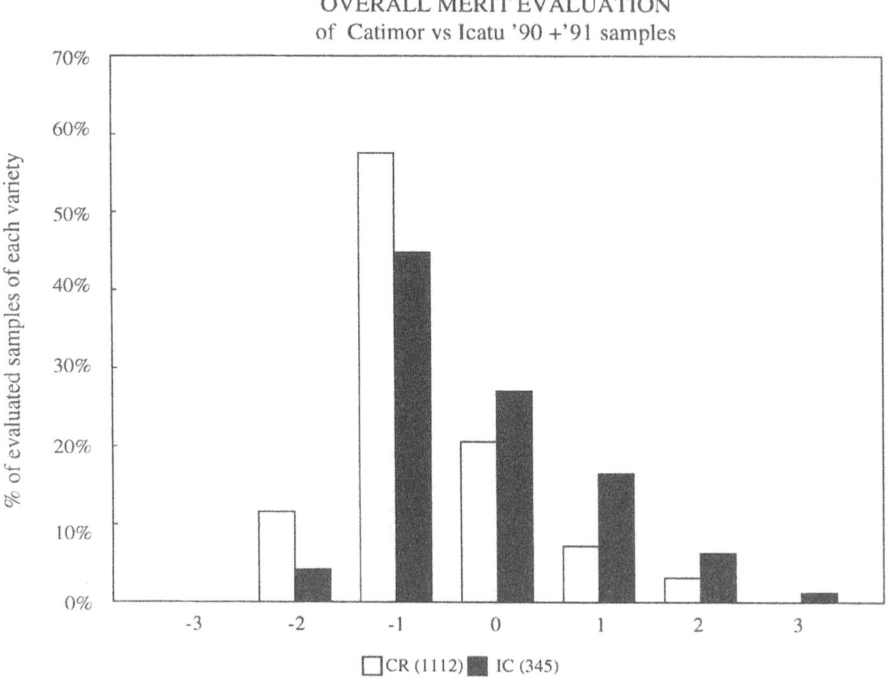

Figure 3. Overal merit evaluation of Catimor vs. Icatu "90+91" samples.

Attention is called, on the other hand, on the gap of samples declared unacceptable without being offensive (merit = -1). This score, as a rule, meant an unbalanced bitter/acidic ratio with lack of aroma and body, and a slight presence of astringency. All these traits could be seen as a peculiarity of a botanical origin and induced by genetic traits of a particular variety.

4. Conclusions

The results provided preliminary evidence that under specific growing and processing conditions as described above, the botanical origin of *C. arabica* plants was important to provide the best desirable properties in Espresso coffee.

5. References

Anzueto F., Bertrand B., Peña M., Marbán-Mendoza N., Villain L. (1995), Desarrollo de una variedad porta-injerto resistente a los principales nematodos de America Central. Paper presented at XVII simpósio sobre la caficultura latinoamericana, San Salvador, El Salvador, p. 7

Guyot B., Gueule D., Manez J.C., Perriot J.J., Giron J., Villain L. (1996), Influence de l'altitude et de l'ombrage sur la qualité des cafés Arabica. *Plantations, récherche, développement* vol.3 (4): 272-280

Illy R., AndIlly F. (1990), From Coffee to Espresso, Mondadori, Milano, Italy, pp.166-188

Illy, A.,Viani R., eds. (1995), Espresso Coffee: The Chemistry of Quality, Academic Press, London

Moreno G., Moreno E., Cadena G. (1995), Bean characteristics and cup quality of the Colombia variety (*Coffea arabica*) as judged by international tasting panels. Proc. XVI colloque ASIC, Kyoto, Japan, pp.574-583

Moschetto D., Montagnon C., Guyot B., Perriot J.J., Leroy T., Eskes A.B., Yapo A. (1995), Résultats récents sur l'amélioration génétique de la qualité à la tasse de *Coffea canephora* en Côte d'Ivoire. Proc. XVI colloque ASIC, Kyoto, Japan, pp.557-564

Osorio Garcia F.O. (1990), Evaluación de la estabilidad del rendimiento de introducciones en un ensaio regional de café. Paper presented at IX reunión regional de mejoramiento de café de PROMECAFE', Managua, Guatemala

Petracco M. (1989), Physico-chemical and Structural Characterisation of Espresso Coffee Brew. Proc. XIII colloque ASIC, Paipa, Colombia, pp.246-261

Roche D. (1995), Coffee genetics and quality. Proc. XVI colloque ASIC, Kyoto, Japan, pp.584-588

Sondahl M.R., Bragin A. (1991), Somaclonal Variation as a Breeding Tool for Coffee Improvement. Proc. XIV colloque ASIC, S. Francisco, USA, pp.701-710

Chapter 33

MYCOTOXIGENESIS IN GRAINS APPLICATION TO MYCOTOXIC PREVENTION IN COFFEE

J. Le Bars and P. Le Bars

Laboratoire de Pharmacologie-Toxicologie I.N.R.A., BP 3, 31931 Toulouse Cedex, France

Running title: Mycotoxic prevention in coffee

1. Introduction

Numerous fungal species may develop all along the food chain, from the plant in the fields during processing steps and storage down to consumer's plate or cup. Fungal development in alimentary substrates can lead to several detrimental effects:
(i) alterations of aspect and technological properties,
(ii) modification of nutritive value,
(iii) impairment of organoleptic features,
(iv) production of toxic compounds. In this regard, it would be important to understand the general mechanisms and conditions conducive to mycotoxic contamination in foods in order to contribute to a logical prevention.

2. Mycotoxins and mycotoxigenesis

2.1. TOXIC PRODUCTS RESULTING FROM FUNGAL DEVELOPMENT IN VEGETAL PRODUCTS

Occurrence of toxic metabolites in foods and feeds resulting from fungal development on plant substrates may result from any of four general mechanisms. Firstly, a saprophyte mould growing on a favourable substrate can directly produce toxic metabolites (e.g. aflatoxins by *Aspergillus flavus*). Secondly, the association of an endophyte with a living plant can lead to toxicosis in animals (e.g. fescue foot disease in cattle with *Neotyphodium coenophialum*). Thirdly, in response to the invasion of phythopathogenic fungi, a living plant can produce diverse metabolites, some of them at

T. Sera et al. (eds.), Coffee Biotechnology and Quality, 355–368.

levels that are toxic (e.g. ipomeamarone in *Ipomea patatas*). Fourthly, atoxigenic species may contribute by bioconversion to transform atoxic compounds to toxic ones (i.e. dicoumarol in mouldy *Melilotus* forage) (Le Bars and Le Bars, 1996).
Only the first two situations correspond to mycotoxin *sensu stricto*, which may be defined as zootoxic metabolites directly produced by micromycetes. They are structurally diverse group of molecular weight compounds mostly lower than 500 Da. In structural complexity, they vary from simple C4-compounds (e.g. moniliformin) to polycyclic ones (e.g. tremorgenic mycotoxins)(Steyn, 1980).

2.2. ORIGIN OF MYCOTOXINS

Mycotoxins are principally characterized as secondary metabolites.

Table 1. Mycotoxins as representative of the Main Biosynthetic Categories of secondary metabolites (Steyn, 1998).

Representative mycotoxins	Structure	Biosynthetic Category
Moniliformin	Di-	Polyketides
Patulin, penicillic acid	Tetra-	Polyketides
Citrinin, diplosporin, ochratoxins	Penta-	Polyketides
Maltoryzine	Hexa-	Polyketides
Rugulosin, viomellein, viriditoxin, xanthomegnin	Hepta	Polyketides
Ergochromes, luteoskirin	Octa-	Polyketides
Citreoviridin, cytochalasin, fumonisins, zearalenone	Nona-	Polyketides
Aflatoxins, austocystins, erythroskirine	Deca-	Polyketides
Cyclopiazonic acid, tenuazonic acid		Tetramic acids
Aspergillic acid, echinulins	Simple	Diketopiperazines
Brevianamides, fumitremorgens, oxaline, roquefortine, sporidesmins, verruculotoxin	Modified	Diketopiperazines
Ergotamine, phomopsins , tryptoquivaline		Peptides
Viridicatumtoxin	Mono-	Terpenes
Trichothecenes	Sesqui-	Terpenes
Aflatrem, penitrems	Di-	Terpenes

The secondary metabolism comprises biochemical pathways whose terminal synthesis products have no pertinent role for the biology of the organism.
In contrast to the primary metabolism, which is fundamentally similar among different living organisms, the secondary metabolism qualitatively depends on the fungal species and quantitatively is often characteristic of each individual strain. In comparison to other organisms, secondary metabolism is very developed in micromycetes and leads to a vast array of molecules (Turner 1971,Turner and Alridge 1983), some of which are toxic to

other organisms: bacteria (antibiotics), animals (mycotoxins). Mycotoxins are elaborated through different metabolic pathways (Steyn, 1980). Polyketides and terpenes are the two largest groups (Table 1).
Secondary metabolites often are synthesized as families of chemically related products. The type and quantity of molecules produced largely depends up on the characteristics of the individual strain as well as environmental conditions (nutrients, physico-chemical parameters, biochemical and biological status of the substrate). The term "toxigenic fungal species" generally means a fungal species for which one or more toxic metabolites can be produced by certain strains (at least under certain conditions) (Le Bars, 1988). Main mycotoxins and their major fungal sources are presented in the Table 2.

2.3. FACTORS INFLUENCING TOXIGENESIS

Conditions permissive to toxigenesis are more limited than those permitting fungal growth. Among the factors leading to the presence of mycotoxins in foods, there are two main categories; intrinsic factors, which are dependent on the fungal species or strain and extrinsic factors which depend on environmental conditions (Le Bars, 1988; Le Bars and Le Bars, 1998).

Table 2. Main mycotoxins (naturally occurring according to substrate, climatic and technological conditions) and their major fungal sources.

Main mycotoxins	Major toxigenic species of fungi
Aflatoxins	*Aspergillus flavus, A. parasiticus*
Cyclopiazonic acid	*A. flavus, Penicillium cyclopium*
Ochratoxin A	*A. ochraceus, P. verrucosum (viridicatum)*
Patulin	*A. clavatus, P. expansum, P. granulatum, Byssochlamys nivea, Paecilomyces varioti*
Sterigmatocystin	*A. versicolor, A. nidulans*
Citrinin	*P. citrinum, P. verrucosum, Monascus purpureus*
Fumonisins, moniliformine, fusarenon X	*Fusarium moniliforme*
Zearalenone	*F. graminearum, F. culmorum*
Trichothecens:	
T2 toxin,	*F. sporotrichoïdes, F. poae*
diacetoxyscirpenol,....	F. culmorum, F. graminearum
Deoxynivalenol, ...	
Sporidesmins	*Pithomyces chartarum*
Phomopsins	*Phomopsis leptostromiformis*
Ergot alkaloïds	*Claviceps, Neotyphodium coenophialum* *

* Endophyte in *Festuca arundinacea.*

2.3.1 Intrinsic factors:

Mycotoxin is rarely produced by only one species or closely related ones, as it is the case for aflatoxins by *Aspergillus flavus* or *A. parasiticus,* fumonisins by *Fusarium moniliforme* (and related species), sporidesmins by *Pithomyces chartarum* (Table 2). More frequently, the same toxin can be synthesized by a diversity of fungal species sometimes belonging to different genera, for example, cyclopiazonic acid by *A. flavus* and *P. camemberti*, ochratoxin A (OTA), patulin and penicillic acid by strains of different species of *Aspergillus* and *Penicillium*. Moreover, one fungal species and strain can produce more or less simultaneously several mycotoxins: OTA and citrinin by *P. verrucosum*, OTA and penicillic acid by *A. ochraceus*, aflatoxins and cyclopiazonic acid by *A. flavus*, PR toxin, roquefortine, mycophenolic acid and patulin by *P. roquefortii*.

In contrast, within a particular species reputed to be toxigenic, every strain does not possess this property; the frequency of toxigenic strains being dependent on the species considered and in some cases the region or substrate from which the strain was isolated. Of greater concern than this frequency is the difference in toxigenic potential within a particular species: among isolates from the same species toxin production can vary by as much as 1000-fold.

Studies on mould-mycotoxin pair on the basis of freshly isolated strains tested in optimal conditions for each toxigenesis showed four schemes of toxigenic potential distribution (TPD): almost 100% toxigenic strains- TPD was close to a normal distribution, inasmuch potential was expressed by the logarithm of the maximal observed concentration of toxin: sterigmatocystin / *A. versicolor*, penicillic acid / *P. verrucosum cyclopium*, cyclopiazonic acid / *P. camemberti*, fumonisin B1 / *F. moniliforme*; a noticeable proportion of non-toxigenic strains (20 to 75%)- the other ones presenting a TPD similar as the preceding one: aflatoxins and cyclopiazonic acid / *A. flavus*, OTA / *A. ochraceus*, deoxynivalenol and zearalenone / *F. graminearum*;
two clear-cut groups of strains, non toxigenic and highly toxigenic ones- satratoxins / *Stachybotrys atra*, sporidesmins / *Pithomyces chartarum*; very rare producing strains- fusarochromanone (TDP1) / *F. roseum* (only one strain out of 200).

Specific biological properties of the fungus may determine the moment and the nature of substrate contamination. Certain fungi considered as pathogens develop within the living plant sometimes symptomless and can lead to mycotoxic contamination of end product, i.e. *Fusarium* sp. in cereals and their mycotoxins. Other species are occasionally pathogenic on specific substrate, such as *P. expansum* on apples conducive to patulin contamination. Other ones are endophytes, strictly associated to a vegetal species, non-pathogenic but producing mycotoxins: *Neotyphodium coenophialum* in *Festuca arundinacea*. Other ones present an opportunist weak pathogenicity, i.e. *A. flavus* penetrating the flower of corn, or invading peanut pods and grains in the field. Lastly, most of the known mycotoxigenic species are only saprophytes. When investigations are undertaken on a toxigenic species or on a mycotoxic contamination, it is important to know (and sometimes to verify) what is such biological situation. Thus, if the mycotoxigenic species is transmitted by the seed (i.e. *F. moniliforme*), the first step to limit fumonisin contamination in corn is to use upstream *Fusarium*-free seeds.

2.3.2. Extrinsic factors:

a) Water availability: Water activity (a_W) influences toxin production considerably, particularly in low hydrated products. Toxigenesis takes place only for a_w slightly higher (i.e. + 0.02) than a_w limit for growth, which is particular for each fungal species (0.80 for A. flavus, 0.76-0.78 for A. ochraceus, 0.85 for numerous Penicillia, 0.90 for most of Fusaria), and then appears proportional to water availability. For higher a_w (i.e. + 0.05), toxin production rate usually increases exponentially with a_w in so far as no other limiting factor interferes (oxygen availability, stability of the toxin, other competitive micro-organisms).

b) Temperature: The optimal temperature for toxigenesis, defined as the temperature at which the rate of toxin production is maximal, is generally slightly lower than the optimal temperature for growth when the fungus is not in competition with other organisms. The example of *F. moniliforme* and fumonisin B1 is presented in Figure 1.

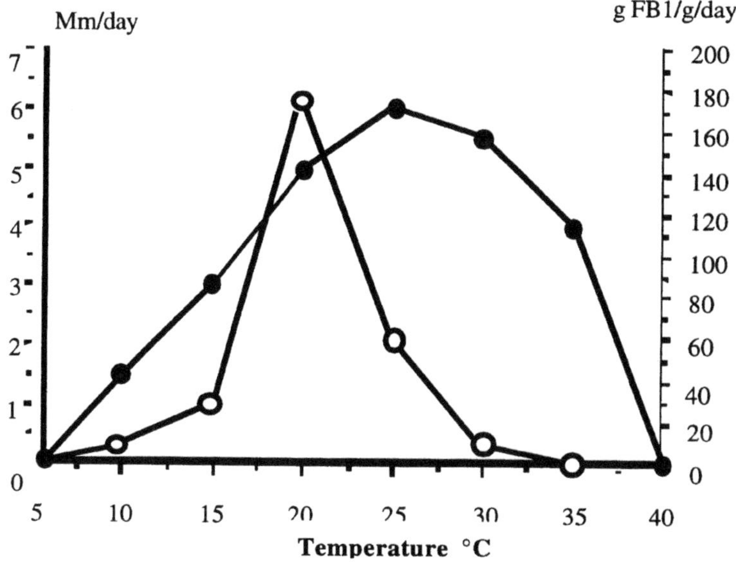

Figure 1. Comparison of growth rate and fumonisin production rate according to temperature for *Fusarium moniliforme* (Le Bars and Le Bars, 1995).

In addition, the temperature span permissive to significant toxigenesis is narrower than that permitting fungal growth. As an example, the array of favourable and limit hydrothermic conditions for ochratoxin A production is presented in Figure 2.

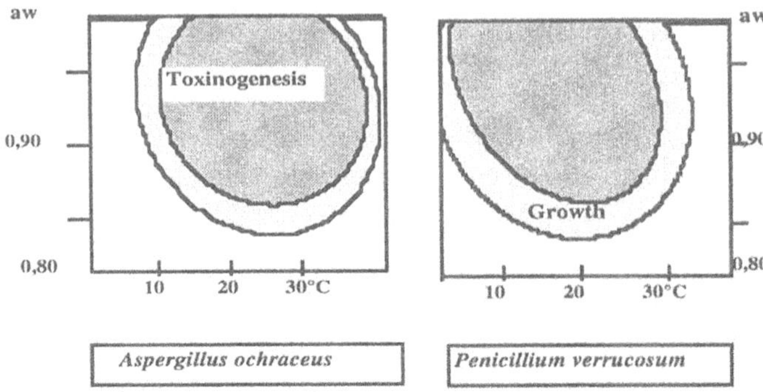

Figure 2. Favorable and limit hydrothermic conditions for growth and ochratoxin A production by *A. ochraceus* and *P. verrucosum* (from Northolt *et al*, 1979).

An analogue general scheme is observed for stable toxins (most of cited ones in Table 2). For other ones, temperature can also affect stability and therefore toxin accumulation in natural substrates. For example, penicillic acid is produced by numerous species of *Penicillium*, notably *P. verrucossum* var. *cyclopium*, at temperatures ranging from 4-30°C, with the maximal rate of production occurring at about 25°C. However, as the temperature increases the rate at which penicillic acid is coupled to -SH radicals also increases, resulting in actual detoxification. As a result, this toxin usually accumulates at relatively low temperatures at which detoxification is more reduced than toxin production (Le Bars, 1982). The situation is similar for other mycotoxins such as patulin, citrinin and PR toxin, which are unstable in many artificial and natural substrates. Thus, for these toxins, the hydro-thermic span for toxin production is much more restricted than for growth.

c) *Hydro-thermic conditions in the field*: In a general way, a stressed plant is more susceptible to opportunist invasion by fungi and consecutive mycotoxinogenesis. Thus, pre-harvest contamination of peanuts by aflatoxins is a complex situation. Invasion of peanuts by *A. flavus* is related to drought stress and soil temperature in the geocarposphere during the last 50 days before harvest (Sanders *et al.*, 1981; Cole *et al.*, 1982; Hill *et al.*, 1983). Neither elevated temperature alone nor drought stress alone cause aflatoxin contamination in sound mature kernels (SMK). The mean threshold geocarposphere temperature required for toxin development during the latter part of the peanut growth cycle is 25-27°C. Aflatoxin was absent from irrigated plots, from drought stressed plots with cooled soil, and from irrigated plots with heated soil, and was present in drought stressed SMK (Hill *et al.*, 1983). Moreover, drought conditions favour insect's proliferation, which increases fungal and subsequent toxin contamination (see below). This is an example of the multifactorial situation leading to mycotoxic contamination.

d) *Gaseous composition:* The reduction of partial pressure of oxygen and above all the rise of carbon dioxide level have a depressive effect that is much greater on toxigenesis than on growth. Aflatoxin production is moderately reduced between 21 and 5% oxygen, but is practically inhibited only when the oxygen concentration is below 1%. An increase in CO_2 level provokes a large drop in aflatoxin formation (Diener and Davis, 1977). In atmosphere enriched with CO_2, ochratoxin A production is reduced when the O_2 concentration is below 20% (Paster *et al.*, 1983). A slightly confined air drastically reduces Fumonisin biogenesis. On the contrary, *Byssochlamys nivea* can produce patulin even in anaerobic conditions (Escoula and Le Bars, 1975). After airtight storage, in which moulds can more or less develop, the introduction of fresh air or ventilation rapidly provokes intense toxigenesis.

e) *Nature of the substrate*: Toxigenesis, at least for most of the known mycotoxins, is much more dependent on the chemical composition of the substrate than fungal growth. In particular, an elevated level of carbohydrates and/or lipids is favourable for toxigenesis. For each fungus-mycotoxin pair, we have searched for the optimal culture medium for toxin production with the aim of purification and crystallization on the gram level. By varying (factorial plans) the culture composition, in particular the percentages of saccharose and yeast extract, it has been found that for a dozen different fungus-toxin pairs studied the most favourable conditions for toxin production generally comprised 16% saccharose for 1-2% yeast extract. Similar results are seen for *in vitro* toxigenesis on a natural substrate: cereals and oleaginous seeds are generally more favourable for toxigenesis than are high-protein substrates. In addition, it is important to consider the substrate/temperature interaction, which is involved in the stability of certain mycotoxins.

f) *Biological factors:* Insects and mites are vectors for spores of moulds, which they introduce into the grains or the fruits by the lesions they generate. The contamination of peanut (Aucamp, 1969), cotton (Ashworth *et al.*, 1971), maize (Widstrom *et al.*, 1976) by *A. flavus* or aflatoxins before harvest is often linked to attack by insects. During storage, the deterring effect of weevils on aflatoxin contamination of corn was clearly demonstrated by Beti *et al.* (1995). More recently, the role of coffee berry borer (*Hypothenemus hampei*) as vector of *A. ochraceus* was observed by Vega and Mercadier (1998) in Uganda and Benin. Among other biological factors, there are specific relations between plant and fungus and evolution of fungal population during storage from xerophilic species to more hydrophilic ones.

3. General data on mycotoxic situation in coffee

The main known mycotoxic contamination in coffee is ochratoxin A (OTA). Let us first remind some data on risk assessment, consider toxin origin and its fate during processing, and then examine the critical points for prevention.

3.1. RISK ASSESSMENT

3.1.1. Natural occurrence:

Levi *et al* 1974 first reported OTA in green coffee. Then it was generally accepted that OTA was found in commercial roast coffee and in coffee brews (Tsubouchi *et al.*, 1988). More recent surveys on 633 samples from different European countries (Van der Stegen *et al.*, 1997) and in UK retail coffees (Patel *et al.*, 1997) allowed evaluation of distribution and mean level of natural contamination in the final product.

3.1.2. Toxicological data:

Animal studies have shown OTA to be a potent nephrotoxin, immune suppressant, teratogen and foetotoxic agent (IARC, 1993). The World Health Organisation / Food and Agricultural Organization Joint Expert Committee on Food Additives (JECFA) has considered the toxicity of OTA and has set a Provisional Tolerable Weekly Intake (PTWI) for OTA of 100 $ng.kg^{-1}$ body weight (WHO, 1996). On the other hand, on the basis of recent surveys (Van der Stegen *et al.*, 1997; Patel *et al.*, 1997), the mean intakes of OTA (from consuming four cups of coffee daily) would be about 2.4 $ng.kg^{-1}$ body $weight.week^{-1}$, that is to say about 2% of the PTWI proposed by JECFA for OTA (Walker, 1997). Despite this reassuring element of risk assessment, evidently an effect has to be made to reduce mycotoxin contamination level.

3. 2. TOXIN ORIGIN AND ITS FATE DURING PROCESSING

3.2.1. Biological origin:

There are two major OA producing moulds, *Aspergillus ochraceus* and *Penicillium viridicatum* (the group of chemotype producing OA was subsequently shown to fit the description of *P. verrucosum*; Frisvad, 1984; Pitt, 1987; Moss, 1996). They differ somewhat in their hydrothermic requirements (Northolt *et al.*, 1979; Ramos *et al.*, 1998). The minimal a_w for growth of *A. ochraceus* is 0.76 and 0.81 *for P. verrucossum*, but the minimum for OA production is 0.85 for both organisms. The temperature range is 10-37°C (optimum 25-30°C) for *A. ochraceus* and 4-30°C for *P.* verrucosum (Figure 2). Such respective ecological characteristics explains, at least partially, the fact that the former one is associated with OA contamination in warm countries, and the second one in cooler ones.

3.2.1. Fate during processing and brewing:

Inconsistent results have been published regarding the influence of roasting on the OTA content in roasted beans and the transfer into the coffee brew. This would be due to the high inhomogeneity of the OTA contamination within one sample (Studer-Rohr *et al.*,

1995; Blanc *et al.*, 1998), a general problem encountered with mycotoxins in grains and fruits. Studer-Rohr *et al* (1995) observed that in naturally contaminated samples no statistically significant difference between OTA content of the green and roasted beans was found. However, in artificially inoculated samples, much more homogeneously contaminated, loss of OTA was estimated from 2-29%. More recently, after analysis of OTA distribution in a naturally contaminated green coffee lot, Blanc *et al.* (1998) obtained a slight elimination of toxin during green coffee cleaning. But the roast and ground coffee contained only 16% of OTA originally present. This reduction was mostly due to thermal degradation (less than 20% accounted by the chaff). Such a result obtained with a roasting cycle of 14 min and a final temperature 223°C, was consistent with the thermostability data of OTA in a dry substrate; half-life times of 12 and 6 min, respectively at 200 and 250°C (Boudra *et al.*, 1995).
In the brewing methods frequently used, 70% to almost complete extraction of OTA occurs (Studer-Rohr *et al.*, 1995, Viani, 1996;Van der Stegen *et al.*, 1997; Blanc et al 1998). Taking into accounts this partial stability (and transfer to the brew) during processing, decrease of contamination frequency and level in the end product is based on prevention of ochratoxigenesis upstream.

3.3. MOST PROBABLE CRITICAL POINTS

Taking in account the preceding data, let us examine the three major possible periods for fungal infection and toxigenesis, in the orchard, during harvesting-processing-drying, and storage.

3.3.1. The orchard:

The ochratoxigenic fungal species are neither phytopathogens nor endophytes, hence direct internal infection of the fruit from the tree is not possible. They are only opportunist saprophytes. Thus, general prevention all along the chain would be logically based on,

a) avoiding fungal spores sources,
b) limiting open entries into the fruit, and
c) reducing, after ripeness, water availability as soon as possible.

In an orchard, clearing away decaying organic material, particularly fallen damaged fruits contributes to reduce fungal spores sources. On tree, infection occurs due to inoculated spores by insects or airborne spores penetrating into the pulp or seed through skin lesions such drought lacerations, insect damages or splitting berries consecutively to rain during fruit ripening stage, similarly as for *A. flavus* infection in figs (Boudra *et al.*, 1994). Partial studies on sound, over ripe, sun scorched and split fruits on the plant indicated that *Aspergillus* and *Penicillium* populations were comparatively high in split fruits (Naidu *et al.*, 1997). Unfortunately, complete identification of species or mycotoxin analysis was not reported. Moreover, in Uganda and Benin, 5-17% of adults coffee berry borer emerging from the beans was infected by *A. ochraceus* (Vega and

Mercadier, 1998). These authors consider that its potential to serve as a vector for this cosmopolitan fungus is high.
It would be important to examine when and how the fungus colonizes the pulp and beans by mycological and histological techniques (passive or active penetration?), and the delay and localization of toxigenesis (in beans or diffusion of the toxin from the pulp?), because we have found different mechanisms in each commodity (figs, walnuts, grains) leading to practical different means for prevention.

3.3.2. Harvesting-Processing-Drying:
As a result of more than thirty years studies on mechanisms of mycotoxic contamination of different grains, seeds, fruits, forages and their prevention, we consider that this "peri-harvest period" (Le Bars, 1990; Le Bars *et al.*, 1994) comprises frequently the most critical steps or points. Maturity stage at harvest, particularly for fruits, determines greatly the risk of physical lesions and consecutive fungal contamination (Boudra *et al.*, 1994). The best compromise results from a close co-operation with the specialists of the commodity. Freshly fallen berries and insect damaged and split fruits have to be collected separately and dried without delay. May be, as for other commodities, mats under the trees would reduce fungal infection of the fruits from the soil. For most of grains and fruits, the intermediary post-harvest period before drying (or other stabilization procedures) has to be as short as possible. Sometimes, when drying process is delayed or transportation is long, volatile fungistatics may be used to prevent fungal development (easily detected by self-heating). The effects of main parameters, along wet or dry processing of coffee on moulds and mycotoxins contamination have to be studied. Such research is difficult because many parameters, variable in time and space, interfere. Usually fruitful investigations comprise two approaches: an analytical one on models, and a synthetic one on pilot scale. They require a very close collaboration between complementary partners (mycologists, technologists, analysts). At each step, it is beneficial to sort out foreign material and damaged berries or grains. Drying should be done as soon as possible to minimize the three main parameters (duration, free water, temperature) effects. For certain commodities, when drying equipment is temporarily insufficient, it is sometimes performed in two steps, a first one to decrease without delay moisture content to about 20%, followed by a slightly delayed one to obtain a water content allowing safe storage.

3.3.3. Storage:
During storage, a commodity is somewhat in equilibrium with inter-particular water vapour. So the notion of Equilibrium Relative Humidity (ERH %) or water activity (a_w) may be used as a determinant parameter for fungal development initiation. The a_w limit for initial growth of *A. ochraceus* is 0.76, which corresponds to 14.2% moisture content (MC) on wet weight basis (w.b.) for green coffee beans (Fig. 3). They are 0.81 a_w and 15.5% MC (w.b.) for P. *verrucosum*. But, for several months storage to avoid any fungal evolution (starting by xerophilous species as *A. glaucus* group conducing

with time to more hydrophilous ones including toxigenic species), it is recommended to store grains at a maximal 0.65 a_w, corresponding to 11.8 % MC (w.b.) for green coffee (Fig.3).

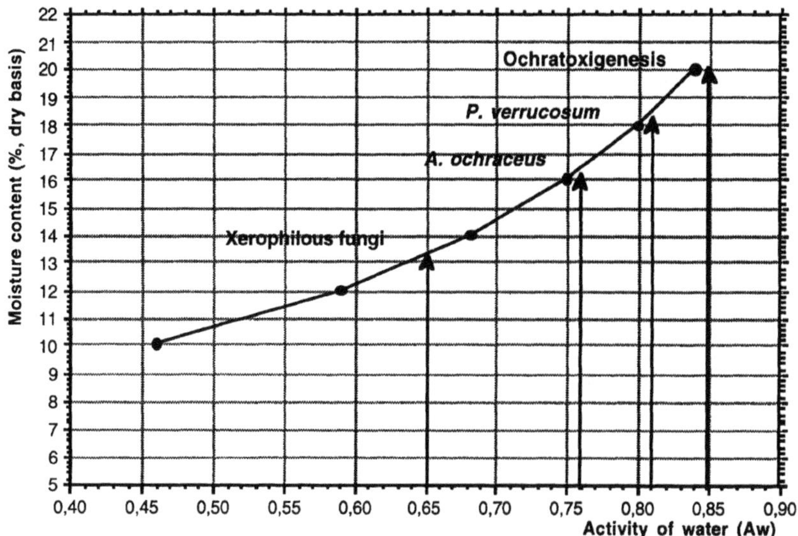

Figure 3. Sorption curve of green coffee (robusta and arabica, 30°C, from Bucheli *et al.*, 1998), and minimal a_w and moisture content for fungal growth and toxigenesis

During experimentation on long term storage (eight months) of green coffee under tropical climate and in different industrial conditions, the average data on MC and a_w fit the predictive model relatively well (Bucheli *et al.*, 1998). The storage conditions were 13.5% MC (dry basis, ca. 11.8% MC w.b.) and 71% ERH. In no case, increase in OTA amount was observed (average OTA of 3.9 $\mu g.kg^{-1}$ at the beginning of storage). Microbiological evaluation and OTA data during the different industrial storage conditions favoured the idea that OTA contamination, as frequently for numerous other mycotoxins in grains and fruits, occurred in the course of steps before storage.

For long term storage of a lot, homogeneity in the physico-chemical characteristics of the commodity and in the hydro-thermic conditions are highly recommended. When storage of a sufficiently dry product goes wrong, it is typically due to heterogeneous conditions. Localized differences in temperature in a lot (i.e. because sunny side) lead to vapour migration and initiation of fungal development (for example, for a 10°C difference between silo sides, the safe storage period of compound feeds for animals is divided by two).

4. Summary and conclusions

Mycotoxins are toxic secondary metabolites produced by fungi. These toxins are low molecular weight compounds derived from polyketo-acids, terpenes or amino acids. A toxin can be synthesized by a diversity of fungal species. Depending on the biology of the fungus and ecological parameters (substrate, hydro-thermic conditions), toxigenesis can occur in the field (i.e. alkaloids from endophytic or parasitic fungi), or during pre-harvest period (i.e. fumonisins from *F. moniliforme*), or in storage (i.e. aflatoxins from *A. flavus*). Moreover, the contaminating toxin can be stable (i.e. aflatoxins, ochratoxin A), or unstable in certain substrate and/or with thermal treatment (i.e. patulin, citrinin). Several parameters have to be considered in the mechanism of a mycotoxic contamination of a commodity. Ochratoxin A is the main reported mycotoxin in green, roasted and soluble coffee.
There is a lot of possible mycotoxic contamination in alimentary products. In each case, biological, technological and environmental conditions are different. So, a reasoned prevention on a specific problem is based on understanding the mechanisms of contamination and on determination of the main parameters. In the particular case of ochratoxin in green coffee, main critical steps are upstream, harvesting and sorting, separation of tissues and drying, storage and transport. Mycological studies all along these steps are necessary. Among fungal contaminants of this commodity, is *A. ochraceus* the alone ochratoxigenic species? Where, when and how does it develop and produce ochratoxin A? Any way, decrease of fungal development would lead not only to improve hygienic quality of coffee but also to maintain organoleptic features, the main value of this commodity.

5. References

Ashworth L.J., Rice R.E., Mc Means J.L., Brown C.M. (1971), The relationship of insects to infection of cotton balls by *Aspergillus flavus*. Phytopathology, 61: 488-493

Aucamp J.L. 1969. The role of mite vectors in the development of aflatoxin in groundnuts. *J. Stored Prod. Res.*, 5: 245-249

Beti J.A., Phillips T.W., Smalley E.B. (1995), Effects of maize weevils on production of aflatoxin B1 by *Aspergillus flavus* in stored corn. *J. Econ. Entomol.*, 88: 1776-1782

Blanc M., Pittet A., Munoz-box Viani R. (1998), Behaviour of ochratoxin A during green coffee roasting and soluble coffee manufacture. *J. Agric. Food Chem.*, 46: 673-675

Boudra H., Le Bars J., Le Bars P., Dupuy J. (1994), Time of *Aspergillus flavus* infection and aflatoxin formation in ropening of figs. *Mycopathologia*, 127: 29-33

Boudra H., Le Bars P., Le Bars J. (1995), Thermostability of ochratoxin A in wheat under two moisture conditions. Appl. Environ. Microbiol., 61: 1156-1158

Bucheli P., Meyer I., Pittet A., Vuataz G., Viani R. (1998), Industrial storage of green robusta coffee under tropical conditions and its impact on raw material quality and ochratoxin A content. *J. Agric. Food Chem.*, 46: 4507-4511

Cole R.J., Hill R.A., Blankenship P.D., Sanders T.H., Garren K.H. (1982), Influence of irrigation and drought stress on invasion by *Aspergillus flavus* of corn kermels and peanut pods. *Dev. Ind. Microbiol.*, 23: 229-236.

Diener U.L., Davis N.D. (1977), Aflatoxin formation in peanuts by *Aspergillus flavus*. *Agric. Exp. Sta.*, Auburn Univ., Bull. n° 493

Escoula L., Le Bars J. (1975), Production de patuline par *Byssochlamys nivea* sur un modèle d'ensilage. *Ann. Rech. Vet.*, 6: 219-226

Frisvad J.C. (1984), Expression of secondary metabolism as fundamental characters in *Penicillium* taxonomy. In Toxigenic Fungi, ed. H. Kurata and Y. Ueno, Elsevier, Amsterdam, 98-106

Hill R.A., Blankenship P.D.,. Cole R.J., Sanders T.H. (1983), The effects of soil moisture and temperature on pre-harvest invasion of peanuts by *Aspergillus flavus* and subsequent aflatoxin development. *Appl. Environ. Microbiol.*, 45: 628-633

I.A.R.C, 1993. Ochratoxin A. Monograph on the evaluation of carcinogenic risks to humans (Lyon: Internat. Agency Research Cancer), 489-521.

Le Bars J. (1982), Facteurs de l'accumulation d'acide pénicillique dans les denrées d'origine végétale. *Sci. Aliments*, 2: 29-33.

Le Bars J. (1988),Toxigenesis as a function of the ecological conditions of the grain/micro-organisms system. *In* "Preservation and storage of grains, seeds and their by-products", J.L.MULTON, Lavoisier pub. New York, Paris, 347-366.

Le Bars J. (1990), Contribution to a practical strategy for preventing aflatoxin contamination of dried figs. *Microbiologie- Aliment -Nutrition*, 8: 265-270.

Le Bars J., Le Bars P., Dupuy J., Boudra H. (1994), Biotic and abiotic factors in Fumonisin B1 production and stability. *J.A.O.A.C Internat.* 77: 517-521

Le Bars P., Le Bars J. (1995), Ecotoxinogénèse du *Fusarium moniliforme*: Enveloppe des risques des Fumonisines. *Crytogamie- Mycologie*, 16: 59-64.

Le Bars J., Le Bars P. (1996), Recent acute and subacute mycotoxicoses recognized in France. *Vet. Res.*, 27: 383-394

Le Bars J., Le Bars P. (1998), strategy for safe use of fungi and fungal derivatives in food processing in Mycotoxins in Food chain", J. Le Bars and P. Galtier (eds), Revue Méd. Vétér., 149, 493-500

Levi C., Trenk H.L. Mohr H.K. (1974), Study of the occurrence of ochratoxin A in green coffee beans. *J.A.O.A.C.*, 57: 866-870

Moss M.O. (1996), Mode of formation of ochratoxin A. *Food Add. Cont.*, 13, (suppl), 5-9.

Naidu R., Daivasikamani S. Nagaraj J.S. (1997), Current status and strategies for management of mycotoxins in India coffees. In 17th Coll. Scientific. sur le café, ASIC (ed), Nairobi. 20-25 /07/1997, p. 69-78.

Northolt M.D., Van Egmond H.P., Paulsch W.E. (1979), Ochratoxin A production by some fungal species in relation to water activity and temperature. *J. Food Protection* , 42:485-490.

Paster N., Lisker N., Chei I. (1983), Ochratoxin A production by *Aspergillus ochraceus* grown under controlled atmospheres. *Appl. Environ. Microbiol.*,45: 1136-1139

Patel S., Hazel C.M., Winterton A.G.M., Gleadle A.E. (1997), Survey of ochratoxin A in UK retail coffees. *Food Add. Cont.*, 14: 217-222

Pitt J.I. (1987), *Penicillium viridicatum, P. verrucosum,* and the production of ochratoxin A. *Appl. Environ. Microbiol.*, 53: 266-269

Ramos A.J., Labernia N., Marin S., Sanchis V., Magan N. (1998), Effect of water activity and temperature on growth and ochratoxin production by three strains of *Aspergillus ochraceus* on a barley extract medium and on barley grains. *Int. J. Food Microbiol.*, 44: 133-140

Sanders T.H., Hill R.A., Cole R.J., Blankenship P.D. (1981), Effect of drought on occurrence of *Aspergillus flavus* in maturing peanuts. *J. Am. Oil. Chem. Soc.*, 58: 9661-9670

Steyn P.S. (1980), The biosynthesis of mycotoxins , 432 pages, Acad. Press.

Steyn P.S. (1998), The biosynthesis of mycotoxins. *Rev. Med. Vet.*, 149: 469-478.

Studer-Rohr J., Dietrich D.R., Schlatter J., Schlatter C. (1995), The occurrence of ochratoxin A in coffee. Fd. Chem. Toxicol., 33: 341-355

Tsubouchi H., Terada H., Yamamoto K., Hisada K., Salabe Y. (1988), Ochratoxin A found in commercial roast coffee. *J. Agric. Food Chem.*, 36: 540-542

Turner W.B. (1971), Fungal metabolites, 426 pages, Acad. Press, London.

Turner W.B., Alridge D.C. (1983), Fungal metabolites II, 631 pages, Acad. Press, London.

Van der Stegen G., Jorissen U., Pittet A., Saccon M., Steiner W., Vincenzi M., Vinkler M., Zapp J., Schlatter C., (1997), Screening of European coffee final products for occurrence of ochratoxin A. *Food Add. Cont.*, 14: 211-216

Vega F.E., Mercadier G. (1998), Insects, coffee and ochratoxin A. *Florida Entomologist*, 81: 543-544

Viani R., (1996), Fate of ochratoxin A during processing of coffee. *Food Add. Cont.* 13, (suppl): 29-33

Walker R., (1997), Quality and safety of coffee. 17th Coll. Scientific Internat. sur le café. ASIC (Nairobi 1997), 51-60.

Widstrom N.W., Lillehoj E.B., Sparks A.N., Kwolek W.F. (1976) Corn earworm damage and aflatoxin on corn ears protected with insecticide. *J. Econ. Entomol.*, 69: 677-679

World Health Organisation, (1996) Evaluation of certain food additives and contaminants. 44th Report of JECFA. WHO Technical Report series N° 859 (Geneva).

Chapter 34

SPECIES RELATED DIFFERENCES IN BRAZILIAN GREEN COFFEE CONTAMINATED BY OCHRATOXIN A

L.V. Soares[1], R.P.Z. Furlani[1], P.L.C. Oliveira[2]

[1]Universidade Estadual de Campinas, Faculdade de Engenharia de Alimentos. CP 6121, 13081-970, Campinas, São Paulo, Brazil; 2 Oliveira e Veiga Consultores Associados, 860202-450 Londrina - PR, Brazil

Running Title: Ochratoxin A in coffee

1. Introduction

Ochratoxin A is a toxic metabolite produced mainly by strains of *Penicillium verrucosum* and *Aspergillus ochraceus. P. verrucosum* seems to predominate in temperate climates and *A. ochraceus* in warm ones (Moss, 1996). Ochratoxin A is acutely toxic to birds, mammals and fish and the kidney is its main target organ followed by the liver. Its teratogenic and immunotoxic action have been described by Busby and Wogan (1981) and it has been shown to be a potent carcinogenic to rats (NTP, 1989). The International Agency for Research on Cancer (IARC) has classified ochratoxin A in the group 2B of substances that means it is possibly carcinogenic to humans (IARC, 1993). Ochratoxin A has been suggested as the causative agent in the endemic human nephropathy affecting villagers in the Balkans (Plestina, 1996). More recently nephropathies have been reported in Algeria and Tunisia, similarly to the case of the Balkans where the toxin was found in bodily fluids and in the food consumed by the population (Crepy *et al.* 1993). The half-life of ochratoxin A in humans was estimated as 35 days (Schlater *et al.*, 1996). In 1995, the FAO/WHO Joint Expert Committee on Food Additives evaluated ochratoxin A and established a Provisional

T. Sera et al. (eds.), Coffee Biotechnology and Quality, 369–376.

Tolerable Weekly Intake (PTWI) of 100 ng per kg of body weight (JEFCA, 1995). The toxin carcinogenic activity was not taken into consideration in this evaluation.
A wide variety of foods, including coffee have been reported contaminated by the toxin (Pohland, 1992). During the brewing of coffee, the toxin is extracted by the boiling water (Tsubouchi *et al.* 1987; Blanc *et al.* 1998). To date, there is still controversy about the losses in the toxin level during roasting with values between 0-80% reported by different authors (Levi *et al.* 1974; Cantàfora *et al.* 1983; Tsubouchi *et al.* 1987; Studer-Rohr *et al.* 1995; Blanc *et al.* 1998). However, risks assessments conducted in several European countries have shown that coffee drinking contributes to less than 10% of the PTWI for ochratoxin A (Stegen *et al.* 1997; Studer-Rohr *et al.* 1995; Jorgensen, 1998; Patel *et al.* 1997).
Brazil is a large coffee producer as well as consumer. A preliminary estimate of the Probable Weekly Intake (PWI) studies in Brazil indicated 6 ng/kg of body weight for the average adult (Soares, 1999). Such intake represents about 6% of the PTWI. Thus, the data for coffee as a source of ochratoxin A in the diet point to it so far not as a public health hazard but as a quality problem for coffee producers and manufacturers. As a result and also due to the fact that ochratoxin A is a natural contaminant the aim should be to keep its incidence and levels as low as technically possible. The production of mycotoxins in plant foods has been and still is extensively researched for in other commodities (Miller and Trenholm, 1994; Jackson *et al.* 1996). In case of ochratoxin A, the contamination in coffee became an issue only recently. The conditions for ochratoxin A production are not understood well yet and a great deal remains to be done on how to avoid or reduce its production in coffee beans.
Two species of the genus *Coffea* dominate international coffee trade: *C. arabica* and *C. canephora*. Surveys of ochratoxin A in green coffee and roasted coffee have been conducted in many countries out of concern with the public health aspects and also in order to allow for risk assessment studies. But among the published results, there is little information on the coffee species of the samples analyzed (Micco *et al.* 1989; Studer-Rohr *et al.* 1995; Jorgensen *et al.* 1997; Nakajima *et al.* 1997; Patel *et al.* 1997; Stegen *et al.* 1997; Trucksess *et al.* 1999). Such information might contribute to increase interest in more resistant varieties. Manufacturing practices may also benefit from such knowledge.

2. Experimental

2.1.SAMPLES

Samples were obtained from the dealers as lots of green coffee was offered for sale to prospective buyers in the cities of Londrina and Santos, Brazil. The samples were dried in an atmospheric oven at 50°C for 16 hours, ground to 20 – 40 mesh and kept in polypropylene flasks until analysis.

2.2. EXTRACTION AND CLEANUP

The samples were extracted and submitted to cleanup according to instructions from the immunoaffinity column suppliers (Ochraprep, Rhône-Diagnostics Technologies). Ten grams of the sample was extracted in a blender for two minutes with 200-ml of 1% $NaHCO_3$ solution. The extract was filtered with Whatman GFB glass filter. An aliquot of 20-ml of filtrate and 20-ml phosphate buffer were mixed and applied to an immunoaffinity column under atmospheric pressure or under enough vacuum to allow a flow of 2-3 $ml.min^{-1}$. The column was washed with 20-ml deionized water and subsequently dried by suction. Ochratoxin A was eluted with 1.5-ml methanol/acetic acid (98:2) followed by 1.5-ml deionized water. The elute was dried and dissolved in methanol/ 9% acetic acid (65:35).

2.3. HIGH PERFORMANCE LIQUID CHROMATOGRAPHY

The HPLC instrument used was comprised by the following modules: Waters pump, model 510; Rheodyne injector with 100 µL loop; Hewlett Packard fluorescence detector, model 1046A; Waters integrator, model 740; Spherisob ODS-2 column, 5 µm, 250 X 4.6 mm. The mobile phase was methanol/ 9% acetic acid (65:35) at 1.0-$ml.min^{-1}$. The excitation and the emission wavelengths were 330 nm and 470 nm, respectively.

2.4. ASSAY AND ANALYTICAL QUALITY CONTROL

Ochratoxin A standard (Sigma) solutions were calibrated according to AOAC (1997). Each batch of samples analyzed included a recovery test. Duplicates were analyzed on different days and results were corrected for the recovery of the batch. RSDs of duplicates were considered acceptable up to 30%.

3. Results and Discussion

The average recovery was 86% and the average RSD for duplicate samples was 13%. The limit of detection for ochratoxin A was 0.7 ng.g^{-1}. Figure 1 shows a chromatogram of a contaminated sample.

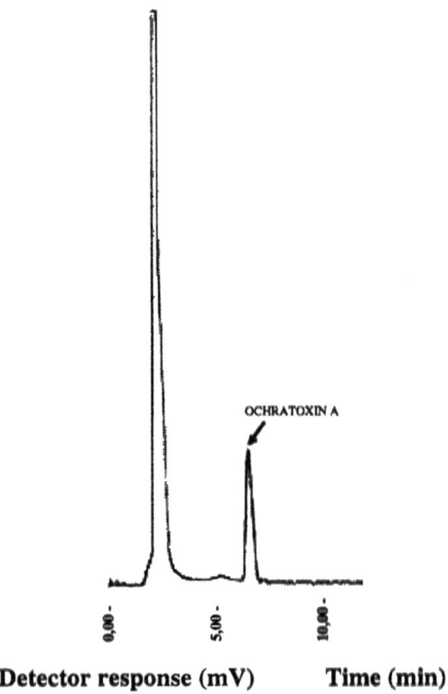

Figure 1. Chromatogram of a contaminated green coffee sample.

Seventy-nine samples of Brazilian green coffee from six producing States were analyzed for ochratoxin A (Table 1). Eleven out of the 70 samples of *C. arabica* analyzed were contaminated with ochratoxin A at levels between 1.7-12.7 ng.g^{-1}. Higher levels of ochratoxin A were found among the samples of *C. canephora* var. Conillon (Robusta). The very fact the Conillon variety commands lower prices in the market plays a part in the agricultural tract this type of coffee may receive in the farms during harvest, drying (fermentation) and storage. This fact certainly is to be expected to have a bearing in the microbial contamination levels of the commodity, and thus the mycotoxin levels as well.

Table 1. Levels of ochratoxin A in Brazilian green coffee from different coffee producing states.

Coffee producing state	Species	Number of samples*	Ochratoxin A (ng/g) in contaminated samples
Minas Gerais	*C.arabica*	3/24	1.7
			6.3
			12.7
Parana	*C.arabica*	7/37	5.7
			8.8
			4.9
			3.6
			10.4
			4.2
			7.3
			2.7
Sao Paulo	*C.arabica*	0/7	-
Espirito Santo	*C.canephora*	0/4	-
Rondonia	*C.canephora*	2/5	5.5
			114.2
Bahia	*C.arabica*	0/1	-

Results represent averages from duplicates.
* Contaminates samples/ Total number of samples analyzed

Two samples out of nine were contaminated with 5.5 and 114.2 ng.g^{-1} of ochratoxin A. It should be noted that modern agricultural methods are currently practised in the major Brazilian coffee producing States. In these areas, *C. arabica* predominates by far. However, more recently coffee was started as a crop in undeveloped areas of the country where agriculture is inching its way in little populated areas and subsistence still based in fishing and hunting. This was the case of the State of Rondonia where exceptionally high level of ochratoxin A (114.4 $ng.g^{-1}$) was found in one sample. This perhaps may find better explanation from harvesting practices, storage and transportation practices rather than in the coffee variety itself (*C. canephora* var. Conillon). This should be further investigated by surveys in areas where the Conillon variety is more extensively planted and a supportive structure for agriculture exists as is the case of the State of Espirito Santo.

4. Conclusions

Ochratoxin A was found in 11 out of 70 samples of *C. arabica* and in two out of nine samples of *C. canephora* var. Conillon (Robusta) in ranges from 1.7-12.17 $ng.g^{-1}$ and from 5.5-114.2 $ng.g^{-1}$ of ochratoxin A, respectively. Some samples of *C. canephora* originated from the areas with poor farming and storage practices, showing that care must be exercised when comparing the data obtained for the two species.

5. Futuristic approach

The behaviour of coffee cultivars toward ochratoxin A production in the field and during fermentation and storage should be further studied. Other biotic and abiotic aspects of ochratoxin A production should also receive attention as well as decontamination procedures in order to keep the contamination level as low as possible. The increase in knowledge in the subject may open the way for the elimination of the contaminant from the commodity, a situation right now not deemed as possible.

6. References

AOAC International (1997), Official Methods of Analysis, Gaithersburg

Blanc M., Pittet A., Munoz-Box R., Viani R. (1998), Behaviour of ochratoxin A during green coffee roasting and soluble coffee manufacture, J. Agric. Food Chem., 46: 673 – 675

Busby Jr.W.F., Wogan G.N. (1981) Ochratoxins. In - Mycotoxins and N-Nitroso Compounds: Environmental Risks, Shank, R. C. (ed.), Vol. II., CRC Press, Boca Raton, pp. 129 – 136

Cantàfora A., Gossi M., Miraglia L., Benelli L. (1983), Determination of ochratoxin A in coffee beans using reversed-phase high performance liquid chromatography, Riv. Soc. Ital. Sci. Alim., 12: 103 – 108

Creppy E., Castegnaro M., Dirheimer G. (1993), Human ochratoxicosis and its pathologies, John Libbey Eurotext, France

Jackson L.S., De Vries J.W., Bullerman L.B. (1996), Fumonisins in Food, Plenum Press, New York

Jefca (1995), Technical Report Series No. 859, Joint FAO/WHO Expert Committee on Food Additives, Geneva

Jorgensen K. (1998), Survey of pork, poultry, coffee, beer and pulses for ochratoxin A, Food Add. Contam., 15: 550-554

IARC (1993), Ochratoxin A, IARC Monographs on the evaluation of carcinogenic risks to humans: some naturally occurring substances, food items and constituents, heterocyclic amines and mycotoxins, International Agency for Research on Cancer, Geneva, pp. 489 - 521

Levi C., Trenk H.L., Mohr H.K. (1974), Study of the occurrence of ochratoxin A in green coffee beans, J. Assoc. Off. .Anal. Chem., 57: 866 – 871

Micco C., Grossi M., Miraglia M.., Brera C. (1989), A study of the contamination by ochratoxin A of green and roasted coffee beans, Food Add. Contam., 6: 333-339

Miller J.D., Trenholm H.L. (1994), Mycotoxins in Grains-compounds other than aflatoxin, Eagan Press, St Paul

Moss M.O. (1996), Mode of formation of ochratoxin A, Food Add. Contam. 13: 5-9

Nakajima M., Tsubouchi H., Miyabe M., Ueno Y. (1997), Survey of aflatoxin B1 and ochratoxin A in commercial green coffee beans by high-performance liquid chromatography linked with immunoaffinity chromatography, Food Agric. Immun., 9: 77 – 83

NTP (National Toxicology Program) (1989), NTP Technical report on the toxicology and carcinogenesis studies of ochratoxin A in F344/N rats, U.S. Department of Health and Human Services, National Institutes of Health, Research Triangle Park, pp. 1- 141

Patel S., Hazel C.M., Winterton A.G.M., Gleadle A.E. (1997), Survey of ochratoxin A in UK retail coffees, Food Add. Contam., 14: 217-222

Plestina R. (1996), Nephrotoxicity of ochratoxin A, Food Add. Contam., 13: 49-50

Pohland A.E. (1992), Ochratoxin A· A review, Pure App. Chem., 67: 1029 – 1046

Schlatter C., Studer-Rohr J., Rasonyi T. (1996), Carcinogenicity and kinetic aspects of ochratoxin A, Food Add. Contam., 13: 43- 44

Soares L.M.V. (1999), Ocratoxinas e aflatoxinas em café no Brasil, Anais III SIBAC, Londrina, in press.

Stegen G.V.D., Jörrissen U., Pittet A., Saccon M., Steiner W., Vicenzi M., Winkler M., Zapp J.A.N., Schlatter, C. (1997), Screening of European coffee final products for occurrence of ochratoxin A, Food Add. Contam., 14: 211 – 216

Studer-Rohr I., Dietrich D. R. Schlatter J., Schlater C. (1995), The occurrence of ochratoxin A in coffee, Fd. Chem. Toxic., 33: 341- 355

Trucksess M.W., Giler J., Young K., White K.D., Page S. (1999), Determination and survey of ochratoxin A in wheat, barley, and coffee – 1997, J. AOAC Int., 82: 85 – 89

Tsubouchi H., Yamamoto K., Hisada K., Sakabe Y., Udagawa S. (1987), Effect of roasting on ochratoxin A level in green coffee beans inoculated with Aspergillus ochraceus, Mycopathology, 97: 111 – 115

Chapter 35

DEVELOPMENT OF BIOPROCESSES FOR THE CONSERVATION, DETOXIFICATION AND VALUE-ADDITION OF COFFEE PULP AND COFFEE HUSK

S. Roussos[1], C. Augur[1], I. Perraud-Gaime[1], D.L. Pyle[2], G. Saucedo-Castañeda[3], C.R. Soccol[4], A. Pandey[4], I. Ferrao[5], M. Raimbault[1]

[1]*Laboratoire de Microbiologie IRD, IFR-BAIM, Université de Provence, ESIL, Case 925, 163 Avenue de Luminy, 13288 Marseille cedex 9, France;* [2]*Department of Food Science and Technology, The university of Reading, Whiteknights, PO Box 226, Reading RG6 6AP, UK;* [3]*Departamento de Biotecnologia, Universidad Autonoma Metropolitana-Iztapalapa, Apto 55 535, 09340 Mexico DF, Mexico;* [4]*Laboratorio de Processos Biotecnologicos, Departamento do Engenharia Quimica, Universidade Federal do Parana, CP 19011, CEP 81531-970, Curitiba-PR, Brazil;* [5]*DG XII/B4 -INCO-DC, Square de Meeus, 8 - Office 1/109, 1050 Brussels, Belgium*

Running title: Biopulca project

1. Introduction

A collaborative effort by research teams from France, the UK, Mexico and Brazil funded by the European Union INCO-DC programme (contract no IC18*CT970185) was initiated with the global objective of recycling the coffee pulp and coffee husk by biotechnological processes. Mexico alone produces more than 300,000 metric tons of

T. Sera et al. (eds.), Coffee Biotechnology and Quality, 377–392.

grain coffee each year using 'wet-processing' technology resulting coffee pulp as the waste residue. The Brazilian State of Parana produced 220,000 metric tons of grain coffee in 1998 using 'dry-processing' technology resulting coffee husk as the waste residue. Considering the great volume of this agro-industrial waste, its high biodegradability and its potential as feed source for animals, a proposal was set-up to stabilize fresh pulp by the lactic acid silage technique, as it is being produced, during the crop season. However, Recalcitrant and Toxic Compounds (RTC) such as caffeine, tannins and polyphenols need to be removed. RTC's actually represent a strong limitation on the use of coffee pulp/husk as a nutritive source for animal feeding purposes (Bressani, 1979). They also cause serious problems of environmental contamination (Violle *et al.*, 1995; Zambrano-Franco *et al.*, 1999). Ensiling of coffee pulp and its further utilization could avoid the rapid degradation of the material, which quickly becomes a very significant source of contamination of water in coffee growing areas (Zuluaga, 1989). The stabilized and detoxified coffee pulp/husk will be further utilized after the crop season as animal feed or for mushroom, fungal metabolites, or enzymes production (Perraud-Gaime, 1995; Ramirez-Martinez, 1999; Pandey and Soccol, 2000).

Recently, several mechanical improvements of coffee processing have been made in Colombia and Central America such as the use of 'dry' (no water added) depulpers and demucilagers and the re-design of water currents for minimal washing of fresh coffee beans (Alvarez-Gallo, 1991, 1995; Avallone, 1999). Thus, a new kind of coffee pulp is produced enriched with soluble compounds, mainly sugars. Those improvements have been shown to save a great deal of process water but they raise questions about the quality of the main product (beans) and the by-product (coffee pulp). The first question to answer involved local observations, mass balances and sampling of products and by-products in order to make a technical assessment of such improvements in terms of future market value of these coffee derivatives, together with an economical appraisal of the impact of water saving techniques. Special reference should be made to the cost of spent water treatment, as compared to the traditional system which involves great amounts of water for de-pulping, fermentation and washing operations (Urquiza, 1989; Viniegra-Gonzalez, 1993; Violle *et al.*, 1995; Zambrano-Franco *et al.*, 1999).

Coffee pulp produced by the new dry process is rich in soluble sugars, rots easily and is difficult to handle. Furthermore, it contains anti-physiological and anti-nutritional compounds such as caffeine, tannins and a large variety of phenolic compounds (Elias,

1978; Bressani, 1979). There is, therefore, a need for a new technology that can handle, detoxify and keep large amounts of coffee pulp using simple farm machinery at a minimal cost, but should also improve the nutritional quality and market value of the final product.

Recent work as described in following sections indicated that such an alternative could be an induced ensiling technique with a mixed inoculum of lactic bacteria and non pathogenic filamentous fungi followed by a two step fermentation process: a) aerobic step (solid-state fungal culture to detoxify and improve the nutritive value of the material), b) anaerobic step (silage fermentation to keep the material in good shape). This way, fungal organisms would be activated under forced aeration and silage could be kept for several months before animal consumption or before further industrial processing (sun or oven drying). Alternatively, pectinase and cellulase crude extracts could be obtained from the fermented mash, leaving a residual material with possible probiotic value for ruminant or fish consumption (Favela *et al.*, 1989; Tapia *et al.*, 1989; Roussos *et al.*, 1998). This new technical alternative for processing and storing coffee pulp ought to be tested at pilot scale in terms of mass balance, engineering feasibility and changes in biochemical and chemical composition of the product (Saucedo-Castañeda *et al.*, 1999).

Castillo *et al.* (1993) showed that the traditional wet process uses 6-23 m^3 of water per ton of fresh cherries, which corresponded to 30-114 m^3 of water per ton of green beans. This water is used mostly in de-pulping, fermentation, washing and transportation of cherries or coffee beans. The new 'dry' process has been shown to reduce water use to the level of only 2 m^3 of water per ton of cherries if transport of materials is mechanical and also if wash water is properly recycled (Vásquez-Morera, 1993; Alvarez-Gallo, 1995). Spent waters can be treated by anaerobic digestion with significant BOD reduction (Jacquet, 1993; Zambrano-Franco *et al.*, 1999), as previously shown by Violle *et al.* (1995) in Mexico. Apparently, the dry process does not affect the final quality of coffee beans (Alvarez-Gallo, 1995) but little published information is available on the fine chemistry of coffee beans and coffee pulp derived from this new process (Avallone, 1999).

Recent work done in the Biotechnology Laboratory of IRD, Montpellier (France) and UAM-I (Mexico) has shown that it could be possible to keep and improve the biochemical quality of coffee pulp by using a mixture of selected strains of lactic bacteria and filamentous fungi. Solid-state fermentation (SSF) of this material yielded a decaffeinated product which could be dried or re-ensiled (Roussos *et al.*, 1989; Perraud-Gaime and Roussos, 1997a). Some HPLC measurements suggested that a major fraction of phenolic compounds was broken down (Perraud-Gaime, 1995). On the other hand, work by Antier *et al.* (1993 a,b) has shown that coffee pulp could be an excellent substrate for pectinase production by selected strains of *Aspergillus niger* (Boccas *et al.*, 1994). The solid residue after such fermentation has been found to have a probiotic effect when assayed *in vitro* (Tapia *et al.*, 1988). This probiotic effect was apparently linked to a water soluble enhancement growth factor present in fungal biomass acting on rumen cellulolytic bacteria (Campos-Montiel and Viniegra-Gonzalez, 1995; Soto-Cruz *et al.*, 1999).

2. Benefits

As with many other tropical agricultural products, green coffee beans agro-industry also suffers large international price fluctuations of the selling product *i. e.* coffee beans. In particular, the period during 1978-92 has been catastrophic for coffee price, which sunk to its lowest levels in 1989. This commercial problem is being worsened by the progressive increase in production costs. There has also been an increase in the payments for the right to use water, associated with the cost of treating wastewater. The commercial crisis has not affected large companies, which are strong and diversified. However, small and medium enterprises, wich are weak and vulnerable, have disappeared (Viniegra-Gonzalez, 1989). The consequences of the crisis have been felt in all coffee sectors and particularly in the research sector. In the latter part of the crisis period (1987-92), the number of coffee research oriented workers decreased in Latin America. Several national research institutes have reduced their size or closed down. For example in Colombia, Centro Nacional de Investigaciones sobre el Cafe (CENICAFE) was reduced to half its previous size, and in Mexico, Instituto Mexicano del Cafe (INMECAFE) totally disappeared. What could be done to cope up with financial entrepreneurial problems and protect the environment at the same time? How can one support the remaining research teams in subjects related to coffee agro-industry? Experts suggest to make a new and improve old 'beneficio' technology. The main thrust is to present alternatives in order to upgrade and diversify the products obtained from all the organic residues derived from agro-industrial cherry transformation. This way, coffee pulp/husk will become one of the selling products from this industry. Also, utilisation of

the upgraded by-products should be carried out in such a manner to help reduce the volume of spent water and the net cost of wastewater treatment. Storage of fresh ensiled and decaffeinated pulp would help in solving its toxicity aspect, together with season to season availability (Porres et al., 1993; Perraud-Gaime and Roussos, 1997b). Thus, the silage product could be kept without immediate drying. Later on during the season, the product could be oven or sun dried without imposing any strains on overloaded machinery and hand labour.
Nutritional tests done on ruminants and fish would help in evaluating the commercial value of the upgraded coffee pulp/husk (Ramirez-Martinez, 1999). Basic studies on the enzymology of silage and SSF could help in selecting the best performing microbial strains to be used for silage and detoxification purposes. It would also be helpful to isolate new strains of lactic acid bacteria, filamentous fungi and edible mushrooms with specific features such as probiotic and enzymatic activities or with other new properties related to a better market value (Denis, 1996; Soares, 1998; Soares *et al.*, 1999, 2000; Brand 1999; Brand *et al.*, 2000; Hakil, 1999; Fan, 1999; Fan et al., 1999a, b, c, 2000a, b; Suzuki-Lopez, 1999). These are some of the planned BIOPULCA project goals.

3. Lactic acid starters for coffee pulp silage

The overall aim of the investigation on lactic acid starters for coffee pulp silage was to study the biodiversity of lactic acid bacteria (LAB) from coffee biotopes and to isolate, characterize and select LAB for the formulation of starters for coffee pulp silage associated with detoxifying fungal enzymes. Furthermore, this work involved co-operation regarding technology associated with the engineering and scale-up basis of controlled coffee pulp silage and detoxification by means of SSF as well as the testing of applications of this bioprocess (Perraud-Gaime et al., 1997a, b).

3.1. ISOLATION AND SCREENING OF LAB FROM COFFEE BIOTOPES

Bacterial strains were isolated from fresh coffee pulp. Fermentation profiles were studied in order to select suitable strains. Each strain was submitted to biochemical tests for precise phenotypic characterization. Among the studied strains, 62% were LAB, out of which 38% were further tested in silage studies. Most of the isolated strains were homo-fermentative with less than 5% being hetero-fermentative. An interesting aspect to point out was that 20% of the isolated strains were able to grow in the presence of tannic acid. It has yet to be determined whether these bacterial strains would be able to degrade hydrolysable tannin (Suzuki-Lopez, 1999).

3.2. LAB SCALE STUDIES OF SILAGE FROM INOCULATED COFFEE PULP

Various silage conditions were tested on laboratory scale with an inoculation of between 300-1000g of fresh coffee pulp. In Mexico, a preliminary assay to inoculate fresh pulp with one homo-lactic natural strain (*Lactobacillus paracasei*) was compared with the natural uninoculated silage. Both the trials were satisfactory, but ethanol and acetic acid were detected besides lactic acid. Inoculated silage caused the production of acetic acid, but did not avoid ethanol accumulation, probably due to acidic tolerant yeast development. New assays would be tried at lab scale in Mexico with other selected strains (*L. plantarum* strains) to improve acidification kinetics of silage and to limit ethanol production (Romano-Machado *et al.*, 1999).

4. Fungal enzymes for coffee pulp silage/detoxification

The production of fungal enzymes as decaffeinases and tannases was obtained in SSF with coffee pulp or impregnated polyurethane foam as solid substrates (Denis *et al.*, 1998; Hakil, 1999; Hakil *et al.*, 1998, 1999). Large-scale enzyme production could be obtained by the development of a continuous solid-state process (Van de Lagemaat *et al.*, 1999). Continuous SSF is carried out in a counter-current reactor adapted for this purpose and a rotating cylinder equipped with mixing baffles (Roussos and Pyle, 1998). The produced enzymes could be useful in the detoxification of the pulp.

Several fungal strains known to be GRAS (generally recognized as safe) were tested for their ability to produce tannases (Van de Lagemaat *et al.*, 1999). Screening of these fungi for tannase was necessary since originally fungi were only screened for caffeinases (Hakil *et al.*, 1999; Aguilar et al., 1999; Romano-Machado *et al.*, 1999). Tannases from the seven most productive strains were isolated on polyacrylamide gels and identified (gel activity assay). *Penicillium frequentans* and *Aspergillus phoenicis* were the most productive strains and would be used for tannase production with continuous reactors. A large-scale purification process is proposed to be developed after sufficient amounts of the enzyme are produced in continuous processes. Monitoring of biomass and tannic acid concentration would be necessary in both batch and continuous fermentations and methods have been developed for this.

4.1. PILOT PLANT PRODUCTION OF FUNGAL ENZYMES BY SSF

Recent experimental evidence indicated that it was possible to degrade caffeine and tannin (polyphenolic compounds) simultaneously by using a single strain of filamentous fungi. It is worth noticing that these results could allow the production of detoxifying enzymes in only one fermentation step (Saucedo-Castañeda *et al.*, 1999).

4.1.1. Process engineering studies for scale-up of continuous bioreactors and process modelling for enzyme production in SSF

Two different types of laboratory scale prototype reactors have been built with the specific aim to develop reactors, which can operate continuously with solid substrates and without inoculation of the feed. In consequence it is important to have sufficient back mixing within the reactor to be able to operate with a sterile feed. The screw reactor ran successfully with sterile feed for around ten days before the experiment had to be stopped (Van de Lagemaat *et al.*, 1999).This device is very promising and a full programme of experimental work began in Autumn 1999. The other design is a rotating baffled cylinder, which ran continuously for 28 days. We believe this period as the longest time a truly continuous sterile SSF has operated work on this design is also continuing. We now have also made considerable progress on modelling these systems. We have developed mathematical models for the solids mixing behaviour and also a new model for the behaviour of the system as a reactor. Together these represent very encouraging progress in our aim to develop and run truly continuous solid substrate reactors.

5. Mushroom cultivation on raw wastes of coffee pulp

Studies were carried out by Fan *et al.* (1999a, b, 2000a) to compare the growth and activity of *Pleurotus* sp. on different residues of coffee industry; viz. coffee husk, coffee leaves and coffee spent ground in solid state fermentation (SSF). It was found that in coffee husk, protein levels increased with the time during fermentation; during first 15 days of fermentation, protein increased by 6.8%. After this period, although there was further increase, it was slow (1.2% during the last 10 days fermentation). Fibre contents decreased 6% during first 15 days of fermentation. This also decreased further with time course of fermentation; the decrease was slow (1.1% during the last 10 days fermentation). Better increase in protein contents were noted when the substrate had 60-65% moisture in comparison to that at 45-55% and 70-75%. With 60-65% moisture maximum protein content was 8.2% and minimum fibre was 5% after 15 days fermentation. Studies on the effect of inoculum size showed that protein increased from 7.6-8.3% when the inoculation rate was 15%. It was significant to note that the concentration of caffeine decreased after colonization and fructification (60 days) up to 60% on average, but caffeine was actually not degraded by the fungus. Tannins also decreased in fermented coffee husk, up to 79% on average.

When coffee leaves were used as substrate, protein increased with time of fermentation. During the first 15 days of fermentation, protein increased by 7.6%. In this case also, the pattern was similar to that of coffee husk and during the last 10 days period, the increase was only 1.7%. The fibre contents also decreased in a similar pattern (decreasing with the time of fermentation by 6.9% during the first 15 days fermentation). A 65-70% substrate moisture was found best than 45-60% and 75% when maximum protein was 10.2% and the content of fibres was minimum. The inoculum rate of 15% was found most suitable in coffee-leaf SSF, which resulted in 9.4% protein after 15 days of fermentation. Correspondingly, the fibre contents decreased to 7%. Higher inoculum size was not effective. SSF using coffee spent ground as substrate also revealed a similar pattern.

SSF was also performed using *Flammulina velutipes*. When coffee husk was used as the substrate, the biological efficiency reached 55%. Experiments carried out to evaluate the efficiency of *F. velutipes* in submerged fermentation using the extract of coffee husk showed that yeast extract ($2g.l^{-1}$) was the best nitrogen source, resulting in 1.8 $mg.ml^{-1}$ biomass after 20 days. During the fermentation, the pH increased from 7.0 to 8.8.

6. Pre-treatment of coffee husk

Thermal hydrolysis of coffee husk was applied as the treatment method to separate efficiently its carbohydrate components (Woiciechowski *et al.*, 1999). Temperature and time of reaction and chemical catalysts were the variables used to optimize the hydrolysis. Temperatures tested were 100, 120 and 140°C for 5, 10 and 15 minutes for dry coffee husk treatment and acid chemical hydrolysis. Results obtained for reducing sugars recovered in the hydrolysate were submitted to a statistical treatment, testing two factors (temperature and time) at three levels. Results showed that the concentration of reducing sugars was almost similar with or without acid catalysis. Thus, in view of the economical and environmental aspects, acid hydrolysis was not recommended. Due to the severe conditions of temperature and acid concentration, during processing sugars was degraded, producing toxic compounds such as furfural. These compounds are harmful for micro-organisms and affect fermentative processes. Best results were obtained with aqueous hydrolysis at 121°C for 15 min, which resulted in 240-g of reducing sugar.kg^{-1} of coffee husk. Further work on using the hydrolysate for fungal cultivation is being carried out.

7. Production of gibberellic acid from coffee husk

Gibberellic acid (GA_3) is a plant hormone widely used in the agro-industry. It is produced as a secondary metabolite by specific fungi. Coffee husk was used as carbon source for its production. Five strains of *Gibberella fujikuroi* and one of its imperfect states, *Fusarium moniliforme* were screened for their efficiency to produce GA_3 in SSF and liquid fermentation (SmF). SmF was carried out using aqueous extract of coffee husk. Results showed the superiority of SSF for GA_3 production. *G. fujikuroi* LPB-06, which gave the best performance was chosen for further studies. Fermentation was carried out in 250-ml Erlenmeyer flasks by inoculating the substrate at 15% v/w. The GA_3 isolation was done by acidifying the fermented mash to pH 2.5 and then extracting with ethyl acetate. The organic layer was separated and analyzed by high performance liquid chromatography. All the studies utilized statistical experimental designs. In order to enhance substrate utilization, different pre-treatments were applied using KOH at different concentrations and varied time of treatment. Best results were obtained when coffee husk was pre-treated with 5g.l^{-1} KOH for 45 min in aqueous solution, which resulted in about 100 mg GA_3.kg^{-1} coffee husk. Further studies to optimize the nutritional conditions of the strain are being carried out (Machado *et al.*, 1999).

8. Nutritional quality of processed coffee pulp

8.1. IN AQUACULTURE

Fish farming could be a way to produce animal protein using locally available feedstuff but taking into account current market limitations. Some work has been done on feeding coffee pulp to Tilapia (Garcia and Baynes, 1974), carp and catfish (*Clarius mossenbicus*) as indicated by Christensen (1981). The level of coffee pulp used in the experimental diets was close to 33% without negative effects on the growth rate and yields of the fish. Lagooning may be also used as a secondary water treatment process, after anaerobic primary treatment of spent waters in the coffee mill.

8.2. RUMINANT NUTRITION

Proximate composition of coffee pulp shows a relative low nutritive value due to high levels of wall materials (60%) and lignin (15%) and also due to the presence of caffeine, tannins and chlorogenic acid. Therefore, the use of raw coffee pulp has been suggested to be lower than 20% in ruminant diets (Ruiz and Ruiz, 1977; Vargas *et al.*, 1982; Abate and Pfeffer, 1986). High raw coffee pulp intake has been associated with negative nitrogen balance because of diuretic effect of caffeine (Cabezas *et al.*, 1977). Coffee pulp silage seems to correct this problem probably because of caffeine leaching in the silage liquor (Cabezas *et al.*, 1976; Ramirez-Martinez, 1999). On the other hand, solid-state culture of fungal organisms such as *Penicillium roquefortii* or *Aspergillus niger* may reduce to less than 1% the level of caffeine in coffee pulp, leaving a probiotic activity in the fungal biomass as indicated above (Tapia *et al.*, 1988; Campos-Montiel and Viniegra-Gonzalez, 1995). Therefore, despite the nutritional limitations of raw coffee pulp, solid-state fungal culture and ensiling may increase the ruminant nutritional and market value of this material. This is an interesting feature, which remains to be tested *in vivo*.

9. Summary and conclusions

The objective of Biopulca project, European INCO-DC project N°IC18*CT970185, is to recycle coffee pulp and coffee husk by biotechnological processes using lactic acid bacteria (LAB) for fresh coffee pulp transformation by silage, followed by the use of selected filamentous fungi capable of producing specific enzymes (caffeinases, tannases) in order to detoxify this tropical agro-industrial waste. It is to be kept in mind that the pulp generated from green coffee production accounts for nearly 50% in volume of the total yearly production. Further objectives are

a) to transform fresh coffee pulp into a stable and detoxified lactic acid silage product,
b) to use lactic acid silage as food or feed purposes and as substrate for solid-state fermentation of fungal metabolites or enzymes production,
c) to use coffee industry residues for mushroom and metabolites production,
d) to transform highly toxic compounds (caffeine, tannins) by biotechnological processes,
e) to diversify people's activity and employment in rural coffee-growing areas.

As seen above, a large number of attempts are being made to eliminate coffee pulp or husk, which, at the present time, are a pollutant in Latin America. As soon as such lab-scale studies show a definitive potential, industrial applications can be attempted. Whether coffee by-products are used for animal feed , for mushrooms, enzymes or metabolites production, the result should ultimately be seen through a diversification of peoples activities through new employment potential in rural coffee-growing areas. It is hoped that in the years to come, coffee by-products, initially discarted, will be transformed into high added value products, turning coffee pulp or husk, for example, into a commodity, rather than a waste.

10. Acknowledgements

We thank the European Union for financial support for the INCO-DC project No IC18*CT970185 (BIOPULCA).

11. References

Abate A., Pfeffer E. (1986), Changes in nutrient intake and performance by goats fed coffee pulp based diets followed by a commercial concentrate. *Anim. Feed Sci. Technol.*, 1: 1-10

Aguilar C., Augur C., Viniegra G., Favela E. (1999), Production of fungal tannase in a model system by solid state fermentation. Paper presented at III SIBAC, 24-28 May, Londrina, Brazil

Alvarez-Gallo J. (1991), Despulpado de café sin agua. Avances Técnicos Cenicafé, N° A64: 1-6

Alvarez-Gallo J. (1995) Evaluacion de la despulpadora vertical PENAGOS 255C, operada sin agua. *CENICAFE*, Chinchina, febrero 3 de 1995

Antier P., Minjares A., Roussos S., Viniegra-Gonzalez G. (1993a), New approach for selecting hyper-producing mutants of *Aspergillus niger* well adapted to solid state fermentation. Biotechnol. Adv., 11: 429-440

Antier P., Minjares A., Roussos S., Raimbault M., Viniegra-Gonzalez G. (1993b), Selective medium for the isolation of pectinase hyper-producing mutants of *Aspergillus niger* C28B25 for solid-state fermentation of coffee pulp. Enzyme Microbial Technol., 15:254-260

Avallone S. (1999), Etude de la fermentation naturelle de *Coffea arabica* L. et des mécanismes de fluidification du tissu mucilagineux. PhD Thesis, Université de Montpellier II, France

Boccas F., Roussos S., Gutierrez-Rojas M., Viniegra-Gonzalez G. (1994), Production of pectinase from coffee pulp in solid state fermentation system: Selection of wild fungal isolates of high potency by a simple three-step screening technique. J. Food Sci. Technol., 31(1): 22-26

Brand D. (1999), Detoxificação biológica da casca de café por fungos filamentosos em fermentação no estado sólido. Tese de Mestrado; Tecnologia de Alimentos da Universidade Federal do Paraná- Curitiba, Brazil

Brand D., Pandey A., Roussos S., Soccol, C.R. (2000), Biological detoxification of coffee husk by filamentous fungi using a solid state fermentation system, Enzyme Microb. Technol., in press

Bressani R. (1979), Factores antifisiológicos de la pulpa de café. In- Pulpa de café: composición, tecnología y utilización. Braham J.E., Bressani R. (eds.), Ottawa, Canada, International Development Research Centre, pp. 143-152

Cabezas M.T., Estrada E., Murillo B., Gonzalez J.M., Bressani R (1976), Coffee pulp and husks. 12. Effect of storage on the nutritive value of coffee pulp for calves. Arch. Latinoamer. Nutr., 2: 203-215

Cabezas M.T., Estrada E., Murillo B., Gonzalez J.M., Bressani R. (1977), Response of calves to coffee pulp, caffeine and tannic acid in the ratio. Mem. Latinoamer. Prod. Anim., 15-21

Campos-Montiel R., Viniegra-Gonzalez G. (1995), Utilization of a consortion of cellulolytic bacteria grown on a continuous anaerobic digesteras a bioassay of probiotic extracts from fungal origin. Biotechnol. Tech., 9: 65-68

Castillo M., Bailly H., Violle P., Pommares P., Sallée B. (1993), Tratamiento de las aguas resisuales de los beneficios húmedos de café, análisis de un proyecto piloto en la cuenca de Coatepec Ver. (México). ASIC, 15e Colloque Scientifique International, Montpellier, France

Christensen M.S. (1981), Preliminary tests on the suitability of coffee pulp in the diets of Common Carp (*Cyprius carpio* L) and catfisf (*Clarias mossambicus* peters). Aquaculture, 25: 235-242

Denis S. (1996), Dégradation de la caféine par *Aspergillus* sp. et *Penicillium* sp. Etude physiologique et biochimique. PhD Thesis, Université de Montpellier II, France

Denis S., Augur C., Marin B., Roussos S. (1998). A new HPLC analytical method to study fungal caffeine metabolism. Biotechnol. Tech., 12: 359-362

Elias L.C. (1978), Composicion quimica de la pulpa de café y otros subproductos. In- Pulpa de café: composición, tecnología y utilización. Braham, J.E. and Bressani, R. (eds.), ICAP, 19-29

Fan L. (1999), Produção de fungo comestível do gênero *Pleurotus* em bioresíduos da agroindustria do café. Tese de Mestrado; Curso de Pós Graduação em Tecnologia de Alimentos da Universidade Federal do Paraná- Curitiba, Brazil

Fan L., Pandey A., Vandenberghe L.P.S., Soccol C.R. (1999a), Effect of caffeine on *Pleurotus* sp. and bioremediation of caffeinated residues. IX European Congr. Biotechnol., July 11-15, Brussels, Belgium, p. 2664- A

Fan L., Pandey A., Soccol C.R. (1999b), Cultivation of *Pleurotus* sp. on coffee residues. Proc. 3rd Internatl. Conf. Mushroom Biology & Mushroom Products, Sydney, Australia, pp 301-310

Fan L., Pandey A., Soccol C.R. (1999c), Growth of *Lentinus edodes* on the coffee industry residues and fruiting body production. Proc. 3rd Internatl. Conf. Mushroom Biology & Mushroom Products, Sydney, Australia, pp 293-300

Fan L., Pandey A., Mohan R., Soccol C.R. (2000a), Comparison of coffee industry residues for production of *Pleurotus ostreatus* in solid state fermentation, Acta Biotechnol., 20(1), in press

Fan L., Pandey A., Soccol C.R. (2000b), Solid state cultivation- an efficient technique to utilize toxic agro-industrial residues. J. Basic Microbiol., 40(3), in press.

Favela E., Huerta S., Roussos S., Olivares G., Nava G., Viniegra-Gonzalez G., Gutierrez M. (1989), Producción de enzimas a partir de la pulpa de café y su aplicación en el benefício húmedo. *in* Memorias I Seminario Internacional de Biotecnología en la Agroindústria Cafetalera, Roussos, S., Licona-Franco, R., Gutierrez-Rojas, M. (eds.), INMECAFE, Xalapa, pp. 144-151

Garcia R.C., Baynes D.R. (1974), Cultivo de *Tilapia aurea* (Staindachner) en corrales de $100 m^2$, alimentada artificialmente con gallinazas y un alimento preparado con 30 % de pulpa de café. Ministerio de Agricultura y Ganaderia, Dir. Gen. de Pecursos Nat. Ren., El Salvador, p. 23

Hakil M. (1999), Dégradation de la caféine par les champignons filamenteux, purification d'une théophylline déméthylase d'*Aspergillus tamarii*. PhD Thesis, Université de Franche Conté, Besançon

Hakil M., Denis S., Viniegra-Gonzalez G., Augur C. (1998), Degradation and product analysis of caffeine and related dimethylxanthines by filamentous fungi. Enzyme Microb. Technol., 22: 355-359

Hakil M., Voisinet F., Viniegra-Gonzalez G., Augur C. (1999), Caffeine degradation in solid state fermentation by *Aspergillus tamarii*: Effects of additional nitrogen sources. Process Biochem., 35: 103-109

Jacquet M. (1993), Alternativas tecnológicas al benefício húmedo en relación con la conservación del medio ambiente. IICA, Boletín PROMECAFE 61: 5-10

Machado C.M.M., Oliveira B.H., Pandey A., Soccol C.R. (1999), Coffee husk as a substrate for the production of gibberellic acid by fermentation Paper presented in III SIBAC, May 24-28, Londrina, Brazil, p. 39

Pandey A., Soccol C.R. (2000), Economic utilization of crop residues for value addition - A futuristic approach. J. Sci. Ind. Res., 59: 12-22

Perraud-Gaime I. (1995), Cultures mixtes en milieu solide de bactéries lactiques et de champignons filamenteux pour la conservation et la décaféination de la pulpe de café. PhD Thesis, Université de Montpellier II, France

Perraud-Gaime I., Roussos S. (1997a), Selection of filamentous fungi for decaffeination of coffee pulp in solid state fermentation prior to formation of conidiospores. In- Advances in solid state fermentation, Roussos, S., Lonsane, B.K., Raimbault, M. and Viniegra-Gonzalez, G. (eds.), Kluwer Acad. Publ., Dordrecht, pp. 209-221

Perraud-Gaime I., Roussos S. (1997b). Preservation of coffee pulp by ensiling: Influence of biological additives. In- Advances in solid state fermentation, Roussos S., Lonsane B.K., Raimbault M. and Viniegra-Gonzalez G. (eds.), Kluwer Acad. Publ., Dordrecht, pp.193-207

Porres C., Alvarez D, Calzada J. (1993), Caffeine reduction in coffee pulp through silage. Biotechnol. Adv., 11 : 519-523

Ramirez-Martinez J. (1999), Pulpa de café ensilada. Produccion, caracterizacion y utilizacion en alimentacion animal. Universidad Nacional Experimental del Taxhira (UNET) San Cristobal, Tachira, Venezuela

Romano-Machado J., Gutierrez-Sanchez G., Perraud-Gaime I., Gutierrez-Rojas M., Saucedo-Castañeda G. (1999), Degradación de cafeina en pulpa de café fresca y ensilada por fermentación en medio sólido : influencia des tratamiento del sustrato y el nivel de inóculo. Paper presented in III SIBAC, Londrina, Brazil, p. 46

Roussos S., Aquiahuatl R.M.A., Cassaigne J., Favela E., Gutierrez M., Hannibal L., Huerta S., Nava G., Raimbault M., Rodriguez W., Salas J.A., Sanchez R., Trejo M., Viniegra G.G. (1989), Detoxificación de la pulpa de café por fermentación sólida. in Memorias I Sem. Intern. Biotecnol. Agroindust. Café (I SIBAC), Roussos, S., Licona-Franco, R., Gutierrez-Rojas, M. (eds.), Xalapa, Mexico, pp. 121-143

Roussos S., Pyle D. L. (1998), Continuous enzymes and fungal metabolite production in solid state fermentation using a counter-current reactor. In- Proc. International Trg. course Solid State Fermentation, Raimbault, M., Soccol, C. R. and Chuzel, G. (eds.), ORSTOM, Montpellier, pp. 21- 35

Roussos S., Perraud-Gaime I., Denis S. (1998), Biotechnological management of coffee pulp. In- Proc. International Trg. course Solid State Fermentation, Raimbault, M., Soccol, C. R. and Chuzel, G. (eds.), ORSTOM, Montpellier, pp. 251-261

Ruiz M.E., Ruiz A. (1977), Effect of intake of green forage on intake of coffee pulp and weight gain of young cattle. Turrialba, 1: 23-28

Saucedo-Castañeda G., Romano-Machado J.M., Gutierrez-Sanchez G., Ramirez-Romero G., Perraud-Gaime I. (1999), Experiencia mexicana na valorizaçao biotecnologica de la pulpa do café. Paper presented in III SIBAC, Londrina, Brazil

Soares M. (1998), Biossíntese de Metabólitos Voláteis Frutais Por *Pachysolen tannophylus* e *Ceratocystis fimbriata* pela Fermentação no Estado Sólido sobre Casca de Café. Tese de Mestrado, Curso de Pós Graduação em Tecnologia de Alimentos da Universidade Federal do Paraná- Curitiba, Brazil

Soares M., Christen P., Pandey A. Soccol C.R. (2000), Fruity flavour production by *Ceratocystis fimbriata* grown on coffee husk in solid state fermentation. Process Biochem., (2000), in press

Soto-Cruz O., Saucedo-Castañeda G., Favela-Torres, E. (1999), Efecto del nivel de inoculo sobre el cultivo por lote de la bacteria ruminal *Megasphaera elsdenii*. Congreso Latinoamericano de Biotecnología y Bioingeniería, Huatulco, Oaxaca, México, 12-17 septembre p. 44

Suzuki-Lopez A. (1999), Isolation and partial characterization of lactic acid bacteria for controlled silage production of coffee pulp. Maestria de Biotecnologia, UAM-Iztapalapa, Mexico

Tapia I.M., Herrera S. R., Viniegra G., Gutierrez R.M., Roussos S. (1989) Pulpa de café fermentada: su uso como aditivo en la alimentaci6n de rumiantes, in Memorias I Sem. Intern. Biotechnol. Agroindust. Café (I. SIBAC), Roussos, S., Licona-Franco, R., Gutierrez-Rojas, M. (eds.), Xalapa, Mexico, pp.153-175

Urquiza G. (1989), Un futuro Café? Notables perspectivas para los cafetaleros. ICYT-Informacion, 11 (154) : 40-43

Van de Lagemaat J., Pyle D.L., Augur C. (1999). The screening of fungi for the production of fungal tannases in solid state fermentation. Paper presented in III SIBAC, Londrina, Brazil

Vargas E., Cabezas M.T., Murillo B., Braham J.E., Bressani R. (1982), Effect of large amounts of dried coffee pulp on growth and adaptation of young steers. Arch. Latinoamer. Nutr., 4: 973-989

Vasquez-Morera R. (1993), Influencia de la recirculaci6n de las aguas del despulpado del café sobre su calidad. IICA-PROMECAFE/IHCAFE, Seminario regional sobre el mejoramiento de la calidad del café, 22-24 September San Pedro Sula

Viniegra-Gonzalez G. (1989), Introducci6n, In- Proc. I Seminario Internacional sobre Biotecnologia en la Agroindustria Cafetalera, Roussos, S., Licona-Franco, R., Gutierrez-Rojas, M. (eds.), INMECAFE, UAM-I and ORSTOM, Xalapa, Ver., Mexico, pp. 13-15

Viniegra-Gonzalez G. (1993), Por qué es necesario Renovar y mejorar la tecnologia del beneficio del café? ICYT, Ciencia y Tecnologia, CONACYT, Mexico: 15: 41-46

Violle P., Pommares P., Vivier D., Castillo M., Sallée B. (1995), Tlapexcatl, unité pilote d'épuration des eaux résiduaires d'une usine à café au Mexique. Plantations, Recherche, Développement, vol. 2 n°3 (Montpellier), pp. 35-44

Woiciechowski A., Soccol C.R., Pandey A., Busato E. (1999), Selective hydrolyses of hemicellulose fraction of coffee husk for production of sugars. Paper presented in III SIBAC, Londrina, Brazil, p. 54

Zambrano-Franco Q.A., Isaza-Hinestroza J.D., Rodriguez-Valencia N., Lopez-Posada U. (1999), Tratamiento de aguas residuales del lavado del café. CENICAFE, Boletin Técnico N° 20 : 1-26

Zuluaga J.V. (1989), Utilización integral de los subproductos del café. In- Proc. I Seminario Internacional sobre Biotecnologia en la Agroindústria Cafetalera, Roussos, S., Licona-Franco, R., Gutierrez-Rojas, M. (eds.), INMECAFE, UAM-I and ORSTOM, Xalapa, Ver., Mexico, pp. 63-76

Chapter 36

MICROBIAL DEGRADATION OF CAFFEINE AND TANNINS FROM COFFEE HUSK

D. Brand[1], A. Pandey[1a], S. Roussos[2], I. Brand[1], C. R. Soccol[1]

[1]*Laboratorio de Processos Biotecnologicos, Departamento Engenharia Quimica, Universidade Federal do Parana (UFPR), 81531-970 Curitiba-PR, Brazil;* [2]*Laboratoire de Microbiologie, IRD, IFR-BAIM, Université de Provence, 13288 Marseille, France;* [a]*currently at Biotechnology Division, Regional Research Laboratory, Trivandrum-695 01, India*

Running title: Coffee husk, caffeine and tannins degradation

1. Introduction

Brazil contributes approximately 25% of the world's coffee production. During 1998, it's production reached two million tons coffee beans. Parana State is one of the most important States for coffee cultivation in the country. Its production was 280,000 tons (representing about 14% of total Brazilian production) during 1998. Only 6% of the fresh grain is utilized for the production of the beverage, the remaining 94% is constituted by water and sub-products of the process (Zuluaga, 1989). In Brazil, coffee cherries are generally sun dried and subsequently the outer layers of husk covering the green coffee are removed by a hulling machine as and when needed. This residue poses a serious environmental concern due to its disposal in rivers, lakes located near the coffee processing regions. In view of its richness in proteins, fibres, carbohydrates and minerals, it has been suggested that it could be used as animal feed and organic fertilizer with suitable bio-treatments. It could also be used as substrate for the production of biogas, enzymes edible mushrooms, etc. (Gaime-Perraud, 1996; Fan *et al.*, 1999a, b, c 2000a, b; Pandey and Soccol, 2000).

T. Sera et al. (eds.), Coffee Biotechnology and Quality, 393–400.

The use of the coffee husk as animal feed has some controversies. The presence of substances that have been termed as anti-physiological factors such as caffeine, polyphenols and a high potassium content could cause problems with the intake of husk (Cabezas *et al.*, 1978). Caffeine has some limitations due to its high nitrogen content and its diuretic and physical stimulant effects causing a decrease in urinary retention and loss of nitrogen. Polyphenols interfere with protein digestibility, thiamine utilization and cause reduced iron intake. The high content of potassium may cause ions unbalance at the tissues. (Velez *et al.*, 1985)

In order to make effective use of the huge amounts of coffee husk, thus, it seems important to remove (or reduce to a reasonably low level) the anti-physiological factors present in it. Use of solid state fermentation (SSF) could be a good alternative to aggregate value in its utilization as animal feed, since there are some micro-organisms that are capable to use caffeine as sole nitrogen source and polyphenols as carbon source (Roussos, 1989.) SSF has great potential to utilize agro-industrial residues for value-addition (Soccol, 1996; Pandey, 1991, 1992, 1994; Pandey and Soccol, 2000; Pandey *et al.* 1999a,b, 2000). The filamentous fungi are the micro-organisms best adapted to SSF processes due to their hyphal mode of growth and their good tolerance for low water activity (aw) (Pandey, 1992, 1994; Pandey *et al.*, 2000). High osmotic pressure conditions make fungi efficient and competitive in natural micro-flora for bioconversion of solid substrates (Raimbault, 1997)

There are a few reports, which describe removal or degradation of toxic substances from crops or crop-residues using filamentous fungi. For example, Wang *et al.* (1969) detoxified the cyanogenic compounds of cassava using filamentous fungi. Some authors have shown that fungi increased the digestibility and protein content of foods (Daubresse *et al.*, 1987; Soccol, 1996; Stertz *et al.*, 1999) and produced anti-carcinogenic substances (Wang *et al.*, 1969). The mould Phanerochaete chrysosporium is known to degrade lignin (Mudgett and Paradis, 1985). Attempts were made to degrade caffeine and tannins present in coffee husk using filamentous fungi belonging to three genera (Brand, 1999; Brand *et al.*, 1999; 2000)

2. Chemical composition of the husk

The chemical composition of the coffee husk has not been studied so extensively as that of coffee pulp, although both appear to have several similarities in their compositions. Table 1 shows a comparative profile of the components present in the coffee pulp and coffee husk. Vasco (1989) described the chemical composition of the pulp. It was interesting to note a difference between the values obtained for protein (total N) determined by Kjeldahl and for True Protein determined by Stutzer method, possibly due to the nitrogen content present in caffeine and other nitrogenous compounds present in the husk. The high content of tannins could be probably because the coffee grains are sun dried which favours the production of these compounds in the coffee husk. The contents of different nutrients of the coffee husk and pulp are reasonably good in comparison to other agricultural products or agro-industrial residues such as oats, rice meal, rice bran and wheat bran, all of each are increasingly being used in the diet of man (Christensen, 1981). However, the contents of caffeine and tannins make coffee husk and pulp different from all of these.

Table 1. Chemical composition of coffee husk and coffee pulp*

Components	Coffee pulp	Coffee husk
Moisture	6.93	11.98
Lipids	2.50	1.50
Fibers	21.00	31.86
Ash	8.3	6`.03
Carbohydrates	59.10	26.5
Protein (N*6,25)	8.25	6.8
True Protein (Stuzer)	-	4.8
Caffeine	0.75	1.2
Tannins	3.70	9.3

*Source-Brand, 1999

3. Selection of micro-organisms

The micro-organisms were selected by radial growth velocity and biomass produced in a coffee husk extract agar medium containing coffee husk extract (Brand, 1999). Eleven strains of *Rhizopus* sp., two strains of *Phanerochaete* and one strain of *Aspergillus* sp. were screened. The radial growth was realized by the inoculation of the micro-organisms in the centre of a Petri dish. Mycelial growth was measured every two hours for *Rhizopus* strains and every 12 hours for the strains of *Phanerochaete* and *Aspergillus* sp. Biomass was measured by the dissolution of the agar and separation of mycelia on filter paper (Table 2) (Brand, 1999; Brand *et al.*, 1999, 2000).

Table 2. Growth of strains on coffee husk extract agar medium*

Strain	Radial Growth velocity (MM.H^{-1})	Biomass (mg.plate^{-1})
R. oryzae LPB 68	2.19 ± 1.12	10,20 ± 0,56
R. oryzae LPB 95	2.05 ± 0.87	8,70 ± 1,20
R. delemar LPB 12	2.13 ± 2.12	10,80 ± 6,00
R. circicans LPB 75	2.09 ± 0.83	9.20 ± 0.70
R. arrhizus LPB 79	2.03 ± 0.34	12.10 ± 2.20
R. arrhizus LPB 25	1.94 ± 0.76	6.6 ± 0.40
R.oryzae LPB 27	1.88 ± 1.15	7.90 ± 4.30
Rhizopus sp. LPB 975	1.78 ± 1.32	2.80 ± 0.20
R. oligosporus LPB 67	1.78 ± 0.87	3.60 ± 0.20
R. formosa LPB 22	0.94 ± 0.65	2.90 ± 0.80
P. chrysosporium HD	0.75 ± 0.42	1.83 ± 0.56
P. chrysosporium BK	1.02 ± 0.34	2.21 ± 0.76
Aspergillus sp.	0.68 ± 0.14	14.83 ± 0.02

*Source-Brand, 1999; Brand *et al.*, 1999, 2000

The growth velocity of *Rhizopus* strains ranged from 0.94-2.19 mm.h^{-1}. The strains were capable to assimilate and metabolize the components present in the coffee husk producing biomass, which was characterized by radial growth. Best results were obtained with *R. arrhizus* 16179. The highest biomass (14.83 mg.plate^{-1} in 92 hours) was produced by *Aspergillus* sp., which presented a radial growth velocity of 0.68 $mm.h^{-1}$, showing great potential to degrade the toxic components of the substrate (Brand, 1999; Brand *et al.*, 2000).

4. Solid state fermentation

The strains of *Rhizopus* and *Phanerochaete* were maintained on PDA medium and that of *Aspergillus* on coffee husk extract agar. To prepare the inoculum, the strains were incubated for 10 days at 32, 35 and 28° C for the strains of *Rhizopus, Phanerochaete* and *Aspergillus* sp., respectively (Brand, 1999).

SSF was carried out in 250-ml Erlenmeyer flasks containing 20-g substrate. The experiments realized with selected strains of *Rhizopus, Phanerochaete* and *Aspergillus* were for the optimization of variables such as initial pH and moisture of the substrate, effect of addition of saline solution and effect of addition of water before autoclaving. For *Aspergillus* sp., which gave the best performance, surface response methodology was adapted to optimize various variables, which consisted of two experimental designs. The first one involved 2^{3-0} factors with four central points and one replicate resulting in a total of 22 experiments and the second one was a 3^{2-0} factors at three levels with one replicate that were significant by the analysis of the first study (Brand *et al.*, 1999, 2000).

SSF with *Rhizopus* LPB 79 showed that the addition of saline solution neither have any influence to degrade caffeine and tannins, nor resulted in better growth of the fungi. A 70 % moisture was the maximal water absorbing capacity of the substrate. The degradation rates obtained were 82 and 62% for caffeine and tannins, respectively under most suitable conditions. The pH of the medium influenced the metabolism of the mould for the growth and degradation of caffeine and tannins. SSF with *Phanerochaete chrysosporium* BK showed no beneficial effects by the addition of saline solution for the degradation of caffeine and tannins, although the growth of the micro-organism appeared to be better. The pH of the substrate influenced fungal metabolism to degrade toxic compounds present in coffee husk and maximum degradation was at pH 5.5 (70.8 and 45%, respectively for caffeine and tannins). The pH of the substrate also influenced the activity of *Aspergillus* sp. to degrade caffeine and tannins and under optimized conditions, a 92% reduction in caffeine content of the coffee husk was achieved (Brand, 1999; Brand *et al.*, 1999, 2000).

5. Summary and conclusions

The strains of filamentous fungi belonging to *Rhizopus, Phanerochaete* and *Aspergillus* sp. were able to remove the anti-physiological factors, *viz.* caffeine and tannins from the coffee husk through their cultivation in SSF. The degree of degradation by each strain varied according to the fermentation conditions and was influenced by physical and chemical parameters during SSF. These experiments were realized without supplementation of any nutrients to the coffee husk medium, which proved that the fungal culture could utilize the nutrients available in the husk along with the toxic components of caffeine and tannins. Maximum caffeine degradation achieved was 92%, leaving only a very little quantity in the substrate. This study clearly demonstrated the potentialities in using filamentous fungal strains to degrade anti-physiological factors present in the coffee husk.

6. Futuristic Approach

There is need to translate and test the laboratory findings on a pilot-plant level. Studies should be carried out to test the feed-value of the fermented coffee husk for cattle and other live-stocks, including poultry and aquaculture. Fermented husk could be used alone as feed or as feed-supplement.

7. Acknowledgements

Financial assistance from the European Union (grant no INCO DC: IC18*CT 970185) and PNP & D/CAFE-EMBRAPA, Brazil (Projeto no 07.1.99.057) is gratefully acknowledged. C. R. Soccol thanks CNPq, Brazil for a scholarship under Scientific Productivity Scheme.

8. References

Brand D. (1999), Detoxificação biológica da casca de café por fungos filamentosos em fermentação no estado sólido. Tese de Mestrado, Tecnologia de Alimentos da Universidade Federal do Paraná- Curitiba, Brazil

Brand D., Pandey A., Roussos S., Soccol C.R. (1999), Biological detoxification of the coffee husk using filamentous fungi. Paper presented in III SIBAC, May 24-28, Londrina, Brazil, p. 38

Brand D., Pandey A., Roussos S., Soccol C.R. (2000), Biological detoxification of coffee husk by filamentous fungi using a solid state fermentation system, Enzyme Microb. Technol., in press

Cabezas M.T., Flores A., Egana J.E. (1978), Uso de la pulpa de café en alimentación de ruminantes. In- Pulpa de Café – Composición, Tecnología y Utilización, J.B. Braham R. Bressani (eds.), INCAP, pp 45-67

Christensen M.S. (1981), Preliminary tests on the suitability of coffee pulp in the diets of common carp (*Cyprinus carpio* L.) and catfish (*Clarias Mossambicus* Peters). Aquaculture, 25, 235-242

Daubresse P., Ntibashirwa S., Gheysen A., Meyer J.A. (1987), A process for protein enrichment of cassava by solid state fermentation in rural conditions. Biotechnol. Bioeng., 29, 962-968

Fan L., Pandey A., Soccol C.R. (1999a), Cultivation of *Pleurotus* sp. on coffee residues. Proc. 3rd Internatl. Conf. Mushroom Biology & Mushroom Products, Sydney, Australia, pp 301-310

Fan L., Pandey A., Soccol C.R. (1999b), Growth of *Lentinus edodes* on the coffee industry residues and fruiting body production. Proc. 3rd Internatl. Conf. Mushroom Biology & Mushroom Products, Sydney, Australia, pp 293-300

Fan L., Pandey A., Vandenberghe L.P.S., Soccol C.R. (1999c), Effect of caffeine on *Pleurotus* sp. and bioremediation of caffeinated residues. IX European Congr. Biotechnol., July 11-15, Brussels, Belgium, p. 2664- A

Fan L., Pandey A., Mohan R., Soccol C.R. (2000a), Comparison of coffee industry residues for production of *Pleurotus ostreatus* in solid state fermentation, Acta Biotechnol., 20(1), in press

Fan L., Pandey A., Soccol C.R. (2000b), Solid state cultivation- an efficient technique to utilize toxic agro-industrial residues. J. Basic Microbiol., 40(3), in press.

Gaime-Perraud I. (1996), Cultures mixtes en milieu solide de bactéries lactiques et de champignons filamenteux pour la conservation et la décaféination de la pulpe de café. PhD thesis, Montpellier II University, France

Mudgett R.E., Paradis A.J. (1985), Solid-state fermentation of natural birch lignin by *Phanerochaete chrysosporium*. Enzyme Microb. Technol., 7, 150-154

Pandey A. (1991), Aspects of fermenter design in solid state fermentation. Process Biochem., 26, 355-361

Pandey A. (1992), Recent developments in solid state fermentation, Process Biochem., 27, 109-117

Pandey A. (1994), Solid state fermentation- an overview. In- Solid State Fermentation, Pandey, A. (ed.), Wiley Eastern Publishers, New Delhi, pp. 3-10

Pandey A., Soccol C.R., Nigam P., Soccol V.T., Vandenberghe L.P.S., Mohan R. (1999a), Biotechnological potential of agro-industrial residues: II Cassava bagasse. Biores. Technol., in press

Pandey A., Soccol C.R., Nigam P., Soccol V.T. (1999b), Biotechnological potential of agro-industrial residues: I Sugarcane bagasse. Biores. Technol., in press

Pandey A., Soccol C.R. (2000), Economic utilization of crop residues for value addition - A futuristic approach. J. Sci. Ind. Res., 59, 12-22

Pandey A., Soccol C.R., Mitchell N. (2000), New developments in solid state fermentation. I- Bioprocesses and products. Process Biochem., in press

Raimbault M. (1997), General and microbiological aspects of solid substrate fermentation. In- Proc. International Trg. Course on Solid State Fermentation. ORSTOM, Montpellier, pp. 1-20

Roussos S. (1989), Detoxificación de la pulpa de café por fermentación sólida. In- Proc I SIBAC, Xalapa, Ver., México. pp 121-143

Soccol C.R. (1996), Biotechnology products from cassava through solid state fermentation, J. Sci. Ind. Res., 55, 358-364

Stertz S.C., Soccol C.R., Raimbault M., Pandey A., Rodriguez-Leon J. A. (1999), Growth kinetics of *Rhizopus formosa* MUCL 28422 on raw cassava flour in solid state fermentation. J. Chem. Technol. Biotechnol., 74(6), 580-586

Velez A.J., Garcia L.A., De Rozo M.P. (1985), Interación *in vitro* entre los polifenoles de la pulpa de café y algunas proteinas. Arch. Latinoamer. Nutr., 35 297-305

Wang H.L., Ruttle D.L., Hesseltine C.W. (1969), Antibacterial compound from a soybean product fermented by *Rhizopus oligosporus*. Proc. Soc. Expt. Biol. Med., 131, 579-583

Zuluaga J.V. (1989), Utilización integral de los subproductos del café. In- Proc. I Seminario Internacional sobre Biotecnologia en la Agroindústria Cafetalera, Roussos, S., Licona-Franco, R., Gutierrez-Rojas, M. (eds.), INMECAFE, UAM-I and ORSTOM, Xalapa, Ver., Mexico, pp. 63-76

Chapter 37

COFFEE HUSK AS SUBSTRATE FOR THE PRODUCTION OF GIBBERELLIC ACID BY FERMENTATION

C.M.M. Machado[1], B.H. Oliveira[2], A. Pandey[1a], C.R. Soccol[1]

[1]Laboratório de Processos Biotecnológicos, Departamento Engenharia Química, [2]Laboratório de Produtos Naturais, Departamento de Química, Universidade Federal do Paraná (UFPR), 81531-970 Curitiba-PR, Brazil; [a]currently at Biotechnology Division, Regional Research Laboratory, Trivandrum-695 01, India

Running title: Gibberellic acid production

1. Introduction

Brazil is the largest producer of coffee in the world and during processing, several residues are generated. These residues represent about 50% of total mass and practically do not find any useful application. Instead, their disposal is a major environmental concern. These products are rich in organic matter, which could make them suitable for bioconversions processes. The principal difficulty for its utilization in fermentations is the presence of inhibitory substances such as phenolic compounds (tannins, chlorogenic acid and caffeic acid) and caffeine that represent about 6-14% of coffee husk dry weight. However, if these inhibitory substances could be removed, industrial products can be produced from these residues by cultivating microbes in submerged fermentation (SmF) or solid state fermentation (SSF) (Pandey and Soccol, 1998, 2000; Roussos and Perraud-Gaime, 1997). Gibberellins (GAs) are a group of diterpenoid acids that function as plant growth regulators influencing a range of developmental processes in higher plants. Among these, gibberellic acid (GA_3) has received the greatest attention. It

T. Sera et al. (eds.), Coffee Biotechnology and Quality, 401–408.

influences stem elongation, germination, elimination of dormancy, flowering, sex expression, enzyme induction and leaf and fruit senescence. GA_3 is a high-value industrially important biochemical with its price depending on the purity and potency. Therefore, its used at present is limited to high-premium crops.

The industrial process currently used for fermentative production of GA_3 is based on SmF technique. In spite of the use of the best process technology, it has been felt that SmF process used today is approaching a saturation point beyond which cost reduction is impossible. This is mainly due to low yields of GA_3 obtained through SmF. The presence of product in dilute form in SmF is considered a major obstacle in economic manufacture of GA_3 mainly due to the consequent high costs of downstream processing and disposal of wastewater. Moreover, the cost of separation of the microbial cells from fermentation broth using centrifugation or microfiltration is reported to involve between 48 and 76% of the total production cost of the microbial metabolite by submerged fermentation (Kumar and Lonsane, 1989).

Table 1. Alternative fermentations for production of GA_3

Substrate	Fermentation technique (reactor)	Production (GA_3)	References
Wheat bran	SSF (50 L pilot scale reactor)	3000 $mg.kg^{-1}$	Bandelier et al., 997
Wheat bran	SSF (fed-batch reactor)	1100 $mg.kg^{-1}$	Kumar & Lonsane, 1990
Cassava flour	SSF (column fermenter)	250 $mg.kg^{-1}$	Tomasini et al., 1997
Polymeric *	SSF(immobilized cells)	210 $mg.kg^{-1}$	Lu, et al., 1995
Molasses	SmF (Erlenmeyer flask)	155 $mg.kg^{-1}$	Cihangir & Aksöza, 1997
Vinasse	SmF (Erlenmeyer flask)	136;57 $mg.kg^{-1}$	Cihangir & Aksöza, 1997
Whey	SmF (Erlenmeyer flask)	120 $mg.kg^{-1}$	Cihangir & Aksöza, 1997
Sugar-beet waste	SmF (Erlenmeyer flask)	73 $mg.kg^{-1}$.	Cihangir & Aksöza, 1997
Fruit pomace	SmF (Erlenmeyer flask)	118,73 $mg.kg^{-1}$	Cihangir & Aksöza, 1997

* fibrous carriers

SSF processes based on agro-industrial residues such as coffee husk could provide potential opportunities for GA_3 production. Attempts were, thus, made for its production in SmF and SSF using five strains of *Gibberella fujikuroi* and one of *Fusarium moniliforme*

2. Experimental

2.1. MICRO-ORGANISMS

Five strains of *G. fujikuroi* and one of *F. moniliforme* from LPB Collection were used. These were maintained on potato-dextrose agar slants at 4°C and were grown two times in succession in Czapek Dox medium on a shaker at 30°C for 48 h for developing the inoculum (Kumar and Lonsane, 1987b).

2.2. PREPARATION OF SUBSTRATES AND FERMENTATION

The substrate for SmF was prepared by adding 1-l distilled water to 200-g coffee husk and autoclaving for one h at 100°C. Resulting extract was filtered through cloth and used as substrate for SmF. The solid residue was dried at 65°C for 24 h for subsequent use as substrate in SSF. Fermentation was carried out in flasks, which were autoclaved at 121°C for 15 min, cooled and inoculated (15% inoculum size) with the liquid culture grown in Czapek Dox. The pH of the substrates was adjusted at 4.5 using HCl (40%). For SSF, the initial moisture of the substrates un-treated coffee husk and hot water treated coffee husk) was 60%. All the flasks were incubated at 29°C for seven days (at 220 rpm for SmF and under static conditions for SSF).

2.3. PRE-TREATMENT OF THE SUBSTRATE

To test the influence of the pre-treatment factors in the production of GA_3, two statistical experimental designs were performed. The parameters studied with $3^{(2-0)}$ experimental design were the concentrations of KOH and different times. Table 2 shows the levels utilized in this study.

The second plan was defined to optimize the conditions of the pre-treatment established by the first experimental plan. For that the results of GA_3 and the statistical analysis that related the most significant factor were observed.

Table 2. Pre-treatment of coffee husk (plan 1)

Level	KOH concentration ($g.l^{-1}$)	Extraction time (min)
-1	5	15
0	20	30
+1	35	45

Table 3 shows the region chosen for this study. Only the KOH concentration was significant at level of 5%, so it was the factor, which was changed.

Table 3. Pre-treatment of coffee husk (plan 2)

Level	KOH concentration ($g.l^{-1}$)	Extraction time (min)
-1	0.5	15
0	2.5	30
+1	5	45

2.3. ESTIMATION OF GA_3

After fermentation, GA_3 was extracted with ethyl acetate at pH 2.5. The extract was concentrated under vacuum to 10-15 ml and was subjected to thin layer chromatography (TLC) for confirmation of the presence of GA_3. The mobile phase was chloroform, ethyl acetate, acetic acid (40:60:5). After evaporation of the solvents, the residue was dissolved in 80% methanol in water (ca. 8 ml) and the mixture applied to a pre-conditioned silica C-18 column (300 mg). The volume of the elute was then made 10 ml.

The samples were analyzed by high performance liquid chromatography (HPLC) using a Varian system composed of a pump (9012Q), a diode array detector (9065) and a autosampler (AI200). The data were processed by a Varian Workstation 5.1. Column was a C18 (5 μm, 4,5x250 mm) and the mobile phase methanol, water (40:60).

3. Results and Discussion

3.1. SCREENING OF THE FUNGI AND SYSTEM OF FERMENTATION

Among the total of six strains screened, *G. fujikuroi* LPB-6 produced the highest quantity of GA_3 after seven days in all the three cases (SmF with coffee husk extract, SSF with un-treated coffee husk and hot water treated husk). The comparative production determined by TLC is showed in Table 4.
Considering that a better fluorescence indicated a higher production of GA_3, the treated coffee husk appeared better substrate for SSF (Table 4).

Table 4. Presence of GA_3 produced by fermentation, analyzed with TLC

Strain	SmF	SSF coffee husk un-treated	SSF coffee husk treated
LPB-1	+++	++++	++++
LPB-2	++++	++	+++
LPB-3	++	+++	+++
LPB-4	+	+	++
LPB-5	+	+	++
LPB-6	++++	++++	+++++

More "+" means better fluorescence.

This could probably be explained by the fact that hot water treatment removed some substances present in coffee husk, which were of inhibitory and water-soluble nature and removed from the husk during the treatment. In un-treated coffee husk and in the

extract of the coffee husk during SSF and SmF, respectively these substances were in the substrates and probably interfered with the microbial growth and activity, resulting in lower yields. It is known that benzoic and cinnamic acids and their derivatives are plant growth inhibitors. These compounds may also inhibit the biosynthesis of gibberellins by *G. fujikuroi.* Coffee beans are known to contain large amounts of caffeic and chlorogenic (cinnamic acid derivatives) and tannic acid (a benzoic acid derivative). Therefore, the removal of these compounds by alkali could improve yields of gibberellic acid in coffee husk fermentation. *G. fujikuroi* LPB-06 and LPB 02 showed best performance in SSF and SmF, respectively.
HPLC results showed the production of 30 $mg.l^{-1}$ GA_3 in SmF, 18.3 $mg.kg^{-1}$ GA_3 in SSF using un-treated coffee husk and 38.4 $mg.kg^{-1}$ GA_3 in SSF with treated coffee husk. These confirmed the TLC results (Machado *et al.*, 1999).

3.2. PRE-TREATMENT OPTIMIZATION

The pre-treatment was optimized with a factorial plan, utilizing the Statistica program. It was found that the concentration of the alkali has the most significant effect on GA_3 production, which increased with the decrease of KOH concentration. Maximum production (99.7 $mg.kg^{-1}$ GA_3) was achieved for a 45 minutes treatment with a solution 5 $g.l^{-1}$ KOH. Further optimization studies using lower concentrations of KOH showed the best results with 2.5 $g.l^{-1}$ KOH (112.6 $mg.kg^{-1}$ GA_3 substrate). However, lowest experimental concentration, i.e. 0.5 $g.l^{-1}$ KOH was found unsuitable as it resulted a lower GA_3 production.

4. Conclusions

The results proved that coffee husk could be a useful substrate for the production of gibberellic acid by SSF with suitable pre-treatment. Under the optimized conditions of fermentation a six-fold increase in the production of GA3 was achieved. The results show promise for commercial exploitation and it would be worth to investigate the laboratory findings at pilot level.

5. Acknowledgements

Financial assistance from the European Union (grant no INCO DC: IC18*CT 970185) and PNP & D/CAFE-EMBRAPA, Brazil (Projeto no 07.1.99.057) is gratefully acknowledged. C. R. Soccol thanks CNPq, Brazil for a scholarship under Scientific Productivity Scheme.

6. References

Agosin E., Maureira M., Biffani V., Perez F. (1997), Production of gibberellins by solid substrate cultivation of Gibberella fujikuroi. In- Advances in solid state fermentation. Roussos S., Lonsane B.K., Raimbault M. and Viniegra-Gnzalez G. (eds.), Kluwer Academic Publishers, Dordrecht, pp. 355-366

Bandelier S., Renaud R., Durand A. (1997), Production of gibberellic acid by fed-batch solid state fermentation in an asseptic pilot-scale reactor. Process Biochem., 32:141-145

Brückner B., Blechschimidt D. (1991), The gibberellin fermentation. Crit. Rev. Biotechnol., 11: 163-192

Cihangir N., Aksöza N. (1997), Evaluation of some food industry wastes for production of gibberellic acid by fungal source. Environ. Technol., 18: 533-537

Kumar P.K.R., Lonsane B.K. (1987a), Extraction of gibberellic acid from dry mouldy bran produced under solid state fermentation. Process Biochem.. 10: 138-143.

Kumar P.K.R., Lonsane B.K. (1987b), Gibberellic acid by solid state fermentation: consistent and improved yields. Biotechnol. Bioeng., 30: 267-271

Kumar P.K.R., Lonsane B.K. (1989), Microbial production of gibberellins: state of the art. Adv. Appl. Microbiol., 34: 26-139

Kumar P.K.R., Lonsane B.K. (1990), Solid state fermentation: physical and nutritional factors affecting gibberellic acid production. Appl. Microbiol. Biotechnol., 34: 145-148

Lu Z.-X., Xie Z.-C., Kumakura M. (1995), Production of gibberellic acid in *Gibberella fujikuroi* adhered onto polymeric fibrous carriers. Process Biochem., 30: 661-665

Machado C.M.M., Olivera B.H., Pandey A., Soccol C.R. (1999), Coffee husk as a substrate for the production of gibberellic acid by fermentation. Paper presented in III SIBAC, May 24-28, Londrina, Brazil, p. 39

Pandey A. (1992), Recent developments in solid state fermentation, Process Biochem., 27: 109-117

Pandey A., Soccol C.R., Nigam P., Soccol V.T., Vandenberghe L.P.S., Mohan R. (1999a), Biotechnological potential of agro-industrial residues: II Cassava bagasse. Biores. Technol., in press

Pandey A., Soccol C.R., Nigam P., Soccol V.T. (1999b), Biotechnological potential of agro-industrial residues: I Sugarcane bagasse. Biores. Technol., in press

Pandey A., Soccol C.R. (1998), Bioconversion of biomass: a case study of lignocellulosic bioconversions in solid state fermentation. Brazilian Arch. Biol. Technol., 40: 379-390

Pandey A., Soccol C.R. (2000), Economic utilization of crop residues for value addition - A futuristic approach. J. Sci. Ind. Res., 59: 12-22

Pandey A., Soccol C.R., Mitchell D. (2000), New developments in solid state fermentation. I- Bioprocesses and products. Process Biochem., in press

Roussos S., Perraud-Gaime I., Denis S. (1998), Biotechnological management of coffee pulp. In- Proc. International Training course on Solid State Fermentation, Raimbault M., Soccol C.R., Chuzel G. (eds.), ORSTOM, Montpellier, pp. 251-261

Soccol C.R., Machado C.M.M., Olivera B.H. (2000), Production of gibberellic acid by solid state fermentation of mixed substrate. Brazilian Patent 000187, 24 February, DEINPI/PR, Brazil

Tomasini A., Fajardo C., Barrioz-González G. (1997), Gibberellic acid production using different solid-stat fermentation systems. World J. Microbiol Biotechnol., 13: 203-206

Chapter 38

HYDROLYSIS OF COFFEE HUSK: PROCESS OPTIMIZATION TO RECOVER ITS FERMENTABLE SUGAR

A.L. Woiciechowski, A. Pandey[a], C.M.M. Machado, E.B. Cardoso, C.R. Soccol

Laboratorio de Processos Biotecnologicos, Departamento Engenharia Quimica, Universidade Federal do Parana (UFPR), 81531-970 Curitiba-PR, Brazil; [a]currently at Biotechnology Division, Regional Research Laboratory, Trivandrum-695 01, India

Running Title: Coffee husk hydrolysis

1. Introduction

The fresh fruits of coffee known as a cherry, is formed by a red skin (exocarp), enclosing the pericarp that is the pulp of the fruit, the mesocarp, called mucilage and the endocarp, a film covering the seeds, normally two. The seed is about 55% of the fruit dry matter. The other part is the waste of the process to obtain the dried coffee grain to the commerce (Clifford and Martinez, 1991).

Two processes are used to separate the seed from the fruit: The 'wet' process, used only in about 2% of the processed coffee in Brazil in which the separation is made using water. This technique generates two kinds of waste, a solid one, mainly the pulp, with a large amount of water, and a liquid with a large concentration of solids, sugar and pectic compounds. The second process used to separate the seed form the fruit is the 'dry' process, used in about 98% of the processed coffee in Brazil in which the harvest is dried in sun or in dryers. In 'dry' process the fruits need to be rinsed, similar to the wet process, producing two fractions, the floating one called "*boia*" and the coffee called "*miudo*". The two fractions are dried separately. The drying process can be natural or with equipment. During natural drying, the grains of fruits are put at the surface, onto earth, bricks or asphalt, and are allowed to dry till the water content of 11-12%. With

T. Sera et al. (eds.), Coffee Biotechnology and Quality, 409–417.

equipment, the grains are dried using forced convection in dryers till the same water content is achieved. After drying the grains are cleaned from dirt such as stones, earth, leaves, etc. The dry and clean grains are broken and peel is separated from the seed (Masson, 1985). Coffee husk represents more or less 50% of the dried fruit, and thus a huge amount is generated every year in Brazil as well as in other countries. Coffee husk is highly biodegradable. Usually it is dumped in watercourses or on the soil leading to serious pollution.
The two varieties of coffee, robusta and arabica are cultivated in Brazil and their husk composition is shown at Table 1.

Table 1. Chemical composition of the coffee husk

Component (%)	Robusta Coffee	Arabica Coffee
Protein	9.2	11.3
Lipid	2.0	17
Cellulose	27.6	13.2
Extracts non nitrogen	57.8	66.1
Tannins	4.5	-
Pectic matter	6.5	-
Reducing Sugar	12.4	-

Source: Tango (1971)

Because of its high biodegradability, high content of organic matter, reducing sugars, protein, nitrogen and minerals, coffee husk could be a potential source for being used in animal feeding, or as substrate for the production of bio-molecules of high commercial value. To be used in animal feeding, the coffee husk must be detoxified (because of its high contents of caffeine, tannins, and phenolics). If to be sued as substrate for fermentation to produce organic acids, enzymes, protein from micro-organisms, etc., coffee husk could be used directly in solid state fermentation (SSF) or in the form of its hydrolysate (Pandey and Soccol, 2000; Woiciechowski *et al.*, 1999a).
In order to obtain the coffee husk hydrolysate, the material must be pre-treated with enzymes or with temperature in water with or without acid catalysis. This could hydrolyse the long chains of carbohydrates such as cellulose and lignin into fermentable sugars. The acid hydrolysis has been typically investigated as a possible treatment of

ligno-cellulosic materials, because acids catalyse saccharification reactions (Pandey and Soccol, 1998; Pandey *et al.*, 1999a,b). Generally, acid treatment is effective to dissolve the hemicellulosic fractions of the biomass (Elander and Hsu, 1995; Woiciechowski *et al.*, 1999b).

Acid hydrolysis consists on putting the material under pressure and temperature for a period of time, using acids and water. The pre-treatment conditions must be optimized, because severe conditions of acid concentrations and temperature can cause the degradation of the sugars, producing furfural and other toxic compounds such as acetic acid and phenolics that inhibit the fermentative processes. On the other hand, light conditions result incomplete extraction of the fermentable sugar from the material.

The main point of our work was to evaluate the hydrolysis conditions with respect to time and temperature of the reaction with or without acid catalyst.

2. Experimental

2.1. SUBSTRATE CHARACTERIZATION

Coffee husk was characterized for its chemical composition (AOAC, 1975) such as a) moisture: direct heating to 105°C till constant weight, b) ash: direct heating till 600°C for four hours, c) fibre: gravimetric method, d) lipid Soxhlet extraction of the solutes in *n*-hexane, e) protein: Kjedhal, f) reducing sugars: Somogyi-Nelson method, g) total sugars: acid hydrolysis and Somogyi-Nelson method.

2.2. ACID HYDROLYSIS

Acid hydrolysis was carried out using 10 g of husk in 50 ml of acid-water mixture (200 $g.l^{-1}$) in 250-ml flask. HCl (1-N) was used as the acid catalyst. Hydrolysis was carried out with water only also (auto-hydrolysis). Reactions were carried out at 100, 120, 140 and 160°C for 15, 30 and 60 minutes. After the hydrolysis the samples were cooled to environmental temperature, filtered and analysed to determine the reducing sugar recovered after the pre-treatment. The reducing sugar was analysed using the Somogyi-Nelson method (Nelson, 1944).

2.3. OPTIMIZATION OF AUTO-HYDROLYSIS

The material was prepared with the same amount of coffee husk as used during the acid hydrolysis (200g of coffee husk/L of water). The auto-hydrolysis optimization was done using an experimental design 3^2. Two experimental factors (pre-treatment time and temperature), distributed in three levels were tested. The data analysis was done using surface response methodology through the program "*statistica*". The response variable was the concentration of reducing sugar in the hydrolysates. The real and coded values are shown in Table 2.

Table 2. Experimental variable values used at the experimental design.

Coded values factors	-1	0	+1
Temperature °C	100	120	140
Time, minutes	5	10	15

3. Results and Discussion

3.1. SUBSTRATE CHARACTERIZATION

Coffee husk contained (%) moisture 12.3, ash 5.57, fibres 26.18, greases 1.66, protein 6.5, reducing sugars 11.24 and total sugars 36.43.

3.2. ACID HYDROLYSIS

Fig. 1 shows the results of reducing sugar obtained in the hydrolysates by the acid hydrolysis using HCl as the catalyst. As can be seen from the figure, the best temperature for the acid hydrolysis was 120°C. Increasing the time of reaction at this temperature from 15-60 minutes resulted increase in reducing sugar concentration from 23.45-28.5 $g.l^{-1}$. However, this long period of time could not justify the increase in reducing sugar.

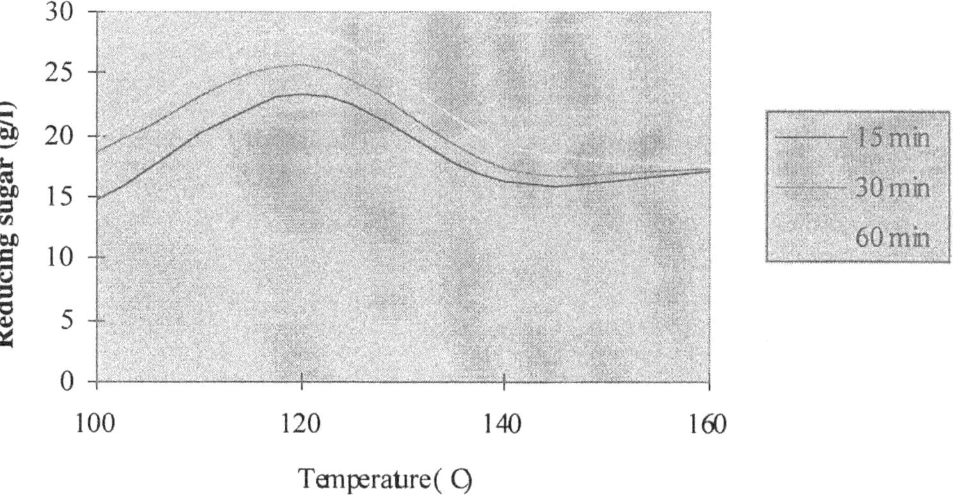

Figure 1. Effect of the temperature and time on hydrolysis of coffee husk using HCl

3.3. AUTO HYDROLYSIS

The reducing sugar determination results obtained with the hydrolysates produced by auto-hydrolysis are shown in Table 3.

Table 3. Chemical composition of the coffee husk used in the experiments

Components	Value (%)
Moisture	12.33
Ash	5.57
Fibre	26.18
Lipids	1.66
Protein	6.59
Reducing Sugar	11.24
Total Sugar	36.43

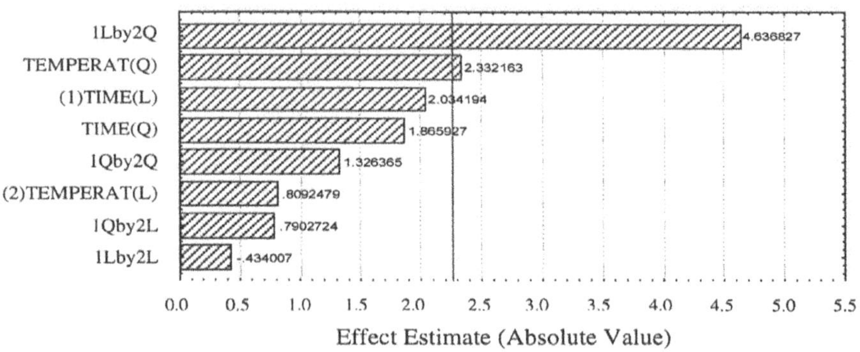

Figure 2. Pareto chart showing reducing sugars concentration in hydrolysates.

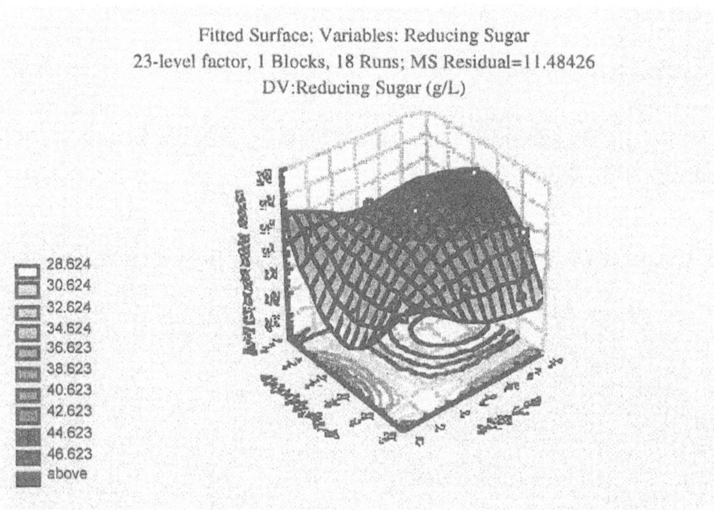

Figure 3. Surface response for reducing sugars concentration in hydrolysates in relation to temperature and hydrolysis time.

The results were submitted for the response surface analysis and statistical analysis. The Pareto graphic as shown in Figure 2 shows the significance levels of each experimental factor tested. From Fig. 2, it is possible to see that the temperature was the most significant factor at the range tested at the significant level of 5% as long as it interacted with the variable time. Considering the amount of reducing sugar recovered, the time of reaction was not significant at this level.

Figures 3 and 4 show the surface response and the contour graphic obtained for the reducing sugar concentration in hydrolysates related to the experimental factors tested. From these, it was possible to see that temperatures between 115-125°C and time of 11-15 minutes were the most favourable conditions to be used to prepare coffee husk hydrolysates.

One litre of hydrolysate prepared from 200 g of coffee husk contained 47.3 g of reducing sugars or 236.5 g of reducing sugars.kg^{-1} of coffee husk. Analysis of data as shown in Table 4, however, showed that the best result was obtained when hydrolysis was carried out at 120°C for 10 minutes.

Table 4. Reducing Sugar concentration (g.l^{-1}) in hydrolysates.

Reaction	Temperature		
	100°C	120°C	140°C
5 minutes	37.6	31.1	39.1
10 minutes	36.8	44.3	40.6
15 minutes	42.8	47.5	35.7

This apparently confusing result about hydrolysis time probably happened due to the fact that this variable (time) was tested in a narrow range and the differences in the response variable according to time were not significant. This was confirmed from the results shown in Fig. 2, which showed that temperature was significant at a level of 5% of significance, but time was not significance factor. Hence, a period of 10-15 minutes could be used with efficiency to recover reducing sugars from coffee husk in liquid phase.

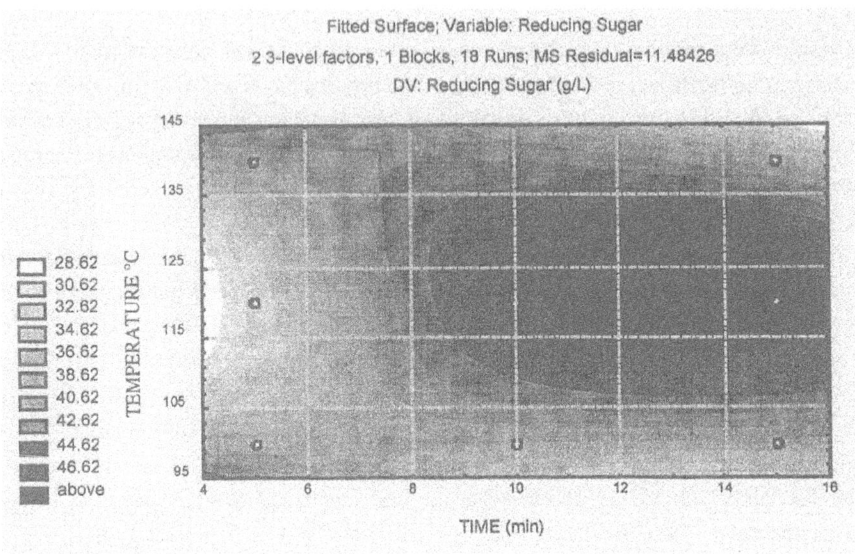

Figure 4. Contour graphic for the response variable (reducing sugar concentration in hydrolysates) related to the temperature and time

The decrease in the concentration of reducing sugar in hydrolysates obtained at 140°C showed that higher temperatures caused the degradation of the reducing sugars. Hence, no further higher temperatures were not tested.

4. Conclusions

Comparing the best result of reducing sugars in the hydrolysates obtained during the hydrolysis with hydrochloric acid (acid catalysis), 28.5 $g.l^{-1}$ and that obtained during the hydrolysis only with water (auto hydrolysis), 47.5 $g.l^{-1}$, both under the similar experimental conditions of temperature and time, it could be concluded that auto hydrolysis was a more effective process with the techo-economical advantages that in this it was not necessary to add any chemical agent.

5. Acknowledgements

Financial assistance from the European Union (grant no INCO DC: IC18*CT 970185) and PNP & D/CAFE- Coordenatior EMBRAPA, Brazil (Projeto no 07.1.99.057) is gratefully acknowledged. C.R. Soccol thanks CNPq, Brazil for a scholarship under Scientific Productivity Scheme.

6. References

AOAC- Association of Official Analytical Chemists (1975), Analysis of foods and other aliments, Official Methods of Analysis, 20th Ed., Horwitz, W. (ed.), Washington

Clifford N.N., Martinez J.R.R. (1991), Phenols and Caffeine in wet-processed coffee beans and coffee pulp. Food Chem., 40: 35-42

Elander R., Hsu T. (1995), Processing and economic impacts of biomass delignification for ethanol production. Appl. Biochem. Biotechnol. 51/52: 463-468

Masson, M.L. (1985), Tecnologia do Café. Boletin CEPPA, 3(2), 1-20

Nelson N.A. (1944), Photometric adaptation of the Somogyi method for the determination of glucose. J. Biol. Chem., 153: 375-378

Pandey A., Soccol C.R., Nigam P., Soccol V.T., Vandenberghe L.P.S., Mohan R. (1999a), Biotechnological potential of agro-industrial residues: II Cassava bagasse. Biores. Technol., in press

Pandey A., Soccol C.R., Nigam P., Soccol V.T. (1999b), Biotechnological potential of agro-industrial residues: I Sugarcane bagasse. Biores. Technol., in press

Pandey A., Soccol C.R. (1998), Bioconversion of biomass: a case study of lignocellulosic bioconversions in solid state fermentation. Brazilian Arch. Biol. Technol., 40: 379-390

Pandey A., Soccol C.R. (2000), Economic utilization of crop residues for value addition - A futuristic approach J. Sci. Ind. Res., 59: 12-22

Tango J.S. (1971), Utilizacao industrial do cafe e dos seus subprodutos. Bol. Inst. Tec. Alim., 28, 49-73

Woiciechowski A.L., Soccol C.R., Pandey A., Bustao E. (1999a), Selective hydrolysis of hemicellulose fraction of coffee husk for production of sugars. Paper presented in III SIBAC, May 24-28, Londrina, Brazil, p. 54

Woiciechowski A.L., Soccol C.R., Ramos L.P., Pandey A. (1999b), Experimental design to optimize the production of L- (+)-lactic acid from steam exploded wood hydrolysate using *Rhizopus oryzae*. Process Biochem., 34 (9): 949-955

Chapter 39

A NOVEL APPROACH FOR THE PRODUCTION OF NATURAL AROMA COMPOUNDS USING COFFEE HUSK

M. Soares[1], A. Pandey[1a], P. Christen[2], M. Raimbault[3], C.R. Soccol[1]

[1]Laboratório de Processos Biotecnológicos, Departamento de Engenharia Química, Universidade Federal do Paraná, Curitiba, Brazil; [2]Laboratoire de Microbiologie IRD, IFR-BAIM, Université de Provence, ESIL, Case 925, 163 Avenue de Luminy, 13288 Marseille cedex 9, France; [3]Laboratoire de Biotechnologie, IRD, BP 5045; 34032 Montpellier cedex 1, France; [a]currently at Biotechnology Division, Regional Research Laboratory, Trivandrum-695 01, India

Running title: Production of aroma compounds

1. Introduction

Flavouring compounds have been traditionally obtained from plant sources but there are strong dependences on factors such as the influence of the weather, plant diseases etc... that are difficult to fully control. As a result, alternative sources were exploited and chemical synthesis was found very effective. Most of the available flavour compounds (84%) are now produced via chemical synthesis, although extraction from natural material continues. However, recent trends have demonstrated that consumers globally prefer foodstuff that can be labelled as natural. A directly viable alternative route for their production can be based on microbial processes. Often, the market prices of such compounds produced from natural materials has been very high. For example, synthetic 4-decalactone (the impact flavour compound of peach) costs US$ 150/kg, while the same substance extracted from a natural source costs about US$ 6000/kg (Janssens *et al.*, 1992). Several microbial species produce volatile fruity aromas. Fungi from the genus *Ceratocystis* produced a large diversity of fruit-like aromas (peach, pineapple, banana, citrus and rose), depending on the strain and culture conditions. Among this genus, *C. fimbriata* seems to be interesting because of its relatively rapid growth, its

T. Sera et al. (eds.), Coffee Biotechnology and Quality, 419–425.

good ability for spore production and the variety of aromas synthesized (Senemaud, 1988; Christen *et al.*, 1994).

One approach can be to use agro-industrial residues such as cassava bagasse, sugarcane bagasse, sugar beet pulp, coffee husk and coffee pulp, etc... (Pandey *et al.*, 1999a, b; Pandey and Soccol, 2000) and cultivate the micro-organisms in solid state fermentation (SSF). SSF has been termed as a potential tool for adding value to agro-industrial residues by production of organic acids, enzymes, secondary metabolites, amino acids, aroma compounds, etc... (Pandey, 1992, 1994; Soccol, 1996; Soccol and Krieger, 1998; Nampoothiri and Pandey, 1996; Pandey and Soccol, 1998; Pandey *et al.*, 1999c, d, e, f, 2000). Coffee husk is a fibrous mucilaginous material obtained during the processing of coffee cherries by dry process. It contains some amount of caffeine, tannins and polyphenols, which makes it relatively toxic. However, it is rich in organic matter, which makes it potentially interesting as a substrate for microbial processes for the production of added value compounds. Several alternative uses of the coffee husk have been tried. These include fertilizers, livestock feed, compost, etc... However, these applications utilize only a fraction of the available quantity and are not technically very efficient. Attempts have been made to detoxify it by degrading caffeine and tannins for its application as feed, and to use as substrate for the production of enzymes, organic acids, mushrooms, etc. (Woiciechowski *et al.*, 1999; Roussos *et al.*, 1994; Brand *et al.*, 1999, 2000). We attempted the production of fruity aroma compounds by SSF using pretreated coffee husk as substrate.

2. Micro-organism, substrate and SSF

A strain of *Ceratocystis fimbriata* CBS 374-83 was used. It was grown and periodically transferred on Potato-dextrose-agar (PDA) slants and stored at 4°C (Soares, 1998; Soares et al., 1999, 2000). Coffee husk, used as substrate, was sieved (0.4-0.8 mm particle size) and treated with hot water (Soares 1998). Fermentation was carried out in 250-ml Erlenmeyer flasks. For all experiments initial pH was 6.0 and moisture of the substrate was adjusted to 70%. The inoculum size was $1x10^7$ spores.g^1 dry matter and incubation temperature 30°C. Some experiments were performed to study the effect of supplementation of this substrate with glucose, leucine, soybean oil, and a nutrient salt solution in order to improve the volatile compound production.

3. Data analysis

Aroma compounds were separated and quantified by GC analysis according to Soares *et al.* (2000). The raw data were integrated in order to calculate the total volatiles (TV) accumulated during the fermentation (Soares *et al.*, 2000). The Gompertz model, a logistic like equation, was used to fit these integrated data, as previously described by other authors (Meraz *et al.*, 1992; Bramorski *et al.*, 1998). This model describes the dynamics of the production with respect to time. Data integration and non-linear Gompertz regression were made with KaleidaGraph program (Abelbeck Software, USA).

4. Fermentation profile of aroma compounds

Experiments using whole coffee husk, hot water treated coffee husk and the liquid extract obtained after the filtration of hot water treatment showed the superiority of the treated coffee husk for both fungus growth and aroma production (Soares *et al.*, 1999). This probably could be due to the removal of anti-physiological factors from the whole coffee husk during the treatment. Addition of glucose in coffee husk resulted in production of strong fruity aroma, which depended on the amount of glucose added (Soares *et al.*, 2000). Experimental results showed that although the fungal culture tolerated as high as 46% glucose concentration, lower concentrations of glucose appeared suitable for its growth and aroma producing activity. At the lowest experimental concentration of glucose (20%), TV production was fastest, and the maximum was about 28 μ mol/l/g DM after 40 h. At 35% glucose concentration, the rate of TV synthesis was slower and the maximum was about 24 μmol/l.g DM after 300 h. This could be due to a decrease in water activity of the substrate in relation to the high glucose concentration. It could also be partially due to catabolic repression due to the carbon source (Soares *et al.*, 1999, 2000). Although these results could not be considered as conclusive, they showed that different aroma compounds with different intensities could be produced with the addition of glucose to coffee husk. Another important aspect of this study was the requirement of a relatively shorter time to produce these compounds in comparison to earlier reports (Chrjsten et al. 1997, Meza *et al.*, 1998) using cassava bagasse or apple pomace (5 and 4 days respectively). The amounts of TV reported in this study are superior to reported by Christen *et al.* (1997) with the same fungus grown on wheat bran or cassava bagasse.

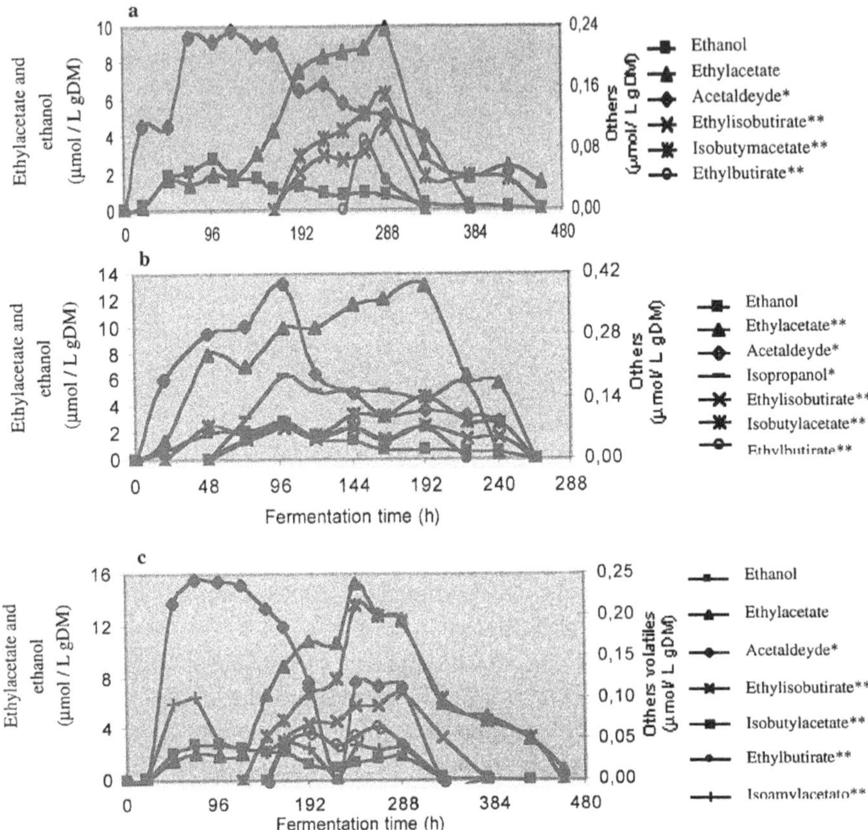

Figure 1. Dynamics of the major compounds in the headspace of the culture with 35% glucose (a), 20% glucose (b) and 35% glucose+leucine (c). * µmol ethanol equivalent.l^{-1}.$g.^{-1}$DM; **µmol ethyl acetate equivalent.l^{-1}.$g.^{-1}$DM)

Studies on the effect of supplementation with soybean oil showed no impact on aroma production, demonstrating that the fungus was not able to use soybean oil. Addition of saline solution drastically decreased the volatile production. Addition of leucine improved TV production (Soares *et al.*, 2000). After the hot water treatment, the coffee husk lost about 30% in weight, which was mostly due to solublization of different minerals and salts present in it. Since hot water treatment or in other words, removal of these salts from the coffee husk improved the TV production, it could be considered logical only to observe such an effect on again supplementing the coffee husk with salts.

5. Compounds produced

Gas chromatography analysis allowed the identification of several compounds (Fig. 1). Under optimized conditions, a total of 13 compounds were produced which included alcohols (ethanol, isopropanol), aldehyde (acetaldehyde), ketones (2-heptanone, 2-octanone) and esters (ethyl acetate, ethyl isobutyrate, isobutyl acetate, ethyl butyrate, isoamyl acetate, propyl acetate, ethyl-3-hexanoate). Ethyl acetate was the prominent compound, followed by ethanol (Soares *et al.*, 1999, 2000).

Figure 1 shows the dynamics of the compounds measured in the headspace of the culture supplemented with glucose (20 and 35%) and with leucine. Addition of glucose (20%) reduced the fermentation time and increased the formation of isopropanol. Addition of leucine although did not alter the evolution of the compounds which was similar to control, it increased ester synthesis (Soares, 1998).

6. Conclusions

Coffee husk can be used as substrate in SSF for the cultivation of fungal strains to produce aroma compounds. Hot water treatment of the husk greatly improved the metabolic activity, particularly the synthesis of esters which is known to be a way for some micro-organisms to avoid a possible inhibition due to acids accumulation in the medium. The fungus was better adapted to this substrate than others mentioned in the literature.

7. Acknowledgements

Financial assistance from the European Union (grant no INCO DC: IC18*CT 970185) and PNP & D/CAFE- Coordenatior EMBRAPA, Brazil (Projeto no 07.1.99.057) is gratefully acknowledged. C. R. Soccol thanks CNPq, Brazil for a scholarship under Scientific Productivity Scheme.

8. References

Bramorski A., Christen P., Ramirez M., Soccol C.R., Revah S. (1998), Production of volatile compounds by the fungus *Rhizopus oryzae* during solid state cultivation on tropical agro-industrial substrates. Biotechnol. Letters, 20: 359-362

Brand D., Pandey A., Roussos S., Raimbault M., Soccol C.R. (1999), Biological detoxification of the coffee husk using filamentous fungi. Paper presented in III SIBAC, May 24-28, Londrina, Brazil, 1999, pp. 38

Brand D., Pandey A., Roussos S., Soccol C.R. (2000), Biological detoxification of coffee husk by filamentous fungi using a solid state fermentation system, Enzyme Microb Technol., accepted.

Christen P., Raimbault M. (1991), Optimization of culture medium for aroma production by *Ceratocystis fimbriata*. Biotechnol. Lett., 13: 521-526

Christen P., Villegas E., Revah S. (1994), Growth and aroma production by *Ceratocystis fimbriata*. Biotechnol. Lett., 16: 1183-1188

Christen P., Meza J.C., Revah S. (1997), Fruity aroma production in solid state fermentation by *Ceratocystis fimbriata*: influence of the substrate type and the presence of precursors. Mycol. Res., 101: 911-919

Janssens L., De-Poorter H.L., Vandamme E.J., Schamp N.M. (1992), Production of flavours by micro-organisms. Process Biochem., 27: 195-215

Meraz M., Shirai K., Larralde P., Revah S. (1992), Studies on bacterial acidification process of cassava (*Manihot esculenta*). J. Sci. Food Agric., 60: 457-463

Meza J.C., Christen P., Revah S. (1998), Effect of added amino acids on the production of fruity aroma by *Ceratocystis fimbriata*, Sci. Alim., 18: 627-636.

Nampoothiri K.M., Pandey A. (1996), Solid state fermentation for L-glutamic acid production using *Brevibacterium* sp., Biotechnol Letts, 16: 199-204

Pandey A. (1992), Recent developments in solid state fermentation, Process Biochem., 27:109-117

Pandey A. (1994), Solid state fermentation- an overview, In- Solid State Fermentation, A. Pandey (ed.), Wiley Eastern Publishers, New Delhi, pp. 3-10

Pandey A., Soccol C.R. (1998), Bioconversion of biomass- a case study of ligno-cellulosics bioconversions in solid state fermentation, Brazilian Arch. Biol. Technol., 41 (1998): 379-390

Pandey A., Nigam P., Vogel M. (1988), Simultaneous saccharification and protein enrichment fermentation of sugar beet pulp. Biotechnol. Letts., 10: 67-72

Pandey A., Soccol C.R., Nigam P., Soccol V.T., Vandenberghe L.P.S., Mohan R. (1999a), Biotechnological potential of agro-industrial residues: II Cassava bagasse. Biores. Technol., in press

Pandey A., Soccol C.R., Nigam P., Soccol V.T. (1999b), Biotechnological potential of agro-industrial residues: I Sugarcane bagasse. Biores. Technol., in press

Pandey A., Benjamin S., Soccol C.R., Nigam P., Krieger N., Soccol V.T. (1999c), The Realm of microbial lipases in biotechnology. Biotechnol. Appl. Biochem., 29: 119-131

Pandey A., Selvakumar P., Soccol C.R., Soccol V.T., Krieger N., Fontana J.D. (1999d), Recent developments in microbial inulinases - Its production, properties and industrial applications. Appl. Biochem. Biotechnol., 81: 35-52

Pandey A., Selvakumar P., Soccol C.R., Nigam P. (1999e), Solid state fermentation for the production of industrial enzymes, Curr. Sci., 77: 149-162

Pandey A., Azmi W., Singh J., Banerjee U.C. (1999f), Fermentation types and factors affecting it. In - Biotechnology: Food Fermentation. Vol. I, V. K. Joshi and A. Pandey (eds.), Educational Publishers & Distributors, New Delhi, pp. 383-426

Pandey A., Soccol C.R (2000), Economic utilization of crop residues a futuristic approach. J. Sci. Ind. Res., 59: 12-22

Pandey A., Soccol C.R., Mitchell D. (2000), New developments in solid state fermentation. I- bioprocesses and products. Process Biochem., in press

Roussos S., Hannibal L., Aquiahuatl M.A., Trejo Hernandez M.R., Marakis S. (1994), Caffeine degradation by *Penicillium verrucosum* in solid state fermentation of coffee pulp: critical effect of additional inorganic and organic nitrogen sources. J. Food Sci. Technol., 31: 316-319

Senemaud C. (1988), Les champignons filamenteux producteurs d'arômes fruités. Etudes de faisabilité sur substrats agro-industriels. Ph.D. Thesis, Université de Bourgogne, France.

Soares M. (1998), Biossintese de metabolitos volateis frutais por *Pachysolen tannophilus* e *Ceratocystis fimbriata* pela fermentacao no estado solido da casca de cafe. Master's Thesis, Universidade Federal do Parana, Curitiba, Brazil

Soares M., Soccol C.R., Christen P., Pandey A., Raimbault M. (1999), Characterization of the fruity aroma compounds produced by *Ceratocystis fimbriata* in solid state fermentation of coffe husk. Paper presented in VI Seminar on Enzymatic Hydrolysis of Biomassa, Maringá, Brazil, 6-10 December, p.130

Soares M., Christen P., Pandey A., Soccol C.R. (2000), Fruity flavour production by *Ceratocystis fimbriata* grown on coffee husk in solid state fermentation. Process Biochem., in press.

Soccol C.R. (1996), Biotechnology products from cassava root by solid state fermentation, J. Sci. Ind. Res., 55, 358-64

Soccol C.R., Krieger N. (1998), Brazilian experiments on valorisation of agro-industrial residues through solid state fermentation, In- Advances in Biotechnology, A. Pandey (ed.), Educational Publishers, New Delhi, pp. 25-36

Woiciechowski A.L., Soccol C.R., Ramos L.P., Pandey A. (1999), Experimental design to optimize the production of L- (+)-lactic acid from steam exploded wood hydrolysate using *Rhizopus oryzae*., Process Biochem., 34: 949-955

Chapter 40

PRODUCTION OF MUSHROOMS ON BRAZILIAN COFFEE INDUSTRY RESIDUES

F. Leifa, A. Pandey[a], C.R. Soccol

Laboratorio de Processos Biotecnologicos, Departamento de Engenharia Quimica, Universidade Federal do Parana, CEP 81531-970, Curitiba-PR, Brazil; [a]currently at Biotechnology Division, Regional Research Laboratory, Trivandrum-695 01, India

Running title: Mushroom cultivation on coffee husk

1. Introduction

Coffee is one of the most important beverages of the world. The crop is cultivated in several countries in Latin America, Asia and Africa among which Brazil is the largest producer. At different stages from harvesting to the processing and consumption, several residues such as coffee pulp/husk, leaves and spent-ground are generated in more than two millions tons quantity yearly (Soccol, 1995; Fan et al. 1999a, b; Pandey and Soccol, 2000). In Brazil coffee cherries are generally processed by the dry method (ICO, 1998). Coffee husk is rich in organic nature and nutrients but also contains compounds such as caffeine, tannins, and polyphenols, which result in toxic nature of the husk. Coffee leaves are mostly collected during harvesting in large volume. Their disposal also difficulties in crop management as they could facilitates epidemic of pathogens and pests. Coffee spent-ground is obtained during the processing of raw coffee powder to prepare 'instant coffee'. It also shows toxic nature as contains caffeine and tannins, although in less quantities when compared with husk. Presence of caffeine, tannins and polyphenols has affected utilization of these residues beneficially and their disposal in the environment poses serious pollution concerns (Fan et al. 1999a, b, c, Pandey and Soccol, 2000).

T. Sera et al. (eds.), Coffee Biotechnology and Quality, 427–436.

Newer advents in biotechnology offer strong potentials for the utilization of agro-industrial residues in economic way (Pandey and Soccol, 2000; Pandey et al., 1999a, b). One of the possibilities to effectively handle these residues could be their utilization as substrate in bioprocesses such as mushroom cultivation, production of enzymes and organic acids, etc. The edible mushroom *Pleurotus* sp. is considered as a good alternative as protein rich food production in tropical countries (Bisaria, 1983; Yang, 1986; Chang, 1989). It generally shows good ability of producing fruiting body and simultaneously reducing or degrading the toxic substances present in the substrate (Yang, 1986; Fan and Ding, 1990; Fan *et al.* 1999c). There are reports on production of *Pleurotus* on the coffee pulp (Martinez and Quirarte, 1984) and coffee-spent ground (Thielke, 1989).

Lentinus edodes is also an important edible mushrooms cultivated world-wide (Yang, 1986; Fan and Ding, 1990). It has excellent organoleptical properties of flavour and aroma and possesses good nutritional and therapeutically values due to the presence of essential amino acids required for humans. It also shows anti-tumoral activity, helps in reducing the cholesterol and strengthening the immunological system. *L. edodes* has also been cultivated on coffee pulp (Mata, 1994).

Flammulina is yet another mushroom cultivar considered important in the category of edible mushrooms. It was produced approximately 143,000 tons during 1990. The production increased to 230,000 tons in 1994, showing a steep hike of 61% (Chang, 1996). During 1995, about 100,000 tons were produced in China (mainland) alone. From these data, it is evident that a faster growth rate, in terms of total production, was being enjoyed globally. In the United States, for example, the production of *Flammulina* has increased at an estimated rate of 25% or more per year for the last four years (Royse, 1995). Production of *Flammulina* is based on synthetic substrate contained in polypropylene bottles. Substrates constitute primarily of sawdust and rice bran at 4:1 ratio. Coffee spent ground has also been used for cultivating a strain of *F. velutipes* (Thielke, 1989).

Not much information is available on the cultivation of edible mushrooms on various residues of coffee industries in Brazil. The enormous amounts of these residues could be used through biotechnological means, for example as a substrate to produce mushrooms in solid state fermentation (SSF), which has been termed as the most potential way to produce mushrooms (Pandey, 1992, 1994; Pandey *et al.*, 2000). As a secondary effect, the amounts of toxic substances within the residues are at least partially degraded which may be, therefore, useful as livestock food in addition to the primary mushroom production process.

2. Selection of strains

Eight strains of *P. ostreatus*, 12 strains of *L. edodes* and one strain of *F. velutipes* were screened for their efficiency to grow was used on a medium containing the extract of coffee husk and agar (pH 7) (Fan et al., 2000a, b). Routinely these strains were preserved on Potato-Dextrose-Agar (PDA) at 4oC.

3. Preparation of spawn and substrate and SSF

The sawdust of *Eucalyptus* sp. (80%) and rice bran (20%) was used for the spawn preparation as described by Yang (1986). After autoclaving (121°C, one h), the spawn medium was inoculated with bits (one disc of one cm in diameter) of mycelia of strains from the PDA slants and then incubated at 24°C in dark. The spawn in the jars was ready for inoculation to the substrate after 20 days growth when the mixture turned totally white (Fan *et al.*, 2000a,b).

The coffee husk, coffee leaves and spent-ground (sun dried) were obtained from the local factories. Leaves were dried and milled in a grinder. These substrates were moistened with water for 4-5 h at 60-65, 60-70 and 50-55% for coffee husk, coffee leaves and spent-ground, respectively before filling in the plastic bags for SSF. The bags (20x35 cm size, 100 g substrate in each bag on dry wt basis) were autoclaved at 121°C for 1.5 h. It was inoculated with the spawn (10%) and mixed thoroughly to facilitate rapid and uniform mycelial growth. The bags were incubated in the dark at 24°C. For *F. velutipes* the substrate was transferred to glass bottle (500-ml) when inoculating for facilitating its fructification later.

After 15 days, the bags of *Pleurotus* were transferred to a lighted environmental chamber (90% relative humidity, 24oC), the plastic was removed to allow stimulation of air, humidity and light to facilitate fruiting body development. The glass bottle was uncovered when some primordia of *Flammulina* appeared after 15 days incubation. For *Lentinus*, the plugs were removed for a little more air to accelerate the transformation of white colour to brown after 25 days incubation. When the colour on the surface of substrate turned to brown, the plastic was removed for fructification

4. Production of *Pleurotus sp.*

Although all the strains grew well on coffee husk extract medium, one of the strains, *viz.* *P. ostreatus* LPB 09 appeared the best strain with high density mycelial growth, resulting the highest radial mycelial growth velocity of 9.68 $mm.day^{-1}$ and biomass 43.4 $mg.plate^{-1}$ in 9 days. The strains of *P. ostreatus* showed better growth, in general, than *P. sajor-caju* (Fan et al. 2000a). Results demonstrated the biological efficiency of *P. ostreatus* LPB 09 on the coffee husk, spent-ground and the mixture of spent ground + leaves (60:40) as 96.5, 90.4 and 76.7%, respectively in 60 days (Table 1). When coffee leaf was used as the substrate, it took five days for full occupation of mycelia under the experimented conditions, but no fruit body was obtained. Although there was increase in the protein content of the fermented matter with the increase in spawn rate from 10-25%, there was not much difference between the results obtained with 10 and 25%. Hence, from economic feasibility point of view, a uniform spawn rate of 10% was considered suitable with all the substrates (Fan *et al.* 2000a).

Table 1. Biological efficiency of different mushroom species on residues of coffee.

Substrate	*P. ostreatus* (60 d)	*L. edodes* (100 d)	*F.velutipes* (50 d)
Coffee husk	96.5	85.8 *	55.8
Spent-ground	90.4	88.6	78.3
Coffee leaves	0	-	-
Mixed substrate	68.5**	78.4***	-

*coffee husk was pre-treated with boiling water for one hour ; **spent-ground 60% + leaf 40% ; ***husk 50% + spent ground 50%

There are a few reports describing the production of *Pleurotus* sp on coffee pulp (Martinez and Quirarte, 1984; Martinez *et al.*, 1985, 1990; Soto et al., 1986, 1987; Lozano, 1990; Bermudez and Traba, 1994; Gaitan and Salmones, 1996) or on coffee pulp mixed with other residues (Bernabe *et al.*, 1993; Galzada *et al.*, 1987).

During SSF, there was increase in the protein content of coffee husk, spent-ground and mixed substrate with the increase in cultivation period. The fibre content decreased simultaneously (Fan *et al.*, 2000a) (Table 2). There are other reports, which have found that *Pleurotus* degraded cellulose and lignin (fibre) (Zadrazil, 1980; Tsang *et al.*, 1987; Valsameda *et al.*, 1991) during SSF.

Table 2. Changes of contents of protein and fibre during SSF by different mushroom species on residues of coffee.

Substrate	*P. ostreatus* (60 d)		*L. edodes* (100 d)		*F. velutipes* (50 d)	
	protein	fibre	protein	fibre	protein	fibre
Coffee husk	+26.08	-30.79	+15.89	-18.15	+24.68	+10.70
Spent-ground	+22.08	-26.17	+24.32	-12.16	+27.05	-7.25
Coffee leaves	+10.06*	-8.79*	-	-	-	-
Mixed substrate	+21.93	-32.00	+18.98	-15.19	-	-

+ increase in % ; - Decrease in %; *25 days SSF

5. Production of *L. edodes*

As was the case with *Pleurotus*, all the strains of *Lentinus* sp. screened grew well on coffee husk extract medium. The strain *L. edodes* LPB 02 showed best growth in general with high density mycelial growth (9.68 $mm.day^{-1}$) and biomass (43.4 $mg.plate^{-1}$) in 12 days. When coffee husk was used as the substrate, it took 20 days for full occupation of mycelia but with the time of growing the mycelia regressed and secreted water so no fruit body was obtained (Fan *et al.*, 2000b). Beaux and Soccol (1996) utilised the coffee husk for growing of *Lentinus* and reported poor mycelial growth in comparison to other substrates. When coffee husk was subjected to a hot water treatment by boiling one-hour in the water, filtering and using the solid residue for cultivating *Lentinus*, the growth was very vigorous. After full occupation of mycelia, the transformation of colour began. First fructification occurred after 60 days of inoculation and the biological efficiency reached at the 85.8% (Table 1). When coffee spent ground was used as the substrate, it also took 20 days for full occupation of mycelia; first fructification occurred after 56 days of inoculation; the biological efficiency reached at the 88.7%. With the mixed substrate, it took 25 days for full occupation of mycelia; first fructification occurred after 65 days of inoculation and the biological efficiency reached at the 78.4% (Fan et al. 2000b). The hot water treatment of the coffee husk probably diminished its soluble contents,

apparently including those, which were exerting toxic effect and inhibiting fungal growth and fruiting body formation.
The trends of protein and fibre contents were similar as observed with *Pleurotus* sp. There was increase in the protein content and decrease in fibre content with increase in time of cultivation.

6. Production of *F. velutipes*

The experimental strain of *F. velutipes* LPB 01 grew well on coffee husk extract medium. The mycelial growth velocity was 7.87 $mm.day^{-1}$ and biomass 45.78 $mg.plate^{-1}$ in 10 days. When coffee husk was used as the substrate, it took 15 days for full occupation of mycelia and the primodia occurred after 25 days of inoculation. The biological efficiency reached at 55.8%. There has not been any work of production of *Flammulina* in the coffee husk and this is the first finding. With spent-ground as substrate, it took 12 days for full occupation of mycelia; first fructification occurred after 21 days and the biological efficiency reached 78.3% at two flush in 45 days. In this case also, the pattern in the change of protein and fibres contents was almost similar to that with *Pleurotus* and *Lentinus* sp. With the different substrates with the increase in cultivation period, the protein content increased and the fibres decreased (Table 2).

7. Contents of caffeine and tannins in the fruit body and fermented husk

Table 3 shows the content of caffeine and tannins in the fruit body of *Pleurotus, Lentinus* and *Flammulina,* and respective fermented residues of husk. The fermented husk showed a decrease of about 61 and 80%, respectively for caffeine and tannins. The fruit body of *Pleurotus* showed 1.25% caffeine and 0.76% tannins, which showed that apparently *Pleurotus* did not degrade caffeine and tannins. The fruit body of *Lentinus* did not contain caffeine or tannins. This could be due to the fact that the husk was pre-treated with hot water. However, in the final residues, decrease in tannins content was only 12.3%. The fruit body of *Flammulina* also did not show any caffeine or tannins and in the fermented residues the caffeine and tannins contents decreased (10.2 and 20.4%, respectively).

Table 3. Contents of caffeine and tannins in the fruit body and fermented substrates.

Analysis	*Pleurotus*		*Lentinus*		*Flammulina*	
	caffeine	tannins	caffeine	tannins	caffeine	tannins
Fruit body (%)	0.2518	0.76	0	0	0	0
Substrate(husk) (% decrease)	61.34	79.76	0	12.33	10.21	20.37

8. Conclusions

Although the increase in the contents of protein and decrease in the fibre was not substantially high, the results were considered significant as these open an avenue for the utilization of these residues. In the laboratory studies, a uniform spawn rate of 10% was considered suitable for all the substrates. However, this must be verified if scale-up is planned to avoid contamination risk. Fermented husk after the production of mushrooms could be used as feed supplement for cattle and other livestock. Results showed the feasibility of using Brazilian coffee industry residues as substrates for cultivation of edible fungus in SSF. These provide one of the first steps in economical utilization of these otherwise unutilised or poorly-utilized residues.

9. Summary

The results showed that the strains of *P. ostreatus, P. sajor-caju, L. edodes and F. velutipes* were able to grow in SSF on different residues produced by coffee industries in Brazil. SSF was carried out by *P. ostreatus* under different conditions of moisture (45-75%) and spawn rate (2.5-25%) using coffee husk, leaves and spent-ground. In general, although 25% spawn rate appeared superior, considering the process economics, the 10% spawn rate was recommended for all the three substrates, as there was not any significant difference in the increase with 10-15%. The ideal moisture for mycelial growth was 60-65 for coffee husk and coffee spent ground, and 60-70 for coffee leaves. With coffee husk and spent-ground the biological efficiency reached about 97 and 90%

after 60 and 50 days, respectively. When coffee leaves were used as the substrate, no fructification was observed even on prolonged cultivation. There was significant decrease in the caffeine and tannins contents (61 and 79%, respectively) of coffee husk after 60 days.

For *L. edodes*, a spawn rate of 10% and moisture level of 55-60% was found suitable for all the substrates. Treatment of the coffee husk with hot water was found useful for its utilization by *L. edodes*. Results showed that there was an increase in the protein content and decrease in the fibre content of the substrates after SSF. Fruiting bodies were obtained from the treated coffee husk, spent ground and mixed-substrate, and the biological efficiency achieved was 85.8, 88.6 and 78.4% for these substrates, respectively. However, no fruiting body was obtained with raw coffee husk was used as the substrate. Results showed that after SSC, there was a decrease of about 27, 40 and 24% in caffeine and about 18, 49 and 12% in tannin contents in the treated coffee husk, coffee-spent ground and mixed substrate, respectively. No caffeine or, tannins were found in fruiting body indicating their degradation by the fungal strain. With *F. velutipes* the biological efficiency reached about 55.8 and 78.3% when coffee husk and spent-ground as substrates, respectively.

10. Acknowledgements

Financial assistance from the European Union (grant no INCO DC: IC18*CT 970185) and PNP & D/CAFE-EMBRAPA, Brazil (Projeto no 07.1.99.057) is gratefully acknowledged. C. R. Soccol thanks CNPq, Brazil for a scholarship under Scientific Productivity Scheme.

11. References

Beaux M.R., Soccol C.R. (1996), Cultivo do fungo comestível *Lentinula edodes* em resíduos agroindustriais do Paraná através do uso de fermentação no estado sólido. Boletim CEPPA, 14:11-21

Bermudez R.C., Traba J.A. (1994), Producción de *Pleurotus sp* cfr. Florida sobre residuales de la agroindústria cafetalera en Cuba. Micol. Neotrop. Apli., 7: 47-50

Bernabe G.T., Dominguez M.S., Bautista S.A. (1993), Cultivo del hongo comestible *Pleurotus ostreatus* var. florida sobre fibra de coco y pulpa de café. Rev. Mexicana Micol., 9: 13-18

Bisaria R., Madan M. (1983), Mushrooms: potential protein source from cellulosic residues, Enzyme Microb. Technol., 5: 251-259

Chang S.T. (1989), Edible mushroom and their cultivation, CRC Press, Inc., Florida

Chang S.T. (1996), Mushroom research and development- equality and mutual benefit. Mush. Biol. Mush. Prod., 2: 1-10

Fan L., Ding C.K. (1990), Handbook of Mushroom Cultivation, Jiangxi Science and Technology Publishing House, Jiangxi, PR China.

Fan L. (1999), Produção de fungo comestível do gênero *Pleurotus* em bioresíduos da agroindustria do café. Tese de Mestrado; Curso de Pós Graduação em Tecnologia de Alimentos da Universidade Federal do Paraná- Curitiba, Brazil

Fan L., Pandey A., Soccol C.R. (1999a), Cultivation of *Pleurotus* sp. on coffee residues. Proc. 3rd Internatl. Conf. Mushroom Biology & Mushroom Products, Sydney, Australia, pp 301-310

Fan L., Pandey A., Soccol C.R. (1999b), Growth of *Lentinus edodes* on the coffee industry residues and fruiting body production. Proc. 3rd Internatl. Conf. Mushroom Biology & Mushroom Products, Sydney, Australia, pp 293-300

Fan L., Pandey A., Vandenberghe L.P.S., Soccol C.R. (1999c), Effect of caffeine on *Pleurotus* sp. and bioremediation of caffeinated residues. IX European Congr. Biotechnol., July 11-15, Brussels, Belgium, p. 2664- A

Fan L., Pandey A., Mohan R., Soccol C.R. (2000a), Comparison of coffee industry residues for production of *Pleurotus ostreatus* in solid state fermentation, Acta Biotechnol., 20(1), in press

Fan L., Pandey A., Soccol C.R. (2000b), Solid state cultivation- an efficient technique to utilize toxic agro-industrial residues. J. Basic Microbiol., 40(3), in press.

Gaitan H.R., Salmones D. (1996), Brief article cultivation and selection of high yielding strains of *Pleurotus sp.* Rev. Mexicana Micol., 12: 107-113

Galzada J.F., Leon R., Arriola M.C., Rolz C. (1987), Growth of mushroom on wheat straw and coffee pulp: strain selection. Biol. wastes, 20: 217-226

ICO-International Coffee Organization (1998), Total production of exporting members. http://www.ico.org/proddoc.htm

Lozano J.C. (1990), Producción comercial del champiñon (*Pleurotus ostreatus*) en pulpa de coffé Fitopatología Colombiana, 14: 42-47

Martinez C.D., Quirarte M. (1984), Prospects of growing edible mushroom on agro-industrial residues in Mexico, Boletin Soc Mexicana Micol., 19: 207-219

Martinez C.D., Soto C., Guzman G. (1985), Cultivo de *Pleurotus ostreatus* en pulpa de café com paja como substrato. Rev. Mexicana Micol., 1: 101-108

Martinez C.D., Morales P., Sobal M. (1990), Cultivo de *Pleurotus ostreatus* sobre bagazo de caña enriquecido com pulpa de café o paja de cebada. Micol. Neotrop. Aplic., 3: 49-52

Mata G., Gaitan Hernandez R. (1994), Advances in shiitake cultivation on coffee pulp Rev. Iberoamer. Microbiol., 11: 90-91

Pandey A. (1992), Recent developments in solid state fermentation, Process Biochem., 27, 109-117

Pandey A. (1994), Solid state fermentation- an overview. In- Solid State Fermentation, Pandey, A. (ed.), Wiley Eastern Publishers, New Delhi, pp. 3-10

Pandey A., Soccol C.R., Nigam P., Soccol V.T., Vandenberghe L.P.S., Mohan R. (1999a), Biotechnological potential of agro-industrial residues: II Cassava bagasse. Biores. Technol., in press

Pandey A., Soccol C.R., Nigam P., Soccol V.T. (1999b), Biotechnological potential of agro-industrial residues: I Sugarcane bagasse. Biores. Technol., in press

Pandey A., Soccol C.R. (2000), Economic utilization of crop residues for value addition - A futuristic approach. J. Sci. Ind. Res., 59: 12-22

Pandey A., Soccol C.R., Mitchell D. (2000), New developments in solid state fermentation. I- Bioprocesses and products. Process Biochem., in press

Royse D.J. (1995), Speciality mushrooms: cultivation on synthetic substrate in the USA and Japan. Interdisciplin. Sci. Rev., 20: 1-10

Soccol C.R. (1995), Aplicações da fermentação no estado sólido na valorização deresíduos agroindustriais, França-Flash Agricultura, 4: 3-4

Soto C. (1986), La producción de hongos comestibles sobre la pulpa de café en la región de Xalapa, Veracruz durante 1985-1986. Rev. Mexicana Micol., 2: 437-440

Soto C., Martinez C.D., Morales P., Sobal M. (1987), La pulpa de café seccada al sol, como una forma de almacenamiento para el cultivo de *Pleurotus ostreatus*. Rev. Mexicana Micol., 3: 133-136

Thielke C. (1989), Cultivation of edible fungi on coffee grounds, Mushroom Sci., 12: 337-343

Tsang L.J. Reid I.D., Coxworth E.C. (1987), Delignification of wheat straw by *Pleurotus sp.* under mushroom growing conditions. Appl. Environ. Microbiol., 53: 1304-1306

Valsameda M., Martinez M.J., Martinez A.T. (1991), Kinetics of wheat straw solid-state fermentation with *Trametes versicolor* and *Pleurotus ostreatus*-lignin and polysaccharide alteration and production of related enzymatic activities. Appl. Microbiol. Biotechnol., 35: 817-823

Yang X.M. (1986), *Cultivation of Edible Mushroom in China,* Agriculture Printing House, Beijing, PR China.

Zadrazil F. (1980), Conversion of different plant waste into feed by Basidiomycetes. European J. Appl. Microbiol. Biotechnol., 9: 243-248

ADDING VALUE TO COFFEE SOLID BY-PRODUCTS THROUGH BIOTECHNOLOGY

I. Perraud-Gaime[1], G. Saucedo-Castañeda[2], C. Augur[1], S. Roussos[3]

[1]IRD Mexico, Calle Ciceron 609, Colonia los Morales, 11530 México DF, Mexico; [2]Departamento de Biotecnologia, UAM Iztapalapa, Mexico DF, Mexico; [3]IRD, Laboratoire de Microbiologie, Université de Provence, IFR-BAIM Case 925, 163 Av de Luminy, 13288 Marseille, Cedex 9, France

Running title: Coffee solid by-products utilization

1. Introduction

Coffee is a universal drink with millions of tons consumed throughout the world. As a result of the brewing of coffee, only 6% by weight of the berry harvested in the coffee field ends up in the cups, leaving 94% as waste, which includes by-products from the process as well as water from the drying of the seeds (Zuluaga, 1989). Although a small amount of these wastes are utilized partially as described below, most of this causes environmental pollution. From 100 g of fresh berries, around 29% of the dry weight constitutes coffee pulp, 12% husk, 4% mucilage and 55% are the coffee seeds (Bressani, 1978). Some typical yields resulting from 100 kg ripe cherries of Canephora (robusta) coffee by different processing methods are: a) wet processing, 22 kg of traded coffee; b) dry processing, 40-45 kg dry berries, 22 kg traded coffee. The by-products include 56-60 kg fresh pulp or 12-15 kg dried pulp, 3-5 kg of parche and 20 kg husk. One hundred kilogram of ripe cherries from arabica coffee using wet process results in 39 kg fresh pulp or 16 kg of dry pulp, 22 kg mucilage, 39 kg wet coffee parche (at 20% moisture) or 20 kg of processed coffee. Whatever the treatment may be, the final yield of processed coffee is around 20% (Zuluaga, 1989, Coste, 1989).

T. Sera et al. (eds.), Coffee Biotechnology and Quality, 437–446.

Potential uses of the husk obtained through dry process include animal feed, composting and extraction of caffeine. Animal feed trials on husk and parches showed that a daily supplement of 10-20% did not affect the feed digestibility. It husk could also be used for caffeine extraction as it contains 0.4-0.7% of the alkaloid. Composting of the by-products would be useful in increasing the carbon and nitrogen contents of soil. At present, these by-products are burned at the sites of production, mainly for economical reasons. The energy value of these products (3,600 kcal.kg^{-1} husk) is high making combustion profitable but only when the combustion is carried out at the production sites. However, the combustion of the husk produces highly corrosive gases and vapours that deteriorate equipment and affect the quality of the seeds. Due to this reason, the husk is used for indirect heating. Besides, dried coffee husk can become a serious pollutant if it re-humidifies accidentally.
The enormous residual volume of by-products generated during coffee processing has sparked a number of research projects geared towards their potential use which include animal and fish feeding, earthworms culture substrate, mushroom *Pleurotus* culture, aerobic compost etc. Roussos et al (1993) investigated biotechnological processes for the production of different metabolites utilizing coffee pulp. High level of humidity (80-85%) of coffee pulp was found inducive for the development of natural micro-flora and its rapid putrefaction (Gaime-Perraud *et al.*, 1993).

2. Chemical composition of coffee pulp

Coffee pulp differs in chemical composition according to the variety of the coffee cultivar, the stage of maturity of the fruit, the type of treatment applied and the production site. Carbohydrates account for 68% of dry matter, fibres 15%, ashes 6.5% and proteins 9% (Elias, 1978). Sixty percent of total nitrogen present in coffee pulp comes from amino acids. Amino acid composition of coffee pulp is very similar to that of soy or cotton flour but contains more valine and lysine and less leucine, tyrosine and phenylalanine than maize flour. The main amino acids present in coffee pulp are lysine, threonine, tyrosine, valine and phenylalanine with methionine and isoleucine as minor constituents (Bressani *et al.*, 1972). There are variations in the percentages of caffeine, tannins, chlorogenic acid and caffeic acid (0.6-1.3, 1.8-8.6, 0.2-3.2, 0.3-2.6 % DM, respectively). Each of these compounds may directly or indirectly have a toxic or anti-physiological effect (Clifford and Ramirez-Martinez, 1991). Condensation of tannins starts a few hours after fruit harvest and intensifies in the presence of water and heat. Condensed tannins increase through polymerization of anthocyanidins. Studies by Velez *et al.* (1985) demonstrated that phenolic compounds from coffee pulp have the capacity to bind proteins with maximal interaction at pH 5. De Rozo *et al.* (1985) observed that phenolic compounds decreased considerably the capacity for iron absorption, even when used as a food supply comprising only 10% of coffee pulp. Coffee pulp may contain

other substances, for example potassium that alone or in synergy with other compounds account for the anti-nutritional effects observed when used as animal feed. The high concentration of potassium could contribute to the anti-nutritional effect of the pulp as animal feed by modifying the ionic equilibrium of tissues (Campabadal, 1987).
Cellulose and lignin contents of coffee pulp suggest that coffee pulp could be a better feed source than citrus fruit skin, presently used as animal feed. However, the fibre content may be an obstacle for use of the fresh pulp as a feed source for monogastrics. D-fructose makes up 50% of the monosaccharides present while D-glucose accounts for 30% of the lyophilized pulp. The remaining 20% are represented by saccharose and galactose. Inositol content is negligible (Zuluaga, 1989).

3. Uses of coffee pulp

3. 1. ANIMAL FEED

Research has primarily been focused on upgrading and utilizing coffee pulp in order to obtain an ensiled or dried product suitable for animal consumption (Bressani *et al*, 1974). Coffee pulp could also serve as a substrate for caffeine extraction with residual products being used as animal feed. Dried pulp accounts for 10% proteins and a little less than 25% fibres. However, its low digestibility allows for only partial substitution as a daily diet. Its high fibre and phenolic compound contents as well as the presence of caffeine also strongly diminish its digestibility, especially with monogastrics, resulting weight hair losses.
Gomez-Brenes (1978) treated coffee pulp with a 1.2-3% calcium hydroxide solution. He observed a decrease in tannin content but without significant change in caffeine, chlorogenic acid or caffeic acid. It was concluded that such a base treatment did not enhance the nutritional value of coffee pulp. Peñaloza *et al.* (1985) grew *Aspergillus niger* in a solid state fermentation (SSF) process and achieved 200% increase in protein content with a significant decrease in fibre content, cellulose and hemicellulose. Despite the fact that tannins and caffeine were unaffected by the fermentation, the nutritional quality of the product was improved when tested as chicken feed. Aquiahuatl *et al.* (1988) isolated 350 fungal strains from soil, leaves or fruit from coffee growing regions. Among these, eight (two *Penicillium* and six *Aspergillus*) presented a high capacity to degrade caffeine in liquid synthetic medium. When tested in SSF using coffee pulp as substrate, all the strains degraded caffeine without any exogenous nitrogen supplementation (Perraud-Gaime, 1995). Roussos *et al.* (1994) and Hakil *et al.* (1998) also reported caffeine degradation by these strains. Boccas *et al.* (1994) reported high levels of pectinases from these strains.

3.2. ORGANIC FERTILIZER

Due to high levels of nitrogen, phosphorus and potassium as well as the presence of organic matter, coffee pulp can be used as soil fertilizer or soil conditioner. Various studies demonstrated that coffee pulp could be a good fertilizer agent, especially on coffee plantations. Such an alternative is presently used on production sites on a small scale. There are, however, two major drawbacks in this: the high water content of the pulp and the price of labour involved. It seems that composting the pulp is necessary to prevent rapid exothermic fermentation of fresh pulp, when stacked at the base of coffee explants. Distribution of the pulp over large plantations is also not cost effective.

3.3. SUBSTRATE FOR BIOGAS PRODUCTION

Coffee pulp has been tested as a substrate in anaerobic fermentation processes for the production of biogas (Calle, 1974a; Blanes, 1982). These studies had the advantage of proposing simple technological solutions, adapted to small size installations within production sites. Pre-treating the pulp by aerating or ensiling along with a good inoculum could be useful to effectively start-up the process. Presence of tannins, caffeine, chlorogenic acid or caffeic acid was termed harmful for production along with the observed drop in pH (Morales and Chacon, 1981).

3.4. MICROBIAL CULTIVATION AND ENZYME PRODUCTION

In view of the fact that citrus fruit used for the industrial production of pectin contain only 1.5-3.5% pectin, high pectin content (33% dry wt.) of the mucilage make it attractive substrate for pectin production. Coffee pulp has potential as a substrate for inducible enzymes production such as pectinases or cellulases (Favela *et al.*, 1989; Boccas *et al.*, 1994). However, up to now, pectic enzyme titres obtained from various fermentations of either coffee pulp itself or from pectin-rich waters derived from the process are lower than those of commercial preparations. Coffee pulp has also been used to cultivate yeast and other micro-organisms (Calle, 1974b; Penaloza *et al.*, 1985).

3.5. PRODUCTION OF EDIBLE FUNGI

Studies on the production of edible fungi (*Pleurotus ostreatus*) grown on coffee pulp have shown satisfactory results with industrial potential (Guzman and Martinez-Carrera, 1985; Rolz *et al.*, 1988; Martinez-Carrera *et al.*, 1989). Residual pulp resulting from the production process could be used as animal feed or as fertilizer (Martinez-*Carrera et al.*, 1989).

3.6. EARTHWORM PRODUCTION

The culture of the earthworm (*Eisenia foetida*, Sav.) on fresh coffee pulp has been envisaged as an alternative (Davila and Arango, 1991). Advantages include a decrease in decomposition time of the pulp, a decrease in its contaminating constituents, an easy set-up with commercially interesting final products. The worms are used to feed fish or chickens and the humus obtained is of better quality than the fertilizing properties of direct fresh coffee pulp addition as previously proposed (Salazar and Mestre, 1991).

4. Microbial flora present in coffee pulp

Coffee pulp (as well as the husk) contains a wide variety of natural micro-flora (bacteria, yeast, and filamentous fungi) which varies in concentration from $7.0x10^5$-$1.1x10^8$ Colony Forming Units (CFU).g^{-1} of dry matter. Samples obtained from the "wet process" contained the highest population of micro-organisms, between 17 and 160 times higher than from samples resulting respectively from the dry process or semi-humid (dry de-pulping). Yeast dominated the population when lyophilized samples (obtained immediately after de-pulping) were analysed. Filamentous fungi predominated samples from coffee husk (Gaime-Perraud, 1995, Roussos *et al.*, 1995).
Natural micro-flora evolves extremely rapidly, resulting in the need for an effective and reliable conservation method. Ensiling would be the method of choice to stabilize coffee pulp through natural micro-flora development of lactic acid bacteria. Bertin and Hellings (1985) advocated a level of 10^5 lactic bacteria per gram of dry matter in order to obtain a satisfactory ensiled product. Anaerobic bacteria from coffee pulp grown on MRS medium accounted for 3.10^4 bacteria per gram of dry matter.

5. Silage: a conservation technique

Ensiling of coffee pulp for its preservation and improvement of feed value is one of the avenues for value-added utilization of coffee pulp. Silage making is based on natural fermentation whereby lactic acid bacteria (LAB) ferment water-soluble carbohydrates to organic acids, mainly lactic acid, under anaerobic conditions. As a result the pH decreases, inhibiting detrimental anaerobes, thereby preserving moist forage. The aim of ensiling is to minimize loss of dry matter as well as nutritious value. It also prevents the development of an undesirable microbial population that would otherwise produce compounds with adverse effects when fed to animals. Other effects such as a better distribution of amino acids is attributed to silage (Weinberg and Muck, 1996). Silage techniques have been developed empirically over the centuries to stock and preserve products of both plant and animal origin. A number of physical, chemical and microbiological factors are of vital importance in obtaining good silage. The substrate to be ensiled should have 30-40% dry matter, compatible to the desired level, amenable for anaerobiosis and contain utilizable sugars in sufficient quantity (up to 13% DM). It also must have the colour, which is nearest to the raw material, fruity aroma and slightly acidic taste. In terms of chemical characteristics and to achieve stability of the organic matter, ensiling should involve a minimum loss of dry matter and the resulting silage should have a pH value lower than 4.5, with over 3% lactic acid, but with less than 0.5 and 0.3% acetic and butyric acids, respectively, and a ratio of ammonical nitrogen over total nitrogen of 10 (Mc Donald *et al.*, 1991). The silage process can be compared to a three-component system: plant substrate, enzymes, and bacteria, in which each player has a key role in the success or failure of the silage (Bertin and Hellings, 1985). LAB develops within the mass of the substrate to be ensiled. They transform soluble sugars, producing lactic acid, a natural preserving agent. These optional anaerobes produce either exclusively lactic acid (strict homo-fermentative bacteria) or lactic acid and acetic acid (optional homo-fermentative bacteria) or lactic acid and acetic acid along with ethanol, butyric acid or CO_2 (strict hetero-fermentative bacteria) (Dellaglio *et al.*, 1994). They are particularly demanding micro-organisms. In addition to fermentable sugars, these bacteria require specific amino acids, vitamins and oligo-elements. LAB can inhibit growth of other micro-organisms through the production of organic acids such as acetic and propionic (Moon, 1983), hydrogen peroxide or bacteriocins such as nisin (Beliard and Thuault, 1989).

Success of a silage depends greatly on the presence of an adequate micro-flora, a quick reduction in pH, and a high production of lactic acid to preserve the substrate by blocking the activity of intracellular enzymes and by inhibiting the proliferation of unwanted micro-organisms (McDonald *et al.*, 1991). Enterobacteria, yeast and fungi represent in general a large proportion of the initial micro-flora present in the substrate to be ensiled (Pahlow, 1991). These micro-organisms, if given a chance, compete with LAB for the fermentable sugar sources. However, anaerobic bacteria such as *Clostridium*, which are strict anaerobes, can multiply rapidly as soon as oxygen becomes scarce, producing toxins (Woolford, 1984). A proposed solution, therefore, could be to add an inoculum of LAB (both homo-fermentative and hetero-fermentative) at the beginning of the silage process in order to inhibit naturally occurring and unwanted micro-organisms.

Plant, bacterial and often fungal enzymes are able to depolymerize the plant cell wall. They liberate soluble sugars within the product being ensiled. These sugars can then be metabolized by LAB. The available literature on the effects of different additives is contradictory. The results depend greatly on the nature of the substrate, its chemical composition as well as on the type of additive, its composition and concentration. The effects of enzyme addition along with lactic acid bacterial starters gives mixed results regarding animal performance (Vanbelle *et al.*, 1994).

5.1. SILAGE OF COFFEE PULP

Most of the reports concerning coffee pulp silage deal with the development of ensiling techniques or with the effect of chemical additives on the processed pulp (Daqui, 1975 ; Murillo, 1978; Carrizales and Ferrer, 1984). In general, an important loss in dry matter has generally been observed. The quality of the silage has also not been satisfactory. Ensiling does exist in tropical countries, in spite of the problems in terms of temperature and humidity. Consequently, the rate of ensiling is slow, putrefaction is common and chemical additives are used (where biological ones should be).
In order to enhance silage quality of coffee pulp, it could be useful to add various types of forage such as those of corn or sorghum depending on availability and price. Coffee pulp has similar nutritious value to those of tropical forage of good quality. Ensiled pulp is better than the dried coffee pulp as far as nutritional value is concerned but shows poor *in vitro* digestibility. The improvement of the acceptability of coffee pulp after silage could be enhanced by a decrease of caffeine and tannin contents. Porres *et al.* (1993) reported caffeine reduction during silage, which most probably was due to its solublization in silage liquids. De Menezes *et al.* (1993) reported that LAB degraded tannins during the fermentation.

5.2. LACTIC ACID STARTERS FOR COFFEE PULP SILAGE

With an aim to study the biodiversity of LAB from coffee biotopes, they were isolated, characterized for preparing the formulation of starters for coffee pulp silage. Due to the presence of toxic compounds such as polyphenols and caffeine, starters could be associated with detoxifying fungal enzymes. Starters were tested on fresh coffee pulp (0.3-1 kg). Samples of arabica coffee pulp were collected at a production site in Coatepec, Veracruz, Mexico during October 1997 until March 1998. This site is equipped with a de-pulping apparatus of the Penagos type. Micro-organisms were isolated from four silage of two kg each of coffee pulp. Two silage resulted from natural endogenous micro-flora found in the pulp and two were prepared by inoculating the pulp using part of previous silage. One hundred and fifty bacterial strains and 20 yeast strains were isolated on MRS and MRS + coffee medium. The selected strains were characterized by HPLC, identified by API, APIZYM and RFLP biodiversity approach.
Fresh pulp was inoculated with one homo-lactic natural strain and the ensiled product was compared to the natural uninoculated silage. Both lab trials were satisfactory, but ethanol and acetic acid were detected besides lactic acid. Inoculation caused the elimination of acetic acid production, but did not avoid ethanol production, probably due to acidic tolerant yeast development. Further studies are underway in Mexico with different homo-fermentative strains associated with a hetero-fermentative strain in order to improve acidification kinetics of silage and to prevent ethanol production.

6. Summary

In tropical countries, coffee industry produces various by-products (coffee pulp, mucilage, parches, husk) which are under-utilized and are a source of environmental pollution. These by-products represent around 50% (dry matter) of the world coffee beans production. Coffee pulp is rich in carbohydrates, amino acids, minerals, and various other nutrients. Nevertheless, the high level of humidity (80-87%) of coffee pulp induces the development of natural micro-flora and causes rapid putrefaction resulting in its transformation from by-product to waste. Coffee pulp offers could be used as solid substrate in biotechnological processes such as animal feed, mushroom and earthworm production, organic fertilizer and micro-organism growth for enzyme production. Due to time constraints during the coffee season, silage of fresh coffee pulp could be the best solution for conservation of the pulp. However, little information exists in inter-tropical regions concerning silage in general, and coffee pulp silage in particular. The global outlook points towards the necessity to elaborate a lactic acid starter specifically for coffee pulp silage.

7. Conclusions

At the present moment, biotechnological innovations and applications offer the greatest potential for agro-industrial by-products. Using a short term outlook, it could be possible to obtain a detoxified product ready for animal feed. It could also be possible to obtain high added value products such as enzymes or secondary metabolites from coffee pulp. However, it must be kept in mind that the utilization of coffee pulp as well as the mucilage depends on a number of factors such as the amount of product, type of treatment, seasonal and regional distribution, humidity, efficient stocking of the product and commercial potential. In recent publications concerning potential industrial applications of coffee by-products, there is an obvious lack of economic and feasibility studies. There is a need to estimate beneficial effects (especially environmentally friendly ones) of the transformed substrates in order to prevent the generation of new sources of pollution.

8. Acknowledgements

Research on coffee pulp silage is funded by a grant from the European Community INCO-DC (BIOPULCA: IC18*CT970185)

9. References

Aquiahuatl M.A., Raimbault M., Roussos S., Trejo, M.R. (1988), Coffee pulp detoxification by solid state fermentation : isolation, identification and physiological studies. Proc.Solid State Fermentation in Bioconversion of Agroindustrial Raw Materials, Raimbault M. (ed.), Montpellier, France, pp. 13-26

Beliard E., Thuault D. (1989), Propriétés antimicrobiennes des bactéries lactiques. In- Microbiologie Alimentaire, Bourgeois C.M., Larpent J.P. (eds.), Lavoisier, Paris, 2, pp. 283-297

Bertin G., Hellings P.H. (1985), Paper presented in Symp L'ensilage: nouveaux aspects biologiques, Paris, 18 Jan., p. 59

Blanes R. (1982), Fermentation anaérobie de la pulpe de café avec production de méthane. Thèse de Doctorat es Science, USTL, Université de Montpellier, France

Boccas F., Roussos S., Gutierrez M., Serrano L., Viniegra G.G. (1994), Production of pectinase from coffee pulp in solid state fermentation system : selection of wild fungal isolate of high potency by a simple three-step screening technique. J. Food Sci. Technol., 31: 22-26

Bressani R., Estrada E., Jarquin R. (1972), Pulpa y pergamino de café. 1. Composición química y contenido de aminoacidos de la proteina de la pulpa. Turrialba, 22 (3): 299-304

Bressani R., Cabezas M.T., Jarquin R., Murillo B. (1974), The use of coffee processing waste as animal feed. In- Proc. Conf. Anim. Feeds Trop. Subtrop. Orig., London: Tropical Products Institute, pp.107-117

Bressani R. (1978), Subproductos del fruto de café. In- Pulpa de café: composición, tecnología y utilización, Braham J.E, Bressani R. (eds.), INCAP, pp. 9-17

Calle V.H. (1974a), Como producir gas combustible con pulpa de café. Boletín Técnico, Cenicafé, 3: 11-17

Calle V.H. (1974b), Processo industrial para propagación de levaduras. Federación Nacional de Cafeteros de Colombia, Centro Nacional de Investigaciones del Café, Chinchina, Caldas, Colombie

Campabadal C. (1987), Utilización de la pulpa de café en la alimentación de animales. III Simposio International sobre utilización integral de los subproductos del café. Guatemala

Carrizales V., Ferrer J. (1984), Recycling agroindustrial waste by lactic fermentations : coffee pulp silage. Paper presented in Symp. Internatl Las fermentaciones lacticas en la industria alimentari, 21-29 Nov., Mexico, p. 26

Clifford M.N., Ramirez-Martinez J.R. (1991), Phenols and caffeine in wet-processed coffee beans and coffee pulp. Food Chem., 40: 35-42

Coste R. (1989), Caféiers et Cafés. Lavoisier, Paris, p. 373

Daqui L.E. (1975), Características químicas y nutricionales de la pulpa de café ensilada con pasto napier y planta de maiz. Tésis de Maestria. Universidad de San Carlos de Guatemala, facultad de Medicina Veterinaria y Zootecnia, INCAP/CESNA, Guatemala

Davila A., Arango B. (1991), Producción de lombrices sobre pulpa de café. In- Memorias II Sem. Intern. Biotecnol. Agroindust. Café (II SIBAC), Manizales, Colombia

Dellaglio F., De Roissart H., Torriani S., Curk M.C., Janssens D. (1994), Caractéristiques générales des bactéries lactiques. In- Bactéries Lactiques, de Roissart H., Luquet F.M. (eds), Lorica, Grenoble, France, 1: 25-116

De Menezes H.C., Samaan F.S., Clifford M.N., Adams M.R. (1993), The fermentation of fresh coffee pulp for use in animal feed. Association Scientifique Internationale du Café (ASIC), 15ème Colloque, Montpellier, France

De Rozo M.P., Velez J., Garcia L.A. (1985) Efecto de los polifenoles de la pulpa de café en la absorción de hierro. Arch. Latinoamer. Nutr., 35: 287-296

Elias L.G. (1978), Composición química de la pulpa de café y otros subproductos. In- Pulpa de café: composición, tecnología y utilización, Braham J.E., Bressani R. (eds.), INCAP, pp. 19-29

Favela E., Huerta S., Roussos S., Olivares G., Nava G., Viniegra G.G., Gutierrez-Rojas M. (1989), Producción de enzimas a partir de la pulpa de café y su aplicación a la indústria cafetalera. In- Roussos S., Licona F.R., Gutierrez R.M. (eds.), I Seminario Internacional de Biotecnología en la Industria Cafetalera. Memorias, Xalapa, Veracruz, pp. 145-151

Gaime-Perraud I., Roussos S., Martinez-Carrera D. (1993), Endogenous microflora of fresh coffee pulp. Mycol. Neotrop. Apl., 6: 95-103

Gaime-Perraud I. (1995), Cultures mixtes en milieu solide de bactéries lactiques et de champignons filamenteux pour la conservation et la decaféination de la pulpe de café. PhD Thesis, Université Montpellier II, France

Gomez-Brenes R.A. (1978), Procesamiento de pulpa de café : tratamientos químicos. In- Pulpa de café: composición, tecnología y utilización, Braham J.E., Bressani R. (eds.), INCAP, pp. 125-142

Guzman G., Martinez-Carrera D. (1985), Planta productora de hongos comestibles sobre pulpa de café. Ciencia y Desarollo, 65: 41-48

Hakil M., Denis S., Viniegra-Gonzalez G., Augur C. (1998), Degradation and product analysis of caffeine and related dimethylxanthines by filamentous fungi. Enz. Microb. Technol., 22: 355-359

Martinez-Carrera D., Morales P., Sobal M. (1989), Producción de hongos comestibles sobre pulpa de café a nivel comercial. In- Proc. I SIBAC, Roussos S., Licona-Franco, R., Gutierrez-Rojas M. (eds.), Xalapa, Mexico, 177-184

McDonald P., Henderson A.R., Heron S.J.E. (1991), The Biochemistry of Silage. 2nd edn., Chalcombe Publications, Marlow, England

Moon N.J. (1983), Inhibition of the growth of acid tolerant yeats' by acetate, lactate and propionate and their synergistic mixture. J. Appl. Bacteriol., 55: 453-460

Morales A.I., Chacon G. (1981), Producción de biogas a partir de pulpa de café. Document CICAFE, San José, Costa Rica, pp. 1-13

Murillo B. (1978), Ensilaje de pulpa de café. In- Pulpa de café : composición, tecnología y utilización, Braham J.E., Bressani R. (eds.), INCAP, pp. 97-110

Pahlow G. (1991), Microbiology of inoculants, crops and silages. Proc Eurobac Symp. Uppsala, pp. 9-18

Peñaloza W., Molina M.R., Gomez-Brenes R., Bressani R. (1985), Solid-State Fermentation: An alternative to improve the nutritive value of coffee pulp. Appl. Environ. Microbiol., 49 (2): 388-393

Porres C., Alvarez D., Calzada J. (1993), Caffeine reduction in coffee pulp through silage. Biotechnol. Adv., 11: 519-523

Rolz C., De Leon R., De Arriola M.C. (1988), Biological pretreatment of coffee pulp. *Biol. Wastes*. 26: 97-114

Roussos S., Gaime I., Denis S., Marin B., Marakis S., Viniegra G. (1993), Biotechnological advances on coffee byproducts utilization. Paper presented in IFCON'93, Mysore, India, 7-12 September, p. 17

Roussos S., Hannibal L., Aquiahuatl M.A., Trejo M., Marakis S. (1994), Caffeine degradation by *Penicillium verrucosum* in solid state fermentation of coffee pulp: critical effect of additional inorganic and organic nitrogen sources. J. Food Sci. Technol. 31: 316-319

Roussos S., Aquiahuatl M.A., Trejo-Hernandez M.R., Gaime-Perraud I., Favela E., Ramakrishna M., Raimbault M., Viniegra-Gonzalez G. (1995), Biotechnological management of coffee pulp - isolation, screening, characterization, selection of caffeine-degrading fungi and natural microflora present in coffee pulp and husk. Appl. Microbiol. Biotechnol., 42: 756-762

Salazar A.J., Mestre M.A. (1991), Utilización del humus de lombriz roja californiana (*Eisenia foetida*, Sav.), obtenido a partir de pulpa de café como sustrato para almacigos de café. in Proc. II SIBAC, Manizales, Colombia

Vanbelle M., Laduron M., Bertin G., Hellings P. (1994), Utilisation d'enzymes dans l'ensilage des fourages. In- Bactéries Lactiques, de Roissart H., Luquet F.M. (eds.), Lorica, Grenoble, France, 2: 271-292

Velez A.J., Garcia L.A., De Rozo M.P. (1985), Interacción *in vitro* entre los polifenoles de la pulpa de café y algunas proteinas. Arch. Latinoamer. Nutr., 35 (2): 297-305

Weinberg Z.G., Muck R.E. (1996), New trends and opportunities in the development and use of inoculants for silage. FEMS Microbiol. Reviews, 19: 53-68

Woolford M.K. (1984), The Silage Fermentation. Marcel Dekker, Inc., New York, NY.

Zuluaga J. (1989), Utilización integral de los subproductos del café. In- Proc. I SIBAC, Roussos S., Licona-Franco R., Gutierrez-Rojas M. (eds.), Xalapa, Mexico, pp. 63-76

Chapter 42

EFFECT OF CONSERVATION METHOD ON CAFFEINE UPTAKE BY *PENICILLIUM COMMUNE* V33A25

G. Gutiérrez-Sánchez[1], I. Perraud-Gaime[2], C. Augur[2], J.M. Romano-Machado[1], G. Saucedo-Castañeda[1]

[1]*Departamento de Biotecnología, Universidad Autónoma Metropolitana, Unidad Iztapalapa (UAMI), A. P. 55-535, C. P. 09340, México, D. F., Mexico;* [2]*Institut de Recherche pour le Développement-México (IRD-México), Cicerón 609, Los Morales, C. P. 11530, México, D. F., Mexico*

Running title: Caffeine degradation

1. Introduction

Coffee pulp is one of the major by-products obtained during the pulping operation of coffee cherries. A fraction of this material is used for compost production for coffee plant nurseries and other part is spilt in rivers or piled up near them. For every two tons of coffee cherry processed, nearly one ton of pulp is generated. During the 1997-98 period, 10.4 million of sacks (60 kg each) of fresh coffee pulp were produced in Mexico (Barreiro, 1999). Coffee pulp is rich in proteins, minerals and fibre that can be used for animal feeding, but utilization of coffee pulp for feed is constrained by anti-physiological (caffeine) and anti-nutritional (polyphenols) compounds (Roussos *et al.*, 1989). These compounds cause adverse effects on the animals that consume coffee pulp. Caffeine (1,3,7-trimethylxanthine, Figure 1) concentration varies according to the coffee variety (0.9-2.4% dry wt. basis). Its physiological role is probably to defend coffee plants from predators and to inhibit growth of other plants. In humans, caffeine is demethylated into three primary metabolites: theophylline, theobromine, and paraxanthine. At 100 $mg.kg^{-1}$ teophylline is toxic to rats. Theobromine has been related with headache, insomnia, restlessness, excitement, mild delirium, muscle tremor, tachycardia and extrasystoles in man and caffeine has been reported to have many other activities including mutagenic, teratogenic and carcinogenic capacities.

T. Sera et al. (eds.), Coffee Biotechnology and Quality, 447–453.

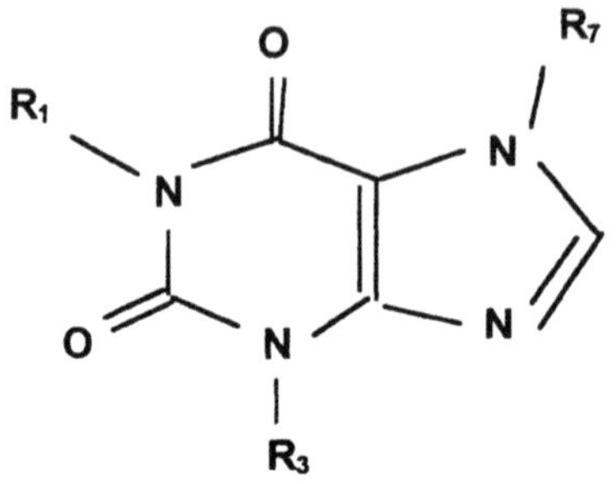

Figure 1. Caffeine structure and related compounds.

Compounds	R_1	R_3	R_7
Caffeine	CH_3	CH_3	CH_3
Theophylline	CH_3	CH_3	H
Paraxanthine	CH_3	H	CH_3
Theobromine	H	CH_3	CH_3
1-metil-xanthine	CH_3	H	H
3-metil-xanthine	H	CH_3	H
7-metil-xanthine	H	H	CH_3
Xanthine	H	H	H

Coffee pulp could be silage for conservation and nutrimental improvement. Silage is an anaerobic fast process, which involves lactic acid bacteria. It has been widely used for forage preservation in regions of moderate climate and enables prevention of forage putrefaction with minimum degradation of the organic material and a reduction in the polyphenols compounds but not in the caffeine content (Perraud-Gaime, 1995). In order to improve the nutritional value, Peñaloza *et al.* (1985), Gómez *et al.* (1985) and Aquiáhualt (1992) carried out solid state fermentation (SSF) studies with coffee pulp using a strain of *Aspergillus niger*. They reported protein enrichment of the substrate but with no significant caffeine elimination.

The removal of caffeine from coffee pulp appears attractive because it could then be used as animal feed. Caffeine elimination could be carried out by two ways, chemically and microbiologically. The first way is not recommended due to the use of heavy salts, solvents and high cost. Microbial caffeine degradation by filamentous fungi such as Penicillium and Aspergillus, which were able to use caffeine as sole nitrogen source has been reported by Hakil *et al.* (1998).

Of the thousands of enzymes a cell is capable of producing, a certain number are always presented in substantial concentration, regardless of what medium in which the organism is growing. Enzyme induction is defined as a relative increase in the rate of synthesis of

a specific enzyme, resulting from exposure to a chemical substance, inducer. Inducible enzymes are necessary when the organism finds itself limited. Induction insures that energy and amino acids are not wasted in making unnecessary enzymes but that, when needed, these enzymes can be formed rapidly (Wang *et al.*, 1979). On the other hand, the ability to preserve successfully a wide range of micro-organisms and cell cultures has been a major achievement in microbiology over the last century that many have taken for industrial development (Smith, 1991). The choice of preservation method depends on many factors. For example, industrial collections place much emphasis on techniques that maintain genetic stability, especially for strains that have special features required for industrial processes such as the production of antibiotics and enzymes.
The aim of this work was to study the effect of the conservation method on caffeine uptake by *Penicillium commune* strain V33A25.

2. Experimental

2.1. MICRO-ORGANISM

Penicillium commune strain V33A25 (IRD-UAMI Collection) was selected because of its ability to use caffeine as nitrogen source (Roussos *et al.* 1995, Denis 1996).

2.2. CULTURE MEDIA

The culture media composition used in this work is shown in Table 1.

2.3. CULTURE CONDITIONS AND STRAIN CONSERVATION

Four methods were used for the evaluation of strain conservation associated to caffeine degradation, which included strain conservation on a) CSA, b) CPA, and lyophilizing the spores harvested from c) PDA, and d) CSA. *P. commune* was sub-cultured three times on CSA medium (except for method b using CPA, where strain was sub-cultured six times on CSA medium) at 30°C for six days and caffeine degradation were evaluated on CS broth according to the methodology reported by Denis (1996).

Table 1. Culture media used for strain conservation, spore harvest and caffeine evaluation (medium g l^{-1})

Component	CPA	CSA	PDA	CS
Saccharose	---	2.0	---	2.0
Milled coffee	---	40.0	---	40.0
Fresh coffee pulp	427.7	---	---	---
$Na_2HPO_4 \cdot 2H_2O$	0.16	0.12	---	0.12
$MgSO_4 \cdot 7H_2O$	0.4	0.3	---	0.3
$CaCl_2 \cdot 2H_2O$	0.4	0.3	---	0.3
Agar	26.7	20.0	---	---
Distilled water (ml)	1000	1000	1000	1000
Potato Dextrose Agar	---	---	39	---

CPA: Coffee pulp-agar medium. **CSA:** Coffee saccharose-agar medium (Aquiáhuatl, 1992; Denis, 1996). **PDA:** Potato Dextrose Agar. **CS:** Coffee saccharose broth.

For coffee-saccharose agar (CSA) and coffee-saccharose broth (CS), commercial coffee (Grand Mère "familial", France) was used. The infusion was filtered using Whatman paper No 41, salts were added and pH was adjusted to 5.5 using KOH (0.1 M). The volume was adjusted to 1-l. In the case of coffee pulp-agar (CPA), fresh coffee pulp was milled in a blender for 5 minutes and then heated until boiling. To this 750 ml distilled water (incorporating salts and agar) were added and pH was adjusted to 5.5 using KOH (0.1 M). The culture media were sterilized at 121°C for 15 minutes.

2.4. CAFFEINE ANALYSIS

Caffeine was extracted from liquid samples (0.5 ml) with 1 ml of a chloroform isopropanol mixture (85:15, v/v), mixing during 30 seconds and recovering the organic phase. The extraction was repeated three times. Solvent was evaporated in a heater and the extract was re-suspended in 1.5-ml de-ionized water. Caffeine was analyzed by High Performance Liquid Chromatography (HPLC), using a modification of the technique reported by Denis (1996). A Sphersorb ODS-2 column, acetonitrile tetrahidrofurane deionized water (5:1:94, v/v) as mobile phase at 1.5 ml min^{-1}, room temperature and diodes array detector at λ=273 nm were used during analysis.

3. Results and discussion

P. commune V33A25 was subjected to four different conservation methods prior to growth in a liquid medium in presence of caffeine. The results are presented in Fig. 2. When spores were conserved on PDA, lag phase of nearly sixty hours before degradation was noted. However, when spores were conserved on CSA, the lag phase was shorter and degradation was observed after 36 hours of cultures. In the case of method c, degradation began later (36 hours). It was evident that the methods c and d were not good alternatives to conserve the capacity of degradation of the caffeine in short periods. These results could be explained probably by the relative damage suffered by the spores during the lyophilization process. When spores were harvested from PDA medium, it was observed that caffeine degradation began after 60 hours, however, the lyophilization process allowed the conservation of strains for long periods of time, as lyophilization allowed the dehydration of fungi to level that halted metabolism (Smith, 1991).

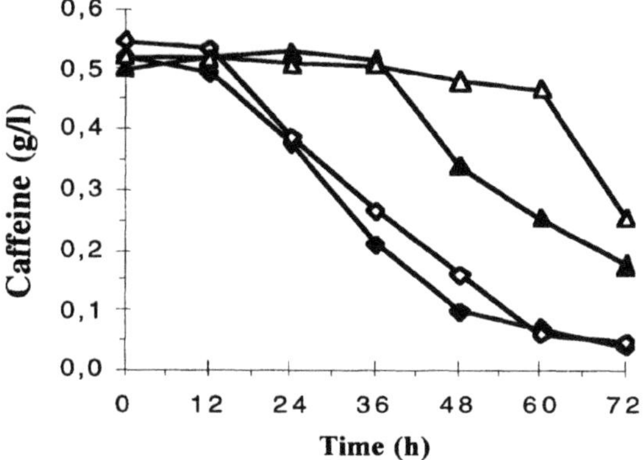

Figure 2. Influence of the conservation method on caffeine degradation ability, using CS broth. Strain conserved on CSA (◆); on CPA (◇); spores harvested from CSA (▲) and from PDA (Δ). Cultures were activated three times on CSA medium, except for CPA were inoculum was sub-cultured six times on CSA medium.

When the strain was conserved with the methods a or b, degradation presented a 12-hour lag phase and degradation rates were closed to 0.01 $g.l^{-1}$ h^{-1}, reaching 92 % degradation after 72 hours in both cases. Method b, however, required more time of adaptation than for the method a. These two methods could be used for short periods of conservation. Hakil *et al.* (1998) reported 90% caffeine degradation in 80 h using an *A. tamarii* strain conserved with the method a. Filamentous fungi (especially *Penicillium* genus) have been reported to possess the ability to use caffeine as sole nitrogen source, yielding teophylline and 3-methylxantine (Hakil *et al.*, 1998). Caffeine degradation pathway begins with two successive demethylation. In the first step teophylline is produced by a

7-demethylation from caffeine, followed by a 1-demethylation leads to 3-methylxantine from teophylline (Fig. 3) (Hakil *et al.*, 1999).
This way it could be established that teophylline is the first step in caffeine degradation (Schwimmer *et al.*, 1971). Results obtained suggested that enzymatic activity responsible for caffeine degradation was inducible due to the fact that degradation was increased considerably after repeated cultivation on CSA medium. Therefore, once enzymatic mechanism for caffeine degradation is expressed, it is important to conserve the strain on CSA medium in order to have a short reactivation period. This is the first report about the influence of the conservation method on the caffeine degradation ability.

caffeine *theophylline* *3-methylxanthine*

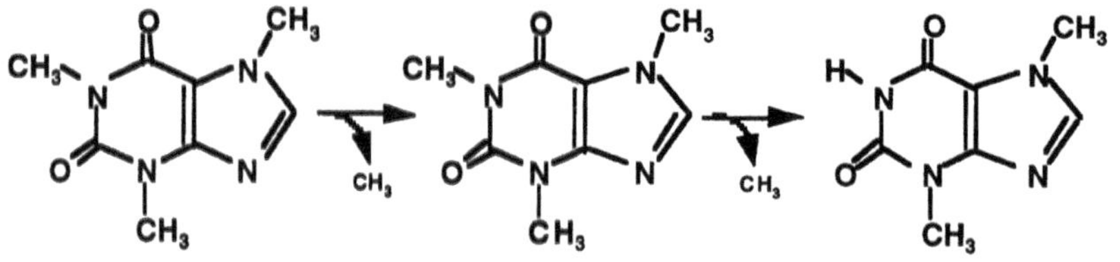

Figure 3. Steps in caffeine degradation by fungi.

4. Summary

Caffeine degradation by *P. commune* V33A25 was evaluated using a liquid culture medium prepared from a coffee infusion with an initial caffeine concentration of 0.5 $g.l^{-1}$. Once caffeinase activity was induced, four different conservation strategies were tested. Two involved the conservation of the mycelium grown either on a coffee saccharose-agar medium (method a) or on a coffee pulp-agar medium (method b). The other two methods involved spore lyophilization, after having grown the strain either on potato dextrose agar (method c) or on coffee saccharose-agar medium (method d). In each case, before evaluating the ability of the strain to degrade caffeine, *P. commune* was sub-cultured on coffee-agar medium. Spores were then harvested and *P. commune* was grown in liquid medium. Caffeine degradation over time was measured. It was observed that using methods c and d, the mould began to degrade caffeine after 60 and 36 hours of culture, respectively. With method 1, the fungus started to degrade caffeine 12 hours after the inoculation, reaching 92% of caffeine degradation in 72 hours; with method b, 39% degradation was observed in 72 hours. The results suggested that enzymatic activity responsible for caffeine degradation was inducible and once caffeine degradation ability was expressed it would be important to conserve the strain in a medium with caffeine.

5. Acknowledgements

This work was supported in part by the National Council of Science and Technology (CONACYT, México), and INCO-DC BIOPULCA Project granted by the European Union (IC 18*CT970185).

6. References

Aquiáhuatl R.M. (1992), Destoxificación de la pulpa de café: morfología, fisiología y bioquímica de hongos filamentosos que degradan cafeína. Tesis de Maestría, Facultad de Ciencias, UNAM

Barreiro Perera M. (1999), Coffee supply and demand balance (Más alla de nuestro campo). Claridades Agropecuarias. Special edition, pp 21-36

Denis S. (1996), Degradation de la caffeine par *Aspergillus commune* et *Penicillium commune* Etude physiologique et biochimique. PhD thesis, Université de Montpellier II, France

Gómez-Bremes R.A., Bendaña G, González J.M., Braham J.E., Bressani, R. (1985), Relación entre los niveles de inclusión de pulpa de café y contenido proteínico en raciones para animales monogástricos. Arch. Latinoam. Nutr. 35,422-437

Hakil M., Denis S., Viniegra-González G., Augur, C. (1998), Degradation and product analysis of caffeine and related dimethylxanthines by filamentous fungi. Enzyme Microb. Technol., 22,355-359

Hakil M., Voisinet G., Viniegra-González G., Augur C. (1999), Characteristics of caffeinases from *Aspergillus tamarii*. III Curso Internacional de Biotecnología Industrial. Biotecnología Enzimática, Abril 7, México, D.F., pp 49-55

Peñaloza W., Molina M.R., Gómez R., Bressani R. (1985), Solid State Fermentation: An alternative to improve the nutritive value of coffee pulp. Appl. Environ. Microbiol. 49,388-393

Perraud-Gaime I. (1995), Culture mixtes en milieu solide de bactéries lactiques et de champignons filamenteux pour la conservation et la décaféination de la pulpe de café. PhD thesis, Université de Montpellier II, Montpellier, France

Roussos S., Aquiahuatl R.M.A., Cassaigne J., Favela E., Gutierrez M., Hannibal L., Huerta S., Nava G., Raimbault M., Rodriguez W., Salas J.A., Sanchez R., Trejo M., Viniegra G.G. (1989), Detoxificación de la pulpa de café por fermentación sólida. In Memorias I Sem. Intern. Biotecnol. Agroindust. Café (I SIBAC), Roussos, S., Licona-Franco, R., Gutierrez-Rojas, M. (eds.), Xalapa, Mexico, pp. 121-143

Roussos S., Aquiáhuatl M.A., Trejo-Hernández M.R., Perraud-Gaime I., Favela E., Ramakrishna M., Raimbault M., Viniegra-González G. (1995), Biotechnological management of coffee pulp: Isolation, screening, characterization, selection of caffeine-degrading fungi and natural microflora present in coffee pulp and husk. Appl. Microbiol. Biotechnol. 42,756-762

Smith D. (1991), Filamentous fungi. Maintenance of filamentous fungi. In- Maintenance of micro-organisms and cultured cells, B. E. Kirsop and A. Doyle (eds.), Academic Press, UK

Wang D.L., Cooney C.L., Demain A.L., Dunill P., Humphrey A.E., Lilly M.D. (1979), Fermentation and enzyme technology. John Wiley & Sons, USA

Chapter 43

SCREENING OF FILAMENTOUS FUNGI FOR THE PRODUCTION OF EXTRA-CELLULAR TANNASE IN SOLID STATE FERMENTATION (SSF)

J. van de Lagemaat[1], C. Augur[2], D.L. Pyle[1]

[1]*Biotechnology and Biochemical Engineering, Department of Food Science and Technology, University of Reading, PO Box 226, Whiteknights, Reading, RG6 6AP, UK;* [2]*IRD, Institut de Recherche pour le Développement. IRD-Mexique, Ciceron 609, Los Morales, 11530 Mexico D.F., Mexico*

Running title: Fungal tannase

1. Introduction

Tannin acyl hydrolase (E.C. 3.1.1.20), commonly called tannase, hydrolyses ester and depside bonds in hydrolysable tannins and gallic acid esters (Dykerhoff and Ambruster, 1933) and is mainly produced by fungal strains belonging to the *Aspergillus* and *Penicillium* genera (Lekha and Lonsane, 1997). Tannase acts on hydrolysable tannins and not on condensed tannins (Dykerhoff and Ambruster, 1933). However, the hydrolysis of condensed tannins such as (-) epicatechin gallate and (-) epigallocatechin-3-galate has been reported (Lekha and Lonsane, 1997).
Commercial tannase is currently produced by submerged fermentation (SmF) (Okamura *et al.*, 1987) and has found practical uses in the chemical, pharmaceutical and food industries. Studies have suggested that fungal tannase production in solid-state fermentation (SSF) could be advantageous over SmF in terms of productivity, the extra-cellular nature and thermo- and pH stability of the enzyme (Lekha and Lonsane, 1994).
Interest has occurred in tannase production by SSF with coffee pulp as a potential tannin-rich and cheap solid substrate. Data for the composition of tannins in coffee pulp are, however, contradictory. Zuluaga (1981) reported the presence of 2% total tannins in

T. Sera et al. (eds.), Coffee Biotechnology and Quality, 455–460.

sun-dried coffee pulp from which 21% were hydrolysable and 79% were condensed tannins while Clifford and Ramirez-Martinez (1991) found only 1-2.7 % condensed tannins depending on the coffee species. Future studies seem necessary in this regard.
Little work has been done on the production of tannase in SSF by different fungal strains. An intensive screening to select a potent strain suitable for an SSF system is necessary since product titres in SmF are usually different from those in SSF (Shankaranand *et al.*, 1992). Since tannic acid seems to be the best substrate for the enzyme (Yamada *et al.*, 1968), a liquid medium containing this component was soaked in polyurethane foam (PUF) and was chosen as the initial SSF system in our study. In contrast to coffee pulp, PUF enables a controlled medium composition to be achieved, offering monitoring of tannase formation and its separation from the inert carrier easily.

2. Experimental

2.1. MICROORGANISMS

Aspergillus niger 11 (IMI 017454) and 137 (IMI 041874), *A. phoenicis* (ATCC 14332), *Fusarium moniliforme* (IMI 350432) and *Penicillium frequentans* (FST 99) were selected as potential producers of tannase in SSF after an initial screening of a large number of fungi. *A. niger* (Aa20) and *P. commune* (V33A25) were obtained from the IRD-UAM culture Collection (IRD, France/Mexico) and used in this study.

2.2. INERT CARRIER AND LIQUID MEDIUM

Polyurethane foam (PUF) type HR 40 was supplied by Wolfson (Ireland) and had a bulk density of 40 kg/m^3. The foam was cut into 5-mm cubes, washed three times with warm water, dried and sterilised for 15 minutes at 121°C. Liquid medium contained (% w/v) 10, 5 or 1 tannic acid, 1 NH_4NO_3, 0.5 KH_2PO_4, 0.25 glucose, 0.1 $MgSO_4.7H_2O$, 0.01 $CaCl_2.2H_2O$, 0.002 $MnCl_2.4H_2O$, 0.001 $FeSO_4.7H_2O$, and 0.001 $Na_2MoO_4.2H_2O$. Fermentation was carried out in conical flasks. (fig.1)

2.3. DETERMINATION OF TANNASE ACTIVITY

The quantity of gallic acid released during hydrolysis of tannic acid in the assay was measured after HPLC fractionation and determination of peak area corresponding to gallic acid (Beverini and Metche, 1990). Tannase activity was expressed in μmole of gallic acid per ml of fermentation extract produced per minute.

2.4. CHARACTERISATION OF TANNASE

Native gels were incubated with tannic acid and subsequently quinine that covered the gel with a white complex. Tannase activity was distinguishable because the enzyme in the gel hydrolysed the substrate into glucose and gallic acid which did not form a white complex with quinine (Aoki *et al.*, 1979).

7 filamentous fungi (*Aspergillus*, *Penicillium*, *Fusarium*)

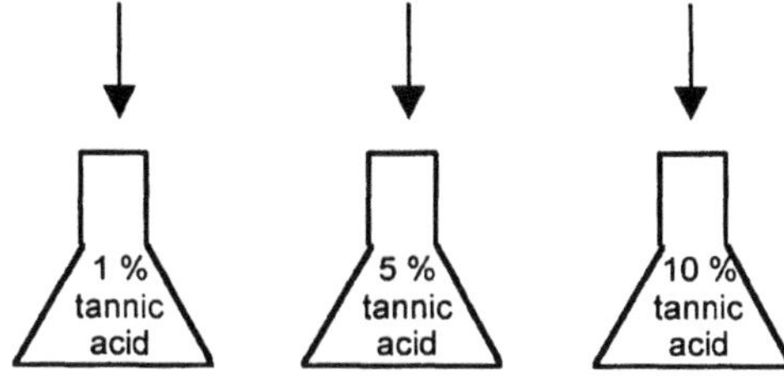

SSF (6 days, 30 °C) on PUF soaked in medium containing different concentrations tannic acid

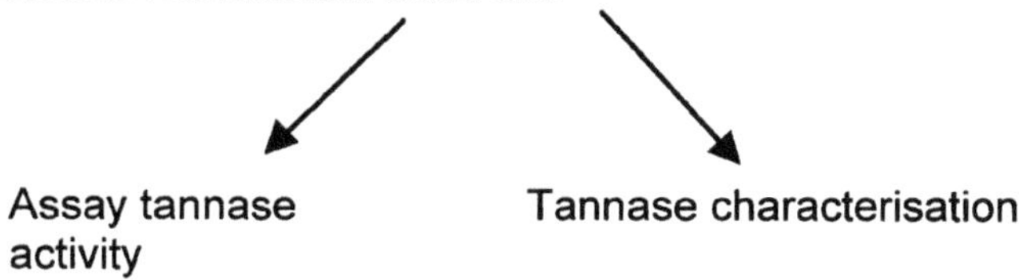

Figure 1. Schematic diagram of Tannase activities for SSF containing a) 1%, b) 5% and c) 10% tannic acid

3. Results and discussion

P. frequentans was the most productive strain in all three fermentations while *F. moniliforme* was least productive in terms of tannase formation (Figs. 2a-c).

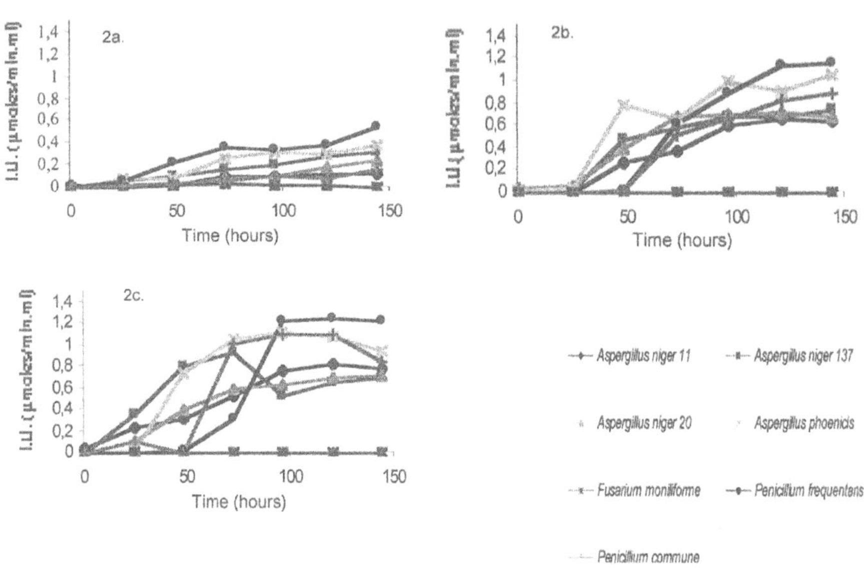

Figure 2a-c. Tannase activities for SSF containing a) 1%, b) 5% and c) 10% tannic acid

A. phoenicis showed an overall high enzyme activity. For the other strains, tannase productivity was very much dependent on the concentration of tannic acid present in the medium. Tannase production was 2-8 times more in media with 5 and 10% tannic acid compared to the 1%. Tannase activities for the 5 and 10% tannic acid fermentations did not differ as much.

Increase in activity can be explained by the fact that the fungus synthesises tannase which hydrolyses the tannic acid to soluble products as gallic acid and glucose. These end products enter into the cell by specific transport pathways and serve as the carbon and energy source for the organism. The decrease in production of tannase may be caused by the lack of substrate in the medium or the secretion of toxic substances. The decrease in tannase activity might be due to the degradation of tannase by proteases but also gallic acid can inhibit the activity (Haslam *et al.*, 1961).

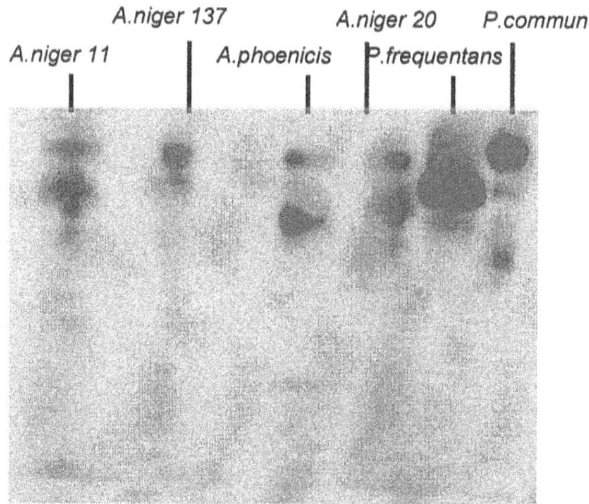

Figure 3. Activity gel with tannase from different fungal strains

Zymograms (Fig. 3) showed that enzyme activity bands differed from strain to strain and that the number of bands varied as well. The fermentation extract from *P. frequentans* showed one clear activity band while the extracts from the other strains gave two or three distinct bands.

4. Conclusions

P. frequentans produced the highest titres of extra-cellular tannase. The fermentation extract from this strain formed a single activity band in the enzyme characterisation studies, which should facilitate purification.

5. Acknowledgements

This study was supported by the European Union INCO-DC No IC18*CT970185 BIOPULCA project.

6. References

Aoki K., Tanaka T., Shinke R., Nishira H. (1979), Detection of tannase in polyacrylamide gels. J. Chromatogr., 170: 446-448

Beverini M., Metche M. (1990), Identification, purification and physicochemical properties of tannase of *Aspergillus oryzae*. Science des Aliments, 10: 807-816

Clifford M.N., Ramirez-Martinez J.R. (1991), Tannins in wet-processed coffee beans and coffee pulp. Food Chem., 40, 191-200

Dykerhoff H., Ambruster R. (1933), Tannase. Z. Physiol. Chem. 292, 38-56

Haslam E., Haworth R.D., Jones K., Rogers H.J. (1961), Gallotannins. Introduction and the fractionation of the tannase J. Chemical Soc., 57: 1829-1835

Lekha P.K., Lonsane B.K. (1994), Comparative titres, location and properties of tannin acyl hydrolase produced by *Aspergillus niger* PKL 104 in solid-state, liquid surface and submerged fermentations. Process Biochem., 29: 497-503

Lekha P.K., Lonsane B.K. (1997), Production and application of tannin acyl hydrolase: State of the art. Adv. Appl. Microbiol., 44: 215-260

Okamura S., Mizusawa K., Mamori K., Takei K., Imai Y., Ito S. (1987), Fermentative manufacture of tannase. *Japan patent* 62: 304 981

Shankaranand V.S., Ramesh M.V., Lonsane B.K. (1992), Idiosyncrasies of solid-state fermentation systems in the biosynthesis of metabolites by some bacterial and fungal cultures. Process Biochem., 27: 33-36

Yamada H., Adachi O., Watanabe M., Sato N. (1968), Studies on fungal tannase Part I Formation, purification and catalytic properties of tannase of *Aspergillus flavus*. *Agric.* Biol. Chem., 32: 1070-1078

Zuluaga J. (1981), Contribution à l'étude de la composition chimique de la pulpe de café (*Coffea arabica* L.). Ph.D. Thesis, University of Neuchatel, Neuchatel, Switzerland.

Chapter 44

INFLUENCE OF CARBON SOURCE ON TANNASE PRODUCTION BY *ASPERGILLUS NIGER* AA-20 IN SOLID STATE CULTURE

C. N. Aguilar[1,2], C. Augur[2,3], G. Viniegra-González[2], E. Favela-Torres[2]

[1]*Food Research Dept. Universidad Autónoma de Coahuila. Unidad Saltillo. P.O. Box 252, 25 000. Coah., México;* [2]*Biotechnology Dept. Universidad Autónoma Metropolitana. Unidad Iztapalapa. P.O. Box 55-535, 09340. México, D.F.;* [3]*Institut de Recherche pour le Développement, IRD-México. Cicerón 609. Los Morales, 11530, México, D.F.*

Running title: Tannase production in SSC

1. Introduction

At present, coffee pulp is one of the most abundant agro-industrial wastes and pollutants generated by the coffee industry. The utilization of coffee pulp as animal feed has been investigated for several years (De Rozo *et al.*, 1985), but its chemical composition is a great limiting factor due to the presence of anti-physiological factors such as caffeine, polyphenolic compounds and tannins (Hakil *et al.*, 1998), which include tannic acid in high concentrations (3-4%). Tannins are water-soluble phenolic compounds with molecular weights ranging from 500 to 3000 Da. These are present in several plants acting structurally as pigments and protecting from microbial attacks (Lekha and Lonsane, 1997).

Coffee pulp has been suggested useful for the production of value-added products such as enzymes. It contains high amounts of soluble sugars (Delgado, 1999). The removal of tannins from coffee pulp could be carried out in three ways, with chemical means, by microbes, or by the direct use of tannin-hydrolysing enzymes. The first option is not recommended due to the use of toxic salts or solvents, which involves high costs. Microbial or enzymatic tannin removal presents several advantages on process costs and

T. Sera et al. (eds.), Coffee Biotechnology and Quality, 461–470.

controls. Several tannase-producing micro-organisms have been reported, which mainly include filamentous fungi such as *Aspergillus, Penicillium, Fusarium* and *Trichoderma* (Iibuchi *et al.*, 1967; Rajakumar and Nandy, 1983; Kawakubo *et al.*, 1991; Lekha and Lonsane, 1994; Bajpai and Patil, 1996; García-Peña *et al.*, 1999), also bacteria (Deschamp *et al.*, 1983) and yeast (Aoki *et al.*, 1976). Thus, the use of tannase-producing fungal strains to degrade tannins might be an alternative in order to use coffee pulp as animal feed. Solid state culture (SSC) might be an attractive possibility for this purpose as this could present two advantages, the removal *in situ* of hydrolyzable tannins, and enzyme production itself.

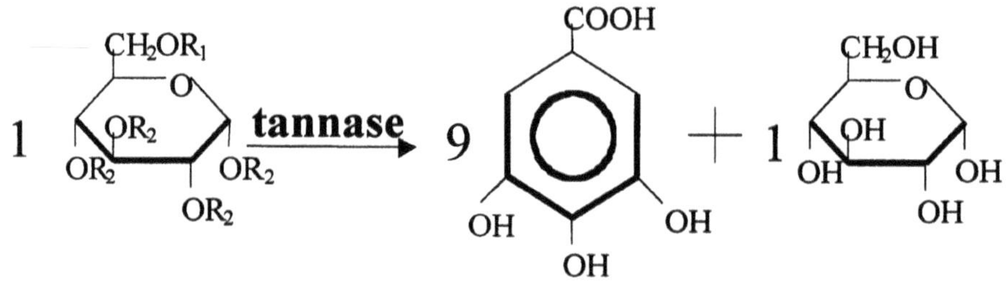

Figure 1. Hydrolysis of tannic acid catalyzed by tannase. R_1 is a galloyl group and R_2 is a di-galloyl group.

Tannase or tannin acyl hydrolase (EC. 3.1.1.20) catalyzes the hydrolysis of ester and depside bonds in hydrolyzable tannins and tannic acid (Fig. 1). Tannase is also used in food and beverage processing (Coggon *et al.*, 1975). Major commercial applications of tannase are in the manufacture of instant tea or acorn wine and the production of gallic acid (Coggon *et al.*, 1975; Chae and Yu, 1983; Pourrat *et al.*, 1985). The latter is a key intermediate required for the synthesis of an anti-bacterial drug, trimethoprim, used in the pharmaceutical industry (Sittig, 1988). Gallic acid is also a substrate for the chemical and enzymatic synthesis of propyl gallate in the food industry. In addition, tannase is used as a clarifying agent in wines, fruit juices and coffee-flavoured soft drinks (Lekha *et al.*, 1993).

Tannase production by *Aspergillus niger* Aa-20 was, therefore, studied in a SSC model in order to determine the effect of tannic acid and glucose concentration on tannase production and tannic acid uptake.

2. Experimental

2.1. MICRO-ORGANISM AND CULTURE MEDIUM

A strain of *Aspergillus niger* Aa-20 (UAM-I Collection) was used in this work. Its spores were stored at -20°C in Cryo-blocks (Bead Storage System, Technical Service Consultants Limited). Inoculum was prepared by transferring the spores to PDA medium, incubated at 30°C for 5 days. The spores were scraped into 0.01% Tween 80 solution and counted in a Neubauer chamber.

Medium for tannase production was the same as reported by Lekha and Lonsane (1994). Salt-containing medium was autoclaved at 121°C for 15 min. Tannic acid (Sigma, U.S.A.) solution was filter-sterilised and added to a final concentration of 12.5, 25, 50 and 100 $g.l^{-1}$. To evaluate the effect of glucose concentration on tannase production by *A. niger* Aa-20, an initial tannic acid concentration of 25 $g.l^{-1}$ and increasing glucose concentrations of 6.25-200 $g.l^{-1}$ were used.

2.2. SOLID STATE CULTURE (SSC)

SSC involved the use of polyurethane foam (PUF) (Expomex, México) as a support to absorb liquid medium. PUF was washed as reported Zhu *et al.* (1994) and then pulverised in a plastic-mill. Column reactors (25 x 180 mm) were packed with 10 g of inoculated PUF (2 x 10^7 $spores.g^{-1}$ of dry inert support). Culture conditions were 30°C incubation temperature, 20 $ml.g^{-1}$ support material per minute aeration rate, 5.5 and 65% initial pH and moisture, respectively. SSC was carried out for 48 h with sample removal at each six hours. For enzyme leaching, the content of each column was mixed with distilled water (10:1 w/v) and vortexed for one minute. Solid particles were filtered using Whatman 41 paper and clear filtrate was assayed for extra-cellular tannase activity. The remaining solids were washed three times with 50 ml distilled water. Intra-cellular enzyme was recovered by freezing the cells in liquid nitrogen and macerating in a chilled mortar. The broken cells were re-suspended in acetate buffer (200 mM, pH 5.5) and the filtered solution was used as intra-cellular tannase extract.

2.3. ANALYTICAL METHODS

Tannase assay was carried out using the HPLC-methodology proposed by Beverini and Metche (1990). One unit of tannase enzyme (IU) was defined as the amount of enzyme able to release one μmol of gallic acid.ml^{-1}.min^{-1}. Biomass formation in SSC was determined by measuring the mycelial protein concentration by the Bradford-microassay (Bio-Rad®) following the technique reported by Córdova-López *et al.* (1996). Tannic acid concentration was evaluated spectrophotometrically (λ = 420 nm) using the phenol-sulphuric acid method. Glucose concentration was evaluated by spectrophotometry (505 nm) using a kit from SPINREACT ™.

3. Results and discussion

In cultures with tannic acid as sole carbon source, tannase production by *A. niger* Aa-20 (both intra- and extra-cellular) in the SSC process was improved when tannic acid concentration was increased. Fig. 2 shows maximum extra-cellular and intra-cellular tannase production.

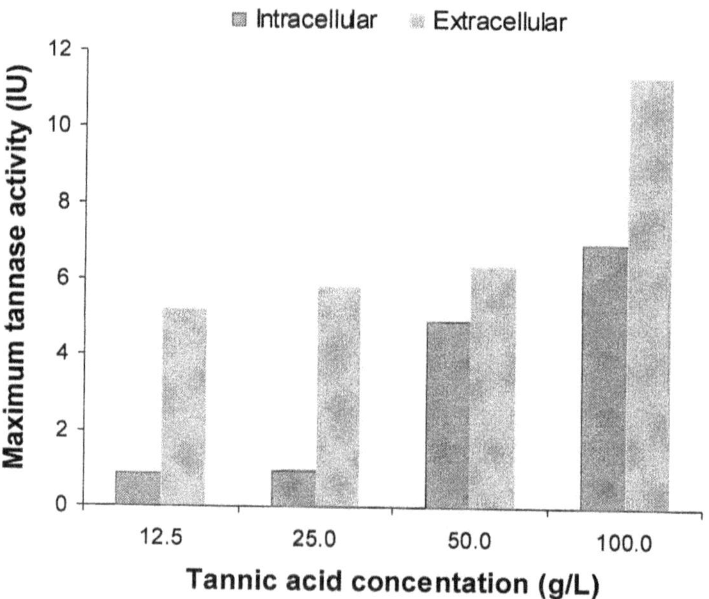

Figure 2. Extra-cellular and intracellular tannase production by *A. niger* Aa-20 in SSC at 48 h of culture with tannic acid as sole carbon source.

A consistent behaviour was found in which the increment in tannic acid concentration was related with increases in tannase activity. Maximum of tannase activity reached with 100 g.l^{-1} of tannic acid after 48 h of culture. The decrease in tannase production (using tannic acid only) by SSC as reported by Lekha and Lonsane (1994) was not observed. Lekha and Lonsane (1994) and García-Peña *et al.* (1999) reported that tannase was wholly extra-cellular during the entire fermentation period as no intra-cellular enzyme was measurable in the fungal cells. This study revealed that the use of higher tannic acid concentrations promoted the expression of intra-cellular activity; which might be due to the fact that the enzyme synthesis rate was higher than the excretion one. Extra-cellular: intra-cellular tannase activity ratios were 1.3:1 and 1.6:1 for initial tannic acid concentrations of 50 and 100 g.l^{-1}, respectively; while with 12.5 and 25 g.l^{-1} tannic acid, the ratios were higher than 6:1.

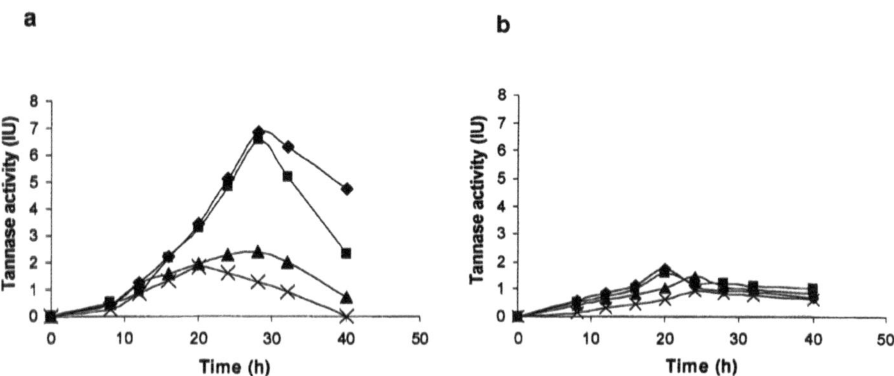

Figure 3. Extra-cellular (a) and intra-cellular (b) tannase production kinetics by *A. niger* Aa-20 in SSC in presence of tannic acid (25 g.l^{-1}) with varying glucose concentrations. 6.25 (◆), 12.5 (■), 50 (_) and 200 (**X**) g.l^{-1}.

Results of tannase production in SSC can be partially explained by those reported for pectin esterase and polygalacturonase by Maldonado and Strasser de Saad (1998). They suggested that the high titres of enzymatic production obtained in SSC were due to the changes in the membrane fatty acids composition provoked by the SSC conditions (low water content and high substrate concentration). Furthermore, it is important to consider that during SSF substrate microgradients on solid matrix could be present giving different tannic acid diffusion profiles and reducing its negative effect on fungal growth and favouring tannase production.

Table 1 shows a summary of the kinetic parameters evaluated in SSC systems. The product yield (Y_p), specific substrate uptake rate (q_s) and specific product formation rate (q_p) were highest for tannic acid concentration of 100 $g.l^{-1}$. The analysis of these kinetic parameters showed that SSC process might be attractive for the degradation of high concentrations of hydrolyzable tannins and for tannase production. *A. niger* Aa-20 was very efficient in degrading tannic acid when grown in SSC. In light of the potential use of the strain to degrade tannins present in coffee pulp, the results presented could be promising.

Table 1. Summary of some kinetic parameters* considered in the tannase production by SSC with tannic acid as sole carbon source.

Tannic acid (g/L)	μ (h^{-1})	$Y_{x/s}$ (g X/g S)	Y_p (IU/mg X)	q_s (g X/g S-h)	q_p (IU/g X h)	Substrate uptake (%)
12.5	0.288	0.306	1.385	0.942	0.399	100
25.0	0.295	0.163	1.443	1.812	0.426	100
50.0	0.223	0.091	1.469	2.440	0.328	100
100.0	0.217	0.074	2.183	2.929	0.474	73.03

* Note: μ, specific growth rate; $Y_{x/s}$, biomass yield; Y_p, product yield; q_s, specific substrate uptake rate; q_p, specific product formation rate; X, biomass and S, substrate.

Glucose addition affected negatively tannase production whether produced extra-cellularly (Figure 3a) or intra-cellularly (Figure 3b). When glucose concentrations lower than 50 $g.l^{-1}$ were used, both substrates were consumed simultaneously (Figure 4).

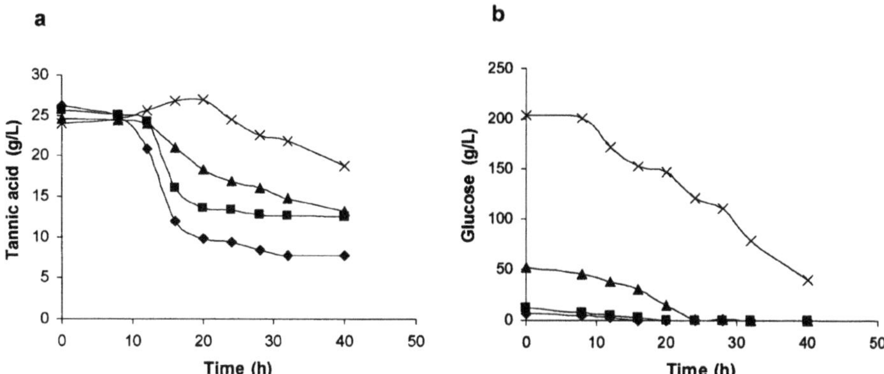

Figure 4. Tannic acid (a) and glucose (b) uptake by *A. niger* Aa-20 in SSC. 6.25 (◆), 12.5 (■), 50 (_) and 200 (X) g.l-1.

At higher glucose concentrations, glucose was used as carbon source preferentially over tannic acid. However, tannase activity decreased in the presence of glucose and tannic acid after 40 h of culture. These results are in agreement with those of Lekha and Lonsane (1994) who reported this behaviour, attributing it to the production of toxic compounds or to scarcity of substrate. In the present case, the observed decrease might be influenced by the increment in carbon source concentration, possibly due to changes in the carbon/nitrogen ratio. When glucose in SSC was used with 25 $g.l^{-1}$ tannic acid, biomass formation increased (Fig. 5) and gallic acid accumulation was not present during the SSC. This is very important since gallic acid has been related with a negative regulation of tannase biosynthesis (Bajpai and Patil, 1997).

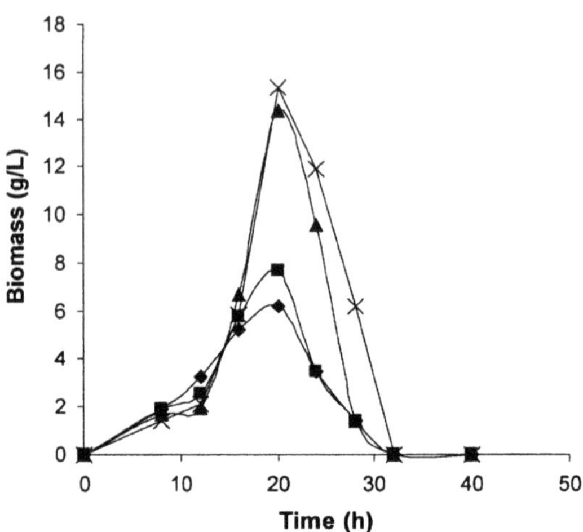

Figure 5. Biomass formation from *A. niger* Aa-20 grown in SSC in presence of 25 $g.l^{-1}$ of tannic acid and glucose at 6.25 (◆), 12.5 (■), 50 (_) and 200 (X) $g.l^{-1}$.

Most of the studies concerning the production of tannase have been carried out in submerged culture (Ganga *et al.* 1978; Pourrat *et al.* 1982; Bajpai and Patil, 1996). Such a system was considered non-suitable for tannase production due to long fermentation times and to the intracellular nature of the enzyme (Lekha and Lonsane, 1997). With respect to tannase production by SSC, the results presented are in agreement with those reported by Lekha and Lonsane (1994) and Chaterjee *et al.* (1996). SSC should be adopted as a tannase production system due to shorter fermentation times, higher activity titres and tolerance to high tannic acid concentrations

4. Summary

Production of tannase by *Aspergillus niger* Aa-20 in SSC using tannic acid (12.5, 25, 50 and 100 $g.l^{-1}$) and glucose (6.25, 12.5, 50 and 200 $g.l^{-1}$) was evaluated. Column reactors were packed with polyurethane foam impregnated with liquid medium inoculated with spores. Tannic acid uptake was measured by the phenol-sulphuric method and extra-cellular and intra-cellular tannase activities were assayed by HPLC. Tannase activity increased with increasing concentration of tannic acid. Maximum extra-cellular and intra-cellular tannase activities (11.35 and 6.95 IU, respectively) were recorded with 100 $g.l^{-1}$ tannic acid. After 48 h, the tannic acid uptake was 100% at concentrations of 12.5, 25 and 50 $g.l^{-1}$, while 74.4% was consumed at its 100 $g.l^{-1}$ concentration. Glucose addition affected negatively tannase production. Total tannase activity decreased from 8.63 to 2.83 IU with increasing glucose concentrations from 6.25 to 200 $g.l^{-1}$. These results demonstrated that high concentrations of tannins could be removed by SSC. Moreover, tannase activity was produced even at high initial glucose concentrations

5. Acknowledgements

C.N. Aguilar thanks CONACYT and UAdeC-DGEPI for financial support. This work was performed as part of a co-operative agreement between the National Council of Science and Technology (CONACYT, México) and the Institut de Recherche pour le Développement (IRD-France) within a specific programme at the Universidad Autónoma Metropolitana (México). Part of the research was funded by the European Community INCO-DC grant (IC 18*CT970185).

6. References

Aoki K., Shinke R., Nishira, H. (1976), Purification and some properties of yeast tannase. Agr. Biol. Chem., 40: 79-85

Beverini M., Metche, M. (1990), Identification, purification and physicochemical properties of tannase of *Aspergillus oryzae*. Sci. Aliments., 10: 807-816

Bajpai B., Patil, S. (1996), Tannin acyl hydrolase (EC, 3.1.1.20) activity of *Aspergillus, Penicillium, Fusarium* and *Trichoderma*. World J. Microbiol. Biotechnol., 12: 217-220

Bajpai B., Patil, S. (1997), Induction of tannin acyl hydrolase (EC, 3.1.1.20) activity in some members of *fungi imperfecti*. Enzyme Microb. Technol., 20: 612-614

Chae S., Yu, T. (1983), Experimental manufacture of acorn wine by fungal tannase. Hanguk Sipkum Kwahakoechi., 15: 326-332

Chaterjee R., Dutta A., Banerjee R., Bhattacharyya B. (1997), Production of tannase by solid-state fermentation. Bioprocess Eng., 14: 159-162

Coggon P., Graham N., Sanderson G. (1975), Cold water-soluble tea. U.K. Pat. 2: 610,533

Córdova-López J., Gutiérrez-Rojas M., Huerta S., Saucedo-Castañeda G., Favela-Torres E. (1996), Biomass estimation of *Aspergillus niger* growing on real and model supports in solid state fermentation. Biotechnol. Tech., 10, 1-6

Delgado F.K. (1999), Variación de la composición química de la pulpa de café sometida a procesos fermentativos. PhD Thesis. Benemérita Universidad Autónoma de Puebla. Pue, México

De Rozo M., Vélez J., García A. (1985), Efecto de los polifenoles de la pulpa de café en la absorción de hierro. Arch. Latinoamer. Nutr., 2, 287-305

Deschamp A., Otuk G., Lebeault G. (1983), Production of tannase and degradation of chestnut tannin by bacteria. J. Ferment. Technol., 61, 55-59

Ganga P.S., Nandy S.C. Santappa M. (1978), Effect of environmental factors on the production of fungal tannase. Leather Sci., 23: 203-209

García-Peña Y., Favela-Torres E., Augur, C. (1999), Purificación parcial de tanasa producida por *Aspergillus niger* en fermentación en medio sólido. In: *Avances en purificación y aplicación de enzimas en biotecnologia*, Prado, L.A., Huerta, S., Rodríguez, G. and Saucedo-Castañeda, G. (eds), Universidad Autónoma Metropolitana Iztapalapa. México, D.F., pp. 247-261

Hakil M., Denis S., Viniegra-González G., Augur, C. (1998), Degradation and product analysis of caffeine and related dimethylxanthines by filamentous fungi. Enzyme Microbiol. Technol., 22, 355-359

Iibuchi S., Minoda Y., Yamada K. (1967), Studies on tannin acylhydrolase of microorganisms. Part III. Purification of enzyme and some properties of it. Agr. Biol. Chem., 32, 803-809

Kawakubo J., Nishira H., Aoki K., Shinke, R. (1991), Screening for gallic acid producing micro-organisms and their culture conditions. Agr. Biol. Chem., 55: 857-877

LekhaP.K., Ramakrishna M. Lonsane B.K. (1993), Strategies for isolation of potent culture capable of producing tannin acyl hydrolase in higher titres. Chem. Mikrobiol. Technol. Lebensmitt., 15: 5-10

Lekha P.K., Lonsane B.K. (1994), Comparative titres, location and properties of tannin acyl hydrolase produced by *Aspergillus niger* PKL 104 in solid-state, liquid surface and submerged fermentations. Process Biochem., 29: 407-503

Lekha P.K., Lonsane B.K. (1997), Production and application of tannin acyl hydrolase: state of the art. Adv. Appl. Microbiol., 44: 215-260

Maldonado M.C., Strasser de Saad, A.M. (1998), Production of pectin esterase and polygalacturonase by *Aspergillus niger* in submerged and solid state systems. J. Ind. Microbiol. Biotechnol., 20: 34-38

Pourrat H , Regerat F., Pourrat A., Jean, D. (1982), Production of tannase by a strain of *Aspergillus niger*. Biotechnol. Lett., 4, 583-588

Pourrat H., Regerat F., Pourrat A., Jean D. (1985), Production of gallic acid from tara tannin by a strain of *A. niger*. J. Ferment. Technol., 63: 401-403

Rajakumar G., Nandy, S. (1983), Isolation, purification and some properties of *Penicillium chrysosporium* tannase. Appl. Environ. Microbiol. 46: 525-527

Sittig M. (1988), Trimethoprim. In: *Pharmaceutical manufacturing encyclopedia*. New Jersey, Noyes Publication, pp. 282-284

Zhu Y., Smith J., Knol W., Bol, J. (1994), A novel solid state fermentation system using polyurethane foam as inert carrier. Biotechnol. Lett., 16: 643-648

Chapter 45

COMMERCIAL PRODUCTION AND MARKETING OF EDIBLE MUSHROOMS CULTIVATED ON COFFEE PULP IN MEXICO

D. Martínez-Carrera, A. Aguilar, W. Martínez, M. Bonilla, P. Morales, M. Sobal

College of Postgraduates in Agricultural Sciences, Campus Puebla, Mushroom Biotechnology, Apartado Postal 701, Puebla 72001, Puebla, Mexico (CONACYT project 28985-B)

Running title: Mushroom cultivation on coffee pulp

1. Introduction

Large-scale utilization and management of coffee pulp around the world still remains as a challenge for the 21st century. Several alternatives have been studied, such as silage, aerobic composting, biogas production, vermiculture, animal feed (cattle, porks, chickens, fishes), and production of ethanol, vinegar, single-cell protein, enzymes, biopesticides, and probiotics (Braham and Bressani, 1979; Adams and Dougan, 1981; Roussos *et al.*, 1991; Soccol *et al.*, 1999). Although these processes are feasible in the laboratory or at small scale, successful technology transfer programmes are limited. Most efforts have focused on coffee growers, who have normally failed in taking advantage of technologies which are quite challenging, time-consuming, and technically very demanding.

We explore another alternative which considers the development of an agro-industry parallel to and connected with coffee production (Martínez-Carrera *et al.*, 1998). In this case, efforts are concentrated on entrepreneurs capable of managing mushroom biotechnology, capital investment from other sectors of the economy, and the market in a regional or world context.

2. Importance of mushroom biotechnology

Mushrooms are fleshy, spore-bearing reproductive structures of fungi. For a long time, wild edible mushrooms have played an important role as a human food. However,

T. Sera et al. (eds.), Coffee Biotechnology and Quality, 471–488.

empirical methods for their cultivation are relatively recent (Martínez-Carrera, 1999). They were independently developed in China about 1,000 years ago for *Auricularia* spp. and *Lentinula edodes* (Berk.)Pegler, and in France about 350 years ago for *Agaricus bisporus* (Lange)Imbach. During the last 50 years, these methods have been significantly improved and modern technologies permit the cultivation of about 20 species at different levels around the world (Chang and Miles, 1989).

Recent figures indicate that commercial production of fresh edible mushrooms is a rapidly-growing industrial activity (Chang and Miles, 1991). During the period 1990-1994, world mushroom production increased by 30.5%, reaching about 4,909 thousand tons in 1994. The global economic value, although difficult to evaluate, has been estimated to be more than 9.8 billion dollars per annum (Chang, 1996). These data include several species of eleven main genera: *Agaricus*, *Lentinula*, *Volvariella*, *Pleurotus*, *Auricularia*, *Flammulina*, *Tremella*, *Hypsizygus*, *Pholiota*, *Grifola*, and *Hericium*. Mushrooms belonging to the genus *Agaricus* are the most widely cultivated, and their total global production in 1994 was 1,846 thousand tons fresh weight.

The biodegradation of lignocellulosic by-products from agriculture or forestry confers ecological importance on mushroom cultivation (Wood, 1985). To achieve this, mushroom hyphae (*i.e.*, mycelium) produce a wide range of extracellular enzymes capable of degrading complex organic material (Martínez-Carrera, 1999). Millions of tons of these by-products, which otherwise would remain unused, are recycled every year as substrates for mushroom growing. The resulting so-called spent substrate is less bulky, and has been traditionally used as a soil conditioner by farmers and gardeners. However, recent studies have shown that spent mushroom compost can also be used for growing containerised nursery plants (Maher, 1991), or as a potential animal feed (Grabbe, 1990). The nutrient needs of maize crops can be satisfied if 200-400 tons/acre of spent mushroom compost are incorporated into the soil, without the negative consequence of adding nitrate to either surface or field drainage water (Flegg, 1991).

Mushroom cultivation is an efficient and relatively short biological process of food protein recovery from lignocellulosic materials. The protein content of edible mushrooms can be considered as their main nutritional attribute. An average value of 19-35% on a dry weight basis has been reported (4% on a fresh weight basis), as compared to 13.2% in wheat, and 25.2% in milk. Edible mushrooms are also a good source of some vitamins and minerals, although fat, carbohydrate, and dietary fibre contents are comparatively low. Recent research work indicates medicinal attributes in several species, such as antiviral, antibacterial, antiparasitic, antitumor, antihypertension, antiatherosclerosis, hepatoprotective, antidiabetic, anti-inflammatory, and immune modulating effects (Wasser and Weis, 1999). Mushrooms are now considered as genuine nutraceuticals, from which nutriceuticals and pharmaceuticals can be developed. Overall value of mushroom products obtained from mushrooms is rapidly increasing, and it has been estimated to be about 3.6 billion dollars per annum (Chang, 1999).

3. Principles of mushroom biotechnology

In general, there are three major fundamental technologies involved : 1) Spawn technology, 2) Mushroom production technology, and 3) Processing technology (Martínez-Carrera, 2000; Fig. 1). A variety of methods and techniques have been developed and described in detail for each technology (Flegg *et al.*, 1985; Chang and Miles, 1989; van Griensven, 1988).

Spawn technology includes the isolation of strains from wild mushrooms growing in nature (Fig. 2a) either by tissue culture or spore culture (Fig. 2b-d). Intensive selection and breeding through classical and molecular genetics is necessary, as wild strains are normally not suitable for commercial cultivation. Genetic improvement is focused on high-yielding strains having additional characteristics, such as disease/chemical resistance, earliness, tolerance to low or elevated temperatures, as well as shape, taste, and colour of fruit bodies (Fig. 2e). When a selected strain is available, spawn preparation is carried out using cereal grains (*e.g.*, wheat, rye, millet, rice, sorghum) or other organic substrates (*e.g.*, coffee pulp, straw, cotton waste, sawdust) sterilized in glass jars or polypropylene plastic bags (Fig. 3a). These organic materials are inoculated, and incubated to be completely colonized by the mushroom mycelium and then used as spawn (Fig. 3b-d).

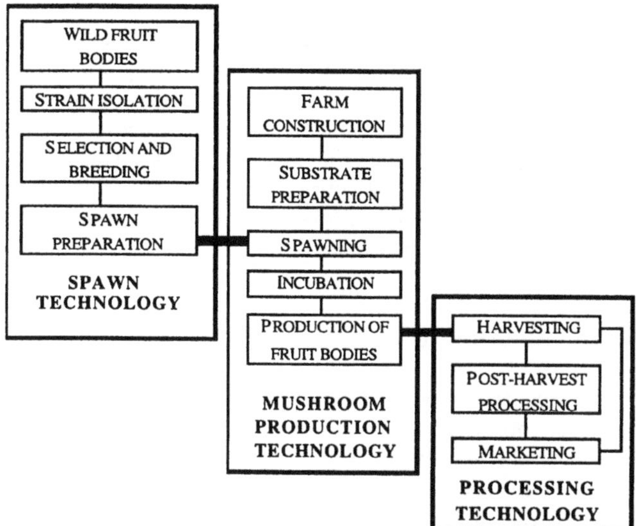

Figure 1. Fundamental principles of mushroom biotechnology.

Mushroom production technology starts with construction of the farm according to local environmental conditions and species requirements. Suitable substrates are then prepared using organic raw materials, easily available at low costs in each region. These substrates, selective for mushroom growing, are spawned and incubated in the farm. Production of fruit bodies varies according to each species, spawn quality, substrate quality, and environmental conditions.

Processing technology is fundamental for commercial production and requires specific facilities. After harvesting, mushrooms are normally cooled down to retard fruit-body metabolism, and then sent to the fresh market. Alternative methods are available, which include cooling, vacuum cooling, cooling with positive ventilation, and ice-bank cooling with positive ventilation. Further processing, *e.g.* canning, drying, or irradiation can also be carried out depending on marketing strategies.

4. Mushroom cultivation on coffee pulp

Historical records suggest that coffee was introduced to Mexico around 1790. Coffee cherries have mainly been processed by the method of wet process ever since, and coffee pulp discarded causing insalubrious conditions and pollution in nearby rivers. Coffee pulp has also been exposed to a variety of microorganisms naturally occurring in the environment. A new ecological niche was then available for native strains of edible oyster mushrooms (*Pleurotus*), primary decomposers having the ability of degrade lignocellulose (Fig. 2a). In 1982, isolation and characterization of *Pleurotus* strains capable of growing on sterilized coffee pulp were reported (Martínez-Carrera and López, 1982; Martínez-Carrera, 1984; Martínez-Carrera *et al.*, 1984). This was followed by a series of studies which have shown that coffee pulp, either as a sole substrate or mixed with other organic materials, is a good substrate for cultivation of the edible mushrooms *Pleurotus*, *Lentinula*, and *Auricularia* (Table 1).

4.1. SUBSTRATE PREPARATION

Fresh coffee pulp produced by wet processing is immediatly subjected to microbial degradation, as yeast (60.6%), fungal (2.4%), and bacterial (37%) populations occur naturally (Gaime-Perraud *et al.*, 1993). Natural fermentation develops rapidly following different pathways (*e.g.*, acetic, lactic, anaerobic, aerobic), depending on physical, chemical, biological, and environmental factors. For these reasons, coffee pulp should be managed appropriately and pretreated in order to be used as substrate for mushroom growing.

Fresh coffee pulp is allowed to drain for 4-8 hours in order to reach 60-80% moisture content, and piled up into long pyramidal heaps (*ca.* 1 m high x 1.5 m wide at the base).

An efficient aerobic fermentation should be promoted by turning the pile every three days (about 4-6 tonnes can be turned in one man-day) [Fig. 4a]. Coffee pulp fermented for up to 10 days has good structure and consistency, and can be used for *Pleurotus* cultivation. After fermentation, coffee pulp is relatively odourless, less bulky, and physically and chemically more homogeneous. Its pH (6.0-7.0) remains suitable for mycelial growth, and mushroom yields are slightly higher (Table 2). Coffee pulp can also be mixed or supplemented with other agricultural by-products to favour aerobic fermentation, such as straw (barley, wheat), maize stubble, and sugar cane bagasse.

Table 1. Edible fungi which can be cultivated on coffee pulp, either as a sole substrate or mixed with other organic materials (Martínez-Carrera, 1987, 1989a-b; Martínez-Carrera *et al.*, 1985a-b, 1990, 1996b; Bernabé-González *et al.*, 1991; Calvo-Bado *et al.*, 1996).

Species	Substrate	Dry weight (g)	Yield average (g)	Biological efficiency (%)[1]
Auricularia	*Inga* sawdust + coffee pulp[2]	159.3	59.2	37.1
fuscosuccinea	Corn cobs + coffee pulp + *Leucaena*[3]	-	-	20.8
Lentinula edodes	*Quercus* sawdust + wheat bran + coffee pulp[4]	232.7	50.9	21.8
Pleurotus sp.	Coffee pulp[5]	999	1,756.5	175.8
cfr. Florida	Coffee pulp + coconut fibre[6]	500	447	89.4
P. ostreatus	Coffee pulp[5]	999	1,598	159.9
	Coffee pulp + sugar cane bagasse[7]	1,350	1,309	96.9
	Coffee pulp + barley straw[6]	1,611	1,607	99.7
P. sajor-caju	Coffee pulp[5]	999	1,280	128.1
P.opuntiae	Coffee pulp [6]	999	1,437	143.8
P. salmoneo-stramineus	Coffee pulp [6]	999	1,549	155

1 Yield of fruit bodies (fresh weight) as a percentage of the dry weight of substrate at spawning (Tschierpe and Hartmann, 1977).

2 Sterile substrate; proportion 1:1 on a dry weight basis.

3 Sterile substrate; proportion 94:3:3 on a dry weight basis.

4 Sterile substrate; proportion 1:1:1 on a dry weight basis.

5 Pasteurized.

6 Pasteurized; proportion 1:1 on a dry weight basis.

7 Pasteurized; proportion 2:1 on a dry weight basis.

Table 2. Effect of aerobic fermentation of coffee pulp on the mushroom yield of *Pleurotus ostreatus* (Martínez-Carrera *et al.*, 1985b).

Treatment	pH	Dry weight substrate (kg)	T[1]	Yield average (g)	Biological efficiency (%)
Fresh	6.0	1.161	34	1,316	113.3
5 days fermentation	6.0	0.999	23	1,320	132.1
10 days fermentation	7.0	0.954	36	1,135	118.9

[1]Average time after spawning to produce the first flush (days).

Caffeine causes adverse effects in animal metabolism. Caffeine content in the coffee pulp varies during mushroom cultivation (Table 3). The highest reduction takes place during substrate preparation by aerobic fermentation and pasteurization (immersion in hot water at 70°C for 15 min).

Table 3. Caffeine content of coffee pulp used as a substrate for the cultivation of *Pleurotus* mushrooms (Martínez-Carrera *et al.*, 1985b).

Coffee pulp treatment	Caffeine content (%)		
	Before pasteurization	After pasteurization[1]	After mushroom cultivation[2]
Fresh	0.99	0.25	0.20
5 days fermentation	0.52	0.20	0.14
10 days fermentation	0.45	0.22	0.20

[1] Immersion in hot water at 70°C for 15 min.
[2] After mycelial growth, fruiting, and harvesting four flushes.

Caffeine reduction during mushroom cultivation (*i.e.*, mycelial growth, fruiting, harvesting) is not significant. This is supported by laboratory experiments, in which mycelial growth on agar plate is gradually inhibited at caffeine concentrations ranging from 0.250-2.0 mg/ml (Martínez-Carrera *et al.*, 1988).

Table 4. Mushroom yields of *Pleurotus ostreatus* cultivated on coffee pulp dried by direct exposure to the sun, stored for different periods of time, and pasteurized (Soto *et al.*, 1987).

Period of storage of the coffee pulp (months)	Dry weight substrate (kg)	Yield average (g)	Biological efficiency (%)[1]
Control (fermented for 5 days)	0.888	1,418	159.6
1	0.888	1,267	142.6
2	0.888	1,298	146.1
7	0.888	1,265	142.4
12	0.888	1,346	152.7
24	0.888	1,290	145.2

1 Yield of fruit bodies (fresh weight) as a percentage of the dry weight of substrate at spawning (Tschierpe and Hartmann, 1977).

After drainage, uniform solar drying of coffee pulp is possible in 4-6 days if environmental conditions are suitable (Fig. 4b). Dry coffee pulp can be stored, and used for *Pleurotus* cultivation even after two years without significant variations in mushroom yields (Table 4). Large-scale artificial drying of fresh coffee pulp is also possible taking advantage of facilities available within coffee regions. Drained coffee pulp is loaded in a commercial coffee drier, having a gas burner and a fan (Fig. 4c). Each load of about 5,000 kg can be dried at 80°C for 30 h. There is no significant difference between fresh and dry coffee pulp in terms of general characteristics (Table 5), as they contain similar amounts of organic matter, nitrogen, phosphorus, potassium, calcium, magnesium, and a pH slightly acid (Martínez-Carrera *et al.*, 1996b).

Dry coffee pulp can be used for mushroom cultivation without previous aerobic fermentation. Dry coffee pulp also offers transportation advantages, as it is less bulky and has a higher water retention capacity. For example, 1 m^3 of dry coffee pulp has about 136 kg, while 1 m^3 of wheat straw (a substrate commonly used for large-scale mushroom cultivation in Mexico) has around 70 kg. This means that one lorry transporting 18,000 kg of dry coffee pulp is equivalent to 4.4 lorries of wheat straw. In addition, dry coffee pulp increases its weight about 275% when rehydrated, while wheat straw increases around 200% (Martínez-Carrera *et al.*, 1996b).

Table 5. Proximate chemical analysis of coffee pulp dried artificially (Martínez-Carrera *et al.*, 1996b).

Component/characteristic	Coffee pulp	
	Fresh	Dry
pH (1:10)	4.4	5.6
Carbon / nitrogen	67.5	50.6
Organic matter	92.9%	91.4%
Nitrogen	0.80%	1.05%
Phosphorus	0.11%	0.13%
Potassium	3.51%	3.99%
Calcium	0.53%	0.63%
Magnesium	0.12%	0.15%

4.2. SUBSTRATE PASTEURIZATION OR STERILIZATION

After fermentation or rehydration, coffee pulp (as a sole substrate, mixed or supplemented with other organic materials) should be pasteurized for the cultivation of *Pleurotus* (Martínez-Carrera, 1987; 1989), while sterilized for growing *Lentinula* and *Auricularia.* Coffee pulp can be pasteurized by immersion in hot water at 70°-90°C for 1-2 hours, a method suitable for rural mushroom cultivation on a small scale (Fig. 4d). For large-scale processing, coffee pulp is placed in an appropriate room or tunnel and pasteurized with steam at 60°-100°C for 6-24 hours. In the case of substrate sterilization, coffee pulp is introduced into polypropylene plastic bags, and autoclaved at 100°-121°C for 1-2 hours (Martínez-Carrera, 1998).

4.3. SPAWNING

Appropriate mushroom strains should be selected according to local environmental conditions, considering that coffee plantations occur at a variety of altitudes (300-1,400 m). Pasteurized coffee pulp is cooled, and homogeneously inoculated with the spawn (*Pleurotus*), either by hand or mechanically, at a rate ranging from 0.5-3% of fresh substrate weight (Fig. 5a-d). In the case of substrate sterilization, the inoculation of supplemented coffee pulp is carried out at a similar spawning rate (*Lentinula, Auricularia*) under aseptic conditions in a laboratory.

4.4. PRODUCTION SYSTEMS

Coffee pulp spawned with *Pleurotus* species is introduced in plastic bags of different sizes, although trays, shelves, vertical plastic sacks, and pressed rectangular blocks may also be used. Inoculated containers (*Pleurotus*, *Lentinula*, *Auricularia*) are placed in growing rooms for incubation and/or fruiting, where temperature (15°-30°C), relative humidity (> 60%), ventilation, and light should be as stable as possible for mushroom cultivation (Martínez-Carrera, 1987; Martínez-Carrera *et al.*, 1992a). Complete colonization of coffee pulp by the mushroom mycelium normally takes 25-30 days for *Pleurotus*, while 60-120 days for *Lentinula* and *Auricularia*. After this incubation period, main environmental factors are managed to promote fruiting. Differentiation starts with formation of small structures called primordia, and complete fruit-body development takes 4-7 days (Fig. 5e-g). Mushroom yields vary according to biological factors, environmental conditions, as well as pests and diseases present during cultivation (Figs. 5h, 6a-b). The biological efficiency, defined as the yield of fresh fruit bodies as a percentage of the dry weight of substrate at spawning (Tschierpe and Hartmann, 1977), varies from 89-175% in *Pleurotus*, from 20-37% in *Auricularia*, and is higher than 21% in *Lentinula*.

In subtropical regions, fresh mushrooms should be cooled down or processed further in order to avoid rapid deterioration before marketing (Fig. 6c). Mushroom canning using local recipes allows to produce a commercial product which is safe, stable, economic, and with good sensory and nutritive properties (Fig. 6d). This processing technology also permits to increase the value added to mushrooms, to standardize mushroom quality, to highlight certain culinary properties of mushrooms by good recipes, and to develop marketing strategies at a national or international level (Martínez-Carrera *et al.*, 1996a).

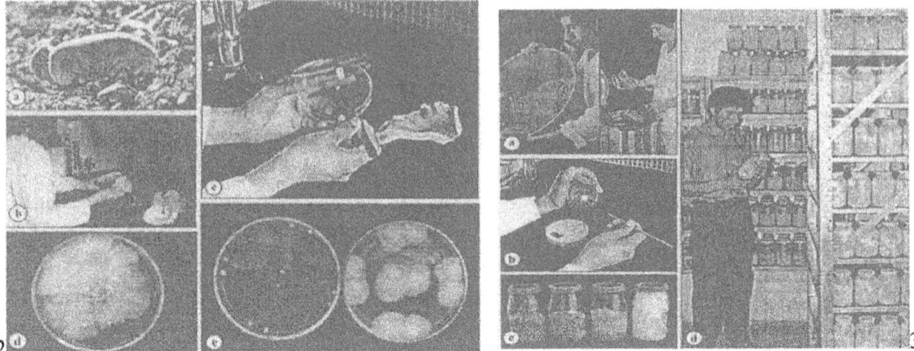

Figures 2 and 3. Several aspects of spawn technology for edible mushroom production. On the left, (a) wild fruit bodies of *Pleurotus* growing on coffee pulp, (b-c) isolation of strains by tissue culture from wild fruit bodies, under aseptic conditions in the laboratory, (d-e) mycelial growth and interstock matings developed in sterile condition. On the right, (a-d) preparation of spawn of cultivated edible mushrooms on wheat kermels in small-scale production.

Figures 4 and 5. On the left, (a-d) coffee pulp preparation as substrate for mushroom cultivation. On the right, (a-b) coffee pulp inoculation and introduction in plastic bags, (c-g) incubation, primordia and fruit-body development, and (h) *Pleurotus* commercial production in Mexico.

Figures 6 and 7. On the left, (a-d) rural production of edible mushrooms (*Pleurotus*) in Cuetzalan, Puebla, Mexico. On the right, (a) spent coffee pulp after mushroom cultivation, and (b) organic fertilizer produced by vermi-composting.

4.5. SPENT COFFEE PULP

After mushroom cultivation, a proportion of about 27% from the original substrate will remain. Approximate chemical composition of spent coffee pulp, after *Pleurotus* cultivation, is shown in Table 6. Carbohydrates (29.9%), crude protein (21.5%), crude fat (1.8%), and crude fibre (31.4%) are main components present. This spent substrate can be composted, either aerobic composting and/or vermicomposting, to produce an organic fertilizer or soil conditioner for crop soils (Fig. 7a-b).

Table 6. Proximate chemical analysis of spent coffee pulp after the cultivation of *Pleurotus ostreatus* (Martínez-Carrera, 1989a).

Component/characteristic	%
Moisture	81.8
Ash	15.2
Crude fat	1.8
Carbohydrates	29.9
Crude protein (N x 6.25)	21.5
Crude fibre	31.4
Tannin (qualitative analysis)	Negative

4.6. SOCIO-ECONOMIC ASPECTS

Social, economic, and ecological benefits can be obtained through mushroom biotechnology using coffee pulp as a growing substrate (Martínez-Carrera, 1989a, 1999b; Martínez-Carrera and Larqué-Saavedra, 1990; Martínez-Carrera *et al.*, 1991a,b; 1993; 1992a, b; 1995; 1998). Main production and operation costs from commercial and rural production of edible mushrooms (Table 7) are salaries (48.3%; five workers), raw materials and energy (33.6%), travelling expenses (5.5%), maintenance (6.8%), and regional marketing (3.3%).

Table 7. Costs (USD) of production and operation in a rural commercial farm from Cuetzalan, Puebla, Mexico (Martínez-Carrera *et al.*, 1998).

Years	Salaries	Raw materials and energy	Administration expenses	Travelling expenses	Maintenance	Marketing
1992-1997	17,845.09	12,418.45	928.78	2,042.04	2,505.22	1,197.01
(%)	(48.3)	(33.6)	(2.5)	(5.5)	(6.8)	(3.3)

A cost-benefit analysis of a mushroom farm operating commercially indicates that this biotechnological process is profitable (Table 8), even under rural conditions (c/b ratio= 1.10). In comparison with other crops and agro-industries, mushroom cultivation is also an efficient process for using and converting energy or water into a human food. Water consumption is considerably higher in mushroom production (*ca.* 97%) than in spawn production (*ca.* 3%). Overall data show that 28 L of water are required for producing 1 kg of fresh oyster mushrooms using rustic technologies, in a considerably short period of time (25-30 days after spawning).

Table 8. Financial analysis (USD) of the commercial mushroom production in a rural farm from Cuetzalan, Puebla, Mexico (Martínez-Carrera *et al.*, 1998).

Years	Production costs ($)	Gross incomes ($)	Profits ($)			Cost-benefit ratio
			Fresh oyster mushrooms	Spawn	Total	
1992-1997	36,936.59	40,576.57	2,536.60	1,103.36	3,639.96	1.10[1]

[1] Average data.

This is a smaller amount in comparison with estimations for other foods or forages, such as potatoes (500 L/kg), wheat and alfalfa (900 L/kg), sorghum (1,110 L/kg), corn (1,400 L/kg), rice (1,912 L/kg), soybeans (2,000 L/kg), broiler chicken (3,500 L/kg), and beef (100,000 L/kg). The production of 1 kg of beef requires 3,571 times more water than the amount needed to produce 1 kg of oyster mushrooms (Table 9).

Table 9. Estimated amount of water required for producing 1 kg of fresh oyster mushrooms using rustic technologies, in comparison with that for other food and forage crops (Martínez-Carrera *et al.*, 1998).

Product	Litres of water/kg	Protein content[a]	Litres of water per gram of protein
Oyster mushrooms (*Pleurotus*)	28	2.7	1.0
Potatoes	500[b]	2.1	23.8
Wheat	900[b]	14.0	6.4
Alfalfa	900[b]	6.0	15.0
Sorghum	1,110[b]	11.0	10.0
Corn	1,400[b]	3.5	40.0
Rice	1,912[b]	6.7	28.5
Soybeans	2,000[b]	34.1	5.8
Broiler chicken	3,500[b]	23.8	14.7
Beef	100,000[b]	19.4	515.4

[a] Composition in 100 g, edible portion (fresh weight) (Watt and Merrill, 1975; Duke and Atchley, 1986; Chang and Miles, 1989).

[b] Data according to Pimentel *et al.* (1997).

Several environmental, economic, and social indicators have been identified and interpreted to assess sustainability of rural mushroom cultivation (Table 10).

Table 10. Environmental, economic, and social indicators which contributed to understand the sustainability of the model for rural production of edible mushrooms using rustic technologies in the coffee growing region of Cuetzalan, Puebla, Mexico, during the period 1992-1997 (Martínez-Carrera et al., 1998).

Category	Indicator	Critical value	Factor(s) evaluated
Environmental	Biological efficiency	≥32%	Spawn, yields, substrates
	Degradation rate	≥31%	Spent substrates, potential organic fertilizer[a]
	Contamination rate	≤20%	Raw materials, growing systems and techonlogy, spawn, environmental conditions, hygiene, labour skills
	Energy–use efficiency	≤ 12%	Energy consumption [b]
	Water-use efficiency	≥ 28 L/kg	Water consumption to produce mushrooms
	Temperature	15°C	Minimum temperature [c]
		33°C	Maximum temperature [c]
	Relative humidity	≥ 70%	Environmental moisture
Economic	Cost-benefit ratio	≥ 1.0	Gross incomes, production costs, profits
Social	Mushroom consumption	≥ 0.914 kg	*Per capita* [d]
		≥ 4.100 kg	per household [d]
	Labour efficiency	≤ 3	Number of workers in the farm [e]
	Market	qe	Potential increase in mushroom production within the farm [f]

qe= Qualitative estimations are usually available, as market trends depend on social, economic, and political circumstances. The market can be local, national or international. National production, imports, exports, real and potential domestic demand are to be considered.

[a] Variations are not significant on a large scale. Data expressed on a dry weight basis.

[b] Proportion as a percentage from total production cost.

[c] Temperatures may be higher or lower, depending on strain tolerance.

[d] Minimum mushroom consumption per year required to maintain the mushroom farm.

[e] Each worker should produce at least 1,689 kg per year.

[f] If enough financial support is available.

5. Future and prospects

Mushroom cultivation is a well-established and profitable biotechnological process carried out worldwide on a large or small scale. Coffee pulp, either as a sole substrate or supplemented with other organic materials, can be used as a substrate for growing the edible mushrooms *Pleurotus*, *Lentinula*, and *Auricularia*. However, utilization of fresh coffee pulp is limited due to: 1) Seasonal availability during the year, 2) Active natural fermentation, 3) Impractical and uneconomic transportation, and 4) Large-scale handling for substrate preparation is more demanding and laborious, in comparison with other agricultural by-products (*e.g.*, straw and corn stubble). Accordingly, at present, the establishment of a mushroom farm within a coffee region, as an independent industry connected with regional agricultural activities (*i.e.*, agro-industry), is a realistic alternative for large-scale utilization of fresh coffee pulp. Appropriate regional adaptations are necessary to design the mushroom farm, considering strain selection, spawn preparation, substrate availability, spawning, production systems, fruiting, and post-harvest processing. The sustainable model for rural production of edible mushrooms represents a strategy that allows large-scale, small-scale, and domestic cultivation to promote regional development (Martínez-Carrera *et al.*, 1998). Associations of coffee growers and mushroom producers may be a productive alternative, as long as enough financial and technical assistance is available and appropriate marketing strategies developed.

Dry coffee pulp can also be used for mushroom cultivation (Martínez-Carrera *et al.*, 1996b). However, solar drying is inefficient for processing large amounts, time-consuming, and dependent on environmental conditions, while artificial drying is more expensive. Dry coffee pulp has several advantages: water retention capacity, homogeneity, physico-chemical structure, practical transportation, and availability throughout the year. Although production costs are relatively higher, in comparison with other agricultural by-products, dry coffee pulp represents a high quality raw material. Enough capital investment is necessary to develop an efficient technology for large-scale processing, in which dry coffee pulp can be stored in appropriate silos to be marketed worlwide as a supplement for substrate formulations in order to improve yields in the mushroom industry.

Further research work may be focused on testing other methods of coffee pulp preservation, such as ensiling, for mushroom cultivation. Coffee pulp can also be studied as a substrate for other cultivated species of edible mushrooms. Increasing importance of organic coffee in the world market will benefit mushroom cultivation, as the coffee pulp produced would permit the production of organic mushrooms.

6. References

Adams M.R., Dougan J. (1981), Biological management of coffee processing wastes. Trop. Sci., 23, 177-196

Bernabé-González T., Domínguez M.S., Bautista S.A. (1991), Cultivo de *Pleurotus floridanus* Singer, sobre fibra de coco y mezclada con pulpa de café. Paper presented at 4th Mexican Congress of Mycology, October 14-18, Tlaxcala, Mexico

Braham J., Bressani R. (1979), Pulpa de café, composición, tecnología y utilización, CIID, Bogota, Colombia.

Calvo-Bado L.A., Sánchez-Vázquez J.E., Huerta-Palacios G. (1996), Cultivo de *Auricularia fuscosuccinea* (Mont.)Farlow sobre sustratos agrícolas en el Soconusco, Chiapas, México. Micol. Neotrop. Apl., 9, 95-106

Chang S.T. (1996), Mushroom research and development - equality and mutual benefit. In-Mushroom Biology and Mushroom Products, D.J. Royse (ed.), Pennsylvania State University, University Park, pp. 1-10

Chang S.T. (1999), Global impact of edible and medicinal mushrooms on human welfare in the 21st century: nongreen revolution. International Journal of Medicinal Mushrooms, 1, 1-7

Chang S.T., Miles P.G (1989), Edible Mushrooms and their Cultivation, CRC Press, Boca Raton, U.S.A

Chang S.T., Miles P.G (1991), Recent trends in world production of cultivated edible mushrooms. Mushroom Journal, 504, 15-18

Duke J.A., Atchley A.A. (1986), CRC Handbook of Proximate Analysis Tables of Higher Plants, CRC Press, Boca Raton, U.S.A

Flegg P.B. (1991), Used compost: a resource or liability? Mushroom Journal, 504, 27

Flegg P.B., Spencer D.M., Wood D.A. (1985), The Biology and Technology of the Cultivated Mushroom, John Wiley & Sons, Chichester, England.

Gaime-Perraud I., Roussos S., Martínez-Carrera D. (1993), Natural microorganisms of the fresh coffee pulp. Micol. Neotrop. Apl., 6, 95-103

Grabbe K. (1990), Upgrading of lignocellulosics from agricultural and industrial production processes into food, feed and compost-based products. In- Advances in Biological Treatment of Lignocellulosic Materials, M. P. Coughlan & M. T. Amaral Collaço (ed.), Elsevier Applied Science Publishers, Barking, pp. 331-342

Maher M.J. (1991), Spent mushroom compost (SMC) as a nutrient source in peat based potting substrates. Mushroom Science, 13, 645-650

Martínez-Carrera D. (1984), Cultivo de Pleurotus ostreatus sobre desechos agrícolas. I. Obtención y caracterización de cepas nativas en diferentes medios de cultivo sólido en el laboratorio. Biotica, 9, 243-248

Martínez-Carrera D. (1987), Design of a mushroom farm for growing *Pleurotus* on coffee pulp. Mushroom Journal for the Tropics, 7, 13-23

Martínez-Carrera D. (1989a), Simple technology to cultivate *Pleurotus* on coffee pulp in the tropics. Mushroom Science, 12, 169-178

Martínez-Carrera D. (1989b), Past and future of edible mushroom cultivation in tropical America. Mushroom Science, 12, 795-805

Martínez-Carrera D. (1998), Oyster mushrooms. In- McGraw-Hill Yearbook of Science & Technology, McGraw-Hill, Inc., New York, pp. 649-652

Martínez-Carrera D. (1999), Mushroom. In- *McGraw-Hill Encyclopedia of Science and Technology, 9th Edition*, McGraw-Hill, Inc., New York, pp. 579-580

Martínez-Carrera D. (2000), Mushroom biotechnology in tropical America. International Journal of Mushroom Science, 3, 9-20

Martínez-Carrera D., Larqué-Saavedra A. (1990), Biotecnología en la producción de hongos comestibles. Ciencia y Desarrollo, 95, 53-64

Martínez-Carrera D., López A. (1982), Obtención, caracterización y comportamiento de cepas de *Pleurotus ostreatus* en cultivo de laboratorio, Paper presented at 1st. Mexican Congress of Mycology, October 26-30, Xalapa, Veracruz, Mexico

Martínez-Carrera D., Aguilar A., Martínez W., Morales P., Sobal M., Bonilla M., Larqué-Saavedra A. (1998), A sustainable model for rural production of edible mushrooms in Mexico. Micol. Neotrop. Apl., 11, 77-96

Martínez-Carrera D., Larqué-Saavedra A., Morales P., Sobal M., Martínez W., Aguilar A. (1993), Los hongos comestibles en México, biotecnología de su reproducción. Ciencia y Desarrollo, 108, 41-49.

Martínez-Carrera D., Soto C., Guzmán G.(1985a), Cultivo de *Pleurotus ostreatus* en pulpa de café con paja como substrato. Rev. Mex. Mic., 1, 101-108

Martínez-Carrera D., Vergara F., Juarez S., Aguilar A., Sobal M., Martínez W., Larqué-Saavedra A. (1996a), Simple technology for canning cultivated edible mushrooms in rural conditions in Mexico. Micol. Neotrop. Apl., 9, 15-27

Martínez-Carrera D., Guzmán G., Soto C. (1985b), The effect of fermentation of coffee pulp in the cultivation of *Pleurotus ostreatus* in Mexico. Mushroom Newsletter for the Tropics, 6, 21-28

Martínez-Carrera D., Quirarte M., Soto C., Salmones D., Guzmán G. (1984), Perspectivas sobre el cultivo de hongos comestibles en residuos agro-industriales en México. Bol. Soc. Mex. Mic., 19, 207-219

Martínez-Carrera D., Sobal M., Morales P. (1988), El efecto de la cafeína sobre el crecimiento e intracruzamiento de *Pleurotus ostreatus* en el laboratorio. Rev. Mex. Mic., 4, 131-135

Martínez-Carrera D., Sobal M.,. Morales P., Larqué-Saavedra A. (1992a), Prospects of edible mushroom cultivation in developing countries. Food Laboratory News, 8(3), 21-33

Martínez-Carrera D., Sobal M., Morales P., Martínez-Sánchez W., Aguilar A., Larqué-Saavedra A. (1995), Edible mushroom cultivation and sustainable agriculture in Mexico. The African Journal of Mycology and Biotechnology, 3, 13-18

Martínez-Carrera D., Morales P., Sobal M. (1990), Cultivo de *Pleurotus ostreatus* sobre bagazo de caña enriquecido con pulpa de café o paja de cebada. Micol. Neotrop. Apl., 3, 49-52

Martínez-Carrera D., Morales P., Sobal M., Larqué-Saavedra A. (1992b), Reconversión en la industria de los hongos? TecnoIndustria, 7, 52-59.

Martínez-Carrera D., Morales P., Sobal M., Chang S.T., Larqué-Saavedra A. (1991a), Edible mushroom cultivation for rural development in tropical America. Mushroom Science, 13, 805-811

Martínez-Carrera D., Morales P., Martínez W., Sobal M., Aguilar A. (1996b), Large-scale drying of coffee pulp and its potential for mushroom cultivation in Mexico. Micol. Neotrop. Apl , 9, 43-52

Martínez-Carrera D., Leben R., Morales P., Sobal M., Larqué-Saavedra A. (1991b), Historia del cultivo comercial de los hongos comestibles en México. Ciencia y Desarrollo, 96, 33-43

Pimentel D., Houser J., Preiss E., White O., Fang H., Mesnick L., Barsky T., Tariche S., Schreck J., Alpert S. (1997), Water resources: agriculture, the environment, and society. Bioscience, 47, 97-106

Roussos S., Gaime-Perraud I., Raimbault M., Viniegra G., Trejo M.R., Aquiahuatl A, Gutiérrez M. (1991), Avances en la valorización biotecnológica de los subproductos del café. Paper presented at 2nd. Internat. Seminar on Biotechnology of Coffee Industry, April 12-15, Manizales, Colombia.

Soccol C.R., Roussos S., Sera T. (1999), Abstracts III International Seminar on Biotechnology in the Coffee Agroindustry, May 24-28, Londrina, Brazil

Soto C., Martínez-Carrera D., Morales P., Sobal M. (1987), La pulpa de café secada al sol, como una forma de almacenamiento para el cultivo de *Pleurotus ostreatus*. Rev. Mex. Mic., 3, 133-136

Tschierpe H.J., Hartmann K (1977), A comparison of different growing methods. Mushroom Journal, 60, 404-416

Van Griensven L.J.L.D. (1988), The Cultivation of Mushrooms, Darlington Mushroom Laboratories, Rustington, England

Wasser S.P., Weis A.L. (1999), Medicinal properties of substances occurring in higher basidiomycetes mushrooms: current perspectives (review). International Journal of Medicinal Mushrooms, 1, 31-62.

Watt B.K., Merrill A.L (1975), Composition of Foods, Agriculture handbook no. 8, United States Department of Agriculture, Washington, D. C., U.S.A

Wood D.A. (1985), Useful biodegradation of lignocellulose. In- Plant Products and the New Technology, K. W. Fuller & J. R. Gallon (ed.), Clarendon, Oxford, pp. 295-309

Chapter 46

COFFEE PULP VERMICOMPOSTING TREATMENT

E. Aranda, I. Barois

Departamento de Biología de Suelos, Instituto de Ecología, A.C. Xalapa, México

Running title: Vermicomposting of coffee pulp

1. Introduction

The concept of vermicomposting began with the knowledge that certain species of epigeic earthworms, which live in the litter and primarily feed on pure organic matter (Bouché, 1977; Lavelle, 1981, Abdul and Abdul, 1994) grow in and consume organic waste materials, converting them into an earth-like soil-building substance that forms a beneficial growing environment for plant roots. Thus, the practice of vermicomposting could be defined as the combination of biological processes, designs and techniques used to systematically and intensively culture large quantities of certain species of earthworms in order to speed up the stabilization of organic waste materials, which are eaten, ground and digested by the earthworms with the help of aerobic and some anaerobic microbiota, and thereby naturally converted into much finer, humidified, microbially active faecal material, where important plant nutrients are held in a form much more soluble and available to plants than those in the parent compound (Aranda *et al.*, 1999). The end product, a decomposed faecal material (earthworm faeces, vermicompost or "castings") consists of very finely structured, uniform, stable and aggregated particles of humidified organic material with excellent porosity, aeration and water holding capacity, rich in available nutrients, hormones, enzymes and microbial populations. Thus, this product is a valuable, marketable and superior plant growth media (Edwards and Bohlen, 1996).

Although most of the vermicomposting information is based on research programmes and technical efforts from temperate countries, there are quite remarkable manuals and experiences from the tropics (Martínez, 1995; Martínez, 1996). Most of the tropical vermiculture literature found is from Colombia (Dávila *et al.*, 1990; Arango and Dávila, 1991; Orozco *et al.*, 1996), Chile (Velázquez *et al.*, 1986; Venegas and Leticia, 1988; López and Ricardo 1989; Azancot and Alvaro, 1989), Costa Rica (León *et al.*, 1992), Cuba (Ramón and Romero, 1993; Werner and Ramon, 1996), Ecuador (Barkdoll, 1994; Landín,

T. Sera et al. (eds.), Coffee Biotechnology and Quality, 489–506.

1994), Perú (Quevedo, 1994 a, b), India (Beena and Sachin, 1993; Bhawalkar, 1993; Senapati, 1994) and México (Aranda, 1988; Barois and Aranda, 1995; Irissón, 1995; Arellano, 1997).
With the aid of earthworms, coffee pulp and many other products can be transformed more quickly into a useful vermicompost for farmland and urban application. Nutrients within vermicomposted material are readily available to plants, and can be added to agricultural land to improve soil structure and fertility or used as a high-quality and marketable additive to potting soil or plant growth media. Coffee pulp vermicomposting treatments offer new eco-technological solutions to take advantage of the main by-product of the coffee agro-industry, providing new opportunities in the commercial farming business.

2. Coffee pulp disposal

Waste disposal is one of the major problems in the traditional coffee-processing units or "beneficios". Approximately three tons of by-products are generated and four tons of water is required to produce one ton of dry de-hulled coffee. Coffee pulp is the largest of these by-products and is produced when coffee beans are extracted from the fresh fruit with "depulpers" in the "wet process". It represents around 39% of the fresh fruit weight, or 28.7% of the weight in dry basis (Zuluaga, 1989). Usually it is dumped on large open piles in gorges or near rivers, causing both water and soil pollution. Efforts have been made to use the fresh coffee pulp as food for animals (Campabadal, 1987; Jarquin, 1989; Braham and Bressani, 1979), but the presence of caffeine, polyphenols, tannins, clorogenic, ferulic and cafeic acids (Roussos *et al.*, 1989) obstruct the feasibility for animal nourishment (Gaime-Perraud *et al.*, 1991a, 1991b). In macro-fauna samplings taken at large coffee pulp piles on open land in Mexico active macro-fauna *of Diptera, Acarida, Oligochaeta, Coleoptera, Thysanoptera* and *Collembola* were found (Barois and Aranda, 1995). The most commonly found earthworm species in coffee pulp piles are *Perionyx excavatus* (Perrier, 1872) in the initial stages or *Amynthas gracilis* (Kinberg, 1867) and *Dichogaster sp.* in advanced stages. *Eisenia fetida* or *E. andrei* have rarely been found freely in the field (Aranda and Arellano, 1995). However, it was observed that in all open piles macro-fauna proliferation reached only the most superficial layers of the piles, around 5-30 cm deep. In the inner layers, which are commonly a bright red colour when fresh, the coffee pulp becomes acidic, hot and yellowish (without any macro-fauna), because of the absence of oxygen. This inactive "ensilaged" material darkens to brown and re-starts its organic transformation when it is manually opened or aerated. At very advanced stages, eight to ten months later, the nearby grasses starts to invade and cover the piles, while the inner layers still remain untouched. Coffee piles should be removed from this location.

3. Vermicomposting process

Different from other traditional composting processes, vermicomposting takes advantage of the biological and physiological capabilities of the composting earthworms in order to enhance the aerobic decomposition of organic materials. Generally speaking, earthworm cultures can perform at the same time three major and useful functions (Sabine, 1983) a) reduce the pollution-potential of the organic residues, b) make good use of organic residues by their bioconversion into casts (a plant growth medium), and c) produce more earthworms in order to extend the vermicomposting areas or as a high- quality protein food suitable for inclusion in various domestic (e.g. livestock) rations. The action of earthworms is performed in various ways, not only by the eating or digesting of the organic matter, which is why they have been called "litter transformers" (Lavelle, 1994). Earthworms derive their nourishment from the micro-organisms that they enhance in the organic material and offer several benefits (Aranda *et al.*, 1999; Edwards and Bohlen, 1996; Edwards and Neuhauser, 1988; Tomati *et al.*, 1987; Satchell, 1983; Van Gansen, 1962) such as:

- While eating, they drill, turn and maintain the substrate like a sponge that is in aerobic condition, ensuring the entrance of O_2 and the release to CO_2.
  -They cover the surface of the beds with their faecal material or casts, reducing bad odours and the presence of unwanted animals such as flies and others.
- They macerate the organic materials through their grinding gizzard, which strongly increases the exposed surfaces and enhances the beneficial action of aerobic micro-organisms.
- The beneficial micro-organisms released from the earthworm gut continue their activity for some period outside the gut because of a favourable polysaccharide-mucoprotein medium produced.
- Different organic materials can be mixed together by earthworms, allowing an optimal combination and composition of the nutrient content, producing a much finer, fragmented and uniform material than by any other composting method.
- There is considerable scientific evidence that plants and human pathogens do not survive the vermicomposting process, so if materials containing pathogens are used, they are for the most part killed in passing through the earthworm gut.
- Small inorganic pieces such as rocks, plastics or glass, hard to collect when mixed with the organics, can be easily sieved after vermicomposting due to the finer size of castings.

Contrary to common belief, earthworms do not have many serious natural enemies, diseases or predators.

4. Coffee pulp vermicomposting treatment

To perform a successful coffee pulp vermicomposting treatment and to produce an efficient and valuable plant growth medium, several components are needed:

i) the proper earthworm species with their biological and demographic properties,
ii) an appropriate coffee pulp quality and its adequate preservation,
iii) the correct interior and exterior environmental conditions, and,
iv) the implementation of efficient design and operation.

4.1. Earthworm Species

Although there are nearly 6,500 described species of earthworms in the world, only a few are known to be suitable for culture in organic waste materials and even fewer have been used on a widespread scale. This means that not just any kind of earthworm is suitable to process organic residues. The earthworms that appear to be the most common, which can be found in the soil of pots or in the garden belong to a different ecological category of "endogeic" earthworm species (Bouché, 1977). Endogeic earthworms are soil feeders "Ecosystem engineers" (Lavelle, 1994) with physiology, habits, distribution and capabilities very different from the "epigeic" or compost earthworms.

The role of earthworms in the breakdown of organic debris on the soil surface and in the soil turnover process was first highlighted by Darwin (1881). Since then it has taken almost a century to appreciate their important contribution in curbing organic pollution and providing topsoil in impoverished lands (Kale, 1998). A clear-cut demarcation of the role of epigeic and endogeic species of earthworms and their limitations have to be made in order to avoid mixed roles of the soil dwelling earthworms from the field (endogeics and anecics) with the epigeics or compost-earthworms in organic residue conversion. Several reports have been published on various aspects of vermiculture (Bouché, 1987; Satchell, 1983; Bonvicini-Pagliai and Omodeo, 1987; Kretzchmar, 1992; Hoerschelmann and Andrés, 1994; Edwards, 1997, 1998). The most commonly used earthworm species world-wide and considered as best for coffee pulp waste management include *Eisenia andrei* (Bouché, 1972), the red earthworm, and the closely related E. *fetida* (Savigny, 1826), the brandling or tiger earthworm (Jaenike, 1982). Both are hardy earthworms, readily handled with wide temperature tolerances and can live within a wide range of moisture contents. In mixed cultures (with other species) *E. andrei* usually becomes dominant (Edwards and Bohlen, 1996; Salazar *et al.*, 1994).

Many vermicomposting sites in México also have naturally-occurring populations of *Perionyx excavatus* (Perrier, 1872), the oriental compost worm, which is believed to be a native species. In the central area of Veracruz, *P. excavatus* is commonly the first species to invade the coffee pulp piles in the field. It can survive field conditions and high temperatures characterized by the initial decomposition of the coffee pulp and have been used satisfactorily

in open field conditions, sometimes withstanding the conditions even better than inoculated populations of *E. andrei. P. excavatus* is a tropical earthworm and is a common species in Asia and Australia (Guerrero, 1983). Other suitable species include *Eudrilus eugeniae* (Kinberg, 1867), the African night-crawler, which has received increased attention as a protein source and waste decomposer in tropical conditions, except that it appears to be less cold temperature tolerant and easily escapes during heavy rains (Neuhauser *et al.*, 1988; Rodriguez *et al*, 1986). Compared to other vermicomposting species, *E. eugeniae* seems to have an initial growth advantage as a potential protein producer. Some basic facts about the life-cycle and reproduction of this species under favourable conditions are well documented in the literature (Viljoen and Reinecke, 1989; Rodríguez, 1996). Some attempts have been made to establish in México the populations of *E. eugeniae* from the State of Quintana Roo (México), Costa Rica and Cuba, but although it develops fairly well in coffee pulp, this species seems to be more attractive to ant invasion (*Solenopsis sp.*), which may cause the loss of the worms.
It is believed that *Lumbricus rubellus,* the red worm, could be suitable for organic waste breakdown, but this has not yet been scientifically sustained. Surveys in USA, Europe, and Australia have shown that the species sold to commercial earthworm farms under the name *Lumbricus rubellus* were *E. fetida* or *E. andrei.*

4.2. BIOLOGICAL AND DEMOGRAPHIC PROPERTIES OF EARTHWORMS

Earthworms have a very simple biological cycle spending their whole time within the organic matter where they live. One of the main reasons why the earthworms reproduce so much is the fact that they are hermaprodithe, where all of the individuals have *clitellum* and both male and female genitals. However, it is required that two individuals copulate. Both individuals then produce cocoons that resemble the size and shape of a water drop. Inside each cocoon can be several individuals (1-5), depending on the growth of the parental earthworm and therefore the size of the cocoon. The apparition of clitellum, fecundity rate, cocoon hatchlings, and even the average mortality rate presented in *E. andrei* and *P. excavatus* in coffee pulp have been discussed by Arellano (1997).
The best way to measure the population density of earthworms in a vermicomposting bed is the quantity (or biomass) of earthworms per square meter. Although 50,000 individuals.m^{-2} (with 3,540 g m^{-2}) of *E. andrei* are commonly obtained from coffee pulp in controlled conditions, for practical values, a good population density around 25,000 earthworm individuals m^{-2} in field conditions is considered suitable, which in terms of worm biomass would be around 2,500-3,000 g m^{-2} (considering an average of 0.110 g per worm). Lower values of their population density are in a particular area and substrate show the bigger size of earthworms. More densely concentrated population diminishes their individual growth, reproduction rate and some other demographic properties. If we consider that more biomass can be attained with more individuals, although they can be less developed, then two

concepts can be explained; one is the substrate carrying capacity, understood as the optimum value of earthworm population density, where a maximum biomass production is obtained in a given area (estimated to 532.4 $g.m^{-2}$ to *P. excavatus*, and 3,371.1 $g.m^{-2}$ to *E. andrei*). This density-dependent relation also implies that to prevent overcrowded and stagnant conditions, the worms should be regularly extracted from the vermicomposting beds (Taboga, 1980; Hartenstein,1981; Tacon *et al.*, 1983; Velásquez *et al.*,1986). Some good experimental tests have been made in Mexico using worms to feed captive animals (García, 1978; Guerra, 1982; Aranda and Aguilar, 1995) and salmon trout, *Oncorhynchus mykiss* as an alternate source of protein.

5. Substrates for earthworm culture

Vermicompost produced from organic wastes depends on the material used. A product with excellent fertilizing qualities cannot be obtained from inferior quality raw material. A wide range of organic waste materials could be used for vermicomposting. These could come from animal (manure, ruminal content), vegetable (plants, plant residues), urban (municipal solid wastes, domestic residues, anaerobically digested effluents, trimmings, supermarkets) and agro-industrial residues (sugarcane processing mill residuals, banana stems, mushroom industry residuals, horticultural and fruit-processing plant by- products, cacao residues). One of the most relevant characteristics for substrates to support earthworm growth is the C/N ratio. The suggested optimum ratio for an efficient organic transformation is 30:1. Some pesticides, heavy metals or other contaminants included in the raw materials can reduce the growth, reproduction and conversion rates of the earthworms. (Hartenstein and Mitchell, 1978; Hartenstein *et al.*, 1980; Ireland, 1983). Both, the presence or high concentration of these substances in raw materials must be avoided.

5.1. EARTHWORM CULTURE USING COFFEE PULP

Coffee pulp has proved to be a dependable source of high quality castings and high quality organic amendment so as to improve soil structure and fertility and plant growth medium (Aranda *et al.*, 1999). However, when compared with other organic substrates, success with coffee pulp vermicomposting appears to be related not only to their well-balanced nutrient content, but to their consistency in purity, texture, effect and the unusually low variable nutrient content of the castings (Scott, 1988). In terms of chemical and physical properties for breeding of earthworms, it possesses high water retention capacity (around four times its own dry weight), a good bulk density (around 0.500 kg l^{-1} when fresh, 0.700 kg l^{-1} when applied to worm beds), a redox potential around 450 mv, pH practically neutral (7.4) and uniform-sized, aggregated particles. In addition, it does not need any additional nutrients.

5.1.1. Substrate preservation:

Due to the natural biodiversity of micro-organisms present in the fresh fruit (Gaime-Perraud *et al.* 1991a, b) when the coffee pulp is dumped on open land, a self-inhibited proliferation of thermophilic micro-organisms occurs within few days. In this, *Aspergillus fumigatus* seems to be dominant. Temperature increases due to fermentation (65-70 °C) and pH becomes low (3.5-4.0). The absence of oxygen (redo potential 199 mv) makes the material to an inactive stage, or "ensilage" (Perraud-Gaime *et al.* 1991c, Irissón, 1995) that delays or even stops its organic transformation. After the initial silage period, the temperature in the coffee pulp descends slowly to reach similar to weather conditions.

Because of the seasonal harvesting of the coffee fruit ensilage is not a problem for vermicomposting. In fact, it seems to be the best way to preserve the optimal nutrient content of the coffee pulp from one crop to the next. This is particularly important in the case of soluble elements such as potassium, because it can either be preserved in the finished product (around 2.14% dry basis) or drained to values as 0.08-0.13% dry basis (if the substrates are excessively soaked in the ensilage stage or in the vermicomposting process).

Studies have been made to conduct and standardize the proliferation of lactic acid bacteria for ensilage process of the coffee pulp (Perraud-Gaime, 1995; Saucedo-Castañeda, 1999; Viniegra-Gonzalez, 1999).

Compared with other common organic residues, coffee pulp could be considered one of the best plant-originated organic residues to raise compost earthworms. This is clear from their ecological and biological parameters, especially from their experimental carrying capacity, estimated to be 3,371 g biomass m^{-3} of *E. andrei* versus a value estimated to 926 g biomass m^{-3} of the same species in cow manure (Arellano, 1997).

5.1.2. Environmental factors:

Optimal conditions for breeding earthworms have been presented for *E. fetid* and *E. andrei* by Lofs-Holmi (1985) and Edwards and Neuhauser (1988) who suggested that these conditions do not differ much from those suitable for the other species. One of the basic limiting factors to overcome in providing wastes for earthworms is the heat generated during the placing of organic matter in the vermicomposting beds. Exposing earthworms to temperatures above 35°C kills them. The reactivated thermophilic micro-organisms can raise temperatures to more than 35°C if the organic matter is placed in layers thicker than 10 cm. Avoiding such overheating requires adding relatively narrow layers of organic material on the surface of the vermicompost bed, and allowing the earthworms to process the new residues under aerobic conditions. In this way the earthworms are always concentrated in the upper layers of waste at around 2-20 cm.

A proper and systematic programme of feeding is essential to keep the vermicomposting process at efficient level. Under experimental conditions, it is necessary to feed the composting bed when the substrate is close to being consumed. Biweekly feeding is the most commonly practised method. At the same time, earthworms need to be kept under wet conditions (around 85%). Prolonged lower moisture (60%) could dehydrate the earthworms or even kill them. However, excessive water reduces the aeration, and thus the worm activity. Periodical stirring of the upper layers is useful when compacting or excessive water is present.

5.1.3. Coffee-vermicompost conversion rate:

If we compare the nutrient content of the raw material (coffee pulp) with that resulting in the vermicompost, we can find that the vermicompost has more or less double the percentage of each element (except for carbon). It can be explained by the volume reduction of the raw material, which could be in the order of 70%. In terms of the fresh weight, the raw material also undergoes a clear diminution of about 60-70%. On dry basis total solids in vermicompost represent about 40% of the total solids of coffee pulp. The main loss of total solids is due to respiration in form of CO_2. Some other small quantities (3-5%) are chemical elements, which are used by the earthworm for its physical growth or drained with the water. Irissón (1995) reported the yield of about 42% of coffee pulp converted to vermicompost on initial dry weight. This was around 12.5% of the initial fresh weight (at 88.6% of water content) if 60% of water content was required in the final product.

5.1.4. Design and operation:

Basic models to grow earthworms in terms of the position of the beds are pits or heaps. Pits under the surface soil level have lower aeration but better water retention. Heaps above soil level can have more lateral aeration, but lower water retention. It is usually easier to work with a heap, but the choice has to be adapted according to specific situations. One of the simplest and clearest examples of vermicomposting can be found with worms working on rabbit farms. The worm bins (or heaps) are situated under the rabbit cages, the droppings (and urine) just fall every day to nourish the worms. When the bins are full, the castings can be harvested and sold.

The first designs of vermicomposting practised in Mexico with coffee pulp were made as single or twin windrows of building blocks in an effort to obtain better control of the worm population. These were constructed about 0.80 m wide by 0.60 m high, varying in length and situated on a bed of gravel. Several variations of this model were also made with local or readily available materials such as wood, wood flanks, bamboo, plastic or metal meshes and even with rubber conveyor belts. In these cases, it is possible to find different kinds of shelter using black plastic meshes, bracken fern (*Pteridium aquilinum*), or also living creeper plants such as "chayote". In larger farms, the most common models used are simple beds or windrows on the ground on open land, sometimes under the shade of native or planted trees such as macadamia and castor-oil plants (*Ricinus* sp.). Some other farms cover the beds making use of sugarcane tips or spent-yute sacks from the coffee industry. With time course, these materials also decompose and incorporate into the product.

Traditional vermicomposting methods of operation in Mexico are mostly based on the addition of successive thin layers of food to the surface of the beds. Stirring the food together with the castings in the beds can increase the colonization of worms and reduce the heat generated. In this basic model, earthworms move upwards to follow the food. About three or four months later, when the windrows are full, the upper layers of material, where the majority of the worms remain, can be momentarily separated to allow the harvesting of the inner layers of castings. When the castings are harvested, the upper layer of food and worms can be shifted in succession to an empty row to restart with a new series of food supplies.
A different and simplified method has been successfully implemented in open land in "San Alfonso", Coatepec, Veracruz which works in agreement with the harvest of coffee, its sowing season and even with the seasonal changes of the year. After the harvesting of the crop the coffee pulp is collected from the ensilaged piles and placed in parallel lines of successive mounds as left on the ground by trucks and then shaped as full-formed windrows with spaces in between. The worms need to be inoculated only one time on the beds once they overcome the thermophilic phase (approximately 3-4 weeks later). Some periodic stirring of the upper layers is needed to permit the worms to move to lower layers for food. When the whole substrate is close to being consumed, another row of mounds in the left spaces is formed with the next coffee pulp production. The habit of escaping of *P. excavatus* (also observed with *E. andrei)* during the rainy season, favouring the first rainy nights makes them gradually move to the neighbouring windrows, which are full of fresh food (coffee pulp collected from the ensilaged piles). A few weeks more is enough to allow the worm-worked windrow to be single-harvested almost without any worms at all. In this way, a single-general harvest of castings is directly used (in the rainy-sowing season) to sow those plants coming from nurseries (also prepared with coffee pulp, vermicompost and sandy soil) to be replanted in their definitive place in field.
One other experience has been successfully attained. Instead of a costly traditional anaerobic digestion treatment of washing waste-waters from the same coffee factory, empirical tests have been made using the beds as "vermifilters"; that is, adding the waste-waters to the vermicomposting beds. This practice was favoured because of the dry season that is present within the coffee cropping season.

5.1.5. Equipment and human resources:

Machinery is not commonly used on Mexican vermicomposting farms, except trucks, tractors or trailers to move residues and end products. The watering of windrows in the field is done by irrigation or with trucks, but some still have garden hoses and the smaller ones simply use watering cans. Effort needs to be done to design cheap and rustic but efficient vermicomposting equipment and tools. The low technological input in vermicomposting is characterized by relatively low costs in labour together with a large labour force and large areas of land. Generally speaking, for field activities one man is needed per 500 square meters of vermicomposting beds.

5.1.6. Vermicompost harvesting:
Prior to being used as a market product or for use by producers themselves, the worm-worked castings need to be harvested, dried, sieved and also bagged. Harvesting is commonly done manually. It must be ensured that it is free of worms. The harvested castings are still muddy and difficult to sieve or handle. One of the most difficult technical obstacles to solve in the production of vermicompost is to dry the end-product, because of the wet and rainy weather where most of the sites are located. At most of the farms which sell castings, the material is simply sun-dried in patios, which is costly and slow to reduce the initial water content of 75-80% to about 55%. Less than 45% moisture makes the casts less efficient with diminished microbial activity, changes their black colour to a dusty grey, disintegrates material, and is difficult to wet again. With the aid of simple wooden or mechanical sieves, the casts are also commonly screened (with a 0.6-0.5 cm mesh) to remove rocks, debris, metal, wood or even glass in order to produce a more uniform, smooth and high quality product. The sieved rejected organic material can be mechanically ground and reincorporated into the main production. Then the product is packed in polypropylene 10-20 or 50-kg sacks.

6. Quality and nutrients content of vermicompost

When the nutrient content of vermicompost is compared with that of a commercial plant growth medium to which inorganic nutrients have been added, earthworm castings usually contain similar quantities of the main nutrients N, P, K and most other mineral elements as needed by the plants. During vermicomposting process most of the nutrients such as nitrate, ammonium exchangeable phosphorous and soluble potassium, calcium and magnesium contained in the waste materials are changed to forms more readily available to plants (Edwards and Bohlen, 1996, Scott *et al.*, 1998). Casts can be rich in available nutrients, allowing not only an immediate supply of plant nutrition, but also a build-up of reserves for future crops. Casts have a superior bioactive potential, presenting plant growth hormones (Tomati *et al.*, 1983, 1987), suppressive effects on some root infecting pathogens (Szczech *et al.*, 1993), and enhanced levels of soil enzymes and high-soil microbial populations. They are weed free and rich in humic compounds, creating an efficient and valuable soil-building medium.
Some authors have examined the influence of chemical composition and mineralization of nitrogen to evaluate how vermicompost derived from differing organic wastes could be compared to other compost. Testing the impact of vermicomposting for different mixtures of pig manure slurries, Dominguez (1997) found that the humic substances showed an increase of 40-60%, which was higher than the value obtained for the composting process. Also the nitrification was 50-65% higher in the earthworm treatments than in the controls. Preliminary data indicated that human pathogens did not survive vermicomposting; faecal

coliform dropped from 39,000 MPN g^{-1} to 0 MPN g^{-1}, and *Salmonella* sp. dropped from <3 MPN g^{-1} to <1 MPN g^{-1}. Penninck and Verdonnck (1987) found only minor differences between earthworm-worked and wastes composted with conventional methods. Haimi and Huhta (1987) considered vermicompost superior to ordinary compost only due to its physical structure.

Humic acids are an important constituent of earthworm-worked material and are natural by-product of the microbial decomposition or alteration of plant or animal residues and of cellular components and products synthesized by soil organisms. Some important and beneficial properties of humus are slow release of plant nutrients, improvement of soil physical properties, enhancement of micro-nutrient element nutrition of plants through chelation reactions, help in the solubilization of plant nutrient elements from insoluble minerals, high adsorptive or exchange capacity for plant nutrient elements, increase in the soil buffer capacity, promotion of heat absorption and earlier spring planting, support of a greater and more variable microbial population which favours biological control including disease suppression, reductions of toxic chemical substances both natural and man made, and increased soil water-holding capacity lant growth regulators belonging to the auxin,giberellin and cytocinin groups and present in the earthworm-worked materials are produced by a large range of soil micro-organisms, many of which live in the guts of earthworms or within the castings (Tomati *et al.*, 1983).

7. Evaluation of coffee vermicompost

A population of *P. excavatus*, *E. andrei* and *E. fetida* was experimentally cultured using coffee pulp as a substrate in order to compare their development (Salazar *et al.,* 1994) and the resulting vermicompost and composts for physical and chemical evaluation (Irissón, 1995). Additionally, a fourth box-treatment was inoculated with the three species together to compare poly- versus mono-specific cultures. Two controls were also used, one with a composting treatment (turned over by shovel every 15th day) and the other with a single initial aeration (Irissón, 1995). The resulting end-product (vermicompost or castings) consisted of finely structured, minute aggregated particles of humidified organic substances with excellent porosity, lightness, aeration and water holding capacity. It was rich in available nutrients, hormones, enzymes, antibiotics, and essential microbial populations. In the composted and untouched treatment, a less efficient transformation took place. The remaining material was hard, dry and mouldy with pieces of unprocessed coffee pulp. After drying in the shade, vermicompost appeared brownish-black colour with 55-60% water content and a bulk density of close to 500 g.l^{-1}. In general, the vermicompost doubled the nutrient content of the organic residues, resulting C/N ratio from 28 to12, with a well-balanced chemical content, and a high nitrogen content, unusual for a vegetal substrate. The earthworms appeared to have stimulated ammonifying bacteria and depressed de-nitrifying ones when compared with the composted treatment. In addition, enzyme activity in the vermicompost was significantly higher than in the case of the controls. Harmful compounds such as caffeine and polyphenols were almost completely decomposed during the process in all the six treatments, but transformed least in the controls (Irissón, 1995). Phosphorous content of the initial coffee pulp was low, which almost doubled during decomposition. Potassium was the most sensitive mineral, because it can either be preserved in the finished product or drained if the substrates are excessively soaked in the stabilization process.

8. Effect of coffee pulp vermicomposting on coffee and other plants

Most of the vermicompost producers use their own product to prepare nurseries of coffee seedlings and to transplant them into the field. Agronomic tests have been carried out on seedlings at nurseries with a divided-plots design with a combination of substrates and fertilization treatments. Three different substrates were used, a) soil alone, b) 60% soil + 40% vermicomposted coffee pulp, and c) 60% soil + 40% composted coffee pulp. Four treatments of chemical fertilization regimes were applied to each of these three substrate, a) no fertilizer (NF), b) soil fertilization (SF; NPK 18-12-06, 3.0 g two times per plant), c) foliar fertilization (FF; "Grow Green", NPK 20-30-10 + micro-nutrients, $5 g.l^{-1}$, two applications), and d) both fertilizers (BF). Although very little statistical significant difference between coffee plants grown on earthworm castings and those grown on composted pulp were found, plants grown in soil with earthworm casts or compost were about 45% higher, 28-30% thicker in diameter. They had 49-52% more leaves, a 100% heavier dry aerial biomass and a 63-83 % heavier dry root biomass, when compared with those grown in soil alone. Soil fertilization of the three different substrates did not result in any positive effect, while foliar fertilization did produce a slight and similar effect on plants grown in all three substrates. Currently, not much scientific work is being done to test the effect of worm-worked coffee pulp on different crops, although much empirical work is underway. However, here we present one successful example because of its potential impact on some agricultural practices.

Many large agricultural companies situated in “The Bajio” plateau region of central México (in the states of Querétaro, Guanajuato, Michoacán, San Luis Potosí and Aguascalientes) are oriented towards the export market and are specialized in producing and selling seedlings grown in large high-tech greenhouses. Large quantities of polystyrene or plastic multi-cell trays are continuously filled with peatmoss-based growing mediums for seedlings to encourage rapid establishment with high yields. In this way, millions of little plants of broccoli, carrots, asparagus, onion, garlic, tomato, chilli, celery among others are consistently produced for the field. Coffee pulp vermicompost was tested on seedlings in greenhouses under normal working conditions. Two similar tests were made on mixes of 100, 75, 50, 25 and 0% of coffee pulp vermicompost, combined with peat moss (Sunshine

3 and Fafard FPM) to grow seedlings of broccoli (Sakata and Peto Seed var. Maratón) in 392 polystyrene cavity-trays in two commercial large high-tech greenhouses. Results showed that with even the addition the smallest portion of coffee pulp vermicompost, vigorous plants with dark green leaves and healthy roots were produced and the transplanting time was reduced from 36 and 37 days on controls to 25 and 26 days with 100% vermicompost. Because these are high value crops, growers and users can only afford and purchase organic ingredients that are highly consistent in texture, content and effect on plants.

9. Current status of vermicompost farms in Mexico

Since the initial work started at the Instituto Mexicano del Café with the Instituto de Ecología in 1989 on vermicomposting project, five covered bins with controlled conditions of food, temperature and moisture to breed and reproduce worms have been established. Over the course of the last 14 years, more than 100 coffee producers, co-operatives or small farmers have begun to work directly with worms supplied from the Institute. Transition and expansion is occurring quickly, because there is a notable public demand for the testing of vermiculture by coffee industries and by other organic producers. The vermicomposting farms promoted by the Institute start with a relatively small amount of worms (10,000-50,000 worms), but with further assistance, they generally increase (8-fold after 120 days). It is estimated that some eight big worm farms (five in the State of Veracruz and three in other states), twenty medium sized and around two hundred small farms are currently active in Mexico.

10. Conclusions

Current results have shown vermicomposting as a good solution for coffee pulp utilization. Since coffee pulp decomposes more rapidly in vermicomposting than in overturned pulp treatment, in long runs it could be a cheap process resulting two products, vermicompost and earthworms. In several less developed countries there are many agro-industrial and market residues that are relatively uncontaminated and could be suitable for composting. These can be used to help solve health, environmental, agricultural and food problems at low cost. To achieve this, however, more work needs to be done to test new residues and determine their viability for processing with vermicomposting technology. The results of research carried out to date along with the demand for the development of further applications indicate that the value of this technology has only begun to be explored.

11. Acknowledgements

We thank IRD, France for kind invitation and financial support to present this subject, and to Judy Shirley for reviewing the English version of this manuscript. This work was partially funded by CONACYT-Mexico Project No. 045-N9108 and by the Instituto de Ecología, A. C. Soil Biology project No. 902-08.

12. References

Abdul R., Abdul M. (1994), Los gusanos de tierra y el medio ambiente. Mundo Científico, 146(14): 408-415

Aranda D.E. (1988), La utilización de lombrices en la transformación de la pulpa de café en Abono Orgánico. Acta Zoológica Mexicana, 27: 21-23

Aranda D.E., Arellano P. (1995), Factores ambientales que influyen en la transformación de la pulpa de café abandonada en al campo. In- Utilización de lombrices en la transformación de la pulpa de café en abono orgánico. Barois, I. and E. Aranda (eds.), Consejo Nacional de Ciencia y Tecnología-Instituto de Ecología. Final Report, pp.1-15

Aranda, D.E., Aguilar, S. (1995), Tasa de extracción de lombrices de *Eisenia andrei* para su utilización como proteína animal In- Utilización de lombrices en la transformación de la pulpa de café en abono orgánico. Barois, I. and E. Aranda (eds.), Consejo Nacional de Ciencia y Tecnología-Instituto de Ecología. Final Report, pp.47-54.

Aranda D.E.,.Barois I, Arellano P., Irissón S., Salazar T., Rodríguez J., Patrón J.C (1999), Vermicomposting in the Tropics. In- Earthworm Management in Tropical Agro-ecosystems. Lavelle, P., L.Brussaard and P.Hendrix. (eds.), CAB International, pp. 241-275

Arango B.L.G., Dávila M.T. (1991), Descomposición de la pulpa de café por medio de la lombriz Roja Californiana. Avances tecnicos. CENICAFE, 161:1-4

Arellano P. (1997), Descomposición de la pulpa de café por *Eisenia andrei* (Bouché, 1972) *y Perionyx excavatus* (Perrier, 1872) *(Annelida, Oligochaeta)*. Tesis de licenciatura, Facultad de Biología, Universidad Veracruzana. Xalapa, Veracruz, México

Azancot R., Alvaro A. (1989), Respuesta del cultivo de ajo (*Allium sativum* L.) a la aplicación de nitrógeno en forma de urea, humus de lombriz y estiercol de bovino. Universidad Católica de Valparaíso, Quillota, Facultad de Agronomía, Chile

Barkdoll A. (1994), South American Operations. Biocycle J. Composting & Recycling, 35(10): 64

Barois I., Aranda E, (eds.). (1995), Utilización de lombrices en la tranformación de la pulpa de café en abono orgánico. Final report, CONACyT, México

Beena P., Sachin Ch. (1993), Vermiculture: Nature's bioreactors for soil improvement and waste treatment. Biotechnol. Development Monitor, 16:8-9

Bhawalkar U.S. (1993), Turning garbage into gold, an introduction to vermiculture biotechnology. Pune Bhawalkar Earthworm Res. Inst. Booklet, India

Bonvicini-Pagliai A.M., Omodeo P. (eds.). (1987), On Earthworms. Selected Symposia and Monographs, U.Z.I. Ed. Mucchi, Modena, Italy

Bouché M.B. (1977), Stratégies lombriciennes. In- Soil organisms as components of ecosystems. Lohm,U. and T. Persson (eds.), Ecol. Bull. (Stockholm), 25: 122-132

Bouché M.B. (1987), Emergence and development of vermiculture and vermicomposting: from a hobby to an industry, from marketing to a biotechnology, from irrational to credible practices. In- On Earthworms. Selected Symposia and Monographs, Bonvicini-Pagliai A.M. and P.Omodeo (eds), U.Z.I. Ed. Mucchi, Modena. 2, 519-532

Braham J.E., Bressani R (1979), Pulpa de café, Composición, Tecnología y Utilización. INCAP. Bogotá, Colombia

Campabadal C. (1987), Utilización de la pulpa de café en la alimentación de animales. In- Utilización Integral de los Subproductos del Café, Memoria del Tercer Simposio Internacional. ANACAFE-ICAITI, Guatemala, C.A. pp. 37-44

Darwin C. (1881), The formation of vegetable mould through the action of worms with observations of their habits. Murray, London, UK

Dávila M.T., Arango L.G, Zuluaga J. (1990), Utilización de la pulpa de café para el cultivo de la Lombriz Roja Californiana (*Eisenia foetida*, Sav.), *Seminario para la Valorización del los Subproductos del café*. Conacyt-Ecuador, CEE, IRCC, JUNAC. Quito, Ecuador

Dominguez J.1997. Testing the impact of vermicomposting. BioCycle J. Composting & Recycling, 38, 58

Edwards C.A. (ed.) (1997), Fifth International Symposium on Earthworm Ecology. Soil Biol. Biochem., 29(3/4), 215-766

Edwards C.A. (ed.) (1998), Earthworm Ecology. St. Lucie Press, USA

Edwards C.A., Neuhauser E. (1988), Earthworms in waste and environmental management. Academic Publishing, The Netherlands

Edwards C.A., Bohlen P. (1996), The Biology and Ecology of Earthworms. 3rd Edition Chapman and Hall, London, UK

Gaime-Perraud I., Zuluaga J., Callies C, Roussos S. (1991a), Microflore naturelle de la pulpe de café. Paper presented in II SIBAC, Manizales, Colombia

Gaime-Perraud I., Ramakrishna M.,. Roussos S (1991b), Etude comparative de la microflore naturelle presente dans la pulpe et la parche de café. Paper presented in II SIBAC, Manizales, Colombia

Gaime-Perraud I, Hannibal L., Trejo M., Roussos S. (1991c), Resultados preliminares sobre el ensilaje de la pulpa de café. Paper presented in II SIBAC, Manizales, Colombia

Gaime-Perraud I. (1995), Cultures mixtes en milieu solide de bacteries lactiques et de champignons filamenteux pour la conservation et la decafeination de la pulpe de café. Thèse de Doctoarat, Université Montpellier II, Sciences et techniques de Languedoc, France

García G. (1978),Utilización de la lombriz roja (*Helodrilus foetidus*) fresca, como sustituto parcial de proteína en la alimentación de gallinas ponedoras. (Profesional thesis) Departamento de Zootecnia, Escuela Nacional de Chapingo. México

Guerra P.L.(1982), Estudio preliminar sobre la utilización de la lombriz de tierra (*Helodrilus foetidus*) en la alimentación del bagre de canal (*Ictolus punctatus*). (Profesional thesis). Universidad Autonoma Agraria. Monterrey, México

Guerrero R.D. (1983), The culture and use of *Perionyx excavatus* as a protein resource in the Philippines. In- Earthworm Ecology, from Darwin to Vermiculture. Satchell J. (ed.), Chapman and Hall, UK, pp. 309-313

Haimi J., Huhta V. (1987), Comparison of composts produced from identical wastes by "vermiestabilization" and conventional composting. Pedobiologia, 30: 137-144

Hartenstein R. (1981), Production of earthworm *Eisenia foetida* as a potentially economical source of protein. Biotechnol. Bioeng., 23: 1812-1997

Hartenstein R., Mitchell M. (1978), Utilization of earthworms and microorganisms in stabilization, descontamination and detoxification of residual sludges from treatments wastewater. Final report to the National Science Foundation, Sony College of Environmental Science and Forestry

Hartenstein R., Neuhauser E., Collier J. (1980), Accumulation of heavy metals in the earthworm *Eisenia foetida*. J. Environ. Quality, 9: 23-26

Hoerschelman H., Andrés H.G. (1994), Wilhelm Michaelsen Gedächtnis Symposion. *Mitteilungen aus dem Hamburgischen Zoologischen Museum und Institut*. Hamburg, Germany

Ireland M.P. (1983), Heavy metal uptake and tissue distribution in earthworms. In- Earthworm Ecology, from Darwin to Vermiculture. J. E. Satchell (ed.). Chapman and Hall, London, UK, pp. 247-265

Irissón N.S. (1995), Calidad del abono y de la lombriz de tierra resultante del lombricompostaje de la pulpa de café, (profesional thesis). Facultad de Química Farmacéutica Biológica, Universidad Veracruzana, Xalapa, Veracruz, México

Jaenike J. (1982), "*Eisenia foetida*" is two biological species. Megadrilogica, 4(1,2): 6-7

Jarquin R.(1989), Alimentación de animales con pulpa de café. In- Utilización Integral de los Subproductos del Café, Memoria del Tercer Simposio Internacional. ANACAFE-ICAITI, Guatemala, C.A, pp. 45-53

Kale R.D. (1998), Earthworms: Nature's gift for utilization of organic wastes. In- Earthworm Ecology. Edwards, C. (ed.), St. Lucie Press, U.S.A. pp. 355-376

Kretzschmar A., Ed (1992), ISEE 4, Fourth International Symposium on Earthworm Ecology. Soil Biol. Biochem., 24(12),1193-1774

Landín C. (1994), Earthworms earn their credit. ILEIA Newsletter, 10(3), 9

Lavelle P. (1981), Stratégies de reproduction chez les vers de terre. Acta Ecológica, (Oecología Generalis), 2: 117-133

Lavelle P. (1994), Faunal activities and soil processes: adaptive strategies that determine ecosystems function. In- Transactions of the of the 15th World Congress of Soil Science. Volume1: Inaugural and State of the Art Conferences. Etchevers, J.D. (ed.). ISSS, Acapulco, México, pp.189-220

León S., Villalobos G., Fraile J., González N. (1992), Cultivo de lombrices *(Eisenia foetida)* utilizando composta y excretas animales. Agronomía Costarricence, Costa Rica 16(1), 23-28

López, A.R., Ricardo J. (1989), Caracterización y evaluación agronómica del proceso de humificación realizado por la lombriz *Eisenia foetida* en guanos de bovinos y caprinos. Tesis Universidad de Chile, Santiago. Escuela de Agronomía, Chile

Martín Focht (1986), Biological properties of soils. Chapter 6, In- Soils for management of organic Wastes and Waste Waters, ASA, CSSA, SSSA, Wisconsin, USA. pp.115-169

Martínez A.A. (1995), A grande e poderosa Minhoca, manual práctico do minhocultor. 3rd Edition, Jaboticabal, FUNEP, SP, Brazil

Martinez C.C. (1996), Potencial de la lombricultura, elementos básicos para su desarrollo. lombricultura Técnica Mexicana, México

Neuhauser E.F., Loehr R.C., Malecki M.R. (1988), The potential of earthworms for managing sewage sludge. In: Earthworms in waste and environmental management. Edwards, C. A. and Neuhauser E. (eds.) Academic Publishing, The Netherlands, pp. 9-20

Orozco F.H., Cegarra J., Trujillo A., Roig A. (1996), Vermicomposting of coffe pulp using the earthworm *Eisenia fetida*: effects on C and N contents and the availability of nutrients. Biol. Fertil. Soils, 22,162-166

Penninck R., Verdonck O. (1987), Earthworms compost versus classic compost in horticultural substrates. In- Compost: Production, Quality and Use. De Bertoldi, M; M.P. Ferranti, P. L'Hermite, and F. Zucconi (eds) pp. 814-817

Quevedo A.G. (1994a), Crecimiento inicial de *Ceiba samauma* trasplantada a campo abierto con aplicación de humus de lombricultura. Folia Amazonica, 6 (1-2), 183-196

Quevedo, A.G. (1994b), Crecimiento inicial de *Guazuma crinita* trasplantada a campo abierto con aplicacion de tres dosis de humus de lombriz y a tres distanciamientos. Folia Amazonica, 6 (1-2),139-151

Ramón C. J., Romero G.A. (1993), La lombricultura, una opción tecnológica. In- Manejo y disposición final de residuos sólidos municipales. Congreso Regional del Sureste. SMISAAC y AIDIS. Mérida, Yucatán, México, pp. 110-117

Rodríguez C., Canetti Ma. E., Sierra A., Reines M., (1986), Ciclo de vida de *Eudrilus eugeniae* (Oligocheta: Eudrilidae) a dos temperaturas. Rev. Biol., 2(2), 45-54

Roussos S., Aquiahuatl A., Cassaigne J., Favela E., Gutierrez M., Hannibal L., Huerta S., Nava G., Raimbault M., Rodríguez W., Salas J.A., Sanchez R., Trejo M., Viniegra-Gonzalez G. (1989), Detoxificación de la pulpa de café por fermentación sólida. Proc. I SIBAC, S. Roussos, R.F. Licona, Gutierrez M. (eds.), ORSTOM-UAM-INMECAFE, Xalapa, Ver. México. pp. 121-143

Sabine J.R. (1983), Earthworms as a source of food and drugs. In- Earthworm Ecology, From Darwin To Vermiculture. Satchell J. (ed.). Chapman and Hall. UK, pp. 285-296

Salazar T., Aranda E., Barois I. (1994), Comparative vermicompost production by *Eisenia andrei, E. fetida* and *Perionyx excavatus* using coffee pulp. Paper presented at Fifth International Symposium of Earthworm Ecology, ISEE5, Ohio, USA

Satchell J.E., (ed.) (1983), Earthworm Ecology, from Darwin to Vermiculture. The University Press, Chapman and Hall, London, UK

Saucedo-Castañeda J. (1999), Valorization of by-products of the coffee agro-industry: Mexican experience. Paper presented in III SIBAC, Londrina, Brazil

Scott M. A. (1988), The use of worm-digested animal waste es a supplement to peat in loamless compost for hardy nursery stock. In- Earthworms in waste and environmental management. Edwards, C.A. and E. Neuhauser (eds.), pp. 221-228

Scott S., Edwards C., Metzeger J. (1998), Comparing vermicompost and compost. In: BioCycle J. Composting & Recycling, 39 (7), 63-66

Senapati B.K. (1994), Science for village and Agrobased industry through vermiculture. Ecology section, School of life sciences, Sambalpur University. Syoti Vihar-768019, Orissa, India

Szczech M., Rondomanski W., Brzeski M.W., Smolinska U., Kotowski J. F. (1993), Suppressive effect of a commercial earthworms compost on some root infecting pathogens of cabbage and tomato. Biol. Agric. Horti., 10: 47-52

Taboga L. (1980), The nutricional value of earthworms for chickens. British Poultry Sci., 21:405-410

Tacon A.G., Stafford E.A., Edwards C.A. (1983), A preliminary investigation of the nutritive value of three terrestrial lumbricid worms for rainbow trout. Aquaculture, 35: 187-199

Tomati V., Grapelli A., Galli E., Nizi W. (1983), Fertilizer from vermiculture as an option for organic wastes recovery. Agrochemestry, 27: 244-251

Tomati, U.,.Grappelli A., Galli E.. (1987), The presence of growth regulators in earthworm-worked wastes. In- Earthworms. Bonvicini-Pagliai A.M and P. Omodeo, (eds.), 2: 423-435

Van Gansen P. (1962), Structures et fonctions du tube digestif du Lombricien *Eisenia foetida* Savigny. J. Microscopie, 1: 1- 363

Velásquez B.L., Herrera C.C., Ibañez B.I. (1986), Harina de Lombriz. I parte: Obtención, composición química, valor nutricional y calidad bacteriológica. Alimentos, 2 (1): 15-20

Venegas M. Leticia del P. (1988), *Caracterización y evaluación agronómica del proceso de humificación por lombriz Eisenia foetida en guanos de cerdo y conejo* Universidad de Chile, Santiago, Escuela de Agronomía, Chile

Viniegra-Gonzalez G. (1999), Biotechnology and the future of coffee production. Paper presented in III SIBAC, Londrina, Brazil

Werner M., Ramon J.C. (1996), Vermiculture in Cuba. BioCycle J. Composting & Recycling, 37 (6): 57-62

Zuluaga V.J. (1989), Utilización integral de los subproductos del café. *In: Memorias del I Seminario Internacional sobre Biotecnología en la Agroindustria Cafetalera.* S. Roussos, R. F. Licona, M. Gutierrez (eds.). ORSTOM-UAM-INMECAFE, Xalapa, Ver. México, pp. 63-76

Chapter 47

COFFEE PULP POLYPHENOLS: AN OVERVIEW

J.R. Ramírez Martínez[1], M.N. Clifford[2]

[1]*Núcleo de Fitoquímica, Decanato de Investigación, Universidad Nacional Experimental del Táchira (UNET) San Cristóbal, Estado Táchira, Venezuela;* [2]*Food Safety Research Group, School of Biological Sciences, University of Surrey, Guildford, Surrey, GU2 5XH, UK*

Running title: Coffee pulp polyphenols

1. Introduction

Polyphenols are compounds present in all kinds of plant materials and in some cases, as in coffee, in significant amounts in fruits and leaves. They are found as simple phenols, phenolic esters, hydroxycinnamic acid derivatives, flavonoids and tannins (Ho, 1992). Polyphenols are generally considered as secondary metabolites not necessary for plant's living. Ingestion of these compounds by human beings and animals in their regular diets is apparently harmless as established experimentally and by daily life observations (Singleton, 1981). However, some of these natural compounds are highly toxic to certain species and their intentional or accidental ingestion above certain levels can produce deleterious effects.

The studies on coffee polyphenols have been done mostly in coffee seeds (Corse *et al.*, 1970; Carelli *et al.*, 1974; Clifford and Wight, 1976; Clifford and Staniforth, 1977; Van der Steegen and Van Duijn, 1980; Clifford, 1985; Clifford *et al.*, 1989) and leaves (El–Hamidi and Wanner, 1964; Zuluaga-Vasco *et al.*, 1971a, b; Orozco-Castano and Casselet, 1974; Amorim *et al.*, 1978). It seems that Robiquet and Bourton (1837) made the first report on coffee polyphenols. Payen (1846 a, b) described a material isolated from coffee green beans which he termed chlorogenic acid in the belief that it was a pure compound. Later studies (Fisher and Dangschat, 1932) showed that the principal component of this material was a caffeic acid conjugate of quinic acid. Presently it is known that coffee bean chlorogenic acid (CGA) is a complex group which includes not

T. Sera et al. (eds.), Coffee Biotechnology and Quality, 507–515.

only monoesters of caffeic acid (5-caffeoylquinic, 4-caffeoylquinic and 3-caffeoylquinic acids) but also esters of other cinnamates (p-coumaric and ferulic acids), as well as dicaffeoylquinic and caffeoylferuloylquinic acids (Clifford, 1979; Clifford *et al.*, 1989). As regards to polyphenols in coffee leaves most of the studies have been made to establish a relationship between susceptibility to specific diseases and abundance of such compounds (Zuluaga–Vasco *et al*, 1971 a, b; Amorim *et al.*, 1978). The results obtained from these studies were inconsistent and the picture was not clear regarding the nature and content of the polyphenols presumably involved.

2. Coffee pulp simple polyphenols

Lopes-Longo (1972) made an extensive and detailed study of flavonoids present in the methanolic extract of berries pulp of the genus Coffea. Using two-dimentional paper chromatography all the spots showing fluorescence under UV light were considered as true polyphenols. It is well known that not all compounds, which emit fluorescence under UV light, are polyphenols, whereas some of these compounds do not fluoresce under UV light. With regard to chlorogenic and caffeic acids contents Molina *et al.* (1974) found values of 2.71% and 0.31% on a dry basis (db), respectively, whereas the values reported by Bressani and Elias (1976) were 2.6% and 1.6%, respectively. El-Hamidi and Wanner (1964) observed that with increasing age of the coffee berry chlorogenic acid diminished in the pulp. Elias (1978) recognised the need for reliable compositional values of the coffee pulp according to coffee cultivars, agricultural practices and processing methods.

At the beginning of the ´80s, the available information on coffee pulp polyphenols was scant and inconsistent. Detailed studies made by paper chromatography on 12 cultivars of *Coffea arabica* L. showed that fresh coffee pulp polyphenols composition followed a similar pattern in both red and yellow berries cultivars (Ramírez, 1987). Differences among cultivars with regard to the presence or absence of a few polyphenols were minimal. Moreover, there were no qualitative differences between cultivars resistant and susceptible to coffee rust (*Hemileia vastatrix).* Among the more abundant simple polyphenols did the dicaffeoylquinic (isoclorogenic) acids follow some of the CGA group, with 5-caffeoylquinic acid being the most conspicuous. Other prominent polyphenols were compounds belonging to the catechins group (flavanols) as well as seemingly leucoanthocyanidins and/or proanthocyanidins (condensed tannins). The use of high performance liquid chromatography (HPLC) allowed not only to verify these observations but also to quantify most of the simple polyphenols extracted from the fresh coffee pulp of 12 cultivars of *C. arabica* L. (Ramïrez-Martínez, 1988). This technique allowed also detecting polyphenols which due to its low concentration could not be detected by paper chromatography, such as protocatechuic and ferulic acids. However, the presence of free caffeic acid in coffee pulp, reported by Molina *et al.* (1974), Bressani and Elias (1976) and Lopes *et al.* (1984), was not confirmed by either paper chromatography or HPLC. Griffin and Stonier (1975) did not find caffeic acid and chlorogenic acid together in the same extract of coffee pulp.

Ramírez-Martínez (1988) reported that 5-caffeoylquinic acid was the simple polyphenols more abundant in fresh coffee pulp and it varied between 0.24-0.88% (db) in the 12 analysed *C. arabica* L. cultivars. Adding the contents of 5-caffeoylquinic acid contents with isochlorogenic acids for each cultivar, the total CGA content was between 0.36-1.41%, which were well below the values reported by Molina *et al.* (1974) and Bressani and Elias (1976). Another abundant polyphenol was epicatechin (0.19-0.44%) with the lowest values in the yellow pulp cultivars. Average composition for the sum of the tentatively identified polyphenols was as follows (%): 5-caffeoylquinic acid, 42.2; epicatechin, 21.6; 3,4-dicaffeoylquinic acid, 5.7; 3,5-dicaffeoylquinic acid, 19.3; 4,5-dicaffeoylquinic acid, 4.4; catechin, 2.2; rutin, 2.2; protocatechuic acid, 1.6; and ferulic acid, 1.0.

In sun-dried coffee pulp from two cultivars of *C. arabica* L. (Red Bourbon and Red Caturra), two hybrids (Timor Hybrid and Catimor) and *C. canephora* the most abundant polyphenol was 5-caffeoylquinic acid (Ramírez-Martínez and Clifford, 1989; Clifford and Ramírez-Martínez, 1991a). Its content was similar in *C. arabica* L. and *C. canephora* (around 0.86%) whereas in the hybrids it was less than half (around 0.35%). Very low contents of 3-caffeoylquinic and 4-caffeoylquinic acids were found in *C. arabica* L. cultivars and the hybrids (around 0.02%) but were relatively significant in *C. canephora* (around 0.14%). The total contents of dicaffeoylquinic acids was 0.49% in *C. canephora*, around 0.16% in *C. arabica* L. cultivars and 0.06% in the hybrids. Protocatechuic acid was relatively high in Timor Hybrid (0.7%), low in *C. canephora* (0.1%) and Red Caturra (0.2%) and intermediate in Catimor and Red Bourbon (0.5%). From Timor Hybrid, a relatively pure compound was isolated by preparative HPLC, which was identified as protocatechuic acid when examined with proton NMR and structure-specific reagents. Protocatechuic content values of 0.02% had been previously reported for Red Bourbon and Red Caturra fresh coffee pulp (Ramírez Martínez, 1988).

There are some reports on the nature of anthocyanins in red coffee berries. Anthocyanins are pigments in which the flavonoid structure is fully conjugated when protonated, and bears one or more sugars. Lopes *et al.* (1984) found two anthocyanins in the coffee pulp of the Red Bourbon cultivar which were identified as cyanidin-3-monoglycoside and cyanidin-3,5-diglycoside. Zuluaga-Vasco (1989) reported also two anthocyanins, presumably from coffee pulp of the cultivar Typica, which were identified as cyanidin-3-glucoside and cyanidin-3-glucorhamnoside. Barboza and Ramírez-Martínez (1991) detected the anthocyanins cyanidin-3-monoglucoside and cyanidin-3-diglucoside in coffee pulp of the cultivar Red Bourbon by paper chromatography and HPLC techniques. It should be mentioned here that anthocyanin identification based only on paper chromatographic techniques could lead to confusion. Harborne and Sherrath (1957) found that arabinose was produced in the isolation of anthocyanins by paper chromatography as a consequence of the action of HCl on the filter paper. Thus, it is advisable to complement this technique with analytic and preparative HPLC and other analytical techniques.

3. Coffee pulp polymeric polyphenols

Condensed tannins (or proanthocyanidins) are outstanding among the plant polymeric polyphenols. Condensed tannins consist of chains of flavan-3-ols units with C-4 to C-8 or C-4 to C-6 linkages (Czochanska *et al.*, 1980). Information on coffee pulp tannins is also scarce and sometimes contradictory. Possibly the first report on coffee tannins goes back to 1937 (as cited by Lopes-Longo, 1972) when Willbaux treated a coffee pulp extract with lead acetate and obtained a blue precipitate which he thought contained `anthocyonosides´. Later on Franco (1939) verified the presence of tannins in the cultivar Typica. Jaffé and Ortiz (1952) observed that the extent of the time the coffee pulp was kept wet before its sun-drying affected the tannins content. Coffee pulp kept wet for 24 hours showed a tannin content of 1.44% whereas in that stored wet for three days (fermented pulp), it decreased to 0.88%. Aguirre (1966) and Molina *et al.* (1974) reported values of 4.5% and 2.4%, respectively for coffee pulp tannins content in Guatemala, whereas Zuluaga-Vasco *et al.* (1975) obtained a value of 3.7% for dehydrated coffee pulp (7% moisture) in Colombia. Later on Zuluaga-Vasco and Tabacchi (1979) found that fresh coffee pulp did not contain condensed tannins and that their formation started a few hours after pulping, especially under humid and hot weather. They also found that in sun-dried pulp 2.07% tannins were formed out of which one fifth were hydrolysable tannins. García *et al.* (1985) reported up to 2.56% alkali-soluble condensed tannins in a Caturra cultivar coffee pulp, a value that included free flavan-3-ols. Clifford and Ramírez-Martínez (1991b) using Porter's improved autooxidative procedure (Porter *et al.*, 1986) and calculating the content of condensed tannins as cyanidin equivalents ($E^{1\%}_{550}$ = 470) reported values around 1.0% for *C. arabica* L., Timor Hybrid and Catimor, whereas for *C. canephora* was 2.7%. Evidence was also obtained for the presence in coffee pulp of significant amounts of `insoluble' condensed tannins. Analyses for `total' tannins revealed that the yield of anthocyanidin pigments (Porter reaction) declines as the tannin ages. Although earlier studies (Zuluaga-Vasco and Tabacchi, 1979) suggested that hydrolysable tannins (gallotannins and ellagitannins) were present in coffee pulp, it was demonstrated unequivocally that coffee pulp tannins consist primarily of proanthocyanidins.

Clifford *et al.* (1991) reported a Sephadex LH-20 batch procedure to extract and fractionate proanthocyanidins from sun-dried coffee pulp and made a comparison of the use of Porter's reagents to quantify tannins with quantitative methods based on substitution with dimethylaminocinnamaldehyde in acidic media. The immediate testing of freshly prepared coffee pulp by direct addition to Porter's reagents yielded a red colour indicating the presence of condensed tannins (Clifford *et al.*, 1993; González de Colmenares *et al.*, 1994) in contrast to the results of Zuluaga-Vasco and Tabacchi (1979) who believed the tannins were formed only by sun-drying. Nevertheless a net production of condensed tannins was observed on drying of the coffee pulp. González de Colmenares *et al.* (1994) for the first time used a purified tannin (quebracho proanthocyanidin) as a calibrant for the quantification of coffee pulp tannins and values in the range 4.57-6.17% were obtained for Red Bourbon, Yellow Catuai and Red Catuai. When these data were compared to those obtained by calculating the corresponding cyanidin equivalents, it was found that the values obtained using quebracho proanthocyanidin as standard were more than 5 times higher. Thus the use of a non-specific tannin as standard could still over or underestimate the true condensed tannin content. Analysis of procyanidin dimer-rich (soluble en ethyl acetate) and procyanidin oligomer-rich (insoluble in ethyl acetate) fractions suggested that while dimers dominate in undried pulp, dimers might be converted to oligomers during drying.

González de Colmenares *et al.* (1998) developed a method to obtain proanthocyanidin-containing preparations from the pulp of ripe *C. canephora* coffee berries. Two proanthocyanidin preparations (PA2 and PA3) were obtained by Sephadex LH-20 column chromatography and characterised by means of infrared and ^{13}C NMR spectroscopy as well as reactions with several functional group-specific reagents. PA2 contains condensed tannins, carbohydrate and some nitrogenous substituent other than protein or an α-amino acid, whereas PA3 is composed of condensed tannins and methoxylated flavonoids, methyl ethers and/or lignans. PA2 at concentrations above 0.5 g litre^{-1} was shown to be an effective inhibitor of the germination of coffee rust uredospores.

Lyophilised PA2 preparation was also used as a standard for the quantification of condensed tannins in wet-processed coffee pulp silage (Ramírez-Martínez *et al.*, 1997; Ramírez-Martínez *et al.*, 1999). Condensed tannins content remained constant during the first three weeks of the coffee pulp silaging, but it decreased by more than 40% after an ensiling period of more than 100 days. It was also found that little amounts of simple polyphenols other than chlorogenic acid were detected in the wet-processed coffee pulp used for the loading of the minisilos and large silos. Recently a proanthocyanidin trimer was isolated from *C. canephora* pulp and its antioxidant activity was determined (González de Colmenares, N., personal communication). This proanthocyanidin trimer showed antioxidant activity against free HO° radicals and not against free ROO° radicals.

4. Summary and futuristic approach

Information on coffee pulp polyphenols is rather scarce, sometimes contradictory and often difficult to interpret because of the use of non-specific analytical methods. Some progress in quantifying and characterising these compounds has been made but much remains to be achieved. With the advent of the dry pulping process, which is gradually replacing the old fashioned wet-processing method, there is a need to analyse the almost intact coffee pulp produced. Work should be done to find new or improved uses for this kind of coffee pulp. Research projects should take into account the final use of these coffee by-products.

It is important that coffee pulp condensed tannins be isolated in sufficient quantities to be used not only as primary standard for quantification purposes but also to pursuit further characterization studies. So far proanthocyanidins have been isolated from *C. canephora* pulp because of its relative abundance. Attempts should to be made to isolate them from *C. arabica* L. cultivars in quantities enough to determine if they are identical to those isolated from *C. canephora* and to establish the pertinence of the use of the later as a primary standard. With the information now available on the nature of coffee pulp proanthocyanidins, research has led the application of tannic acid as inducer and standard in the production of enzymes to degrade coffee pulp tannins. Such esterases are only likely to hydrolyse chlorogenic acids and not the true condensed tannins, thus, necessitating a careful evaluation for their relevance to the commercial exploitation. The studies on the biological activity of the coffee pulp proanthocyanidins as antioxidants, fungicides and bactericides could be useful.

Finally, the study of coffee pulp polyphenols is far from being finished. Further subject of study could be, for example, the localisation of the simple and polymeric polyphenols in the intact coffee pulp by chemical analysis and histochemical techniques. If they are confined to the epidermal cells, then a physical or chemical peeling of the coffee berries before pulping could be useful to remove them. Another subject could be the isolation of specific chlorogenic acids and proanthocyanidins in large scale from the coffee pulp for commercial purposes.

5. References

Aguirre B.F. (1966), La utilización industrial del grano de café y de sus productos. Instituto Interamericano de Investigación y Tecnología Industrial, Investigaciones Tecnológicas del ICAITI N° 1, Guatemala, p. 43

Amorim H.V., Alvarez M.L.M., Lopes C.R., Carvalho A., Monaco, L.C. (1978), Análise de compostos fenólicos em folhas de caffeiros resistentes e susceptiveis a *Hemileia vastatrix*. Turrialba, 28, 109-117

Barboza C.A., Ramírez-Martínez J.R. (1991), Antocianinas en pulpa de café del cultivar Bourbon Rojo. *14ème Colloque Scientifique International sur le Café*, ASIC, Paris, pp. 272-276

Bressani R., Elias L.G. (1976), Utilización de desechos de café en alimentación de animales y materia prima industrial. Documento presentado en la Exposición Pecuaria del Istmo Centroamericano (EXPICA) 76. San Salvador, El Salvador, May 3-8. INCAP, Guatemala, p.25

Carelli M.L.C., Lopes C.R., Monaco L.C. (1974), Chlorogenic acid content in species of *Coffea* and selections of *C. arabica*. Turrialba, 24: 398-400

Clifford M.N., Wight J. (1976), The measurement of feruloylquinic acids and caffeoylquinic acids in coffee beans. Development of the technique and its preliminary application to green coffee beans. J. Sci. Food Agric. 27, 73-84

Clifford M.N., Staniforth P.S. (1977), A critical comparison of six spectrophotometric methods for measuring chlorogenic acids in green coffee beans. *Huitième Colloque International sur la Chemie des Cafés Verts*, ASIC, Paris, pp 109-113

Clifford M.N. (1979), Chlorogenic acids – Their complex nature and routine determination in coffee beans. Food Chem. 4: 63-71

Clifford M.N. (1985), Coffee bean dicaffeoylquinic acids. Phytochemistry, 25: 1767-1769

Clifford M.N., Ramírez-Martínez J.R. (1991a), Phenols and caffeine in wet-processed coffee beans and coffee pulp. Food Chem. 40: 35-42

Clifford M.N., Ramírez-Martínez J.R. (1991b), Tannins in wet-processed coffee beans and coffee pulp. Food Chem. 40, 191-200

Clifford M.N., de Colmenares N.G., Ramírez-Martínez J.R., Adams M.R., de Menezes H. C. (1991), Tannins in the sun-dried pulp from the wet-processing of Arabica coffee beans. *14ème Colloque Scientifique International sur le Café*, ASIC, Paris, pp. 230-236

Clifford M.N., González de Colmenares N., Ramírez-Martínez J.R., Aldana J.O. (1993), Progress in the analysis of proanthocianidins in freshly prepared coffee pulp. *15ème Colloque Scientifique International sur le Café, ASIC*, Paris, pp. 884-886

Clifford M.N., Kellard B., Birch G.G. (1989), Characterisation of caffeoylferuloyl- quinic acids by simultaneous isomerisation and transesterification with tetramethyl- ammonium hydroxide. Food Chem. 34: 81-88

Corse J., Layton L.L., Patterson D.C. (1970), Isolation of chlorogenic acids from roasted coffee. J. Sci. Food Agric. 21: 164-168

Czochanska Z., Foo L.Y., Newman R.H., Porter L.J. (1980), Polymeric proanthocyanidins stereochemistry. Structural units and molecular weight. J. Chem. Soc. Perkin Trans. I: 2278-2286

El-Hamidi A., Wanner H. (1964), The distribution pattern of chlorogenic acid and caffeine in *Coffea arabica.* Planta, 61: 90-96

Elias L.G. (1978), Composición química de la pulpa de café y otros subproductos. In *–Pulpa de café: composición, tecnología y utilización*, Braham, J.E. and Bressani, R. (eds.), CIID, pp. 152

Fischer H.O.L., Dangschat G. (1932), Konstitution der Chlorogensäure (3. Mitteil über Chinasäure und Derivate). Berichte, 65: 1037-1040

Franco C.M. (1939), Sobre compostos fenólicos no café. Instituto Agronómico, Campinas (Brasil), Bol. Tec. 64: p. 14

García L.A., Vélez A.J., de Rozo M. (1985), Extracción y cuantificación de los polifenoles de la pulpa de café. Arch. Latinoamer. Nutr., 35: 492-495

González de Colmenares N., Ramírez-Martínez J.R., Aldana J.O., Clifford M.N. (1994), Analysis of proanthocyanidins in coffee pulp. J. Sci. Food Agric., 65: 157-162

González de Colmenares N., Ramírez-Martínez J.R., Aldana J.O., Ramos-Niño M.E, Clifford M.N., Pékerar S., Méndez B. (1998), Isolation, characterisation and determination of biological activity of coffee proanthocyanidins. J. Sci. Food Agric., 77: 368-372

Griffin B., Stonier T. (1975), Studies on auxin protectors. XII. Auxin protectors and polyphenol oxidase in developing coffee fruit. Physiol. Plantarum 33: 157-160

Harborne J.B., Sherrath H.S.A. (1957), Variations in the glycosidic patterns of anthocyanins. Experientia 13: 486

Ho C.T. (1992), Phenolic compounds in food. An overview. In-*Phenolic Compounds in Food and their Effect on Health II. Antioxidants and Cancer Prevention* (ACS Symposium Series 507), Huang, M.T., Ho, C.T. and Lee, C.Y. (eds.), p.207

Jaffé W., Ortiz D.S. (1952), Notas sobre el valor alimenticio de la pulpa de café. .Agro 23, 31-37

Lopes-Longo C.R. (1972), Estudo dos pigmentos flavonóides e sua contribuição à filogenia do gênero *Coffea.* Tese de doutoramento apresentada á Escola Superior de Agricultura "Luis de Queiroz", da Universidade de São Paulo, Brazil

Lopes C.R., Musche R., Hec M. (1984), Identificação de pigmentos flavonóides e ácidos fenólicos nos cultivares Bourbon e Caturra de *Coffea arabica* L. Rev. Brasil. Genet. VII 4: 657-669

Molina M.R., de la Fuente G., Gudiel H., Bressani R. (1974), Pulpa y pergamino de café. VIII. Estudios básicos sobre la deshidratación de la pulpa de café. Turrialba, 24: 280-284

Orozco-Castaño F.J., Cassalet-Dávila C. (1974), Estudio cromatográfico de un híbrido interespecífico en café. Cenicafé, 23: 65-77

Payen S. (1846a), Untersuchung des Kaffees. Annalen, 60: 286- 294

Payen S. (1846b), Memoire sur le café (3° Part). Comptes Rendus, 23:244-251

Porter L.J., Hritisch L.N., Chen B.G. (1986), The conversion of procyanidins and prodelphinidins to cyanidin and delphinidin. Phytochemistry, 25: 223-230

Ramírez J. (1987), Compuestos fenólicos en la pulpa de café. Cromatografía de papel de pulpa fresca de 12 cultivares de *Coffea arabica* L. Turrialba, 37:317-323

Ramírez-Martínez J.R. (1988), Phenolic compounds in coffee pulp: quantitative determination by HPLC. J. Sci. Food Agric., 43, 135-144

Ramírez-Martínez J.R., Clifford M.N. (1989), Acidos fenólicos en pulpa de café. *13ème Colloque Scientifique International sur le Café*, ASIC, Paris, pp. 205-210

Ramírez-Martínez, Bautista E.O., Clifford M.N., Adams M.R. (1997), Production and nutritional evaluation of coffee pulp silage. *17ème Colloque Scientifique International sur le Café*, ASIC, Paris, pp. 695-702

Ramírez-Martínez J.R., Pernia R.D., Bautista E.O., Clifford M.N., Adams M.R. (1999), Production and characterisation of coffee pulp silage. Trop. Sci., 39: 107-114

Robiquet , Boutron (1837), Ueber den Kaffee von Robiquet und Boutron. Ann. Pharmacie, 23: 93-95

Singleton V.L. (1981), Naturally occurring food toxicants: phenolic substances of plant origin common in foods. Adv. Food Res., 27: 149-242

Van der Stegen G.H.D., Van Duijn J. (1979), Analysis of chlorogenic acids in coffee. *Neuvième Colloque Scientifique International sur le Café*, ASIC, Paris, pp. 107-112

Zuluaga-Vasco J., Valencia G., González J. (1971a), Contribución al estudio de la resistencia del cafeto a *Ceratocystis fimbriata* (Ell.Halst) Hunt. Cenicafé, 22, 43-68

Zuluaga-Vasco J., Valencia G., González J. (1971b), Contribución al estudio de la resistencia del cafeto a *Ceratocystis fimbriata* (Ell. Halst) Hunt. II. Cenicafé, 22, 109-125

Zuluaga-Vasco J., Bonilla V.M., Quijano R. (1975), Contribución al estudio y utilización de la pulpa de café. *Septième Colloque International sur la Chemie des Cafés Verts,* ASIC, Paris, pp. 233-242

Zuluaga-Vasco J., Tabacchi R. (1979), Contribution á l'étude de la composition chimique de la pulpe de café. *Neuvième Colloque Scientifique International sur le Café*, ASIC, Paris, pp. 335-344

Zuluaga-Vasco J. (1989), Utilización integral de los subproductos del café. En- *I Seminario Internacional sobre Biotecnología en la Agroindustria Cafetalera*, México, 12-15 April, pp. 63-76

Chapter 48

BIOTECHNOLOGICAL POTENTIALITIES OF COFFEE AND SIMILAR WITH OLIVE, TWO MODELS OF AGROINDUSTRIAL PRODUCTS RICH IN POLYPHENOLIC COMPOUNDS: A REVIEW.

M. Labat[1,4], C. Augur[2], B. Rio[1], I. Perraud-Gaimé, S. Sayadi[3]

[1]IRD (ex-ORSTOM) Institut de Recherche pour le Développement, Laboratoire de Microbiologie, Université de Provence, CESB/ESIL, Case 925, 163 avenue de Luminy, F-13288 Marseille cedex 9. FRANCE e-mail: labat@esil.univ-mrs.fr e-mail: brio@esil.univ-mrs.fr2 [2] UAM-I, Depto. Biotecnologia; Ave. Michoacán-Purísima, Iztapalapa, C.P. 09340, A.P. 55 535, México D.F. e-mail: augur@xanum.uam.mx e-mail: isa@xanum.uam.mx3 [3]CBS Centre de Biotechnologie de Sfax, BP"K", 3038 Sfax, Tunisie e-mail : sami.sayadi@cbs.rnrt.tn

1. Introduction

Micro-organisms are unique « cell factories » able to valorize agricultural by-products instead of only degrading or mineralizing them for depollution or methanisation processes. The difficulties encountered in treating such compounds often result from high concentration of pollutants and/or high toxicity to the microflora. Liquid effluents with high chemical oxygen demand (COD) (> 100 g COD/litre) are often rich in aromatic compounds.

Olive mill wastewater (OMW) is an exemple of high polluting industrial by-product especially rich in polyphenolic compounds. Difficulties in the treatment of such a polluting compound are due both to the high concentration of organics which are mainly composed of phenolic compounds toxic to the microflora. Indeed the polluting charge is exceptionnally high, generally over 120 g and can reach 200g of COD/litre of OMW. Polyphenolics of OMW span from monoaromatic to high molecular-mass polyphenols.

Coffee husk and coffee pulp represent another example of agro-industrial co-products rich in polyphenolic compounds. Information is available essentially for coffee pulp, where

T. Sera et al. (eds.), Coffee Biotechnology and Quality, 517–531.

flavonoids were first extensively studied with two-dimensional paper chromatography techniques (Lopes-Longo, 1972). Values of total phenolics from lyophilized pulp (6.3 %) and sun dried pulp (6.6 %) were reported (Zuluaga, 1981), and new phenolic structures progressively identified. HPLC techniques enable the identification and quantification of acid phenolics from fresh coffee pulp (Ramirez-Martinez, 1988). Polyphenolic compounds are homo or heterocyclic aromatic structures, where hydroxylic functions are substituted to the benzenic structure. Thus hydroxylated benzenic and cinnamic acids derivatives (acid phenolics), hydroxylated phenols, flavonoids, anthocyans and tannins are polyphenolic compounds. Microbial degradation occurs aerobically or anaerobically. Hydroxylated aromatic compounds are metabolizable in pure anaerobic or aerobic cultures but anaerobic degradation is restricted mostly to sufficiently oxidated aromatics. As an example, methane production from benzoate is achieved only in mixed cultures (Ferry and Wolfe, 1976), as no organisms able to undergo such a methanisation in pure culture were isolated. Moreover, some biotopes or hindguts are specialized in the degradation of specific, usually poorly degradable aromatics. As an example, termite species possess in their hindgut, bacteria that enable the degradation of aromatic compounds with methane production (Brauman *et al.*, 1992). Degradation of complex aromatic compounds is often more efficient aerobically, even if anaerobic degradation was demonstrated with oligognols, or compounds generated from aerobic degradation of lignin (Crawford *et al.*, 1983), and effective with tannins (Field and Lettinga, 1987), and various other complex compounds.
In this review, analytical information is given concerning polyphenolics reported in coffee arabica, a plant species especially rich in polyphenolics and largely cultivated in Latin America. Coffee husk and coffee pulp represent two models of solid effluents containing polyphenolic compounds. Besides olive (*Olea europeae* L.), rather poorly cultivated and consumed in latin America (except Argentina), emphasis is on coffee (*Coffea arabica* L.). These two examples of agricultural products or by-products containing aromatics are then presented together with known biological activities attributed to these aromatics.

2. Potentialites of phenolic compounds found in olive and coffee

2.1 PHENOLIC CHEMICALS IN OLIVE PLANT (*OLEA EUROPAEA L.*)

Various phenolic chemicals are reported from different parts of the olive plant. Hydrophilic phenolic compounds are estimated between 20 and 500 ppm in olive (Léger, 1999), and o-diphenolics between 60 to 100 mg/l of olive oil (Ryan and Robards, 1998). Other cyclic compounds, like cyclo-olivil (branches), cinchonine (leaf) are not phenolic chemicals (N heterocyclic compounds). Sterols (beta-sitosterol-glucoside reported in olive leaf), carotenoids (1.8 - 8.3 ppm beta-carotene reported in olive fruit) and cyclic compounds without phenolic alcohol are excluded from this list; but verbascoside (reported in olive fruit), esculin (reported in olive stem) or estrone (reported in olive seed) are listed as they exhibit a true phenolic function.

Table 1. Part A - List of phenolics reported in olive plant and by-product (alphabetical order).

Chemical name of the phenolic compound Chemical name of similar phenolic compound	Part of the olive plant
d-acetoxypinoresinol	4
d-1-acetoxypinoresinol-4"-o-methyl-ether	4
d-1-acetoxypinoresinol-4'-beta-d-glucoside	4
apigenin (pulp absent)	2, 8
apigenin-7-di-o-xyloside	2
apigenin-7-glucoside	2, 6, 8
caffeic-acid (oil 0.0-1.0 ppm, OMW 90 ppm)	1, 8, 9
1-caffeyl-glucose	1
catechin	1
catechol-melanin	1
chlorogenic acid	2
cinnamic acid	8
cornoside	6
p-coumaric acid (oil 0.0-0.6 ppm)	1, 2, 8, 9
o-coumaric acid	6, 8
cyanidin-3-monoglycoside	2, 5, 6
cyanidin-3-rutinoside	1
cyanidin-3-rhamnosylglucosylglycoside	1
3,4-dihydroxyphenylethanol (oil 0.0-351.2 ppm)	1, 2, 6, 8, 9
linked with dialdehydic form of elenolic acid	8
3,4-dihydroxyphenylethanol-4-diglucoside	1
3,4-dihydroxyphenylethanol-4-monoglucoside	1
3,4-dihydroxyphenylpropane	2
elenolic acid	2, 6, 8
elenolic acid glucoside (leaf absent)	6,8
esculin	4
esculetin	4
estrone	7
ferulic acid (oil 0.0-2.4 ppm)	6, 8, 9
gallic acid	1, 6, 8, 9
hesperidin	2, 6

Table 1. Part B - List of phenolics reported in olive plant and by-product (alphabetical order)

Chemical name of the phenolic compound	Chemical name of similar phenolic compound	Part of the olive plant
4-hydroxybenzoic acid		8
4-hydroxyphenylacetic acid (oil 0.2-2.8 ppm, OMW 145 ppm)		6, 8, 9
	3,4-dihydroxyphenylacetic acid	1
4-hydroxyphenylethanol (oil 0.1-123.1 ppm)		1, 6, 8, 9
	4-hydroxyphenylethanol glucoside	6
	4-hydroxyphenylethanol derivative (oil 0.0-113.4 ppm)	8
linked with dialdehydic form of elenolic acid (oil 0.0-79.8 ppm)		8
d-1-hydroxypinoresinol		4
	d-1-hydroxypinoresinol-4"-o-methyl-ether	4
kaempferol		1, 4
	glycosylated kaempferols	1, 4
ligstroside (oil 25 ppm)		2, 6, 8
luteolin		1, 2, 6
	luteolin-4'-glucoside	2, 6
	luteolin-5-glucoside	1
	luteolin-7-glucoside	2, 6
	luteolin-7-rutinoside	2
	luteolin-tetraglucoside	2
nuezhenide		7
	nuezhenide oleosid	7
oleuropein (oil 67 ppm)		1, 2, 6, 8
	demethyl-oleoeuropein	1
	oleuropeic acid	3
	deacetoxyoleuropein aglycon	1
	oleuropein aglycon isomer (oil 0.0-83.5 ppm)	8
	oleuroside	2
	oleoside	
olivin		2
olivin-4'-diglucoside		2
protocatechuic acid		1, 8, 9
quercitin		2, 4, 5, 6
	quercitin-3-rhamnoside	5
	quercitin-3-rutinoside (or rutin)	2, 5, 6
quinone		1

Table 1. Part C - List of phenolics reported in olive plant and by-product (alphabetical order)

Chemical name of the phenolic compound Chemical name of similar phenolic compound	Part of the olive plant
salidroside	7
sinapic acid	1, 8
syringic acid (oil 0.0-2.3 ppm, OMW 710 ppm)	1, 6, 8, 9
3,4,5-trimethoxybenzoic acid (OMW 80 ppm)	9
alpha-tocopherol (oil 40-130 ppm)	8
beta- tocopherol (oil 10-20 ppm	8
gamma-tocopherol (oil 13 ppm)	8
veratric acid (OMW 500 ppm)	6, 9
verbascoside	1,6

1 = fruit, 2 = leaf, 3 = root, 4 = stem, 5 = pericarp, 6 = pulp, 7 = seed, 8 = olive oil, 9 = liquid by-product (olive mill wastewater or OMW).

Phenolic compounds listed above were detected in olive fruit or other parts of the plant (if specified), and in olive oil and its liquid by-product (OMW). This list includes free and glycosylated monomeric (p-coumaric acid; protocatechuic acid; quinone) and polymeric (catechin, cyanidin-3-rutinoside, tannin) phenolics.
This list of phenolic chemicals with ppm (or mg/kg) values is adapted from various bibliographical sources: Balice and Cera, 1984; Salvemini, 1985; Nefzaoui, 1991; Duke, 1992a and 1992b; Labat *et al.*, 1996 and 1997; Baldioli *et al.*, 1996, Léger, 1999; Montedoro *et al.*, 1993; Ryan and Robards, 1998. Chemical names are mentioned with reported concentrations when available. Few compounds are not true phenolic compounds (but associated with). As olive oil generates liquid by-products most of these values in ppm are re-calculated from % f.w. (fresh weight) of OMW, and few from mg/l of oil (with d=0.91). For this reason, this table cannot be directly compared with values in ppm calculated from solid by-products (coffee pulp).

2.2 PHENOLIC CHEMICALS IN COFFEE PLANT (*COFFEA ARABICA L.*)

Various phenolic compounds are reported from different parts of the coffee plant or solid by-products (pulp and husk). Total phenolic compounds are often estimated in mg gallic acid per 100 mg of dry weight (d.w.) of pulp. Values between 6.3 (lyophized pulp) and 6.6 (sun dried pulp) were reported (Zuluaga, 1981). With HPLC techniques, mean value of total phenolics was 1.27 % (d.w.) (Ramirez-Martinez, 1988).

Table 2. Part A - List of phenolics reported in coffee plant and by-product (alphabetical order).

Chemical name of the phenolic compound / Chemical name of similar phenolic compound	Part of the coffee plant
caffeic acid (pulp 3,100-16,000 ppm)	3, 4
caffeol	2
caffesterol (or coffeasterol)	5
caffetannic acid (seed 84,600 ppm)	3
catechin (flavanol) (pulp 2200 ppm)	4
epicatechin (pulp 1900-4400 ppm)	4
chlorogenic acid (or caffeoyl-quinic acid ester or CGA) (seed 50,000-100,000 ppm, pulp 3,600-27,000 ppm)	3, 4
3-caffeoyl-quinic acid (pulp 200-1,400 ppm)	3, 4
4-caffeoyl-quinic acid (pulp 200-1,400 ppm)	3, 4
5-caffeoyl-quinic acid (pulp 2400-8800 ppm)	3, 4
p-coumaric acid	4
p-coumaric ester derivative	5
p-cresol	5
o-cresol	5
m-cresol	5
cyanidin-3-glycoside (or cyanidin-3-monoglycoside)	4
cyanidin-3,5-diglycoside	4
cyanidin-3-diglycoside	4
cyanidin-3-glycorhamnoside	4
3,4-dicaffeoylquinic acid (3,4-isochlorogenic acid) (pulp 5,700 ppm)	3, 4
3,5-dicaffeoylquinic acid (3,5-isochlorogenic acid) (pulp 19,300 ppm)	3, 4
4,5-dicaffeoylquinic acid (4,5-isochlorogenic acid) (pulp 4,400 ppm)	3, 4

Table 2. Part B - List of phenolics reported in coffee plant and by-product (alphabetical order).

Chemical name of the phenolic compound Chemical name of similar phenolic compound	Part of the coffee plant
4-ethylphenol	5
2-methoxy-4-ethylphenol	5
2-ethylphenol	5
eugenol	5
isoeugenol	5
ferulic acid (pulp 1,000 ppm)	4
ferulic ester derivative	5
caffeoylferuloylquinic acid	3
guaiacol	5
2,4-methylenephenol	5
4-methoxy-4-vinylphenol	5
protocatechuic acid (pulp 200-7,000 ppm)	4
rutin (or quercitin-3-rutinoside) (pulp 2200 ppm)	4
sinapic acid	1
tannin (seed 90,000 ppm, pulp 18,000-86,000 ppm)	3, 4
gallotannins* (hydrolysable tannins)	4
ellagitannins* (hydrolysable tannins)	4
flavonoids (pulp 28,000-40,000 ppm)	3, 4
leucoanthocyanidins (condensed tannins)	4
proanthocyanidins (condensed tannins)	4
alpha-tocopherol	1
beta-tocopherol	1
gamma-tocopherol	1
2,3,5-trimethylphenol	5
p-xylenol (dimethylphenol)	5
o-xylenol	5

* = not confirmed, 1 = plant, 2 = leaf, 3 = seed, 4 = solid by-product (coffee pulp), 5 = plant part not mentioned

Compounds such as caffeine (seed 600 - 32,000 ppm); theobromine (leaf and seed 18 ppm) and theophylline (leaf and seed) are not phenolics (N heterocyclic compounds) and are not listed below. Sterols, carotenoids and cyclic compounds without phenolic alcohol are excluded of this list but tocopherols are listed as they exhibit a true phenolic function.

This list of phenolic chemicals with ppm (or mg/kg) values is adapted from various bibliographical sources: Bressani *et al.*, 1972; Bressani et Elias, 1976; Clifford *et al.*, 1989 and 1993; Duke, 1992a and 1992b; Molina *et al.*, 1974; Ramirez-Martinez, 1988; Ramirez-Martinez and Clifford, 2000; Zuluaga *et al.*, 1975. Chemical names are mentioned with reported concentrations when available. As coffee plant generates solid by-products (pulp and husk), most of these values in ppm are calculated from % d.w. (dry weight). For this reason, this table cannot be directly compared with values in ppm calculated from liquid by-products (OMW).

3. Potentialites of three major phenolic compounds found in coffee

Coffee (*Coffea arabica* L.) is a plant species especially rich in polyphenolics. Three polyphenolic compounds found within this plant (seed, leaf and other parts) are chosen for their high potential in terms of biological activities. Chlorogenic and caffeic acids as monoaromatic models and tannin as polyaromatic model of polyphenols, is described.

3.1. CHLOROGENIC ACID CONTENT IN COFFEE AND OTHER PLANTS

Table 3. Plant species reported to contain high amount of chlorogenic acid (> 100 ppm)

Plant name	Concentration	Plant part
Coffee (*Coffee arabica* L)	50,000 - 100,000 ppm	seed
Sunflower (*Helianthus annuus* L.)	1,900 - 28,000 ppm	seed
Damask rose (*Rosa damascena* M.)	15,000 ppm	pollen or spore
Blueberry (*Vaccinium corymbosum* L.)	3,000 ppm	fruit
Coriander (*Coriandrion sativum* L.)	305 - 320 ppm	plant
Wall germander (*Teucrium chamaedrys* L.)	200 ppm	plant

To our knowledge, *Coffea arabica* is the plant species which contains the highest concentration of chlorogenic acid, when compared with other plants known to synthetise this compound. Contrary to expectations coming from the name of the plant itself, caffeic acid is not reported to be in high amount, as a free phenolic compound, in coffee. This compound is reported to be in the highest amount in the tuber of jalap (Ipomoea purga) reaching up to 40,000 ppm (Duke, 1992a). Coffee is not within the 30 plant species with highest amount. Caffeic acid is found only after hydrolysis, where the caffeic structure is issued from chlorogenic acid. The amounts of chlorogenic and caffeic acids depend (1) on the maturity of coffee beans, (2) on the drying method and (3) on the species of *Coffea* studied (Balyaya and Clifford, 1995).

3.2. KNOWN BIOLOGICAL ACTIVITIES ATTRIBUTED TO CHLOROGENIC ACID, A MODEL OF DIMERIC PHENOLIC COMPOUND FOUND IN COFFEE

This aromatic compound, an ester of caffeic acid and quinic acid, is a dimer of a phenolic and a cyclic structure respectively. This polyphenolic structure possesses one carboxylic and five hydroxylic functions (two from the caffeic and three from the quinic acid structure). Its hydrolysis produces diphenolic (caffeic) and tetrahydroxylic (quinic) structures. Different related structures have been identified in coffee (caffeoyl-quinic acid esters in position 3-, 4- or 5-, and dicaffeoyl-quinic acid esters in position 3,4-, 3,5 or 4,5 and derivatives), belonging to CGA (chlorogenic acid) group or ICGA (isochlorogenic acid) group (cf. Table 2).

Table 4. Compilation of 51 reported biological activities of chlorogenic acid (adapted from Duke, 1992a)

Biological activity	Biological activity	Biological activity
Allelochemic	Allergenic	Analgesic
Anti HIV	Anti tumor promoter	Anti EBV
Anticancer (colon, forestomach, liver and skin)	Anticarcinogenic	Antifeedant
Antigenotoxic	Antigonadotropic	Antihemolytic
Antihepatotoxic	Antiherpetic	Antihypercholesterolemic
Antiinflammatory	Antinitrosaminic	Antimutagenic
Antioxidant	Antiperoxidant	Antipolio
Antiradicular	Antiseptic	Antisunburn
Antithyroid	Antitumor	Antiulcer
Antiviral	Bactericide	Cancer preventive
CNS active	CNS stimulant	Cholagogue
Choleretic	Clastogenic	Collagen sparing
Diuretic	Fungicide	Hepatoprotective
Histamine inhibitor	Immunostimulant	Insectifuge
Interferon inducer	Juvabional	Larvistat
Leukotriene inhibitor	Lipoxygenase inhibitor	Metal chelator
Ornithine decarboxylase inhibitor	Oviposition stimulant	Sweetener, Vulnerary

As a consequence CGA was reported to exibit various physiological and biological activities.To our knowledge, up to 51 biological activities were attributed to CGA. They are listed below, but still await an unambiguous demonstration (Table 4).

3.3. KNOWN BIOLOGICAL ACTIVITIES ATTRIBUTED TO CAFFEIC ACID, A MODEL OF MONOMERIC PHENOLIC COMPOUND FOUND IN COFFEE

Caffeic acid is a monoaromatic compound which possesses an o-diphenolic structure (3,4-dihydroxybenzoic acid), resulting in strong antioxydant properties which continues to attract considerable research. Free caffeic acid was reported in coffee (seed and pulp) but a large part was included in CGA (cf. Table2). Like CGA, caffeic acid was shown to exhibit various physiological and biological activities (up to 65). They are listed below but still remain to be unambiguously demonstrated (Table 5).

Table 5 Compilation of 65 reported biological activities of caffeic acid (adapted from Duke, 1992a).

Biological activity	Biological activity	Biological activity
Allergic	Analgegic	Anti-HIV
Anti-tumor-Promoter	Antiadenoviral	Antiaggregant
Anticancer	Anticarcinogenic	Antiedemic
Antiflu	Antigonatotropic	Antihemolytic
Antihepatotoxic	Antiherpetic	Antihypercholesterolemic
Antiinflammatory	Antimutagenic	Antinitrosamic
Antiophidic	Antioxidant	Antiperoxidant
Antiprostaglandin	Antiradicular	Antiseptic
Antispasmodic	Antistomatic	Antisunburn
Antithiamin	Antithyroid	Antitumor
Antiulcerogenic	Antivaccinia	Antiviral
Bactericide	CNS-Active	Cancer-Preventive
Carcinogenic	Cholagogue	Choleretic
Clastogenic	Cocarcinogenic	Collagen-Sparing
Cytoptotective	Cytotoxic	DNA-Active
Diuretic	Fungicid	Hepatoprotective
Hepatotropic	Histamine-Inhibitor	Immunostimulant
Leucotriene-Inhibitor	Lipoxigenase-Inhibitor	Lyase-Inhibitor
Metal-chelator	Ornithine decarboxylase-In	Prooxidant
Prostaglandigenic	Sedative	Spasmolytic
Sunscreen	Tumorigenic	Viricide, Vulnerary

3.4. KNOWN BIOLOGICAL ACTIVITIES ATTRIBUTED TO TANNIN, A MODEL OF POLYMERIC PHENOLIC COMPOUND FOUND IN COFFEE

This aromatic compound, is a model of polymeric and polyphenolic structure. Both condensed and hydrolyzable tannins exist in coffee plant, but it was demonstrated that in coffee pulp, tannins were mostly condensed and consisted primarily of proanthocyanidins (Ramirez-Martinez and Clifford, 2000). Information on calculation of the content of tannins in coffee is sometimes contradictory. The data regarding concentration of hydrolysable tannins (gallotannins and ellagitannins), condensed tannins or both, depends on the drying method and the extraction method. For example, the amounts of condensed tannins were calculated in our laboratory with three storage techniques of coffee pulp :

1- fresh-frozen and crushed,
2- frozen after storage and crushed in N_2, and
3- lyophilized before crushing.

These coffee pulps were both extracted with methanol:H_2O (80:20) and quantified with the method of Swain and Hillis (1959), also known as the ferrous-butanol-HCl method (autodepolymerisation of proanthocyanidins). The results were given as equivalent of mimosa tannin powder and results were respectively 2,000; 6,000 and 11,200 ppm (fresh weight) of condensed tannins .

This gives a 5 times ratio between technique 1 and technique 3. With these samples 100g f.w. (fresh weight) represented 21g d.w (dry weight), which gives 9,500; 28,600 and 53,300 ppm (d.w.) of condensed tannins respectively. Total tannins were previously reported between 18,000 and 86,000 ppm in coffee pulp and condensed tannins (as flavonoids) between 28,000 and 40,000 ppm (cf. Table 2). Our results are comparable with previous reports, and show that lyophilization permits high tannins recovery.

Total tannins were previously reported between 18,000 and 86,000 ppm in coffee pulp (Table 2) and condensed tannins (as flavonoids) between 28,000 and 40,000 ppm

As with CGA and caffeic acid, tannin was reported to exhibit various physiological and biological activities. To our knowledge, up to 33 biological activities were attributed to tannin (as a generic name). They are listed below, but still remain, as previously, to be unambiguously demonstrated (Table 6).

Table 6. Compilation of 33 reported biological activities of tannin (adapted from Duke 1992a)

Biological activity	Biological activity	Biological activity
Anthelminthic	Anti tumor promoter	Anti HIV
Anticancer	Anticariogenic	Antidiarrheic
Antidysenteric	Antihepatotoxic	Antihypertensive
Antilipolytic	Antimutagenic	Antinephritic
Antiophidic	Antioxidant	Antiradicular
Antirenitic	Antitumor	Antiulcer
Antiviral	Bactericide	Cancer preventive
Chelator	Cyclooxygenase inhibitor	Glucosyl transferase inhibitor
Hepatoprotective	Lipoxygenase inhibitor	MAO inhibitor
Ornithine decarboxylase inhibitor	Psychotropic	Viricide
Xanthine oxidase inhibitor	Carcinogenic	Immunosuppressant

4. Conclusion

In this paper, we provide evidence that high biotechnological potential exists with various identified polyphenolic compounds found in liquid by-product like OMW or with solid by-product like coffee pulp. The large lists of examples of phenolic compounds from olive and coffee plant species opens researchpossibilities on other agroindustrial by-products specifically rich in such polyphenolics. Chlorogenic acid, caffeic acid and tannin represent three examples of bioactive compounds identified in olive and coffee by-products. These three compounds along with a few other compounds which are listed, could be valuable as food additive, cosmetic food or nutrient for human health.

5. Abstract

Actual and future use of biotechnologies implies research focusing not only on agricultural production but also on the valorization of by-products generated by the agro industry. Micro-organisms are unique « cell factories » able to valorize agricultural by-products instead of only degrading them in depollution or methanisation processes. This paper gives two exemples of agricultural products and by-products that contain phenolic structures. Coffee pulp and olive mill wastewater (OMW) represent models of respectively solid and liquid by-products rich in polyphenolic compounds. Polyphenolics are homo- or hetero-cyclic aromatic compounds, where hydroxylic functions are substituted to at least one of the cyclic carbonic structures. Hydroxylated phenols, flavonoids, anthocyans, tannins and lignins are polyphenolic compounds. Coffee pulp contains simple polyphenols including acid phenolics and caffeoyl derivatives, and polymeric polyphenols including tannins. OMW possess similar chemical structures with each by-product containing specific polyphenolic compounds. OMW is probably the agro-industrial liquid by-product exhibiting the highest carbon oxygen demand (COD), with values up to 200g of COD/litre. The difficulties encountered in treating such compounds often result from high concentration of pollutants and/or high toxicity to the microflora. Coffee (*Coffea arabica* L.) and olive (*Olea europeae* L.) release after processing a polyphenolic rich by-product where known biological activities are attributed to a large extent to these aromatics. Examples of three polyphenolic structures with high biotechnological potential, are given. Chlorogenic and caffeic acids as monoaromatic models and tannin as polyaromatic model of polyphenols, is described. These compounds are both found in coffee and olive plants. *Coffea arabica* is the plant which contains the highest concentration of chlorogenic acid, when compared with other plant species known to synthetise this compound. Up to 52 different names of biological activities are reported with chlorogenic acid. A tentatively exhaustive list of the various phenolic compounds which are detected and reported from olive and coffee is added in this paper, showing the large potential of valorisation of these two agro-industrial plants.

6. Aknowledgments

For olive results, the financial assistance from the EU DGXII-B in part of the project EC/ORSTOM/CBS, Contract CEE CI1*CT92-0104 is gratefully acknowledged. For coffee results, the financial assistance from the EU DGXII-B4 in part of the project BIOPULCA, Contract CEE IC18*CT97-0185 is gratefully acknowledged.

7. References

Baldioli M., Servili M., Perreti G., Montedoro G.F. (1996), Antioxidant activity of tocopherols and phenolic compounds of virgin olive oil. J.A.O.C.S., 73 (11): 1589-1593

Balice V., Cera O. (1984), Acidic phenolic fraction of the olive vegetation water determined by a chromatographic method. Grasas y Aceites, 35 (3): 178-180

Balyaya K.J., Clifford M.N. (1995), Individual chlorogenic acids and caffeine contents in commercial grades of wet and dry processed indian green Robusta coffee. J. Food Sci. Technol., 32(2), 104-108

Brauman A., Kane M.D., Labat M. Breznak A. (1992), Genesis of acetate and methane by gut bacteria of nutritionally diverse termites. Sciences, 257: 1384-1387

Bressani R., Estrada E. Jarquin R. (1972), Pulpa y pergamino de café. 1. Composicion quimica y contenido de aminoacidos de la proteinadelapulpa. Turrialba, 22 (3): 299-304

Bressani R Elias, L.G. (1976), Utilizacion de desechos de café en alimentacion de animales y materia prima industrial. Exposicion Pecuria del Istmo Centroamericano. EXPICA 76, San Salvador, El Salvador, INCAP Doc., Guatemala, 25 pp

Clifford M.N., Kellard B., Birch G.G. (1989), Characterisation of caffeoylferuloyl-quinic acids by simultaneous isomerisation and transesterification with tetramethyl-ammonium hydroxide. Food Chem., 34: 81-88

Clifford M.N., Gonzàlez de Colmenares N., Ramirez-Martinez J.R., Aldana, J.O. (1993), Progress in the analysis of proantocianidins in freshly prepared coffee pulp. *15ème Coll. Scient. Intern. sur le Café*, ASIC, Paris, 884-886

Crawford D.L., Pometto A.L., Crawford R.L. (1983), Lignin degradation by *Streptomyces viridisporus* : Isolation and caracterisation of a new polymeric degradation intermediate. Appl. Environ. Microbiol., 45: 898-904

Duke J.A. (1992a), CRC Handbook of Phytochemical Constituents in GRAS Herbs and other Economic Plants. CRC Press, Boca Raton, Fl., 654 pp

Duke J.A. (1992b), CRC Handbook of Biologically Active Phytochemicals and Their Activities. CRC Press, Boca Raton, Fl., 183 pp

Ferry J.G., Wolfe R.S. (1976), Anaerobic degradation of benzoate to methane by a microbial consortium. Arch. Microbiol., 107, 33-40

Field J.A., Lettinga G. (1987), The methanogenic toxicity and anaerobic degradability of a hydrolysable tanin. Wat. Res., 20: 367-374

Labat M., Sayadi S., Gargouri A., Zorgani F., Jaoua M., Zekri S., Ellouz R. (1996), Procédé aérobie-anaérobie pour le traitement biologique des résidus liquides de l'industrie oléicole. In: "Journées Industrielles sur la Digestion Anaérobie", JIDA-96, Narbonne, 8 pp

Labat M., Sayadi S., Gargouri A., Garcia J.L. (1997), Development of biological process for treatment of olive mill waste water. Molecular studies of the ligninolytic system by *Phanerochaete chrysosporium. Final Progress Report* CEE DGXII-B, Contrat n° CI1-CT92-0104, feb. 1997, Marseille, 278 pp

Léger C.L. (1999), Co-produits de l'huilerie d'olive: les composés phénoliques et leurs propriétés biologiques. O.C.L., 6 (1), 60-63.

Molina M.R., De La Fuente G., Gudiel H. Bressani R. (1974), Pulpa y pergamino de café. 8. Estudios bàsicos sobre la deshydratacion de la pulpa de café. Turrialba, 24 (3) 280-284

Montedoro G.F., Servili M., Baldioli M., Selvaggini R., Miniati E. Macchioni A. (1993), Simple and hydrolyzable phenolic compounds in virgin olive oil. Note 3. Spectroscopic characterization of the secoroidoid derivatives. J. Agric. Food Chem., 41: 2228-2234

Nefzaoui A. (1991), Valorisation des sous-produits de l'olivier. Options Méditerranéennes, 16: 101-108

Ramirez-Martinez J.R. (1988), Phenolic compounds in coffee pulp: quantitative determination by HPLC. J. Sci. Food Agric., 43: 135-144

Ramirez-Martinez J.R., Clifford M.N. (2000), Coffee pulp polyphenols: an overview, in T. Sera, C.R. Soccol, A. Pandey and S. Roussos (eds.) *Coffee Biotechnology and Quality*, Kluwer Academic Publishers, Dordrecht, pp 501-509

Ryan D., Robards K. (1998), Phenolic compounds in olives. Analyst, 123: 31-44

Salvemini F. (1985), Composizione chimica e valutazione biologica di un mangime ottenuto essicando tercamente le acque di vegetazione delle olive. Riv. Delle Sostanze Grasse, 112: 559-564

Swain T., Hillis W.E. (1959), The phenolic constituents of *Prunus domestica*. I.- The quantitative analysis of phenolic constituents. J. Sci. Food Agric., 10: 63-68

Zuluaga J., Bonilla, C., Quijano, R.M. (1975), Contribucion al estudio y utilizacion de la pulpa de café. 7th Intern. Col. Chim. Coffee, Hamburg, ASIC Ed, Paris, 233-242

Zuluaga J. (1981) Contribution à l'étude de la composition chimique de la pulpe de café (*Coffea arabica* L.). Thesis, Neuchatel Univ., Swiss, 93pp

Subject index CBQ

www.ingramcontent.com/pod-product-compliance
Ingram Content Group UK Ltd.
Pitfield, Milton Keynes, MK11 3LW, UK
UKHW021901190726
13853UKWH00003B/1371
* 9 7 8 9 4 0 1 7 1 0 6 9 5 *